# Computational Methods in Catalysis and Materials Science

*Edited by*
*Rutger A. van Santen*
*and Philippe Sautet*

## *Further Reading*

Jensen, F.

**Introduction to Computational Chemistry**

2007

ISBN: 978-0-470-01186-7

Dronskowski, R

**Computational Chemistry of Solid State Materials**

**A Guide for Materials Scientists, Chemists, Physicists and others**

2006

ISBN: 978-3-527-31410-2

Van Santen, R. A., Neurock, M.

**Molecular Heterogeneous Catalysis**

**A Conceptual and Computational Approach**

2006

ISBN: 978-3-527-29662-0

Ertl, G., Knözinger, H., Schüth, F., Weitkamp, J. (Eds.)

**Handbook of Heterogeneous Catalysis**

**Second, Completely Revised and Enlarged Edition**
**8 Volumes**

2008

ISBN: 978-3-527-31241-2

Morokuma, K., Musaev, D.

**Computational Modeling for Homogeneous and Enzymatic Catalysis**

**A Knowledge-Base for Designing Efficient Catalysts**

2008

ISBN: 978-3-527-31843-8

# Computational Methods in Catalysis and Materials Science

*Edited by*
*Rutger A. van Santen and Philippe Sautet*

WILEY-VCH Verlag GmbH & Co. KGaA

**The Editors**

***Prof. Dr. Rutger A. van Santen***
Schuit Institute of Catalysis
Eindhoven University of Technology
Den Dolech 2
5612 AZ Eindhoven
The Netherlands

***Dr. Philippe Sautet***
Université de Lyon
Institut de Chimie de Lyon
Laboratoire de Chimie
Ecole Normale Supérieure
de Lyon et CNRS
46 Allée d'Italie
69364 Lyon Cedex 07
France



**Library of Congress Card No.:** applied for

**British Library Cataloguing-in-Publication Data**
A catalogue record for this book is available from the British Library.

**Bibliographic information published by the Deutsche Nationalbibliothek**
The Deutsche Nationalbibliothek lists this publication in the Deutsche Nationalbibliografie; detailed bibliographic data are available on the Internet at ⟨http://dnb.d-nb.de⟩.



**Composition** Laserwords Private Ltd., Chennai, India
**Printing** Betz-Druck GmbH, Darmstadt
**Bookbinding** Litges & Dopf GmbH, Heppenheim

Printed in the Federal Republic of Germany
Printed on acid-free paper

**ISBN**: 978-3-527-32032-5

# Contents



*Computational Methods in Catalysis and Materials Science.* Edited by Rutger A. van Santen and Philippe Sautet

ISBN: 978-3-527-32032-5

# Preface

This book contains a collection of chapters based on lectures presented at the IDECAT graduate summer school "Computational methods and applications in catalysis and material science" held September 2007 at the island Porquerolles in France.

IDECAT stands for "Integrated design of catalytic materials"; it is an EU Network of Excellence launched in 2005. It includes 37 laboratories from 17 institutions gathering over 500 researchers with a broad multidisciplinary expertise covering most of the aspects of catalysis.

Computational catalysis is a rapidly developing essential sub discipline of catalysis. The summer school brought together approximately 50 Ph.D. students and post-doctoral students with widely varying backgrounds.

Whereas often such summer schools mainly focus on the use of computational methods in a wide variety of catalytic applications, we decided that we should concentrate on an introduction to the methods. Applications could then be treated as illustrations.

We are very happy that most of the participating lecturers have been able to find the time not only to present their lectures, but also to write a chapter for this book based on their presentations.

The book is organized in four parts:

- The first part introduces the basic methodologies that are currently used in electronic structure calculations. Hartree–Fock and electronic correlation methods are followed by basic aspects of Density Functional Theory (DFT) and by the description of excited states in the framework of time-dependent DFT. This part is completed by the approach of the electronic structure of periodic systems and by a presentation of the concepts of ab initio molecular dynamics.
- The second part contains chapters discussing statistical mechanical approaches, based on force-fields or on kinetic parameters, which are useful to study complex systems and processes occurring at longer time and length scales. Course graining techniques as well as kinetic Monte-Carlo methods or reactive force fields are introduced.

*Computational Methods in Catalysis and Materials Science.* Edited by Rutger A. van Santen and Philippe Sautet

ISBN: 978-3-527-32032-5

- The third part aims at the calculation of physico-chemical properties. Two chapters deal with applications to a particular spectroscopic technique while one introduces to the description of crystal structures. Chapters dealing with the calculation of physical properties as adsorption, diffusion and transport conclude this part.
- The final part opens to applications of theory to problems in homogeneous and heterogeneous catalysis. We especially aim to show how molecular insights are bringing conceptual understanding to many important catalytic aspects.

  This is especially the case for domains at the interface between organometallic chemistry, surface chemistry and heterogeneous catalysis. This part also contains a concluding chapter that summarizes current status and perspectives of computational catalysis.

This book would not have been possible without the pleasant and efficient support of Judith Wachters- and Ad kolen as well as from the Wiley-VCH editorial office.

Eindhoven
Lyon
January 2009

*Rutger A. van Santen*
*Philippe Sautet*

# List of Contributors

***Emilio Artacho***
University of Cambridge
Department of Earth Sciences
Downing Street
Cambridge CB2 3EQ
United Kingdom

***Evert Jan Baerends***
Vrije Universiteit Amsterdam
Section Theoretical Chemistry
De Boelelaan 1083
1081 HV Amsterdam
The Netherlands

***Marie-Laure Bocquet***
Université de Lyon
Institut de Chimie de Lyon
Laboratoire de Chimie
Ecole Normale Supérieure de Lyon et CNRS
46 Allée d'Italie
69364 Lyon Cedex 07
France

***Marie-Laure Bonnet***
Université de Lyon
Institut de Chimie de Lyon
Laboratoire de Chimie
Ecole Normale Supérieure de Lyon et CNRS
46 Allée d'Italie
69364 Lyon Cedex 07
France

***Mark E. Casida***
Universite Joseph Fourier (Grenoble I)
Institut de Chimie Moleculaire de Grenoble
(ICMG, FR2607)
301 rue de la Chimie, BP 53
38041 Grenoble Cedex 9
France

***Christophe Copéret***
CPE Lyon
Laboratoire de Chimie
Organométallique de Surface
43, Bd du 11 Novembre 1918
69622 Villeurbanne Cedex
France

***Alain Dedieu***
Université Louis Pasteur
Institut de Chimie
Laboratoire de Chimie Quantique
UMR 7177 CNRS/ULP
Rue Blaise Pascal
67000 Strasbourg
France

*Computational Methods in Catalysis and Materials Science.* Edited by Rutger A. van Santen and Philippe Sautet

ISBN: 978-3-527-32032-5

**Françoise Delbecq**
Université de Lyon
Institut de Chimie de Lyon
Laboratoire de Chimie
Ecole Normale Supérieure de Lyon et CNRS
46 Allée d'Italie
69364 Lyon Cedex 07
France

**Adri van Duin**
California Institute of Technology
Force Field & Simulation Technology
Beckman Institute
Pasadena, CA 91125
USA

**Michael H. Eikerling**
National Research Council of Canada
Institute for Fuel Cell Innovation
4250 Wesbrook Mall
Vancouver, BC V6T 1W5
Canada
and
Simon Fraser University
Department of Chemistry
8888 University Drive
Burnaby, BC V5A 1S6
Canada

**Boris Le Guennic**
Université de Lyon
Institut de Chimie de Lyon
Laboratoire de Chimie
Ecole Normale Supérieure de Lyon et CNRS
46 Allée d'Italie
69364 Lyon Cedex 07
France

**Klaus Hermann**
Fritz-Haber-Institut der MPG and
Collaborative Research Center
Theory Department
Faradayweg 4–6
14195 Berlin
Germany

**Marcella Iannuzzi**
Laboratory for Reactor Physics
Paul Scherrer Institut
5232 Villigen
Switzerland

**A.P.J. Jansen**
Eindhoven University of Technology
Department of Chemistry
P.O.Box 513
5600 MB Eindhoven
The Netherlands

**Mikaël Képénékian**
Université de Lyon
Institut de Chimie de Lyon
Laboratoire de Chimie
Ecole Normale Supérieure de Lyon et CN
46 Allée d'Italie
69364 Lyon Cedex 07
France

**Mikael Leetmaa**
Stockholm University
Fysikum
AlbaNova University Center
106 91 Stockholm
Sweden

**Hervé Lesnard**
Université de Lyon
Institut de Chimie de Lyon
Laboratoire de Chimie
Ecole Normale Supérieure de Lyon et CN
46 Allée d'Italie
69364 Lyon Cedex 07
France

**Mathias Ljungberg**
Stockholm University
Fysikum
AlbaNova University Center
106 91 Stockholm
Sweden

***Nicolas Lorente***
Université de Lyon
Institut de Chimie de Lyon
Laboratoire de Chimie
Ecole Normale Supérieure de Lyon et CNRS
46 Allée d'Italie
69364 Lyon Cedex 07
France

***Kourosh Malek***
National Research Council of Canada
Institute for Fuel Cell Innovation
4250 Wesbrook Mall
Vancouver, BC V6T 1W5
Canada

***A. J. (Bart) Markvoort***
Eindhoven University of Technology
Department of Biomedical Engineering
Den Dolech 2
5600 MB Eindhoven
The Netherlands

***Martijn Marsman***
University of Vienna
Faculty of Physics
1090 Vienna
Austria

***Serge Monturet***
Université Paul Sabatier
Laboratoire Collisions
Agrégats, Réactivité
IRSAMC
118 route de Narbonne
31062 Toulouse Cédex
France

***Antonio J. Mota***
Universidad de Granada
Departamento de Química Inorgánica
Facultad de Ciencias
Campus de Fuentenueva
18071 Granada
Spain

***Anders Nilsson***
Stockholm University
Fysikum
AlbaNova University Center
SE 106 91 Stockholm
Sweden
and
Stanford Synchrotron Radiation Laboratory
P.O.B. 20450
Stanford, CA 94309
USA

***Lars G.M. Pettersson***
Stockholm University
Fysikum
AlbaNova University Center
106 91 Stockholm
Sweden

***Vincent Robert***
Université de Lyon
Institut de Chimie de Lyon
Laboratoire de Chimie
Ecole Normale Supérieure de Lyon et CNRS
46 Allée d'Italie
69364 Lyon Cedex 07
France

***Jean-Baptiste Rota***
Université de Lyon
Institut de Chimie de Lyon
Laboratoire de Chimie
Ecole Normale Supérieure de Lyon, CNRS
46 Allée d'Italie
69364 Lyon
France

***Rutger A. van Santen***
Schuit Institute of Catalysis
Eindhoven University of Technology
Den Dolech 2
5612 AZ Eindhoven
The Netherlands

***Philippe Sautet***
Université de Lyon
Institut de Chimie de Lyon
Laboratoire de Chimie
Ecole Normale Supérieure de Lyon, CNRS
46 Allée d'Italie
69364 Lyon
France

***Berend Smit***
University of California
Department of Chemical Engineering
201 Gilman Hall
Berkeley, CA 94720-1462
USA

***Thijs J.H. Vlugt***
Delft University of Technology
Process and Energy Laboratory
Leeghwaterstraat 44
2628CA Delft
The Netherlands

***Raphael Wischert***
Université de Lyon
Institut de Chimie de Lyon
Laboratoire de Chimie
Ecole Normale Supérieure de Lyon et CN
46 Allée d'Italie
69364 Lyon Cedex 07
France

# Part I
# Electronic Structure Calculations

# 1
# From Hartree–Fock to Electron Correlation: Application to Magnetic Systems

*Vincent Robert, Mikaël Képénékian, Jean-Baptiste Rota, Marie-Laure Bonnet, and Boris Le Guennic*

## 1.1
## Introduction

At the beginning of last century, quantum mechanics broke out and the famous Schrödinger's and Dirac's equations were derived and constituted tremendously important milestones. Even though they aim at describing the nanoscopic correlated world, it is known that the analytical solution is limited to the two-particle system, a prototype of which being the H atom. In particular, the description of a simple system as $H_2$ necessarily relies on approximations. One may first consider electrons as independent particles moving in the field of fixed nuclei. The appealing strategy of a mean field approximation was thus suggested along with the important picture of screened nuclei. How much the fluctuation with respect to this description dominates the physical properties has been a widely debated challenging issue.

This review will be organized as follows. First, the different methods traditionally used in quantum chemistry are briefly recalled starting from the Hartree–Fock description to the introduction of correlation effects. Since quantum chemistry aims at describing the interactions between atomic partners, the one-electron functions (so-called molecular orbitals, MOs) are derived from one-electron atomic basis sets localized on the atoms (atomic orbitals, AOs). However, it is known that a major drawback in this single determinantal description of the wavefunction is its inability to properly account for bond breaking. The $H_2$ case is used as a pedagogical example in Section 1.2.2.2 to exemplify the need for multireference SCF algorithms. For the study of homolitic breaking of such a single bond, it is recalled that both bonding and antibonding MOs must be introduced to incorporate the so-called nondynamical correlation effects. In this hierarchical construction of the wavefunction, the *Complete Active Space Self-Consistent Field* (CASSCF) [1, 2] procedure is described (see Section 1.2.3.1). Such methodology is particularly efficient since along bond stretching, two electrons become strongly correlated and the CASSCF treatment tends to localize one electron in each atom. The important dynamical correlation effects are then exemplified deriving the $H_2$–$H_2$ interactions,

*Computational Methods in Catalysis and Materials Science.* Edited by Rutger A. van Santen and Philippe Sautet

ISBN: 978-3-527-32032-5

and the short distance behavior ($1/R^6$) of the van der Waals potential is recovered (see Section 1.3.1).

In the last section, the machinery and efficiency of *ab initio* techniques are demonstrated over selected examples. A prime family is represented by magnetic systems which have attracted much attention over the last decades considering their intrinsic fundamental behaviors and possible applications in nanoscale devices. Chemists have put much effort to design and fully characterize new families of systems which may exhibit unusual and fascinating properties arising from the strongly correlated character of their electronic structures. From a fundamental point of view, high-$T_c$ superconducting copper oxides [3–5], and colossal magnetoresistant manganite oxides [6–11] are such families which cannot be ignored in the field of two- and three-dimensional materials. One-dimensional chains [12–15] as well as molecular systems mimicking biological active centers [16,17] have more recently been considered as promising targets in the understanding of dominant electronic interactions. In such materials, a rather limited number of electrons are responsible for the observed intriguing properties. Reasonably satisfactory energetics description of such systems can be obtained by the elegant broken-symmetry (BS) method [18–21]. Let us mention that, in particular, BS *density functional theory* (DFT) calculations have turned out to be very efficient in the determination of magnetic coupling constants and EPR parameters (see [22–29] and references therein). On the other hand, the DFT methodology has been extensively used in surface science to follow at a microscopic level the reactant transformation leading to products. Nevertheless, this description has shown to suffer from an unrealistic description of physisorption [30]. Thus, a combined approach based on the periodic DFT method with $MP_2$ correction has been proposed to overcome this intrinsically methodological drawback [31,32].

These examples aim at shedding light over a selected number of systems in materials science, catalysis, and enzymatic activity which may call for explicitly correlated calculations.

## 1.2
## Methodological Aspects of the Electronic Problem

### 1.2.1
### The Electronic Problem

Physical properties of molecules take their origin in electron assembly phenomena. To understand these properties, one has to investigate the electron distributions and interactions. This information is contained in the electronic wavefunction governed by Schrödinger's equation:

$$\hat{H}\Psi = E\Psi, \tag{1.1}$$

which is to be solved, defining the $N$-electron eigenfunction $\Psi$ and eigenvalue $E$ of the Hamiltonian $\hat{H}$. The nonrelativistic Hamiltonian is written as a sum of different kinetic and potential contributions arising from interacting electrons and nuclei:

$$\hat{H} = \hat{T}_N + \hat{T}_e + \hat{V}_{Ne} + \hat{V}_{ee} + \hat{V}_{NN}. \tag{1.2}$$

Since the nuclei are much heavier than the electrons, their kinetic energy is much smaller and, consequently, can be considered as motionless. In the study of the electronic problem, the nuclei positions are parameters for the motion of the electrons, and the problem is solved by considering only the electronic part of the Hamiltonian (so-called the Born–Oppenheimer approximation [33]). Thus, the electronic Hamiltonian using atomic units reads

$$\hat{H}_{\text{elec}} = \sum_i -\frac{1}{2}\Delta_i - \sum_i \sum_A \frac{Z_A}{r_{iA}} + \frac{1}{2}\sum_i \sum_{j \neq i} \frac{1}{r_{ij}}. \tag{1.3}$$

While the first two terms are monoelectronic in nature, the third one is the electron–electron repulsion which excludes any analytical resolution of the many-body problem.

Traditionally, one looks for a step-by-step procedure to incorporate the important physical contributions in a hierarchical way. A reasonable zeroth-order wavefunction is accessible within the Hartree–Fock scheme. Such treatment relies on a meanfield approximation where each electron moves in the field generated by the nuclei and the average electronic distribution arising from the $N-1$ other electrons (see Section 1.2.2.1). It was rapidly understood that such single determinantal strategy fails to properly describe bond breakings. As a matter of fact, as a bond is stretched, the independent electron approximation breaks down as the electrons tend to localize in a concerted way one on each nuclei. To overcome this failure and incorporate the so-called static correlation, the CASSCF procedure has been proposed [1,2]. Along this procedure, the wavefunction becomes intrinsically multireference (see Section 1.2.3.1). Finally, contributions which tend to reduce the electron–electron repulsion account for the dynamical correlation. Its main effect is the digging of the Coulomb hole to increase the probability of finding two electrons in different regions of space, distinguishing radial and angular correlations. This concept has been widely used in the understanding of DFT approaches.

As both static and dynamical correlations are turned on top of a Hartree–Fock solution, electrons are allowed to occupy arbitrarily (respecting spin and space symmetries!) all the MOs, introducing other electronic configurations which may be necessary to describe the physical state of interest. In a sense, the expansion of the wavefunction as a linear combination of Slater determinants (*configuration interaction*, CI) tends to recover the physical effects absent in the initial orbital approximation.

### 1.2.2
### Finding a Solution

Let us start from an infinite set of MOs, $\phi_i$, and a zeroth-order approximation to the $N$-electron problem. The MOs are split into two sets, either doubly occupied

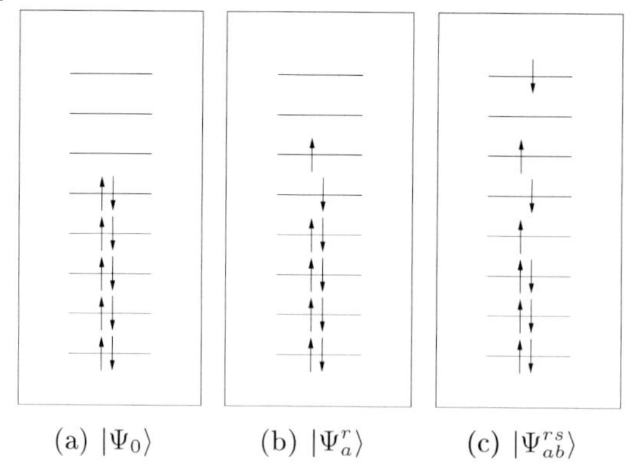


**Figure 1.1** (a) $|\Psi_0\rangle$, (b) single, and (c) double excited determinants.

or empty (referenced as $(a, b, c, \ldots)$ and $(r, s, t, \ldots)$, respectively), defining $|\Psi_0\rangle$ as $|a\bar{a}b\bar{b}c\bar{c}\rangle$. The wavefunction can be developed upon $|\Psi_0\rangle$ and the electronic configurations built from $|\Psi_0\rangle$ by successive excitations (see Figure 1.1),

$$|\Psi\rangle = c_0|\Psi_0\rangle + \sum_{ar} c_a^r|\Psi_a^r\rangle + \sum_{\substack{a<b\\ r<s}} c_{ab}^{rs}|\Psi_{ab}^{rs}\rangle + \sum_{\substack{a<b<c\\ r<s<t}} c_{abc}^{rst}|\Psi_{abc}^{rst}\rangle + \cdots, \tag{1.4}$$

where $|\Psi_a^r\rangle$ represents single excited determinants, $|\Psi_{ab}^{rs}\rangle$ double excited, and so on.

Solving the electronic problem consists in the determination of (i) the MOs, and (ii) the amplitudes of different electronic configurations ($c_0$, $\{c_a^r\}$, $\{c_{ab}^{rs}\}, \ldots$). The first task is achieved along the Hartree–Fock procedure, while the second calls for numerical demanding methods which are constantly under intense investigations.

#### 1.2.2.1 Hartree–Fock Approximation

The goal is to find a set of MOs sustaining the reference determinant $|\Psi_0\rangle$. These orbitals should form an orthonormal basis of one-electron functions. Under these constraints, the Hartree–Fock equations are easily derived by minimizing the expectation value of $\hat{H}$ and $|\Psi_0\rangle$:

$$\left[\hat{h} + \sum_{a,occ}(J_a - K_a)\right]\phi_i = E_i\phi_i, \tag{1.5}$$

where $J_a$ and $K_a$ represent the Coulomb and the exchange operators, respectively.

The eigenfunction problem(s) must be solved iteratively (self-consistent field procedure, SCF) since the Fock operator $\hat{f} = \hat{h} + \sum_{a,occ}(J_a - K_a)$ is constructed on the occupations of its own eigenvectors. $\hat{h}$ is the sum of the kinetic energy and nuclei–electron interactions, while the sum defines the Hartree–Fock potential that averages the interelectronic repulsion so as to give a monoelectronic operator.

#### 1.2.2.2 Example

In order to clarify the Hartree–Fock SCF framework, let us concentrate on the quantum chemist's "swiss army knife" system, namely $H_2$ in a minimal AO basis set $\{a, b\}$. From symmetry consideration, one can build two MOs, symmetric (g)

$$g = \frac{1}{\sqrt{2}}(a + b) \tag{1.6}$$

and antisymmetric ($u$)

$$u = \frac{1}{\sqrt{2}}(a - b). \tag{1.7}$$

Evidently, the Hartree–Fock solution for the ground state is

$$|\Psi_0\rangle = |g\bar{g}| \tag{1.8}$$

or returning to the AOs,

$$|\Psi_0\rangle = \frac{1}{\sqrt{2}}\left(\frac{|a\bar{a}| + |b\bar{b}|}{\sqrt{2}} + \frac{|a\bar{b}| + |b\bar{a}|}{\sqrt{2}}\right) = \frac{1}{\sqrt{2}}\left(|\Psi_{\text{ion}}\rangle + |\Psi_{\text{neutral}}\rangle\right).$$

$|a\bar{a}|$ and $|b\bar{b}|$ are referred to as the ionic forms since the two electrons are localized on the same atomic center. This is to be contrasted with the combination $\frac{1}{\sqrt{2}}(|a\bar{b}| + |b\bar{a}|)$ which is the neutral singlet. Thus, $|\Psi_0\rangle$ consists of an equal weight of ionic and neutral forms. While the $H_2$ molecule should clearly dissociate into $H^\bullet + H^\bullet$ (see Figure 1.2), the Hartree–Fock procedure overestimates the weight of the ionic forms. As a matter of fact, the latter should physically become vanishingly small as the bond is stretched. This is a major pitfall of the Hartree–Fock theory which is being taken care of in a multireference approach.

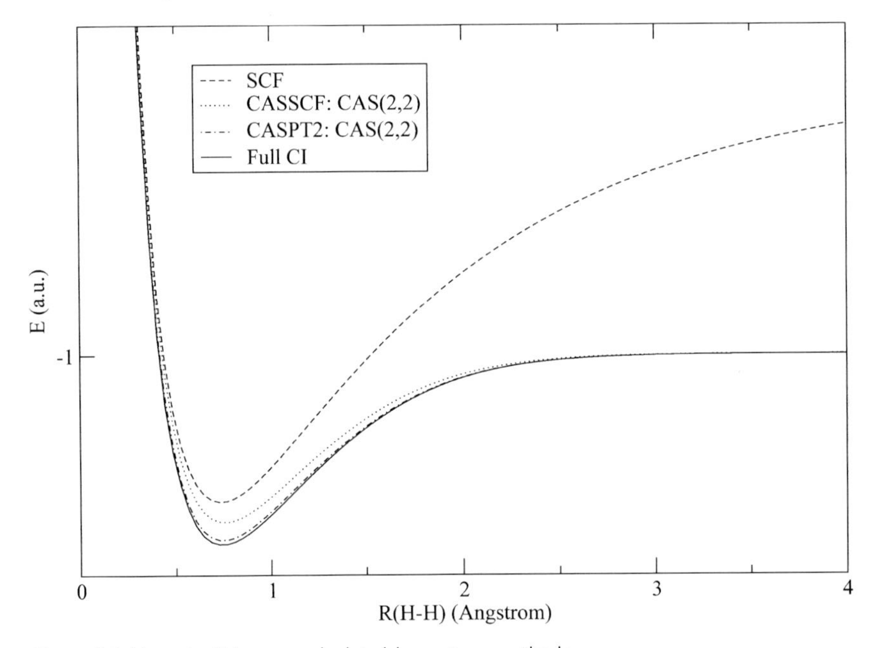


**Figure 1.2** $H_2$ potential curve calculated by various methods.

#### 1.2.2.3 Beyond Hartree–Fock Treatment: Electronic Correlation

As mentioned previously, the main effort is to calculate the coefficients in the wavefunction expansion. This calculation gives a correlated wavefunction, the energy of which defines the correlation energy as

$$E_{\text{corr}} = E - E_{\text{Hartree-Fock}}.$$

Practically the expansion cannot be carried out upon an infinity of excitations and a selection of excited determinants must be made in the configuration interaction treatment. In practice, this procedure cannot lead to the exact solution of the many-electron problem for two main reasons:

- It is impossible to handle infinite one-electron basis sets. Therefore, the constructed determinants $\{|\Psi_i\rangle\}$ cannot form a complete $N$-electron function basis set.
- Even for relatively small basis sets, the number of determinants to be considered may become extremely large. Thus, in practice, one will not take into account all determinants (*full*-CI procedure) but only a small part of these (*truncated*-CI), in single and/or double (CIS, CISD) calculations.

Traditionally, one distinguishes static and dynamical correlations in the CI approach. In the next section, we will clarify these notions using the $H_2$ example.

### 1.2.3 Correlation Energy

#### 1.2.3.1 Static Correlation: CASSCF Approach

Let us concentrate on the problem of bond breaking of $H_2$. In the g orbital, the maximum of electronic density is in the middle of the bond. Conversely, the $u$ function displays a nodal plane and the maximum of density is concentrated on the nuclei.

To overcome the major failure of the Hartree–Fock description, one may introduce in a multireference expansion other determinants. Clearly, $g$ and $u$ become quasidegenerate in the long-distance regime. Thus, $|u\bar{u}|$ may as well significantly participate in the two-electron wavefunction. By allowing the occupations of two MOs by two electrons, leaving all the other orbitals either doubly occupied (inactive) or vacant (virtual), one performs a Complete Active Space Self-Consistent Field CAS(2,2)SCF calculation (see Figure 1.3). From a physical point of view, this procedure consists in treating exactly the correlation in the active space and let the inactive orbitals react to the field generated by different configurations built in the active space. This point constitutes the major difficulty of the CASSCF calculation. Indeed, since the active space is the only part of the system where the correlation is treated with fine accuracy, it has to include the necessary configurations to describe the property of interest.

One defines the best set of MOs under this constraint. The inactive MOs respond to the occupations of the active MOs, treating democratically the $|g\bar{g}|$ and $|u\bar{u}|$

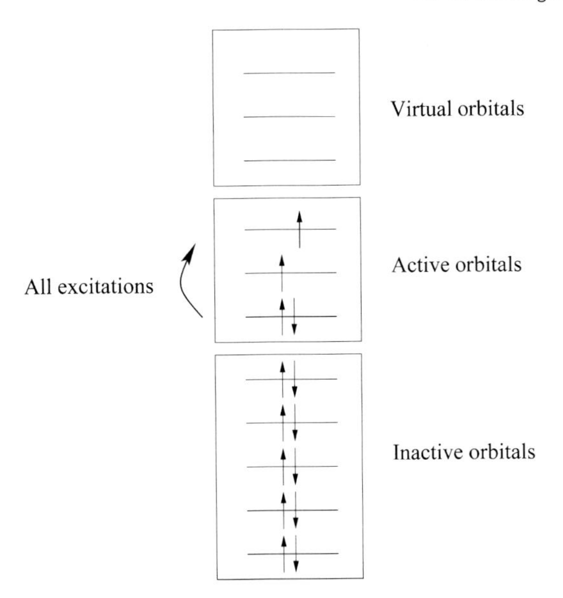


**Figure 1.3** CASSCF configurations.

configurations. The comparison between $H_2$ and $F_2$ systems is instructive since the latter holds such inactive shells. The CAS(2,2) inactive MOs of $F_2$ will be the best compromise between the double occupancy of $g$ and $u$.

For $H_2$ in a minimal basis set, the correlation energy is analytical from the $2 \times 2$ matrix diagonalization (see Ref. [34] for derivations). Writing $|\Psi\rangle = |g\bar{g}| + c|u\bar{u}|$ and $\Delta = \langle u\bar{u}|\hat{H}|u\bar{u}\rangle - \langle g\bar{g}|\hat{H}|g\bar{g}\rangle$

$$E_{\text{corr}} = \Delta - \sqrt{\Delta^2 + K_{gu}^2} \tag{1.9}$$

$$c = \frac{K_{gu}}{-\Delta - \sqrt{\Delta^2 + K_{gu}^2}}. \tag{1.10}$$

These results call for two important comments. First the correlation energy is negative as a result of the flexibility offered to the wavefunction. Then, the amplitude $c$ of $|u\bar{u}|$ being negative reduces the weight of the ionic forms. Eventually, as $R \to \infty$, it can be shown that $c \to -1$ and the wavefunction reduces to $\frac{1}{\sqrt{2}}\left(|a\bar{b}| + |b\bar{a}|\right)$. The electrons are no longer independent, they are said to be correlated. The reduction of the ionic forms stresses the demand of atoms to recover their neutral character. The nondynamical correlation strikes back again the delocalization preference arising from the Hartree–Fock scheme. Along the CASSCF procedure one introduces the leading physical contributions in a multireference wavefunction. This allows one to treat on the same footing quasidegenerated electronic configurations given in a predefined active space (so-called CAS). Typically, the dissociation of $H_2$ can be properly discussed using a CAS(2,2)SCF calculation (see Figure 1.2).

#### 1.2.3.2 Dynamical Correlation

In the light of the previous considerations, let us again concentrate on $H_2$ close to the equilibrium distance. Consequently, the $\Delta$ value is large whilst $c$ is almost negligible. $\Psi$ is almost monoreference. A statistical analysis of the wavefunction shows that the electrons spend much time in the $g$ orbital and sometimes explore the $u$ one. In this case, the correlation is a fluctuation of the electronic density around an average value. This is part of the origin for dynamical terminology. The dynamical correlation brings a correction to the energy and wavefunction, but the qualitative results of the Hartree–Fock approach are not deeply changed.

More generally, on top of the CASSCF wavefunction one traditionally performs either second-order perturbation theory treatment (CASPT2) [35, 36] or variational CI such as the so-called first-order CI which incorporates in a variational way all the single excitations on the CAS determinants. These contributions account for the electronic relaxations which respond to the instantaneous field modifications or spin polarization in the active space.

In this respect, the *Difference Dedicated* CI (DDCI) methodology [37–39] has shown to provide impressive results in magnetically coupled systems [40–42]. The conceptual guideline is the quasidegenerated perturbation theory (QDPT) developed by Bloch [43]. For a two-electron/two-MO system one looks for the singlet–triplet energy difference $2J$, $J$ being the one-parameter model Heisenberg Hamiltonian $\hat{H} = -2J\hat{S}_1\hat{S}_2$ ($S_1 = S_2 = 1/2$). The model space consists of two neutral forms $|a\bar{b}|$ and $|b\bar{a}|$ upon which the QDPT defines an effective Hamiltonian $\hat{H}_{\text{eff}}$. At the second order of perturbation theory, the off-diagonal element of $\hat{H}_{\text{eff}}$ is precisely $J$ and reads

$$\langle a\bar{b}|\hat{H}^2_{\text{eff}}|b\bar{a}\rangle = K_{ab} + \sum_{\alpha} \frac{\langle a\bar{b}|\hat{H}|\alpha\rangle\langle\alpha|\hat{H}|b\bar{a}\rangle}{E_0^{(0)} - E_\alpha^{(0)}},$$

$|\alpha\rangle$ being outer-space determinants, including ionic forms $|a\bar{a}|$ and $|b\bar{b}|$. If the sum is restricted to $\alpha = a\bar{a}, b\bar{b}$, then $J$ reads

$$J = K_{ab} + \frac{\langle a\bar{b}|\hat{H}|a\bar{a}\rangle\langle a\bar{a}|\hat{H}|b\bar{a}\rangle}{E_0^{(0)} - E_{a\bar{a}}^{(0)}} + \frac{\langle a\bar{b}|\hat{H}|b\bar{b}\rangle\langle b\bar{b}|\hat{H}|b\bar{a}\rangle}{E_0^{(0)} - E_{b\bar{b}}^{(0)}}$$

$$J = K_{ab} + 2\frac{t^2}{-U}$$

with $t = \langle a\bar{a}|\hat{H}|a\bar{b}\rangle$ and $U = E_0^{(0)} - E_{a\bar{a}}^{(0)}$. One recovers the famous competition between ferromagnetic and antiferromagnetic contributions. For $|\alpha\rangle$ to be simultaneously coupled to $|a\bar{b}|$ and $|b\bar{a}|$, it should not defer by more than two spin orbitals (Slater's rule). Thus, the determinants are traditionally listed according to the number of holes ($h$) and particles ($p$) generated on the model space. As soon as this space is enlarged to the full valence space (i.e., including the ionic forms), it can be shown that $1h$, $1p$, $1h + 1p$, $2h$, $2p$, $2h + 1p$, and $2p + 1h$ participate in the hierarchical organization of the singlet and triplet.

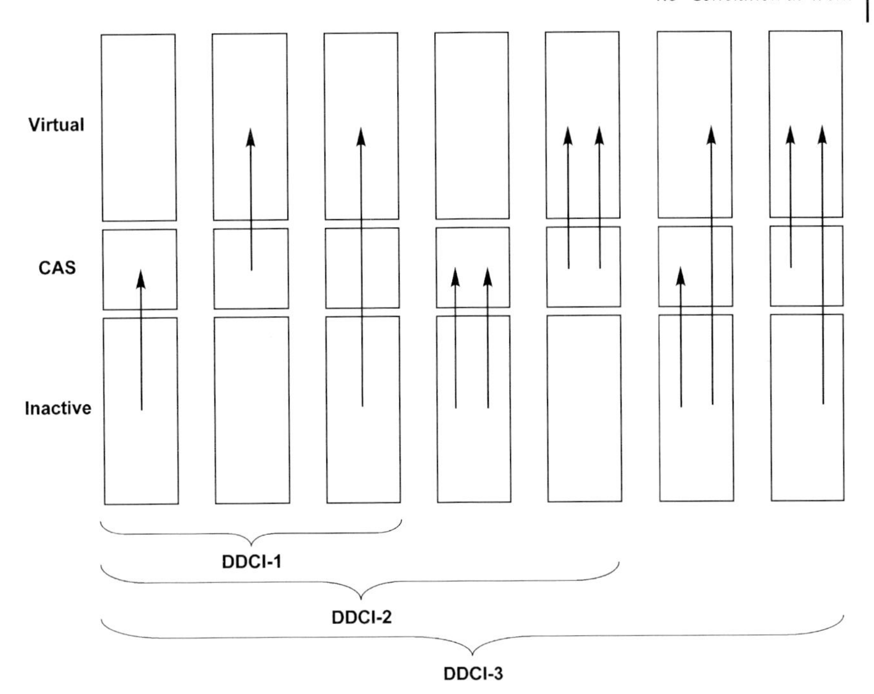


**Figure 1.4** Successive determinants considered in the DDCI approach.

Indeed the purely inactive excitations $2h + 2p$ simply shift the diagonal matrix elements. As shown in Figure 1.4, the selection gives rise to DDCI-1, -2, and -3 levels of calculations. Being a truncated-CI methodology, DDCI suffers from intrinsic size-consistency issue which has been elegantly corrected in the so-called *Size-Consistent Self-Consistent* $(SC)^2$ framework [44]. The physical effects (spin polarization, dynamical correlation) have been clarified by considering different levels of calculations [45–47].

In order to remedy to this size-consistency problem, alternative approaches have been proposed and coupled pairs methodology turned out to be very efficient [48]. Unfortunately, the cost of such calculations does not allow one to handle even moderate size systems. Nevertheless, the CASPT2 method [35, 36] offers a remarkable compromise, introducing at second-order of perturbation theory the correlation effects. The corresponding atomic effects are properly incorporated in this contracted treatment of correlation effects. Such methodology has proven to be remarkably efficient in the inspection of magnetic properties of molecular and extended systems.

## 1.3 Correlation at Work

Over the past decades, a huge amount of experimental data carried out on a wide panel of systems has received much attention from both CI- and DFT-based frameworks. For the present purpose, we limit our inspection to a selection of architectures of various dimensionalities. Over the years, the possibilities of generating magnetic systems using versatile ligands coordinated to different metallic centers have been much considered in the light of the porphyrin-like molecules activity. Thus, the traditional scenario involving open-$d$ shells in the environment of closed-shell magnetic couplers (see Section 1.3.3) has been revisited based on both experimental and theoretical works (see Section 1.3.2). Nevertheless, we shall first investigate prototypes of weak interactions arising in the $(H_2)_2$ dimer (see Section 1.3.1). The van der Waals forces are of prime importance in physisorption phenomena which are likely to control catalyzed reactions. These effects have a purely quantum origin as they correspond to instantaneous charge fluctuations.

### 1.3.1
### Dipoles Interactions: Example of $(H_2)_2$

Let us consider two $H_2$ molecules well separated in space ($l \ll L$, see Figure 1.5).

If $a, b, c$, and $d$ refer to the AOs, one can built the $g$ and $u$ MOs on each $H_2$ fragment (see Figure 1.6).

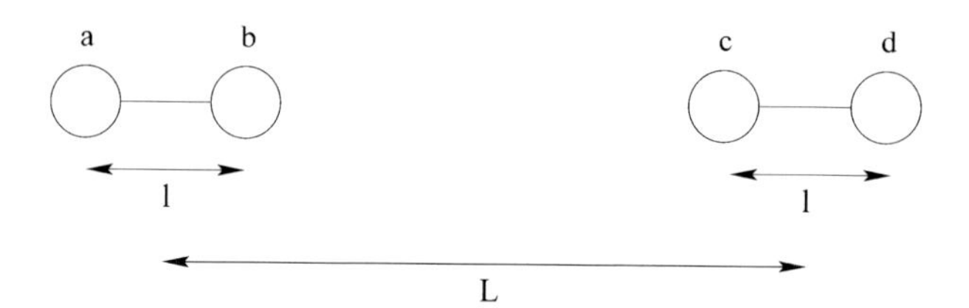


Figure 1.5 Schematic representation of the $(H_2)_2$ dimer.

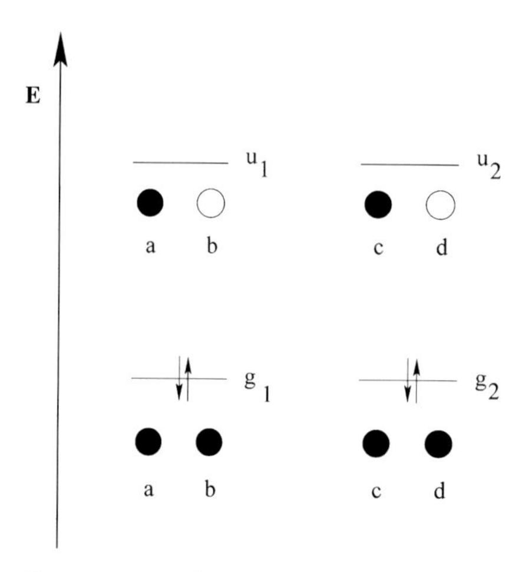


Figure 1.6 $|\Psi_0\rangle$ for $(H_2)_2$.

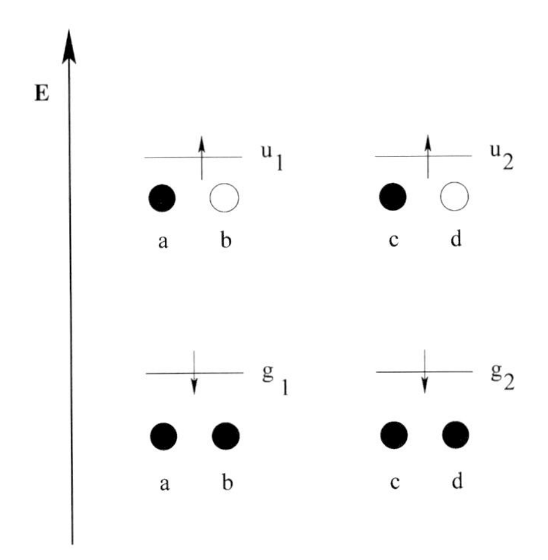


**Figure 1.7** $|\Psi_{g_1g_2}^{u_1u_2}\rangle$ for $(H_2)_2$.

Thus, a zeroth-order wavefunction is given by

$$|\Psi_0\rangle = |g_1\bar{g_1}g_2\bar{g_2}| = \tfrac{1}{4}\left|\left(a\bar{b} + b\bar{a} + a\bar{a} + b\bar{b}\right)\left(c\bar{d} + d\bar{c} + c\bar{c} + d\bar{d}\right)\right|.$$

One can observe in the development of $|\Psi_0\rangle$ that the doubly ionic structures "$H^+H^-H^+H^-$" and "$H^+H^-H^-H^+$" hold equal weights, in disagreement with naive electrostatic argument. However, the double excitation $g_1g_2 \to u_1u_2$ (see Figure 1.7) enhances the former and reduces the latter thanks to configurations interaction:

$$\begin{aligned}\langle\Psi_0|\hat{H}|\Psi_{g_1g_2}^{u_1u_2}\rangle = \langle g_1g_2|\hat{H}|u_1u_2\rangle &= \tfrac{1}{4}(a^2 - b^2, c^2 - d^2)\\ &= \tfrac{1}{4}(J_{ac} + J_{bd} - J_{bc} - J_{ad}).\end{aligned}$$

The bielectronic Coulomb integrals can be approximated as the inverse of interatomic distances,

$$J_{ac} = J_{bd} \approx \frac{1}{L}$$

$$J_{ad} \approx \frac{1}{L+l}, \quad J_{bc} \approx \frac{1}{L-l}.$$

Thus, a second-order development ($l \ll L$) gives

$$\langle g_1g_2|H|u_1u_2\rangle \approx -\frac{l^2}{2L^3}.$$

Using second-order perturbation theory to evaluate the correlation energy, the $L^{-6}$ dependence of the dispersion energy is recovered.

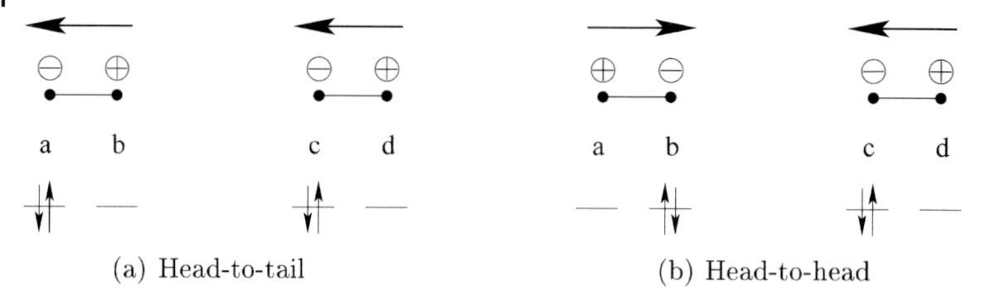


**Figure 1.8** Head-to-tail dipole interactions are favored from correlation effects.

The origin of the dispersion energy is clear in this procedure. Indeed, the development of the doubly excited determinant $|\Psi_{g_1g_2}^{u_1u_2}\rangle$ on the atomic orbitals $a$, $b$, $c$, and $d$ (see Figure 1.8) exhibits the role of correlation between the fluctuations of the positions of the electrons in the two bonds. When the electrons move from $b$ to $a$, then the probability of a concerted displacement from $d$ to $c$ is larger than the one of a movement from $c$ to $d$.

### 1.3.2
### Open-Shell Ligands: Noninnocence Concept

Considering the possibility of generating high oxidation states ions (in iron chemistry for instance, let us mention notable examples of Fe(IV) [49, 50], Fe(V) [51–53] and Fe(VI) [54]), much synthetic effort has been devoted to the preparation of specific multidentate ligands. The use of such ligands, known as *noninnocent*, has opened up the route to original synthetic materials, involving open shells on both metal and ligands partners [55–61]). The spectacular *excited-state coordination* chemistry concept in which a ligand coordinates in an excited electronic state to a metal center has emerged from this class of compounds [62]. The generation of radical ligands in coordination compounds has given rise to a promising route to magnetic materials.

From the theoretical point of view, DFT as well as CI calculations have been undertaken to scrutinize the electronic structures of such *noninnocent* ligand-based systems [58–60, 62, 63]. In particular, the comparison between experimental and calculated exchange-coupling constants and the analysis of the magnetic interactions has been the subject of intense work. While DFT has sometimes failed to fully account for the low-energy spectroscopy, the wavefunction-based DDCI method has elucidated the unusual behavior of several complexes [58, 62]. Among those, a striking example is given by the Fe(gma)CN complex containing the glyoxalbis(mercaptoanil) (gma) ligand (see Figure 1.9) [22]. Even though the *noninnocent* character of the gma ligand was clearly demonstrated both experimentally and theoretically, DFT calculations were only partially successful in the description of the electronic structure of the full complex [62]. The magnetic susceptibility and zero-field Mössbauer measurements clearly favored a doublet ground state. Nevertheless, DFT calculations did not provide any clear evidence in that sense, the $M_s = 1/2$ solution exhibiting a low-spin Fe(III) ($S_{\mathrm{Fe}} = 1/2$) coupled to a closed shell gma ligand ($S_{\mathrm{gma}} = 0$). Clearly, for a good

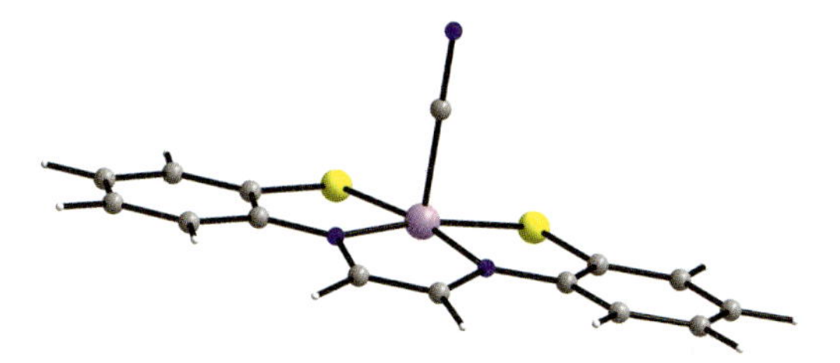

**Figure 1.9** Structure of Fe(gma)CN. Fe, S, N, C, and H are represented in purple, yellow, blue, gray, and white, respectively.

description of the electronic structure of such system, DFT and its monodeterminantal character is not appropriate and correlated *ab initio* calculations might be desirable.

Based on this statement, correlated *ab initio* calculations on this particular system by means of DDCI-2 calculations on the top of the CAS(5,5)SCF wavefunction were performed [22]. Interestingly, the active orbitals consist of three metal-centered and two ligand-centered MOs (see Figure 1.10) [62]. The calculations showed that the low-energy spectrum exhibits a 200 $cm^{-1}$ quartet–doublet gap, in agreement with different experiments, and that the observed strong antiferromagnetic is due to important ligand-to-metal charge transfer (LMCT). The resulting ground-state wavefunction which exhibits an intermediate magnetic/covalent character is rather strongly correlated and is dominated by local ($S_{Fe} = 3/2$ and $S_{gma} = 1$) electronic configuration. Finally, whereas the gma ligand is clearly a closed-shell singlet

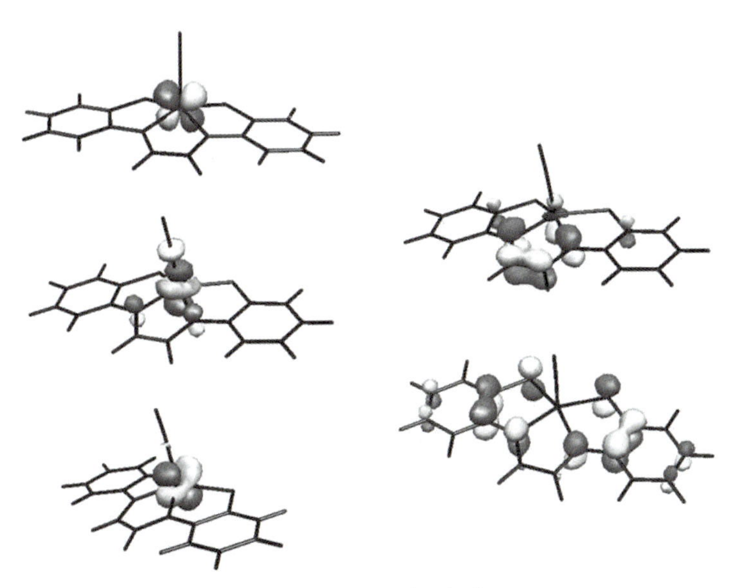

**Figure 1.10** Optmized active average MOs for the lowest doublet and quartet state of Fe(gma)CN.

when considered alone, it is likely to be a triplet when coordinated to the iron center. The multiconfigurational nature of the wavefunction has been identified in this example and makes this class of compounds still very challenging for theoreticians. It has been recently suggested that the energetics of low-lying states of coordination complexes based on porphyrins and related entities may not be accessible by means of DFT methodology (see Ref. [23] and references therein). More troublesome is the dependence of the spin density maps on the functional choice.

### 1.3.3
### Growing 1D Materials: Ni-Azido Chains

With the generation of magnetic properties goal in mind, experimentalists have prepared higher dimensionality materials. One of the main challenges in the synthesis of extended 1D systems is to prevent the local magnetic moments from canceling out. In the presence of most frequent antiferromagnetic interactions, pioneer approaches were devoted to regular heterospin ferrimagnetic chains [12] holding alternating spin carriers, coupled through a unique exchange constant. Another strategy consists in varying the magnetic exchange constants between homospin carriers [64, 65]. Finally, the use of strong anisotropic metal ions to reduce the magnetization relaxation has generated the promising field of the single-chain magnet (SCM) [13–15].

In this respect, the azido ligand turned out to be extremely appealing in linking metal ions and a remarkable magnetic coupler for propagating interactions between paramagnetic ions. The structural variety of the azido complexes ranges from molecular clusters to extended 1D to 3D materials [66–71]. An interesting prototype of such a system has been recently synthesized where a single azido unit bridges in an alternating End-On (EO) and End-to-End (EE) way the Ni(II) ions (see Figure 1.11) [72]. The system can be considered from the chemical point of view as a quasi-1D chain. However, based on magnetic susceptibility measurements, it was suggested that the system should be described from the magnetic point of view as isolated dimers. Indeed, the introduction of a second magnetic interaction was shown to be irrelevant. Therefore, the question of the nature and amplitude of the magnetic interactions between the nearest Ni(II) ions deserved special attention. The alternation of EO and EE units strongly suggested the presence of two magnetic exchange pathways which can be accessible through $Ni_2$ dimers spectroscopy analysis. Thus, CAS(5,6)SCF/DDCI-2 calculations were performed on the molecular EE and EO fragments extracted from the available crystal structure.

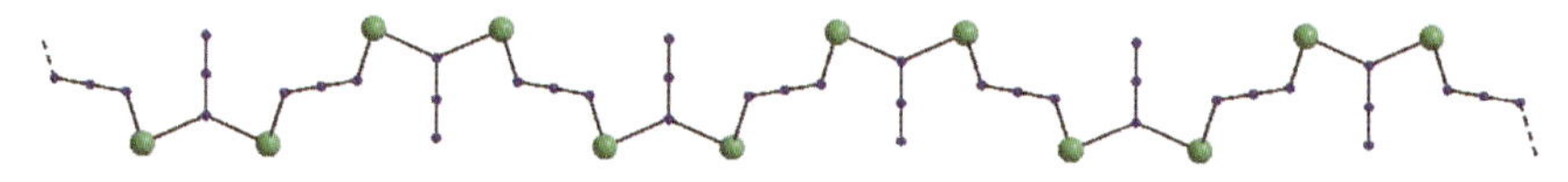

**Figure 1.11** Nickel(II) chain $\{Ni_2(\mu_{1,1}\text{-}N_3)(\mu_{1,3}\text{-}N_3)(L)_2(MeOH)_2]\}_n$ with alternating End-On/End-to-End single azido bridges.

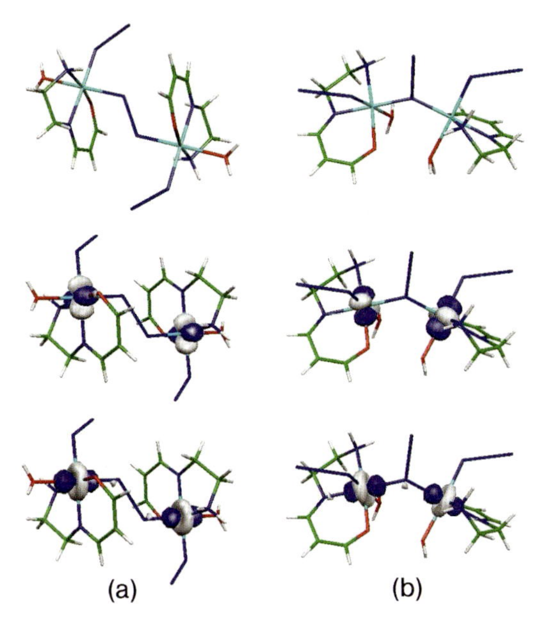


**Figure 1.12** Molecular EE (a) and EO (b) fragments and in-phase active metallic MOs. For the sake of simplicity, the out-of-phase combinations are not shown.

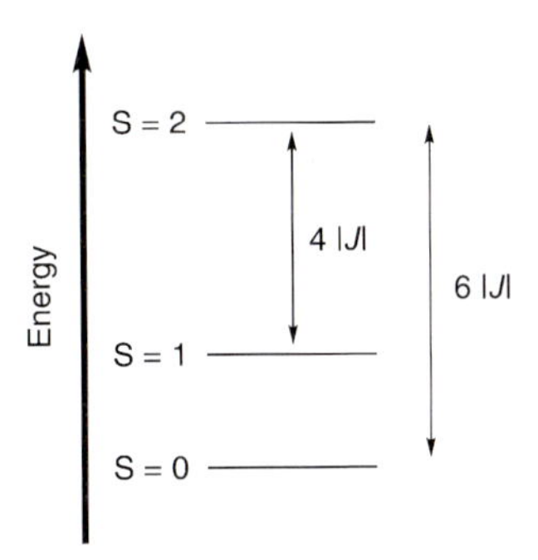


**Figure 1.13** Energy spectrum of a two-center Heisenberg $S = 1$ Hamiltonian.

The active orbitals consist of the in-phase and out-of-phase linear combinations of the $d_{z^2}$ and $d_{x^2-y^2}$ metallic AOs (see Figure 1.12) and the nonbonding MO of the $N_3^-$ bridge.

Since the Ni(II) ion is formally $d^8$, it is expected that exchange interactions between $S = 1$ ions should give rise to three spin states in the $Ni_2$ units, namely singlet ($S$), triplet ($T$), and quintet ($Q$) states. In a Heisenberg picture $\hat{H} = -2J\hat{S}_1\hat{S}_2$ ($S_1 = S_2 = 1$), the energy separations are $6|J|$ and $4|J|$ between the quintet and singlet, quintet and triplet states, respectively (see Figure 1.13). Within the EE unit, a relatively large antiferromagnetic exchange constant ($J_{EE} \sim -50\ \mathrm{cm}^{-1}$) was calculated in good agreement with the unique value extracted from experiment

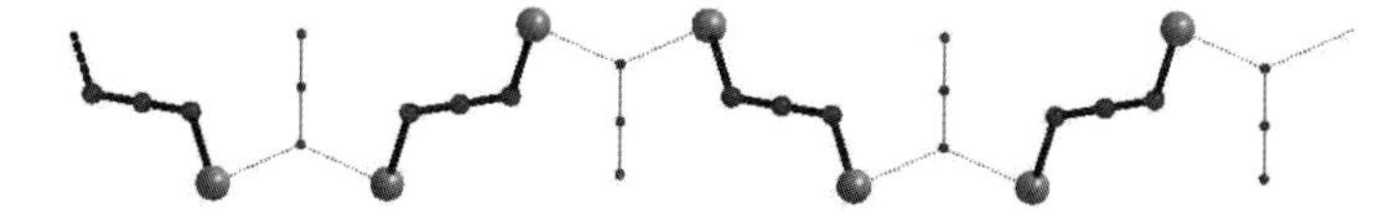

**Figure 1.14** Schematic representation of the Ni-azido chain resulting from the isolated EE dimer picture.

($\sim -40\ \text{cm}^{-1}$). This is to be contrasted with the EO $Ni_2$ unit, which exhibits a negligibly small magnetic interaction ($\alpha = J_{\text{EO}}/|J_{\text{EE}}|$ ratio $\sim$0.02) (see Figure 1.14).

The correlated calculations not only confirmed the isolated dimers picture, but also associated the leading antiferromagnetic exchange pathway with the EE bridging mode. In the light of the calculated $(E_Q - E_S)/(E_Q - E_T)$ ratio, let us mention that the deviation from a pure Heisenberg picture is negligible (less than 2%) ruling out the speculated participation of quadratic terms. The attempt to generate high-enough ferromagnetic interactions between $S = 1$ sites looked very promising since the antiferromagnetic coupling between the resulting $S = 2$ units through EE bridges might have resulted in a Haldane chain with vanishingly small spin gap [73, 74]. The versatility of the azido magnetic coupler should still be considered to generate synthetic models for theoretical physics analysis.

## 1.4
## Discussion and Concluding Remarks

Quantum chemical calculations have become valuable means of investigation which cannot be ignored. As spectroscopy accuracy can be reached down to several tens of wavenumbers, *ab initio* techniques have the ability to rationalize interactions in magnetic materials. Interestingly, the different contributions to energy splitting are accessible and the underlying physical phenomena can be interpreted. The information which is conveyed by the wavefunction is crucial in the characterization of model Hamiltonians. Undoubtedly, significant efforts must be devoted to extract the relevant parameters in a "boil down" procedure of the *ab initio* information. Even though certain CI methodologies might be very demanding when dealing with large systems such as enzyme active sites, they allow one to manipulate symmetry and spin-adapted eigenstates of the exact Hamiltonian. The impressive demand for catalyzed reactions interpretation has led to a spectacular developments of DFT-based tools dedicated to surface-type issues. Popular codes take advantage of the crystal periodicity by introducing plane waves rather than localized atomic orbitals. It is noteworthy that some recent works have suggested that *aposteriori* corrections should be performed on the reaction site cluster embedded in a periodic environment. Such methodology has opened up new routes to important issues involving biological systems. Nevertheless, some specific systems including open-shell compounds are the concern of explicitly correlated calculations which allow an efficient treatment of both nondynamical and dynamical correlations.

## References


1 Roos, B. O. *Adv. Chem. Phys.* **1987**, *LXIX*, 399.
2 Olsen, J., Roos, B. O., Jorgensen, P., Jensen, H. J. A. *J. Chem. Phys.* **1988**, *89*, 2185.
3 Bednorz, J. G., Müller, K. A. *Z. Phys. B: Condens. Matter* **1986**, *64*, 189.
4 Dagotto, E. *Rev. Mod. Phys.* **1994**, *66*, 763.
5 Nagaosa, N. *Science* **1997**, *275*, 1078.
6 Kiryukhin, V., Casa, D., Hill, J. P., Kelmer, B., Vigliante, A., Tomioka, Y., Tokura, Y. *Nature* **1997**, *386*, 813.
7 Asamitsu, A., Tomioka, Y., Kuwahara, H., Tokura, Y. *Nature* **1997**, *388*, 50.
8 Fäth, M., Freisem, S., Menovsky, A. A., Tomioka, Y., Aarts, J., Mydosh, J. A. *Science* **1999**, *285*, 1540.
9 Kimura, T., Tokura, Y. *Annu. Rev. Mater. Sci.* **2000**, *30*, 451.
10 Yoo, Y. K., Duewer, F., Yang, H., Yi, D., Li, J.-W., Xiang, X.-D. *Nature* **2000**, *406*, 704.
11 Dagotto, E., Hotta, T., Moreo, A. *Phys. Rep.* **2001**, *344*, 1.
12 Kahn, O. *Molecular Magnetism;* New York, Wiley-VCH, **1993**.
13 Caneschi, A., Gatteschi, D., Lalioti, N., Sangregorio, C., Sessoli, R., Venturi, G., Vindigni, A., Rettori, A., Pini, M. G., Novak, M. A. *Angew. Chem. Int. Ed.* **2001**, *40*, 1760.
14 Coulon, C., Miyasaka, H., Clérac, R. Single-Chain Magnets: Theoretical Approach and Experimental Systems. In *Single-Molecule Magnets and Related Phenomena*, Vol. 122; Winpenny, R., Ed.; Springer: Berlin, Heidelberg, **2006**.
15 Lescouëzec, R., Toma, L. M., Vaissermann, J., Verdaguer, M., Delgado, F. S., Ruiz-Pérez, C., Lloret, F., Julve, M. *Coord. Chem. Rev.* **2005**, *249*, 2691.
16 Yoon, J., Mirica, L. M., Stack, T. D. P., Solomon, E. I. *J. Am. Chem. Soc.* **2004**, *126*, 12586.
17 Yoon, J., Solomon, E. I. *Coord. Chem. Rev.* **2007**, *251*, 379.
18 Noodleman, L., Norman, J. G. *J. Chem. Phys.* **1979**, *70*, 4903.
19 Noodleman, L. *J. Chem. Phys.* **1981**, *74*, 5737.
20 Noodleman, L., Case, D. A. *Adv. Inorg. Chem.* **1992**, *38*, 423.
21 Noodleman, L., Peng, C. Y., Case, D. A., Mouesca, J.-M. *Coord. Chem. Rev.* **1995**, *144*, 199.
22 Messaoudi, S., Robert, V., Guihéry, N., Maynau, D. *Inorg. Chem.* **2006**, *45*, 3212.
23 Ghosh, A. *J. Biol. Inorg. Chem.* **2006**, *11*, 712.
24 Pierloot, K., Vancoillie, S. *J. Chem. Phys.* **2006**, *125*, 124303.
25 Caballol, R., Castell, O., Illas, F., de P. R. Moreira, I., Malrieu, J. P. *J. Phys. Chem. A* **1997**, *101*, 7860.
26 Kahn, O., Martinez, C. J. *Science* **1998**, *279*, 44.
27 Christou, G., Gatteschi, D., Hendrickson, D. N., Sessoli, R. *MRS Bull.* **2000**, *25*, 66.
28 Gatteschi, G., Sessoli, R. *Angew. Chem. Int. Ed.* **2003**, *42*, 268.
29 Bogani, L., Sangregorio, C., Sessoli, R., Gatteschi, D. *Angew. Chem. Int. Ed.* **2005**, *44*, 5817.
30 Ruzsinszkky, A., Perdew, J. P., Csonka, G. I. *J. Phys. Chem. A* **2005**, *109*, 11015.
31 Tuma, C., Sauer, J. *Chem. Phys. Lett.* **2004**, *387*, 388.
32 Tuma, C., Sauer, J. *Phys. Chem. Chem. Phys.* **2006**, *8*, 3955.
33 Born, M., Oppenheimer, R. *Annal. Phys.* **1927**, *84*, 457.
34 Szabo, A., Attila, N. S. *Modern Quantum Chemistry;* New York, Dover, **1982**.

35 Neese, F. ORCA – An *Ab Initio*, Density Functional and Semi-Empirical Program Package, Max-Planck Institute for Bioinorganic Chemistry, Mülheim an der Ruhr, Germany, **2005**.
36 Cabrero, J., Ben Amor, N., de Graaf, C., Illas, F., Caballol, R. *J. Phys. Chem. A* **2000**, *104*, 9983.
37 Miralles, J., Daudey, J.-P., Caballol, R. *Chem. Phys. Lett.* **1992**, *198*, 555.
38 Cabrero, J., de Graaf, C., Bordas, E., Caballol, R., Malrieu, J.-P. *Chem. Eur. J.* **2003**, *9*, 2307.
39 de P. R. Moreira, I., Illas, F., Calzado, C. J., Sanz, J. F., Malrieu, J.-P., Ben Amor, N., Maynau, D. *Phys. Rev. B* **1999**, *59*, R6593.
40 Suaud, N., Lepetit, M.-B. *Phys. Rev. B* **2000**, *62*, 402.
41 Munoz, D., Illas, F., de P. R. Moreira, I. *Phys. Rev. Lett.* **2000**, *84*, 1579.
42 Petit, S., Borshch, S. A., Robert, V. *J. Am. Chem. Soc.* **2002**, *124*, 1744.
43 Lindgren, I., Morrison, J. *Atomic Many-Body Theory;* Berlin, Springer-Verlag, **1982**.
44 Daudey, J.-P., Heully, J.-L., Malrieu, J.-P. *J. Chem. Phys.* **1993**, *99*, 1240.
45 de Loth, P., Cassoux, P., Daudey, J.-P., Malrieu, J.-P. *J. Am. Chem. Soc.* **1981**, *103*, 4007.
46 Calzado, C. J., Cabredo, J., Malrieu, J. P., Caballol, R. *J. Chem. Phys.* **2002**, *116*, 2728.
47 Calzado, C. J., Cabredo, J., Malrieu, J. P., Caballol, R. *J. Chem. Phys.* **2002**, *116*, 3985.
48 Bartlett, R. J., Musial, M. *Rev. Mod. Phys.* **2007**, *79*, 291.
49 Collins, T. J., Uffelman, E. S. *Angew. Chem., Int. Ed. Engl.* **1989**, *28*, 1509.
50 Kostka, K. L., Fox, B. G., Hendrich, M. P., Collins, T. J., Rickard, C. E. F., Wright, L. J., Munck, E. *J. Am. Chem. Soc.* **1993**, *115*, 6746.
51 Meyer, K., Bill, E., Mienert, B., Weyhermüller, T., Wieghardt, K. *J. Am. Chem. Soc.* **1999**, *121*, 4859.
52 Grapperhaus, C. A., Mienert, B., Bill, E., Weyhermüller, T., Wieghardt, K. *Inorg. Chem.* **2000**, *39*, 5306.
53 Wasbotten, I., Ghosh, A. *Inorg. Chem.* **2006**, *45*, 4910.
54 Berry, J. F., Bill, E., Bothe, E., George, S. D., Mienert, B., Neese, F., Wieghardt, K. *Science* **2006**, *312*, 1937.
55 Dutta, S. K., Beckmann, U., Bill, E., Weyhermüller, T., Wieghardt, K. *Inorg. Chem.* **2000**, *39*, 3355.
56 Chaudhuri, P., Verani, C. N., Bill, E., Bothe, E., Weyhermüller, T., Wieghardt, K. *J. Am. Chem. Soc.* **2001**, *123*, 2213.
57 Herebian, D., Bothe, E., Bill, E., Weyhermüller, T., Wieghardt, K. *J. Am. Chem. Soc.* **2001**, *123*, 10012.
58 Herebian, D., Wieghardt, K. E., Neese, F. *J. Am. Chem. Soc.* **2003**, *125*, 10997.
59 Bachler, V., Olbrich, G., Neese, F., Wieghardt, K. *Inorg. Chem.* **2002**, *41*, 4179.
60 Sun, X., Chun, H., Hildenbrand, K., Bothe, E., Weyhermüller, T., Neese, F., Wieghardt, K. *Inorg. Chem.* **2002**, *41*, 4295.
61 Beckmann, U., Bill, E., Weyhermüller, T., Wieghardt, K. *Inorg. Chem.* **2003**, *42*, 1045.
62 Ghosh, P., Bill, E., Weyhermüller, T., Neese, F., Wieghardt, K. *J. Am. Chem. Soc.* **2003**, *125*, 1293.
63 Ray, K., Petrenko, T., Wieghardt, K., Neese, F. *Dalton Trans.* **2007**, 1552.
64 Vicente, R., Escuer, A., Ribas, J., Solans, X. *Inorg. Chem.* **1992**, *31*, 1726.
65 Julve, M., Lloret, F., Faus, J., De Munno, G., Verdaguer, M., Caneschi, A. *Angew. Chem. Int. Ed.* **1993**, *32*, 1046.
66 Ribas, J., Escuer, A., Monfort, M., Vicente, R., Cortés, R., Lezama, L., Rojo, T. *Coord. Chem. Rev.* **1999**, *193–195*, 1027.
67 Ribas, J., Escuer, A., Monfort, M., Vicente, R., Cortes, R., Lezama, L., Rojo, T., Goher, M. A. S. In *Magnetism: Molecules to Materials II: Molecule-Based Materials;* S., M. J., Drillon, M., Eds.; Wiley-VCH: Weinheim, **2002**.
68 Manson, J. L., Arif, A. M., Miller, J. S. *Chem. Commun.* **1999**, 1479.

69 Liu, F.-C., Zeng, Y.-F., Jiao, J., Bu, X.-H., Ribas, J., Batten, S. R. *Inorg. Chem.* **2006**, *45*, 2776.
70 Monfort, M., Resino, I., Ribas, J., Stoeckli-Evans, H. *Angew. Chem. Int. Ed.* **2000**, *39*, 191.
71 Mautner, F. A., Cortes, R., Lezema, L., Rojo, T. *Angew. Chem. Int. Ed.* **1996**, *35*, 96.
72 Bonnet, M.-L., Aronica, C., Chastanet, G., Pilet, G., Luneau, D., Mathonière, C., Clérac, R., Robert, V. *Inorg. Chem.* **2008**, *47*, 1127.
73 Haldane, F. D. M. *Phys. Rev. Lett.* **1983**, *50*, 1153.
74 Yamashita, M., Ishii, T., Matsuzaka, H. *Coord. Chem. Rev.* **2000**, *198*, 347.

# 2
# Basic Aspects of Density Functional Theory[1)]

Evert Jan Baerends, Philippe Sautet, and Rutger van Santen

## 2.1
## Introduction

The popularity of density-functional-theory-based computational electronic structure methods [1] is that it can be applied to large systems with energy accuracy typically between 10 and 40 kJ mol$^{-1}$.

The essential feature that has opened this possibility is the discovery of the two Hohenberg–Kohn theorems [2], which demonstrated that the electronic energy of any system is defined by its electron density. This contrasts with the Hartree–Fock-based methods, discussed in Chapter 1, which require the (approximate) solution of the Schrödinger equation to obtain a multielectron wavefunction. Equation (2.1) represents the approximate wavefunction of the Hartree–Fock model, a single determinant of occupied orbitals. Its expectation value for the total Hamiltonian, $\langle \Phi_0^{\mathrm{HF}} | \hat{H} | \Phi_0^{\mathrm{HF}} \rangle = E_0^{\mathrm{HF}}$, is necessarily, according to the variation theorem, an upper bound to the exact total energy.

*Hartree–Fock model: orbitals*

$$\Phi_0^{\mathrm{HF}} = |\varphi_1(1)\varphi_2(2)\cdots\varphi_N(N)| \; E_0^{\mathrm{HF}} \geq E_0^{\mathrm{exact}} \tag{2.1}$$

$$\Psi_0 = c_0\Phi_0^{\mathrm{HF}} + \sum_{i,a} c_i^a \Phi_i^a + \sum_{ij,ab} c_{ij}^{ab}\Phi_{ij}^{ab} + \cdots. \tag{2.2}$$

The Hartree–Fock model neglects electron correlation. This is apparent from the one-electron equations that are obeyed by the orbitals (one-electron wavefunctions) in the Hartree–Fock determinantal wavefunction: the Fock operator describes the potential of an electron in an *average* field of the other electrons, and does not depend on the instantaneous positions of the other electrons, with a probability distribution that depends on the position of the reference electron; correlated motion of the electrons is thus neglected.

1) A summary of the lectures by E.J. Baerends and P. Sautet written by R.A. van Santen.

*Computational Methods in Catalysis and Materials Science.* Edited by Rutger A. van Santen and Philippe Sautet

ISBN: 978-3-527-32032-5

The exact solution of the many-electron Schrödinger equation is found by expanding the many-electron wavefunction into a linear combination of Slater determinants (Eq. (2.2)) that describe all possible excitations of the system in terms of electron transitions in the space of the Hartree–Fock orbitals. This is the configuration interaction (CI) method (see Chapter 1). The simultaneous excitation of two electrons leads to the determinants that are most important for the improvement of the Hartree–Fock wavefunction. The Brillouin theorem states that the variation principle that leads to the Hartree–Fock equations implies that matrix elements between the HF determinant and all singly excited determinants are zero. This holds, since the Hartree–Fock wavefunction represents a variational minimum, in the sense that a small variation of an orbital in the HF determinant (an occupied orbital), orbital $\varphi_i$ say, by admixing an infinitesimal amount of an unoccupied orbital, $\varphi_a$ say, should lead to a zero energy change to first order. This condition is obeyed if the Hamiltonian matrix element of the HF determinant and the singly substituted determinant $\Phi_i^a$ is zero. The Brillouin condition leads to small coefficients (vanishing in first order) for the singly excited determinants in the expansion (2.2). Nevertheless, they cannot be neglected, and CI calculations including all single and double excitations are the lowest order CI calculations that are routinely used. Already these calculations are fairly time-consuming, and any improvement by incorporating higher excitations makes the resulting computational problem highly demanding for large systems.

An essential simplification in the computational methods is afforded by the Hohenberg–Kohn theorems. The first Hohenberg–Kohn theorem states that all ground state properties of an $N$-electron system are uniquely determined by the ground state electron density of the system. Hohenberg and Kohn proved that different systems, which are characterized by different "external" potentials for the electrons, must have different ground state electron densities and different ground state wavefunctions. The ground state densities and ground state wavefunctions are therefore in one-to-one correspondence. (For molecules and solids, the external potential is usually just the potential field of the nuclei of the system.) The theorem does not depend on the precise form of the electron–electron interaction, it may be the familiar distance dependence $1/r_{ij} = 1/|\mathbf{r}_i - \mathbf{r}_j|$, or some other $r_{ij}$ dependent form, or just zero. The last case will prove to be important, see below. The most important property to which this theorem is applied is the ground state energy, as well as its components, the kinetic and potential energy terms: they are all functionals of the electron density. In this formulation, the nuclei are kept fixed (they generate the potential field), which is the so-called Born–Oppenheimer approximation. The kinetic energy therefore just consists of the kinetic energy of the electrons (expectation value of $\hat{T}_e$), and the potential energy has electron–nuclear interaction energy and electron–electron interaction energy components (expectation values of $\hat{V}_{\text{ne}} = \sum_i v(\mathbf{r}_i)$ and $\hat{V}_{\text{ee}} = \sum_{i<j} 1/r_{ij}$, respectively).

Writing the total Hamiltonian as

$$\hat{H}_{\text{el}} = \hat{T}_{\text{e}} + \hat{V}_{\text{ne}} + \hat{V}_{\text{ee}}, \tag{2.3}$$

we can write the total ground state energy in terms of functionals of the densities belonging to the set of ground state densities $\rho_0$,

$$E_0[\rho_0] = T_e[\rho_0] + E_{ne}[\rho_0] + E_{ee}[\rho_0]. \tag{2.4}$$

Each of these terms is defined as an expectation value of the corresponding operator with respect to the unique ground state wavefunction belonging to $\rho_0$, $\Psi[\rho_0]$. Hohenberg and Kohn have also proven a variational principle: if one keeps the external potential fixed, and searches through the set of ground state densities, then the minimum value of the energy is achieved for precisely the ground state density corresponding to that external potential:

$$E[\rho_{\text{trial}}] \geq E_0 \text{ and } E[\rho_0] = E_0. \tag{2.5}$$

Here again $E[\rho_{\text{trial}}] = \langle\Psi[\rho_{\text{trial}}]|\hat{H}|\Psi[\rho_{\text{trial}}]\rangle$. This is, however, not a practical scheme, since there is not an easy way to find the ground state wavefunction $\Psi[\rho_{\text{trial}}]$ belonging to $\rho_{\text{trial}}$. Although the functional $E_{\text{ne}}[\rho_{\text{trial}}] = \langle\Psi[\rho_{\text{trial}}]|\hat{V}_{\text{ne}}|\Psi[\rho_{\text{trial}}]\rangle = \int \rho_{\text{trial}}(\mathbf{r})v(\mathbf{r})d\mathbf{r}$ is easily calculated, the kinetic energy and the electron–electron energy cannot be obtained to any accuracy as functionals of $\rho_{\text{trial}}$. Kohn and Sham made an important step forward by considering the possibility to solve the problem using just one-electron equations [3]. Kohn and Sham made the following ansatz:

For any system of interacting ($\hat{V}_{\text{ee}} = \sum_{i<j} 1/r_{ij}$) electrons moving in a local potential $v(\mathbf{r})$ there exists a noninteracting electron system in a local potential $v_s(\mathbf{r})$, described by the one-electron equations

$$\left(-\frac{1}{2}\nabla^2 + v_s(\mathbf{r})\right)\psi_i^s(\mathbf{r}) = \varepsilon_i\psi_i^s(\mathbf{r}), \tag{2.6}$$

with the characteristic property that its electron density (from the $N/2$ lowest doubly occupied orbitals) is exactly equal to the density of the "real" interacting system. We note that the Hohenberg–Kohn theorem when applied to a noninteracting system assures us that there is only one such potential (potential and density are in one-to-one correspondence also for a noninteracting system). It is a special feature of the Kohn–Sham (KS) potential $v_s(\mathbf{r})$ that the energy of the highest occupied Kohn–Sham orbital is exactly equal to the negative of the ionization potential (IP) (if the potential is required to go to zero at infinity, which fixes the arbitrary constant that can always be added to a potential without consequences). This property results from the long-range behavior of the electron density of that orbital. Since in the asymptotic limit ($|\mathbf{r}| \to \infty$) only the slowest decaying density, which is the one of the occupied orbital with the highest orbital energy, survives, that orbital density has to decay in the same way as the total density. The orbital therefore must have – IP as its one-electron energy. This property is analogous to Koopmans's theorem in the Hartree–Fock theory. This theorem states that the negative of the orbital energies of the Hartree–Fock orbitals are *approximately* equal to ionization energies (errors are typically in the order of 1 eV). The HOMO orbital energy within the Kohn–Sham theory is *exactly* equal to minus the first ionization potential.

So we have the properties:

- $v_s(\mathbf{r})$ yields the *exact* $\rho$: $\rho_s(\mathbf{r}) = \sum_{i=1}^{N} |\psi_i^s(\mathbf{r})|^2 = \rho^{\text{exact}}(\mathbf{r})$
  (NB. the Hartree–Fock density is NOT exact)
- $v_s(\mathbf{r})$ is unique (apply HK theorem to system with $\hat{V}_{ee} = 0$)
- If: $v_s(\mathbf{r}) \xrightarrow[r\to\infty]{} 0$ : $\quad \varepsilon_{\text{HOMO}} = -\text{IP}_{\text{exact}}$

Kohn and Sham then proposed to write the *exact* total energy in the following form:

$$E = T_s[\rho] + \int \rho(\mathbf{r})v(\mathbf{r})d\mathbf{r} + \frac{1}{2}\int \frac{\rho(\mathbf{r}_1)\rho(\mathbf{r}_2)}{r_{12}}d\mathbf{r}_1 d\mathbf{r}_2 + E_{\text{xc}}[\rho]. \tag{2.7}$$

Here $T_s[\rho]$ is the kinetic energy of the noninteracting electrons with density $\rho$, which is precisely the kinetic energy of the one-electron wavefunctions, or Kohn–Sham orbitals, since the noninteracting electron system has the determinantal wavefunction with Kohn–Sham orbitals, $\Psi_s = |\psi_1^s(\mathbf{r}_1\alpha)\psi_1^s(\mathbf{r}_2\beta)\ldots\psi_H^s(\mathbf{r}_{N-1}\alpha)\psi_H^s(\mathbf{r}_N\beta)|$, as exact wavefunction:

$$T_s[\rho] = \sum_{i=1}^{N/2} 2\left\langle \psi_i^s(\mathbf{r}) \left| -\frac{1}{2}\nabla^2 \right| \psi_i^s(\mathbf{r}) \right\rangle. \tag{2.8}$$

($H = N/2$ is the number of the highest occupied molecular orbital, the HOMO, assuming double occupancy of all orbitals). The second and third terms in Eq. (2.7) are the electron–nuclear Coulomb energy and the electron–electron classical Coulomb interaction. These two energies will be the same as for the exact energy, since the Kohn–Sham density and the exact density are equal. The last term, $E_{\text{xc}}[\rho]$, is just a rest term: it is defined as the remaining energy after the three other terms have been subtracted from the exact energy. It is loosely called the "exchange-correlation energy," to distinguish it from the other terms which have simple, classical interpretations. It is, however, not the same as the traditional exchange and correlation energies of quantum chemistry. To make the distinction clear, we write the exact total energy in the conventional way,

$$E = T + \int \rho(\mathbf{r})v(\mathbf{r})d\mathbf{r} + \frac{1}{2}\int \frac{\rho(\mathbf{r}_1)\rho(\mathbf{r}_2)}{r_{12}}d\mathbf{r}_1 d\mathbf{r}_2 + W_{\text{XC}}. \tag{2.9}$$

Here $T$ is the exact kinetic energy, the electron–nuclear and electron–electron Coulomb terms are the same as before, and $W_{\text{XC}}$ is the correction on the electron–electron Coulomb energy, which has to be added to obtain the total electron–electron interaction energy, and which has been traditionally called the sum of the exchange and correlation energies,

$$W \equiv \langle \Psi | \hat{V}_{ee} | \Psi \rangle = \frac{1}{2}\int \frac{\rho(\mathbf{r}_1)\rho(\mathbf{r}_2)}{r_{12}}d\mathbf{r}_1 d\mathbf{r}_2 + W_{\text{XC}}. \tag{2.10}$$

By putting $E$ from Eq. (2.9) and Eq. (2.7) equal, it is now immediately apparent that the $E_{\text{xc}}$ of DFT contains, in addition to the traditional exchange and correlation effects, the correlation correction to the Kohn–Sham kinetic energy, $T_c = T - T_s$:

$$E_{\text{xc}} = T_c + W_{\text{xc}}, \tag{2.11}$$

where $E_{\text{xc}}$ is the great unknown of Kohn–Sham DFT. It is a functional of the density, as are the other terms in Eq. (2.7). However, this functional is unknown. But Kohn–Sham DFT has provided a strong incentive to try to develop good approximations to $E_{\text{xc}}[\rho]$, and the success of DFT is due to the development of rather good approximations, which go under names like LDA (local-density approximation) and GGA (generalized gradient approximation).

## 2.2 The Exchange-Correlation Potential

The Kohn–Sham potential that determines the energies and shape of the Kohn–Sham orbitals, according to Eq. (2.6), consists of three terms:

$$v_s(\mathbf{r}) = v(\mathbf{r}) + v_{\text{Coul}}(\mathbf{r}) + v_{\text{xc}}(\mathbf{r}). \tag{2.12}$$

Here $v(\mathbf{r})$ is the external (nuclear) potential, and $v_{\text{Coul}}(\mathbf{r})$ is the potential at point $\mathbf{r}$ of the electronic charge distribution $\rho(\mathbf{r}')$, $v_{\text{Coul}}(\mathbf{r}) = \int \rho(\mathbf{r}')/|\mathbf{r} - \mathbf{r}'|d\mathbf{r}'$. These are straightforward to calculate. The exchange-correlation potential $v_{\text{xc}}$ is the functional derivative of the exchange-correlation energy,

$$v_{\text{xc}}(\mathbf{r}) = \frac{\delta E_{\text{xc}}[\rho]}{\delta \rho(\mathbf{r})}. \tag{2.13}$$

Any difference between the Hartree–Fock orbitals and the Kohn–Sham orbitals has to come from differences between this local potential and the exchange operator of Hartree–Fock. Since $E_{\text{xc}}[\rho]$ is not known, Eq. (2.13) does not help at all to obtain further insight. However, it is possible to break down $v_{\text{xc}}(\mathbf{r})$ in three physically meaningful components, each of which can be determined from the wavefunction.

$$v_{\text{xc}}(\mathbf{r}) = v_{\text{xc}}^{\text{hole}}(\mathbf{r}) + v_{c,\text{kin}}(\mathbf{r}) + v^{\text{resp}}(\mathbf{r}). \tag{2.14}$$

The first term tends to dominate for two-electron systems. It describes the potential due to the exchange-correlation hole that an electron creates around itself in the total charge density when it is at position $\mathbf{r}_1$ in the system. An electron traveling through a molecule does not see just the total (time-averaged) charge density of $N$ electrons, with potential $v_{\text{Coul}}(\mathbf{r}_1) = \int \rho(\mathbf{r}_2)/r_{12}d\mathbf{r}_2$, but it is surrounded by a hole $\rho^{\text{hole}}(\mathbf{r}_1; \mathbf{r}_2)$ so that it sees a charge density at position $\mathbf{r}_2$ of magnitude $\rho(\mathbf{r}_2) + \rho^{\text{hole}}(\mathbf{r}_1; \mathbf{r}_2)$, see Figure 2.1. The hole is different at different positions $\mathbf{r}_1$ of

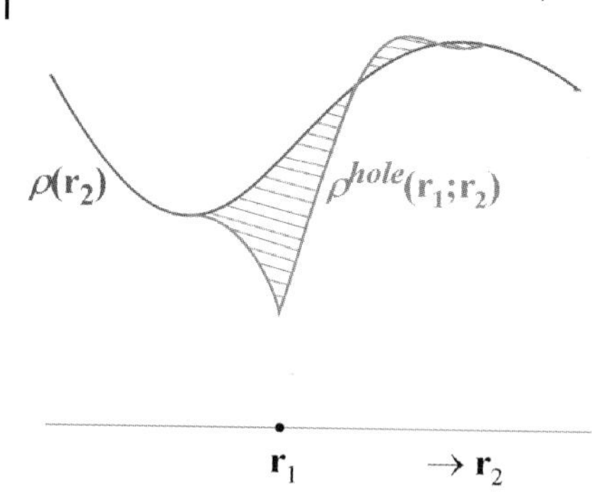


**Figure 2.1** The Coulomb field at the position $\mathbf{r}_1$ of an electron traveling through a molecule is not due to the total charge density $\rho(\mathbf{r}_2)$, but the electron is surrounded by a hole density that cancels the spurious self-repulsion present in $\rho(\mathbf{r}_2)$ and embodies electron correlation effects.

the electron. At any $\mathbf{r}_1$, the hole integrates to $-1$ electron: $\int \rho^{\mathrm{hole}}(\mathbf{r}_1; \mathbf{r}_2) d\mathbf{r}_2 = -1$, so the electron always sees $N-1$ other electrons. This is necessary: the electron cannot have repulsion with itself. This is called the self-interaction correction: The Coulomb potential of the total charge density $\rho(\mathbf{r}_2)$ incorporates the field of the electron itself and therefore entails a self-interaction error.

The self-interaction correction is already present in the Hartree–Fock model. The hole of this model is commonly denoted as *Fermi hole*. The total hole is not equal to the Fermi hole, since electron correlation effects change the shape of the hole. This correlation part of the hole is also called the *Coulomb hole*.

The Kohn–Sham exchange-correlation potential has a hole-potential contribution, which includes the Coulomb hole. This is a crucial difference with the Hartree–Fock theory. This can be explained very clearly with the prototype two-electron bond such as is present in the $H_2$ molecule. In the Hartree–Fock theory, the hole is just the Fermi hole or exchange hole, which in the case of a two-electron system (one occupied orbital) is only a self-interaction correction. The Fermi hole is just minus the occupied orbital squared, $-|\sigma_g(\mathbf{r}_2)|^2$. So it subtracts from the total density of the two electrons, $2|\sigma_g(\mathbf{r}_2)|^2$, the part which comes from the electron itself. Clearly, the Fermi hole is independent of the position $\mathbf{r}_1$ of the reference electron. It is static, i.e., does not move with the electron. In Figure 2.2, the exact Fermi hole is shown for two distances of the $H_2$ molecule, equilibrium distance and elongated. The reference electron is in the figure at a position $\mathbf{r}_1$ close to the right nucleus $b$. The Fermi hole would be, however, exactly the same whatever the position of the reference electron. The Coulomb hole, on the other hand, is very much dependent on the position of the reference electron. It has its greatest depth in the neighborhood of the reference electron, and is positive at larger distance. It has the effect of "polarizing" charge density away from the instantaneous position of the electron. When the bond length increases, we see this effect becomes very large: the Coulomb hole deepens the Fermi hole around its own nucleus and cancels the Fermi hole around the other nucleus. The net effect is that the total hole becomes

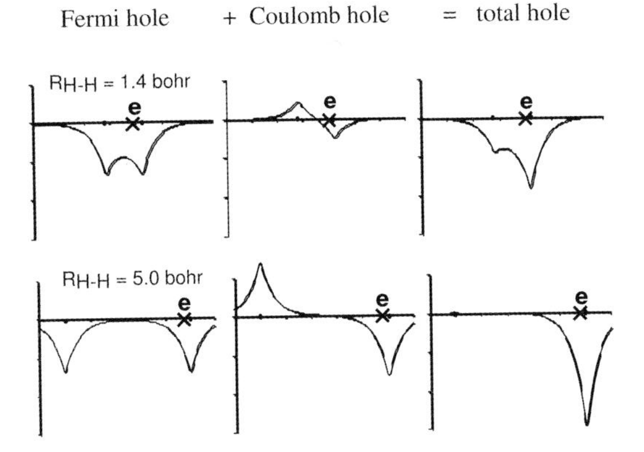


**Figure 2.2** The Fermi hole density, the Coulomb hole density, and the total hole density at equilibrium distance in $H_2$ and at a long bond length, for an electron at position indicated with x.

$-|1s_b(\mathbf{r}_2)|^2$: When an electron (our "reference electron") is close to nucleus $b$, the hole density $-|1s_b(\mathbf{r}_2)|^2$ is added to the total density $\rho(\mathbf{r}_2) \approx |1s_a(\mathbf{r}_2)|^2 + |1s_b(\mathbf{r}_2)|^2$, so that the electron close to the nucleus $b$ only sees a charge of $|1s_a(\mathbf{r}_2)|^2$ (the other electron) around the other nucleus, and sees its own nucleus without any screening charge from the other electron being present. This is exactly what we expect for a dissociating $H_2$ system that is yielding two H atoms.

It is well known that the (restricted) Hartree–Fock energy behaves erroneously at long distances of a dissociating electron pair bond, becoming much too high (above the energy of the fragments). We can explain this from the behavior of the Fermi hole. The Fermi hole subtracts at long bond length just $|\sigma_g(\mathbf{r}_2)|^2 = (1/2)|1s_a(\mathbf{r}_2)|^2 + (1/2)|1s_b(\mathbf{r}_2)|^2$ from the total density, so the reference electron close to $b$ experiences a field from the remaining density $(1/2)|1s_a(\mathbf{r}_2)|^2 + (1/2)|1s_b(\mathbf{r}_2)|^2$. So the reference electron at atom $b$ still sees an electron density $(1/2)|1s_b(\mathbf{r}_2)|^2$ on its own atom, instead of seeing the other electron completely at atom $a$. This gives too much electron–electron repulsion; hence the total energy becomes much too repulsive at long distance. In the Hartree–Fock model, the motion of the reference electron is not properly correlated with that of the other electron.

A consequence of the too repulsive nature of the Hartree–Fock potential is that the orbitals and the total electron density are not sufficiently contracted toward the nucleus, i.e., the Hartree–Fock densities are too diffuse. This yields quite substantial errors in energy terms like the kinetic energy (too diffuse orbitals, i.e., too low gradients, lead to too low kinetic energy) and the electron–nuclear energy (not negative enough). The Kohn–Sham orbitals build, by definition, the exact density. This can be understood from the fact that the Kohn–Sham potential not only incorporates the Fermi hole potential but also the Coulomb hole potential. In this sense the Kohn–Sham orbitals have better shapes than the Hartree–Fock orbitals. There is no reason not to trust the Kohn–Sham orbitals for qualitative MO reasoning. They are rather similar to Hartree–Fock orbitals since the major part of the potentials are the same (nuclear potential and Coulomb potential of the total electron density), and where the potentials differ (in the exchange *versus* exchange-correlation part), the Kohn–Sham potentials are physically more sensible. This also translates into good quality of the orbital energies, see the next section.

The local-density approximation uses at each point $\mathbf{r}_1$ of the reference electron the hole that belongs to the homogeneous electron gas with uniform density equal to the local density $\rho(\mathbf{r}_1)$ of the atom or molecule. This hole is localized around the reference electron, due to the spatial homogeneity of the uniform electron gas. In this sense, it is better than the Fermi hole of Hartree–Fock, which is spread out over the atoms participating in a bond. However, the properties of the homogeneous electron gas differ so much from real molecules that the shape of the homogeneous gas hole is rather deficient. Bond energies are therefore improved over Hartree–Fock values, which are typically 50% too low (not bonding enough), but the LDA model tends to overshoot, yielding too large bond energies. Better approximations are found when density functionals are used, which not only use the local density $\rho(\mathbf{r}_1)$ but also its gradients $\nabla\rho(\mathbf{r}_1)$, yielding the so-called generalized gradient approximations (GGAs). Even mixtures of Hartree–Fock type exchange and GGA functionals can be used that can give results comparable to Hartree–Fock MP2 configuration interaction methods (see Chapter 4).

## 2.3
## Physical Interpretation of Kohn–Sham Orbital Energies

When the exact electron density is known (for small systems this can be exactly calculated using independent methods) it is possible to generate in an iterative fashion the unique local potential which yields precisely that exact density in a Kohn–Sham calculation. This is by definition the exact Kohn–Sham potential. It can be used to solve for the Kohn–Sham orbitals and orbital energies. In contrast to the repeated statement in the literature that Kohn–Sham orbital energies, except for the highest occupied one, do not have any physical meaning, one also finds that the other occupied orbitals have energies that correspond very closely to ionization energies [4]. The upper valence orbitals have energies that on average deviate only a few tenths of an electron volt from the experimental IP, while in contrast the Hartree–Fock orbital energies typically deviate on average by ca. 1 eV. The deep valence levels and particularly the core levels show larger deviations (of the same order of magnitude as Hartree–Fock but with different sign). In Table 2.1, the results of (exact) Kohn–Sham calculations are compared to experimental IPs to illustrate this point. The table also shows that calculations with a KS potential derived from LDA or a GGA functional (in the table the Becke–Perdew GGA is used) yield orbital energies that are too high by ca. 5 eV. However, the shift with respect to the exact Kohn–Sham values is rather uniform, as can be seen by comparing the orbital energies that have been uniformly shifted by −4.83 eV from their GGA-BP values to the Kohn–Sham ones. The property that occupied orbital

**Table 2.1** CO orbital energies (eV) computed with the Hartree–Fock and Kohn–Sham methods (both exact Kohn–Sham and using the Becke–Perdew functional) compared with experimental ionization energies.

| CO MO | HF $-\varepsilon_i$ | GGA-BP $-\varepsilon_i$ | + Uniform shift −4.83 | KS $-\varepsilon_i$ | Experiment $I_i$ |
|---|---|---|---|---|---|
| $5\sigma$ | 15.12 | 9.18 | (14.01) | 14.01 | 14.01 |
| $1\pi$ | 17.42 | 11.95 | (16.78) | 16.80 | 16.91 |
| $4\sigma$ | 21.94 | 14.27 | (19.10) | 19.37 | 19.72 |
| $3\sigma$ | 41.47 | 29.47 | (34.29) | 34.70 | 38.3 |
| $2\sigma$ | 309.17 | 272.50 | (277.33) | 279.27 | 296.21 |
| $1\sigma$ | 562.36 | 513.53 | (518.37) | 519.92 | 542.55 |

energies are closely related to ionization energies is very useful and is generally assumed in qualitative molecular orbital considerations.

We next comment on the virtual orbital energies. The Hartree–Fock virtual orbitals are solutions in the full Coulomb potential of $N$ electrons, there is not a hole integrating to $-1$ that corrects the full density $\rho(\mathbf{r}_2)$. As a consequence, the orbital energies of the virtual orbitals are rather high, and virtual minus occupied orbital energy differences are not good approximations to excitation energies. The "bandgap" in the Hartree–Fock systems is too large because the excited electron moves in the full repulsive electrostatic field of all the ground state electrons, due to the absence of a proper description of the Fermi hole. This is very different for the Kohn–Sham virtual orbitals. All Kohn–Sham orbitals are solutions in the same local potential, also the virtual ones. The virtual orbitals therefore "see" $N-1$ electrons instead of $N$ electrons, just as the KS electrons in occupied orbitals do. The virtual KS orbitals are just excited one-electron states in the same potential. Therefore, the virtual minus occupied orbital energy differences are good approximations to excitation energies in the KS case.

These statements hold for the exact Kohn–Sham potential. The relation between the exact Kohn–Sham potential and approximate KS potentials, such as LDA or GGA potentials, can be illustrated with Figure 2.3 for the potentials in the $N_2$ molecule, plotted along the internuclear axis. The bond midpoint is at $z = 0$, and the N nucleus is at $z = 1$ bohr. It is evident that both the LDA potential and the GGA potential are shifted upward with respect to the KS potential. The difference between the exact potentials and approximate potentials is close to a constant (ca. 5 eV) over most of the molecular domain, in particular in the bonding region and in the outer tail of the molecular density. This causes a uniform shift of the approximate Kohn–Sham eigenvalue energies, as was observed in Table 2.1. However, if one considers occupied and virtual valence orbitals, which have their main amplitude in the molecular region, then both will experience the same shift. Therefore, orbital energy differences between virtual and occupied valence orbitals are rather similar for LDA or GGA on one hand and exact Kohn–Sham on the other hand. Valence excitation energies are usually described quite reasonably in LDA or GGA calculations. However, Rydberg orbitals are very diffuse and have an

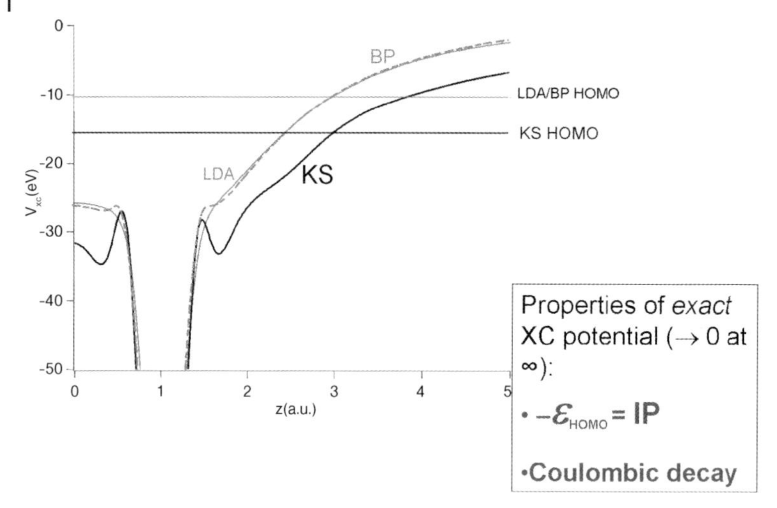


**Figure 2.3** Exchange correlation potentials for $N_2$ along the bond axis.

orbital energy close to zero, both in the LDA or GGA and the exact Kohn–Sham case. Therefore, the orbital energy difference for a Rydberg excitation suffers from the full error in the occupied orbital energy when it is calculated with LDA or GGA. The Kohn–Sham orbital energies, when they are calculated in exact Kohn–Sham calculations or with a good approximation, yield excellent Rydberg excitation energies. For a more detailed discussion of the computation of excitation energies we refer to Chapter 3.

As a general conclusion, we may state that the Kohn–Sham orbitals are physically meaningful and very suitable for electronic structure explanations in terms of qualitative MO considerations. The density is in principle exact, in practice rather accurate, which makes arguments concerning charge control of reactivity valid. The occupied and virtual orbital energies are physically meaningful, giving credibility to arguments based on orbital interaction schemes (orbital control of reactivity). The frequent allegation in the literature that the Kohn–Sham orbitals are unphysical, being only meant for and only suitable for describing the total density, and that the Kohn–Sham orbital energies have no meaning, is totally unfounded. The reverse is true.

## References


1 For a detailed introduction to density functional theory, see W. Koch, and M.C. Holthausen, *A Chemist's Guide to Density Functional Theory*, Wiley-VCH, Weinheim, **2000**.

2 P. Hohenberg and W. Kohn, *Phys. Rev.* 136(**1964**) B864.

3 W. Kohn and L.J. Sham, *Phys. Rev.* 140(**1965**) A1133.

4 D.P. Chong, O.V. Gritsenko, and E.J. Baerends, *J. Chem. Phys.* 116(**2002**) 1760.

# 3
# TDDFT for Excited States

*Mark E. Casida*

## 3.1
## Introduction

It may come as a surprise to younger researchers that DFT (density functional theory) was once considered a "four-letter word" by many traditional *ab initio* quantum chemists. The same younger researchers may also be surprised to learn that, even while DFT was being increasingly accepted in the early 1990s, it was still widely regarded as a firmly grounded principle that DFT could not handle excited states. Those attitudes have mostly changed. Typical modern quantum chemistry practice is now to use DFT where once Hartree–Fock (HF) was used to model the energy landscape and electronic properties of chemical systems, only carrying out HF calculations when needed as a starting point for sophisticated and costly *ab initio* many-body calculations. Moreover, time-dependent DFT (TDDFT) has become, with few exceptions, the primary single-reference method for treating excited states in medium- and large-sized molecules, leading to a veritable explosion of publications over the past decade. The main objective of this chapter is to answer the question "What is TDDFT?" by describing the formalism and mathematics behind modern "conventional" TDDFT. Some illustrations of the performance of the method will be given for the small molecule oxirane.

The use of TDDFT to treat electronic excited states is only about a decade old. Although several ideas had been put forth to deal with the excited-state problem [1], the fact remained that the original Hohenberg–Kohn theorems were for the ground state only and there was not yet any widely accepted way to treat excited states in quantum chemistry using DFT. This changed rapidly in the mid-1990s due in large part to an article entitled: "Time-dependent density-functional response theory for molecules" [2]. That article sent important messages to both the then still fairly separate DFT quantum chemistry and *ab initio* quantum chemistry communities. It explained to the DFT community that linear response theory could be used to reformulate TDDFT to resemble configuration interaction (CI) and hence to handle automatically configuration mixing in excited states. It explained to the *ab initio* community about the history and formal underpinnings of TDDFT, particularly the Runge–Gross theorems [3], and hence justified the adaptation

*Computational Methods in Catalysis and Materials Science.* Edited by Rutger A. van Santen and Philippe Sautet

ISBN: 978-3-527-32032-5

of recently developed technology for DFT second analytic derivatives to calculate excitation spectra using a formalism which very much resembled linear response time-dependent Hartree–Fock (LR-TDHF, also known as RPA for random phase approximation). Our own paper [4] and that of Bauernschmitt and Ahlrichs [5] soon confirmed that the method could give excellent excitation energies. Since then, LR-TDDFT has become part of virtually all quantum chemistry programs and many quantum physics programs around the world.

That TDDFT is an "unfinished subject" is attested to by the many variants which continue to appear, including but not limited to time-dependent current density functional theory [6, 7], propagator corrections [8–10], spin-flip theory [11–13], range-separated hybrids [14–17], and subsystem theory [18–21]. In line with these developments, many review articles on TDDFT have now appeared [2, 22–35, 35–39], each emphasizing a somewhat different aspect of TDDFT. Given the vastness of the subject, some choices have had to be made as to what aspects to present here. I have decided to emphasize those aspects which I think are of most interest to quantum chemists, namely the calculation of excitation spectra and the treatment of excited-state potential energy surfaces (PESs).

## 3.2 Formalism

TDDFT has its roots in ordinary ground-state DFT. This section presents the basic formalism of both exact DFT and exact TDDFT and says a few words about approximate functionals.

### 3.2.1 Ground-State Formalism

Before looking at the time-dependent problem, it is useful to begin with the time-independent (or static) problem. After the usual Born–Oppenheimer separation, the $N$-electron Hamiltonian, $\hat{H}$, of the time-independent Schrödinger equation,

$$\hat{H}\Psi_I = E_I\Psi_I; \quad E_0 \leq E_1 \leq \cdots, \tag{3.1}$$

can be separated, $\hat{H} = \hat{T} + \hat{V}_{ee} + \hat{V}_{ext}$, into a kinetic energy term, $\hat{T}$, the electron–electron repulsion, $\hat{V}_{ee}$, and the external potential, $\hat{V}_{ext} = \sum_{i=1}^{N} v_{ext}(\mathbf{r}_i)$, representing the interaction of the electrons with the electric field of the nuclei and/or some other applied potential.

The objective of DFT is to avoid having to solve the $N$-electron Schrödinger equation (3.1) and working with complicated $N$-electron wavefunctions, by working with a simpler entity, namely the ground-state charge density,

$$\rho(\mathbf{r}_1) = N \int \int \cdots \int |\Psi_0(\mathbf{x}_1, \mathbf{x}_2, \mathbf{x}_3, \ldots, \mathbf{x}_N)|^2 \, d\sigma_1 d\mathbf{x}_2 d\mathbf{x}_3 \cdots d\mathbf{x}_N, \tag{3.2}$$

where $\mathbf{x}_i = (\mathbf{r}_i, \sigma_i)$ is shorthand for the space and spin coordinates of electron $i$. That we can do this at all is the somewhat surprising content of the two Hohenberg–Kohn theorems [40].

The first Hohenberg–Kohn theorem (HK1) tells us that the ground-state charge density determines the external potential up to an arbitrary additive constant: $\hat{V}_{ext} + C \leftarrow \rho$. (If desired, this constant can be fixed after the fact for finite systems by requiring that the potential goes to zero at infinite distance.) Since integrating the charge density gives the number of electrons, HK1 tells us that the Hamiltonian is also determined up to an arbitrary additive constant: $\hat{H} + C \leftarrow \rho$. So the ground-state charge density fixes just about everything, including the excitation energies,

$$\hbar\omega_I = E_I - E_0, \tag{3.3}$$

and their associated oscillator strengths,

$$f_I = \frac{2m_e\omega_I}{3\hbar} \sum_{q=x,y,z} |\langle\Psi_0|q|\Psi_I\rangle|^2. \tag{3.4}$$

That is, $\rho$ determines the molecule's stick spectrum.[1] Thus both ground- and excited-state properties are functionals of the ground-state density, but we do not know these more general functionals.[2]

The second Hohenberg–Kohn theorem (HK2) tells us that there is a variational principle for the ground-state energy,

$$E_0 \leq F[\rho] + \int v_{ext}(\mathbf{r})\rho(\mathbf{r})\, d\mathbf{r}, \tag{3.5}$$

with equality only if $\rho$ is a ground-state charge density (degenerate ground states are allowed). The functional, $F$, is universal in the sense that it is independent of $v_{ext}$. It is *not* unknown because it may be written explicitly as

$$F[\rho] = \min_{\Psi \rightarrow \rho} \frac{\langle\Psi|\hat{T} + \hat{V}_{ee}|\Psi\rangle}{\langle\Psi|\Psi\rangle}. \tag{3.6}$$

This constrained variational form [41] is important for a number of reasons, including the facts that it eliminates the original Kohn–Sham $v$-representability assumption (i.e., that the trial densities can be realized as the ground state of some system) and applies to degenerate ground states. However, it is far from practical since it implies carrying out a calculation even more difficult than full

1) Molecular spectra are not stick spectra because they include broadening for a number of reasons, including but not limited to vibrational structure. However, gas phase oscillator strengths can be related to Beer's law extinction coefficient, $\epsilon(\omega)$, by $f_I = [(m_e c \ln(10))/(2\pi^2 N_A e^2)] \int \epsilon(\omega)\, d\omega$, in Gaussian electromagnetic units, where the integration is only over spectral features arising from the electronic excitation to the $I$th electronic excited state.

2) A functional is just a function of a function. It is designated with square brackets to distinguish it from ordinary functions of numbers which are designated by parentheses.

configuration interaction (CI). Thus simplifying approximations are still required if the Hohenberg–Kohn variational principle is to become practical.

These simplifications are aided by the Kohn–Sham reformulation of DFT [42]. A fictitious system of noninteracting electrons,

$$\left[-\frac{\hbar^2}{2m_e}\nabla^2 + v_s(\mathbf{r})\right]\psi_i(\mathbf{r}) = \epsilon_i\psi_i(\mathbf{r}), \tag{3.7}$$

is considered with $N$ occupied orthonormal orbitals which generate the same charge density as that of the interacting system: $\rho(\mathbf{r}) = \sum_i n_i|\psi_i(\mathbf{r})|^2 \leftarrow \Phi_s$. Here the $n_i$ are orbital occupation numbers, and $v_s$ and $\Phi_s$ are, respectively, the Kohn–Sham single-particle potential and the Kohn–Sham determinant made from the $N$ occupied orbitals. The total energy may now be written in a form reminiscent of the HF model,

$$\begin{aligned} E_{\text{KS}} = & -\frac{\hbar^2}{2m_e}\sum_i n_i\langle\psi_i|\nabla^2|\psi_i\rangle + \int v_{\text{ext}}(\mathbf{r})\rho(\mathbf{r})\,d\mathbf{r} \\ & + \frac{1}{2}\int\int\frac{\rho(\mathbf{r}_1)\rho(\mathbf{r}_2)}{r_{12}}\,d\mathbf{r}_1 d\mathbf{r}_2 + E_{\text{xc}}[\rho], \end{aligned} \tag{3.8}$$

where the exchange-correlation (xc) energy is given by

$$\begin{aligned} E_{\text{xc}}[\rho] = & \min_{\Psi\to\rho}\frac{\langle\Psi|\hat{T}+\hat{V}_{\text{ee}}|\Psi\rangle}{\langle\Psi|\Psi\rangle} - \min_{\Phi_s\to\rho}\frac{\langle\Phi_s|\hat{T}|\Phi_s\rangle}{\langle\Phi_s|\Phi_s\rangle} \\ & - \frac{1}{2}\int\int\frac{\rho(\mathbf{r}_1)\rho(\mathbf{r}_2)}{r_{12}}\,d\mathbf{r}_1 d\mathbf{r}_2, \end{aligned} \tag{3.9}$$

and contains not only exchange and correlation but also the difference between the kinetic energies of the real interacting and fictitious noninteracting systems. Minimizing the Kohn–Sham energy subject to the constraint of orthonormal orbitals gives Eq. (3.7), but with $v_s(\mathbf{r}) = v_{\text{ext}}(\mathbf{r}) + v_{\text{xc}}(\mathbf{r})$, where the xc-potential is the functional derivative[3)] of the xc-energy,

$$v_{\text{xc}}(\mathbf{r}) = \frac{\delta E_{\text{xc}}[\rho]}{\delta\rho(\mathbf{r})}. \tag{3.10}$$

The success of the applied DFT is due to the quality of available approximations for the xc-energy. Reviewing all of these approximations would be a chapter in and of itself and so we refer the reader to other works for a more complete treatment [44, 45]. However, we do need to say a few words here about some of the approximations because of their use in TDDFT calculations. (Indeed, TDDFT has proven to be one of the driving horses behind developing new functionals.)

**3)** The functional derivative of the functional $F[\rho]$ is defined by $F[\rho+\delta\rho] - F[\rho] = \int (\delta F[\rho]/\delta\rho(\mathbf{r}))\,\delta\rho(\mathbf{r})\,d\mathbf{r}$ for arbitrary infinitesimal variations, $\delta\rho(\mathbf{r})$. This definition may be compared with the more familiar definition of a partial derivative from multivariable calculus, $f(\mathbf{v}+d\mathbf{v}) - f(\mathbf{v}) = \sum_i(\partial f(\mathbf{v})/\partial v_i)dv_i$. The essential difference is that the sum over discrete indices has been replaced by an integral over a continuous index.

**Table 3.1** Jacob's ladder for functionals [43].

| Quantum chemical heaven | | |
|---|---|---|
| Double-hybrid | – | $\rho_\sigma(\mathbf{r}),\ x_\sigma(\mathbf{r}),\ \tau_\sigma(\mathbf{r}),\ \psi_{i\sigma}(\mathbf{r}),\ \psi_{a\sigma}(\mathbf{r})^{a}$ |
| Hybrid | – | $\rho_\sigma(\mathbf{r}),\ x_\sigma(\mathbf{r}),\ \tau_\sigma(\mathbf{r}),\ \psi_{i\sigma}(\mathbf{r})^{b}$ |
| mGGA | – | $\rho_\sigma(\mathbf{r}),\ x_\sigma(\mathbf{r}),\ \tau_\sigma(\mathbf{r})^{c}$ |
| GGA | – | $\rho_\sigma(\mathbf{r}),\ x_\sigma(\mathbf{r})^{d}$ |
| LDA | – | $\rho_\sigma(\mathbf{r})$ |
| Hartree World | | |

[a] Unoccupied orbitals.
[b] Occupied orbitals.
[c] The local kinetic energy $\tau_\sigma(\mathbf{r}) = \sum_i n_{i\sigma}\psi_{i\sigma}(\mathbf{r})\nabla^2\psi_{i\sigma}(\mathbf{r})$.
[d] The reduced gradient $x_\sigma(\mathbf{r}) = |\vec{\nabla}\rho_\sigma(\mathbf{r})|/\rho_\sigma^{4/3}(\mathbf{r})$.

Improvements in functionals come in two different but complementary ways. One is to make the functional form more exact without changing the variables on which the functional depends. The other approach is to increase the number of variables on which the functional depends. This latter approach consists in climbing Jacob's ladder shown in Table 3.1. While adding more variables should, in principle, make it easier to design more reliably accurate functionals, all that is really guaranteed in climbing the ladder is that calculations will become more expensive. Hence, it is also important to improve the functionals at the lower ends of the ladder.

The very first expansion of the variable list has been to include a dependence on the two densities $\rho_\uparrow$ and $\rho_\downarrow$, rather than the total density, $\rho$. This extension is so commonplace that DFT is commonly understood to mean spin DFT and we will follow this practice. The local density approximation (LDA) and generalized gradient approximations (GGAs) are examples of pure density functionals since they are orbital dependent. They were the traditional workhorses of DFT until Axel Becke proposed the use of hybrid functionals [46] including a fraction of orbital-dependent Hartree–Fock exchange on the basis of adiabatic connection theory and to improve the accuracy of DFT for thermochemistry. Hybrids are often found to lead to increased accuracy in TDDFT calculations, but are computationally more expensive. Meta-GGAs (mGGAs) are a step down the ladder and in the level of computational difficulty in that the orbital dependence is limited to calculating the local kinetic energy. Recently, Stephane Grimme and Frank Neese have moved up the ladder by recommending a double-hybrid of the form

$$E_{\text{XC}}^{\text{double-hybrid}} = a_x E_x^{\text{HF}} + (1 - a_x) E_x^{\text{GGA}} + a_c E_c^{\text{GGA}} + (1 - a_c) E_c^{\text{MP2}}, \qquad (3.11)$$

for applications in TDDFT, where $E_x^{\text{HF}}$ is the HF exchange energy, $E_x^{\text{GGA}}$ and $E_c^{\text{GGA}}$ are, respectively, GGA exchange- and correlation-energy functionals, and $E_c^{\text{MP2}}$ is the second-order Møller–Plesset (MP2) energy [47].

A problem with hybrid functionals is that the xc-potential is no longer a "simple" multiplicative function as intended in the original Kohn–Sham theory. Following Sharp and Horton [48] and Talman and Shadwick [49], insight into the behavior of the exact exchange potential can be obtained by seeking the optimized effective potential (OEP) whose orbitals minimize the HF energy. The answer [48] turns out to be that this is equivalent to asking that the linear response of the DFT charge density to the perturbation $(\hat{\Sigma}_x^\sigma - v_x^\sigma)$ be zero. While exact solutions are possible [49], Krieger, Li, and Iafrate, following up on a footnote in the Sharp–Horton paper [48], found a very successful approximate solution [50]:

$$v_x^\sigma(\mathbf{r}) = \frac{\sum_i n_{i\sigma}\psi_{i\sigma}^*(\mathbf{r})\hat{\Sigma}_x^\sigma\psi_{i\sigma}(\mathbf{r})}{\rho_\sigma(\mathbf{r})} + \frac{\sum_i n_{i\sigma}(\epsilon_{i\sigma}^{\mathrm{DFT}} - \epsilon_{i\sigma}^{\mathrm{HF}})|\psi_{i\sigma}(\mathbf{r})|^2}{\rho_\sigma(\mathbf{r})}, \tag{3.12}$$

where $\rho_\sigma$ is the spin $\sigma$ charge density, the HF exchange operator, $\hat{\Sigma}_x^\sigma$, acts by $\hat{\Sigma}_x^\sigma\phi(\mathbf{r}_1) = -\int(\gamma_\sigma(\mathbf{r}_1,\mathbf{r}_2)/r_{12})\phi(\mathbf{r}_2)\,d\mathbf{r}_2$, and $\gamma_\sigma(\mathbf{r}_1,\mathbf{r}_2) = \sum_i \psi_{i\sigma}(\mathbf{r}_1)n_{i\sigma}\psi_{i\sigma}^*(\mathbf{r})$, is the one-electron reduced density matrix (1-RDM or just "density matrix.") The first term in Eq. (3.12) is Slater's form of $v_x^\sigma$ while the second term is a derivative discontinuity term which leads to sudden rigid jumps in the potential when new orbitals are occupied. At large $r$ only the HOMO of each spin contributes, leading, after a bit of algebra, to

$$v_x^\sigma(\vec{r}) = -\frac{1}{r} + \epsilon_{\mathrm{HOMO}}^{\mathrm{DFT}} - \epsilon_{\mathrm{HOMO}}^{\mathrm{HF}}. \tag{3.13}$$

We now see that the HOMO energy must be the same in DFT and HF if the x-potential vanishes at infinity. This is a reflection of the general theorem that $-\epsilon_{\mathrm{HOMO}}^{\mathrm{DFT}}$ must be the ionization potential when the xc-functional is exact. The exact xc-potential must fall off as $-1/r$ at large distances.

Figure 3.1 shows orbital energies obtained from our own exchange-only OEP calculations [51]. It turns out that minus the Kohn–Sham orbital energy is a remarkably good approximation to the corresponding ionization potential. In fact, the graphic shows a dramatic improvement in Koopmans' theorem when the HF exchange operator is transformed into a localized exchange potential by the OEP procedure. The behavior of the unoccupied orbitals can be understood by using a result of Gonze and Scheffler obtained from an exchange-only time-dependent OEP theory [52]. (See also my own article [10] for an alternative approximate demonstration.) The result is that

$$\epsilon_a^{\mathrm{DFT}} = \epsilon_a^{\mathrm{HF}} - (\epsilon_i^{\mathrm{HF}} - \epsilon_i^{\mathrm{DFT}}) - (aa|f_H|ii) - (ai|f_{\mathrm{xc}}^{\uparrow,\uparrow}(\epsilon_a - \epsilon_i)|ia). \tag{3.14}$$

Here the indices include spin and

$$(pq|f|rs) = \int\int \psi_p^*(\mathbf{r})\psi_q(\mathbf{r})f(\mathbf{r},\mathbf{r}')\psi_r^*(\mathbf{r}')\psi_s(\mathbf{r}')\,d\mathbf{r}d\mathbf{r}', \tag{3.15}$$

where $f$ can be either the Hartree kernel,

$$f_H(\mathbf{r}_1,\mathbf{r}_2) = 1/r_{12}, \tag{3.16}$$

or the xc-kernel which is given by

$$f_{\rm xc}^{\sigma,\tau}(\mathbf{r}_1,\mathbf{r}_2) = \frac{\delta^2 E_{\rm xc}[\rho_\uparrow,\rho_\downarrow]}{\delta\rho_\sigma(\mathbf{r}_1)\delta\rho_\tau(\mathbf{r}_2)}, \tag{3.17}$$

in the adiabatic approximation. Ignoring the small frequency dependence of $f_{\rm xc}$ and choosing $i$ to be the HOMO, we obtain

$$\epsilon_a^{\rm DFT} = \epsilon_a^{\rm HF} - (aa|f_H|ii) - (ai|f_{\rm xc}^{\uparrow,\uparrow}|ia). \tag{3.18}$$

This latter equation has a simple interpretation. Unoccupied HF orbitals see $N$ electrons while unoccupied KS orbitals see the same potential as occupied KS orbitals, hence $(N-1)$ electrons. To go from the HF case to the DFT case, it is therefore necessary to remove a coulomb integral and an exchange integral. Practical experience suggests that the answer should be roughly independent of the occupied orbital $i$ as long as the orbital is not too localized.

Potentials from approximate pure density functionals do not behave like Eq. (3.12). They do not show the sudden jump (or "derivative discontinuity") which arises from the second term when a new orbital is populated and the long-range behavior of the xc-potentials is also different. Since charge densities fall off exponentially at large $r$, the LDA exchange potential ($v_x^\sigma = -C_x\rho_\sigma^{4/3}$) also falls off exponentially at large $r$ which is too rapid in comparison with the correct $1/r$ behavior. Also since the exchange part dominates the correlation part, this leads to underbinding not only in the LDA but also in GGAs, hence to HOMOs which are too small in magnitude by as much as 5 eV in typical small molecules. One way to correct this is to introduce model potentials [53–55] which fall off asymptotically as $1/r$ for exponentially decaying charge densities. However, the resulting xc-potentials are no longer the functional derivative of an xc-energy functional. Hybrid functionals do lead to "potentials" with improved asymptotic behavior, behaving

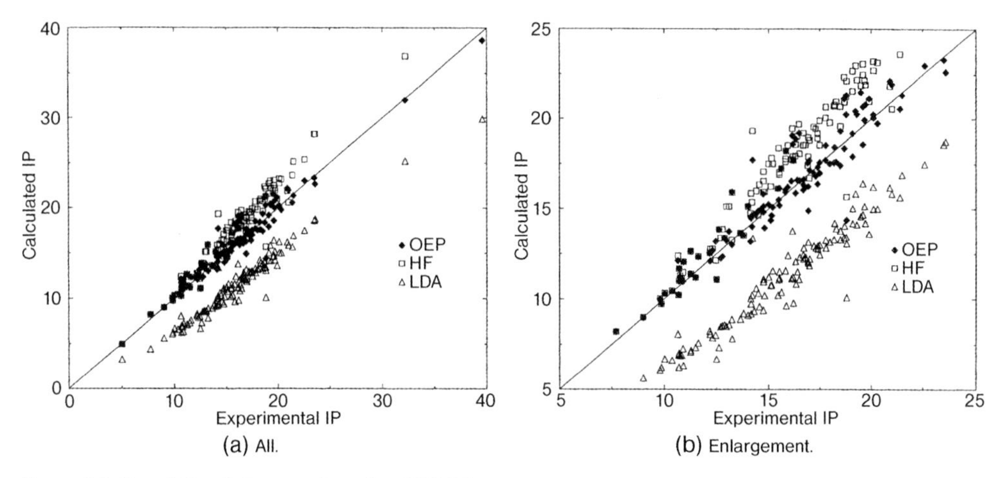


**Figure 3.1** Correlation between minus the OEP (◇), HF (filled square), and LDA (△) orbital energies and experimental outer valence ionization potentials for 26 small molecules and a total of over 100 ionization potentials. (Reprinted from Ref. [51], with permission from Elsevier.)

as a constant times $1/r$ at long distances. The short-range/long-range separated hybrid functionals appear to give even better results in this respect [14–17].

### 3.2.2
### Time-Dependent Formalism

We now turn our attention to the time-dependent version of Eq. (3.1), namely

$$\hat{H}\Psi(t) = i\hbar\frac{\partial}{\partial t}\Psi(t). \tag{3.19}$$

Modern TDDFT is based upon two theorems of Runge and Gross (RG1 and RG2) [3], which are the analogs of the two Hohenberg–Kohn theorems (HK1 and HK2).

The first theorem is fundamentally a theorem about the current density,

$$\mathbf{j}(\mathbf{r}_1, t) = \frac{N\hbar}{m_e}\mathrm{Im}\left\{\int\int\cdots\int\left[\nabla_1\Psi(\mathbf{x}_1, \mathbf{x}_2, \ldots, \mathbf{x}_N, t)\right]\right.$$
$$\left.\times\,\Psi^*(\mathbf{x}_1, \mathbf{x}_2, \ldots, \mathbf{x}_N, t)\, d\sigma_1 d\mathbf{x}_2\cdots d\mathbf{x}_N\right\}. \tag{3.20}$$

The current density satisfies the continuity equation,

$$\frac{\partial\rho(\mathbf{r}, t)}{\partial t} + \nabla\cdot\mathbf{j}(\mathbf{r}, t) = 0. \tag{3.21}$$

In the case of a single-determinant wavefunction, $\mathbf{j}(\mathbf{r}, t) = (\hbar/m_e)\mathrm{Im}\sum_{i=1,N}[\nabla\psi_i(\mathbf{r}, t)]\psi_i^*(\mathbf{r}, t)$.

RG1 states that the time-dependent charge density determines the external potential up to an additive function of time: $v_{\mathrm{ext}}(\mathbf{r}, t) + C(t) \leftarrow \rho(\mathbf{r}, t)$. It is assumed that the external potential can be expressed as a Taylor series in time, $v_{\mathrm{ext}}(\mathbf{r}, t) = \sum_{k=0}^{\infty} c_k(\mathbf{r})(t - t_0)^k$ with $c_k(\mathbf{r}) = (1/k!)\left[\partial^k v_{\mathrm{ext}}(\mathbf{r}, t)/\partial t^k\right]_{t=t_0}$. Most physical potentials can be approximated arbitrarily closely by such a function. The equation of motion,

$$i\hbar\frac{\partial\langle\Psi(t)|\hat{j}_q|\Psi(t)\rangle}{\partial t}(t) = \langle\Psi(t)|\left[\hat{j}_q, \hat{H}(t)\right]|\Psi(t)\rangle, \tag{3.22}$$

is used to show that two external potentials generating the same current density cannot differ by more than an additive function of time. Then the continuity equation (3.21) is used to show from this result that two external potentials generating the same time-dependent charge density cannot differ by more than an additive function of time. This involves the vanishing of a certain integral over a boundary surface which will always be the case for normal electric fields generated by finite sets of charges. A corollary to RG1 is that the time-dependent charge density, $\rho(\mathbf{r}, t)$, fixes the number of particles, $N$, and the external potential up to an arbitrary additive function of time, $v_{\mathrm{ext}}(\mathbf{r}, t) + C(t)$. Hence, $\rho(\mathbf{r}, t)$ determines the time-dependent Hamiltonian up to an additive function of time: $\hat{H}(t) + C(t) \leftarrow \rho(\mathbf{r}, t)$. This means that the equation of motion for the wavefunction can always be integrated provided we know the initial wavefunction, $\Psi_0$, at time $t_0$ to obtain $\Psi(t) = \Psi[\rho, \Psi_0](t)e^{i\phi(t)}$, where the phase factor is given by $\phi(t) = \int_{t_0}^{t} C(t')\,dt'$. If our system is in its ground state at time $t_0$, then we can use eHK1 to remove the dependence on $\Psi_0$ to obtain $\Psi(t) = \Psi[\rho](t)e^{i\phi(t)}$.

Thanks to RK1 we have that the external potential of the real system is a functional of the density. This is also true for the fictitious system of noninteracting electrons, so we can write down a time-dependent Kohn–Sham equation,

$$\left[-\frac{1}{2}\nabla_1^2 + v_{\text{ext}}(\mathbf{r}_1, t) + \int \frac{\rho(\mathbf{r}_2, t)}{r_{12}}\,d\mathbf{r}_2 + v_{\text{xc}}[\rho](\mathbf{r}_1, t)\right]\psi_i(\mathbf{r}_1, t) = i\hbar\frac{\partial\psi_i(\mathbf{r}_1, t)}{\partial t}. \tag{3.23}$$

The goal of RG2 was to propose a stationary action principle in analog to the variational principle of HK2. This will not be discussed here except to note that the Dirac–Frenkel action, $A = \int_{t_0}^{t_1} \langle\Psi(t')|i\hbar\frac{\partial}{\partial t'} - \hat{H}(t')|\Psi(t')\rangle\,dt'$, originally proposed by Runge and Gross proved to be inadequate. It was replaced by Robert van Leeuwen with a more appropriate Keldysh action formalism [27].

Almost all applications of TDDFT make the adiabatic approximation which assumes that the xc-potential reacts instantaneously and without memory to any temporal change in the charge density. Then

$$v_{\text{xc}}[\rho](\mathbf{r}, t) = \frac{\delta E_{\text{xc}}[\rho_t]}{\delta\rho_t(\mathbf{r})}, \tag{3.24}$$

where $\rho_t(\mathbf{r})$ is a function of $\mathbf{r} = (x, y, z)$ obtained by fixing $t$ in the function $\rho(\mathbf{r}, t)$. The result is expected. The time-independent Kohn–Sham equation has been transformed into a time-dependent Kohn–Sham equation by replacing the orbital energy with a time derivative, making the orbitals time dependent and inserting the time-dependent charge density wherever the density appears.

Only a little is known about how to go beyond the adiabatic approximation. Perhaps the most successful approach has been the Vignale–Gross formalism which includes nonadiabatic effects through the current density [6, 7]. Another approach involves a comoving Lagrangian reference frame [56]. More recently, work has been carried out to extract the nonadiabatic behavior of the xc-kernel from the Bethe–Salpeter equation via a polarization propagator formalism [8–10].

## 3.3 Technology

At this point the reader may be wondering how we are going to extract excitation energies from TDDFT. The answer is that we are going to make a textbook application of linear response theory which will completely eliminate time in favor of excitation energies. This section describes important mathematical "technology" needed to go from the formal theory described in the previous section to obtain excited-state quantities of interest in quantum chemistry.

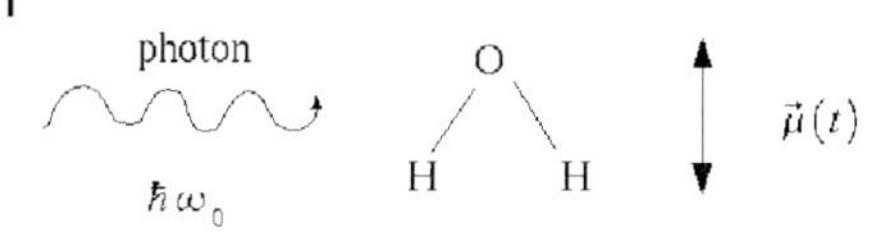


**Figure 3.2** A water molecule perturbed by a photon. The photon is modeled by a classical time-dependent electric field. We are interested in the induced dipole moment which is also a function of time.

### 3.3.1
### Formal Response Theory

Consider a time-dependent perturbation applied to a molecule initially in its ground stationary state. We would like to express the response of a property to this perturbation in terms of the states of the unperturbed system. An important example is shown in Figure 3.2. Other examples include NMR chemical shifts and circular dichroism spectra.

We assume that the exact solution is known for the Schrödinger equation for the unperturbed system (molecule). The time-independent equation was given in Eq. (3.1). The corresponding time-dependent equation is

$$\hat{H}\Psi_I(t) = i\hbar\frac{\partial}{\partial t}\Psi_I(t), \tag{3.25}$$

with $\Psi_I(t) = \Psi_I e^{-iE_It/\hbar}$. (Careful! The notation is compact: $\Psi_I(t) \neq \Psi_I$.)

Now apply the time-dependent perturbation, $\hat{b}(t)$. The equation governing the time evolution of the perturbed system is $\left(\hat{H} + b(t)\right)\Psi_0(t) = i\hbar\partial\Psi_0(t)/\partial t$. Without loss of generality, $\Psi_0(t) = \left(\Psi_0 + \delta\Psi_0(t) + \cdots\right)e^{-iE_0t/\hbar}$, which allows us to deduce that the linear response of the ground state, $\delta\Psi_0(t)$, satisfies the first-order equation,

$$\left(i\hbar\frac{\partial}{\partial t} - \hat{H} + E_0\right)\delta\Psi(t) = \hat{b}(t)\Psi_0. \tag{3.26}$$

After an appropriate Fourier transform, this first-order equation is just,

$$\left[E_0 - \hat{H} + \hbar\omega\right]\delta\Psi_0(\omega) = \hat{b}(\omega)\Psi_0, \tag{3.27}$$

which is very good because this equation is now in the usual form for applying Rayleigh–Schrödinger perturbation theory. We can immediately write down that

$$\delta\Psi_0(\omega) = \sum_{I\neq 0}\Psi_I\frac{\langle\Psi_I|\hat{b}(\omega)|\Psi_0\rangle}{\hbar(\omega - \omega_I)}, \tag{3.28}$$

where $\hbar\omega_I$ is the $I$th excitation energy of the unperturbed system (Eq. (3.3)).

In order to go further, we assume that the perturbation is monochromatic, so $\hat{b}(t) = b\cos(\omega_0 t)$ and $\hat{b}(\omega) = \pi b\left[\delta(\omega + \omega_0) + \delta(\omega - \omega_0)\right]$. Inserting into Eq. (3.28) and back Fourier transforming gives

$$\delta\Psi_0(t) = \left[\sum_{I\neq 0}\Psi_I\frac{\omega_I\langle\Psi_I|\hat{b}|\Psi_0\rangle}{\hbar(\omega_0^2-\omega_I^2)}\right]\cos(\omega_0 t)$$
$$-i\left[\sum_{I\neq 0}\Psi_I\frac{\omega_0\langle\Psi_I|\hat{b}|\Psi_0\rangle}{\hbar(\omega_0^2-\omega_I^2)}\right]\sin(\omega_0 t). \tag{3.29}$$

The linear response, $\delta\langle\hat{a}\rangle(t) = \langle\Psi_0|\hat{a}|\delta\Psi_0(t)\rangle + \langle\delta\Psi_0(t)|\hat{a}|\Psi_0\rangle$, of an observable, $a$, is given by

$$\delta\langle\hat{a}\rangle(t) = \left[\sum_{I\neq 0}\frac{2\omega_I\mathrm{Re}(\langle\Psi_0|\hat{a}|\Psi_I\rangle\langle\Psi_I|\hat{b}|\Psi_0\rangle)}{\hbar(\omega_0^2-\omega_I^2)}\right]\cos(\omega_0 t)$$
$$+\left[\sum_{I\neq 0}\frac{2\omega_0\mathrm{Im}(\langle\Psi_0|\hat{a}|\Psi_I\rangle\langle\Psi_I|\hat{b}|\Psi_0\rangle)}{\hbar(\omega_0^2-\omega_I^2)}\right]\sin(\omega_0 t). \tag{3.30}$$

This result is very powerful. We see that the response to a perturbation at frequency $\omega_0$ is at the same frequency. The phase is also the same if the operators $\hat{a}$ and $\hat{b}$ are both real (polarizability) or both imaginary (NMR). The phase is $\pi/2$ if one of $\hat{a}$ and $\hat{b}$ is real and the other is imaginary (circular dichroism).

We finally arrive at a point where we will specialize to the case of the electric polarizability. The dynamic polarizability is the proportionality tensor between the linear response of the dipole moment, $\mu$, and the applied field, $\mathcal{E}_q(t) = \mathcal{E}_q\cos(\omega t)$, namely

$$\mu_q(t) = \mu_q + \sum_{q'=x,y,z}\alpha_{q,q'}(\omega)\mathcal{E}_{q'}\cos(\omega t) + \cdots\ ; q = x, y, z. \tag{3.31}$$

For us, the dynamic polarizability will be a way to access excited states. At optical frequencies, the variation in the electric field is too rapid for the nuclei to follow, so we may consider them clamped in place. The response of the dipole moment is then entirely electronic and we may write

$$\delta\mu_q(t) = -e\langle\Psi_0|q|\delta\Psi_0(t)\rangle - e\langle\delta\Psi_0(t)|q|\Psi_0\rangle$$
$$= \sum_{q'=x,y,z}\left[\sum_{I\neq 0}\frac{2e^2\omega_I\Re(\langle\Psi_0|q|\Psi_I\rangle\langle\Psi_I|q'|\Psi_0\rangle)}{\hbar(\omega_I^2-\omega^2)}\right]\mathcal{E}_{q'}\cos(\omega t). \tag{3.32}$$

The quantity in square brackets is $\alpha_{q,q'}(\omega)$. As we often do not know the orientation of the molecules, it is the average dynamic polarizability,

$$\alpha(\omega) = \sum_{I\neq 0}\frac{e^2 f_I}{m_e(\omega_I^2-\omega^2)}, \tag{3.33}$$

which is important. The oscillator strengths, $f_I$, are defined in Eq. (3.4). Equation (3.33) is known as the sum-over-states (SOS) theorem and relates the dynamic response of the density to the excited states.

### 3.3.2
### LR-TDDFT

So far, we have assumed that we know the exact wavefunction solution of the unperturbed problem for the interacting system. This is not known in DFT, so some modifications are necessary. The result are the basic equations for linear-response TDDFT (LR-TDDFT). The approach taken here is based upon density matrices [2]. In what follows, "DFT" means "pure DFT." As the result for linear-response time-dependent Hartree–Fock (LR-TDHF) is also given, it is trivial to write down the corresponding generalization for hybrid functionals.

In HF and in DFT the density matrix of the unperturbed system is

$$\gamma_\sigma(\mathbf{r},\mathbf{r}') = \sum \psi_{p\sigma}(\mathbf{r}) P_{pq\sigma} \psi_{p\sigma}(\mathbf{r}'); \quad P_{pq\sigma} = n_{p\sigma}\delta_{p,q}. \tag{3.34}$$

Its response to a time-dependent electric field is $\delta\gamma_\sigma(\mathbf{r},\mathbf{r}') = \sum \psi_{p\sigma}(\mathbf{r})\delta P_{pq\sigma}(\omega)\psi^*_{q\sigma}(\mathbf{r}')$. As

$$\delta\psi_{i\sigma}(\mathbf{r}\omega) = \sum_p \psi_{p\sigma}(\mathbf{r}) \frac{\langle\psi_{p\sigma}|\hat{b}(\omega)|\psi_{i\sigma}\rangle}{\omega - (\epsilon_{p\sigma} - \epsilon_{i\sigma})}, \tag{3.35}$$

and as $\delta\gamma_\sigma(\mathbf{r},\mathbf{r}') = \sum_i \delta\psi_{i\sigma}(\mathbf{r}) n_i \psi^*_{i\sigma}(\mathbf{r}') + \sum_i \psi_{i\sigma}(\mathbf{r}) n_i \delta\psi^*_{i\sigma}(\mathbf{r}')$, then

$$\delta P_{pq\sigma}(\omega) = \frac{n_{q\sigma} - n_{p\sigma}}{\omega - (\epsilon_{p\sigma} - \epsilon_{q\sigma})} \langle\psi_{p\sigma}|\hat{b}_{\text{eff}}(\omega)|\psi_{q\sigma}\rangle, \tag{3.36}$$

for $\hat{b}(t) = \hat{b}\cos(\omega t)$.

The reason for writing $\hat{b}_{\text{eff}}(\omega)$ and not just $\hat{b}(\omega)$ is that $\hat{b}_{\text{eff}}(\omega)$ is the perturbation felt by the HF or DFT orbitals and not the applied field. The difference between these two perturbations is the response of the self-consistent field, $\hat{b}_{\text{eff}}(\omega) = \hat{b}(\omega) + \delta\hat{v}_{\text{SCF}}(\omega)$. In terms of matrices,

$$b^{\text{eff}}_{pq\sigma}(\omega) = b_{pq\sigma}(\omega) + \sum K_{pq\sigma,rs\tau}\delta P_{rs\tau}(\omega), \tag{3.37}$$

where the coupling matrix is

$$K_{pq\sigma,rs\tau} = \frac{\partial v^{\text{SCF}}_{pq\sigma}}{\partial P_{rs\tau}} = \begin{cases} (pq|f_H|rs) - \delta_{\sigma,\tau}(pr|f_H|sq) \;; & \text{HF} \\ (pq|f_H|rs) + (pq|f^{\sigma,\tau}_{\text{xc}}|rs) \;; & \text{DFT} \end{cases}. \tag{3.38}$$

Note that we have made the adiabatic approximation even if this is not strictly necessary [2]. Thus we have that

$$\delta P_{pq\sigma}(\omega) = \frac{n_{q\sigma} - n_{p\sigma}}{\omega - (\epsilon_{p\sigma} - \epsilon_{q\sigma})} \left[ b_{pq\sigma}(\omega) + \sum K_{pq\sigma,rs\tau}\delta P_{rs\tau}(\omega) \right], \tag{3.39}$$

or

$$\sum_{rs\tau}^{n_{p\sigma} \neq n_{q\sigma}} \left[ \delta_{\sigma,\tau}\delta_{p,r}\delta_{q,s} \frac{\omega - (\epsilon_{p\sigma} - \epsilon_{q\sigma})}{n_{q\sigma} - n_{p\sigma}} - K_{pq\sigma,rs\tau} \right] \delta P_{rs\tau}(\omega) = b_{pq\sigma}(\omega). \tag{3.40}$$

We can solve this equation at each frequency $\omega$ and so calculate the response of each property $a$ at that frequency. But we would like to go further and have an SOS-type formula so that we can extract excitation energies and their associated oscillator strengths.

After some algebra and assuming occupation numbers equal to 0 or 1 and real orbitals, it can be shown [2] that Eq. (3.40) can be rewritten as

$$\left\{ \omega \begin{bmatrix} -1 & 0 \\ 0 & +1 \end{bmatrix} - \begin{bmatrix} \boldsymbol{A} & \boldsymbol{B} \\ \boldsymbol{B}^* & \boldsymbol{A}^* \end{bmatrix} \right\} \begin{pmatrix} \delta \mathbf{P}(\omega) \\ \delta \mathbf{P}^*(\omega) \end{pmatrix} = \begin{pmatrix} \mathbf{b}(\omega) \\ \mathbf{b}^*(\omega) \end{pmatrix}, \tag{3.41}$$

where

$$\begin{aligned} A_{ia\sigma,jb\tau} &= \delta_{\sigma,\tau}\delta_{i,j}\delta_{a,b}(\epsilon_{a\sigma} - \epsilon_{i\sigma}) + K_{ia\sigma,jb\tau} \\ B_{ia\sigma,jb\tau} &= K_{ia\sigma,bj\tau}\,. \end{aligned} \tag{3.42}$$

(Matrices have been distinguished from vectors by putting the matrices in bold italic.)

At an excitation frequency, the response of the density matrix is infinite even if the perturbation is finite. This means that the excitation frequency must satisfy the pseudo-eigenvalue problem,

$$\begin{bmatrix} \boldsymbol{A} & \boldsymbol{B} \\ \boldsymbol{B} & \boldsymbol{A} \end{bmatrix} \begin{pmatrix} \mathbf{X}_I \\ \mathbf{Y}_I \end{pmatrix} = \omega_I \begin{bmatrix} +1 & 0 \\ 0 & -1 \end{bmatrix} \begin{pmatrix} \mathbf{X}_I \\ \mathbf{Y}_I \end{pmatrix}. \tag{3.43}$$

One way to solve this equation is to rewrite it as a true eigenvalue problem, $\frown \mathbf{F}_I = \omega_I^2 \mathbf{F}_I$, where $\frown = (\boldsymbol{A} - \boldsymbol{B})^{1/2}(\boldsymbol{A} + \boldsymbol{B})(\boldsymbol{A} - \boldsymbol{B})^{1/2}$ and $\mathbf{F}_I = (\boldsymbol{A} - \boldsymbol{B})^{-1/2}(\mathbf{X}_I + \mathbf{Y}_I)$. This still leads to eigenvalue problems which rapidly become too large to solve by ordinary means. However, the lowest several eigenvalues and eigenvalues may be solved using the iterative block Davidson Krylov space method [57–60]. These iterations are usually plainly evident in the output of programs performing LR-TDDFT calculations. Oscillator strengths may be calculated using the formula [10]

$$f_I = \frac{2}{3} \sum_{q=x,y,z} |\mathbf{q}^\dagger (\boldsymbol{A} - \boldsymbol{B})^{1/2} \mathbf{F}|^2. \tag{3.44}$$

### 3.3.3 TDA-TDDFT

The Tamm–Dancoff approximation (TDA) to the linear response equation is

$$\boldsymbol{A}\mathbf{X}_I = \omega_I \mathbf{X}_I. \tag{3.45}$$

The TDA often gives results which are a good approximation to full LR-TDDFT results around the equilibrium geometry of the molecule with somewhat less computational effort (especially for hybrid functionals). However, we lose the polarizablity sum rule (Eq. (3.33)) and the Thomas–Reiche–Kühn (TRK) sum rule, which says that the oscillator strengths sum to the number of electrons in the system $\sum_I f_I = N$. These are perhaps not such a great loss since we rarely have a complete set of oscillator strengths (even in a finite basis calculation) to use in the sum-over-states expression for the dynamic polarizability. Also the TRK sum rule is only strictly valid in the limit of a complete basis set, and TRK basis set requirements are not the same in practice as those needed to calculate polarizabilities and excitation spectra [4].

Most importantly, the TDA actually seems to give *better* excited-state potential energy surfaces than does a full linear-response calculation [30, 61, 62]. While this may seem strange for an "approximation," the reason for this better behavior is that the quality of the excitation energies obtained in response theory depends upon the quality of the description of the ground-state problem which in DFT depends in turn on the quality of the xc-functional and there can be problems with the xc-functional for the ground state. In particular, broken symmetry solutions should not occur for ground states which are closed-shell singlets when the xc-functional is exact but do occur for approximate xc-functionals. In the latter case, there is a theorem [30] which says that one of the triplet LR-TDDFT excitation energies will go to zero and then become imaginary at geometries where symmetry breaking occurs. The TDA circumvents this "triplet instability problem" by decoupling the excited-state problem from the ground-state problem. Exactly how this happens is difficult to see in TDDFT except by direct calculation but is easy to understand in TDHF. That is because TDA-TDHF is the same as CIS which, as a variational method, forbids collapse of the excitation energies to unphysical values.

### 3.3.4 Analytic Gradients

An important aspect of quantum chemistry is the ability to calculate analytic gradients, permitting automatic geometry optimizations and providing on-the-fly forces for *ab initio* molecular dynamics calculations. Developing the mathematics necessary to implement these for conventional quantum chemistry methods has been a considerable *tour de force* which is described at length in Ref. [63]. This work has been paralleled for DFT and the basic methodology for the computation of TDDFT analytic gradients has now been implemented in a number of codes [64–70]. I describe the basic ideas briefly here for TDA-TDDFT calculations which in any event is the method I recommend for calculating excited-state PESs.

We want to take analytic derivatives of excited-state energies with respect to a parameter which we will call $\eta$, and the most important example of $\eta$ is a geometric parameter associated with a force. The excited-state energy can be written as $E_I = E_0 + \hbar\omega_I$; hence $\partial E_I/\partial\eta = \partial E_0/\partial\eta + \hbar\partial\omega_I/\partial\eta$. Thus we need to be able to take analytic derivatives of both the ground state and the TDA excitation energy. We will not separate space and spin in this subsection.

In quantum chemistry calculations, the molecular orbitals (MOs), $\psi_s$, are expanded in a basis of atomic orbitals (AOs), $\chi_\mu$: $\psi_s(\mathbf{r}) = \sum_\mu \chi_\mu(\mathbf{r})c_{\mu,s}$. (In this subsection, the MO indices include spin.) When the parameter $\eta$ varies, the AOs, $\chi_\mu$, and the MO coefficients in the AO basis, $c_{\mu,s}$, both vary. We would like to separate these two types of $\eta$-dependent variations and, if possible, eliminate any derivatives with respect to $c_{\mu,s}$ since these are costly to calculate. To carry out our program we develop $\partial\psi_s/\partial\eta = \psi_s^\eta + \sum_\mu \psi_r U_{r,s}^\eta$. In general, the superscript $\eta$ is reserved for a derivative over AOs at constant $c_{\mu,s}$ giving so-called core or skeleton terms. So, $\psi_s^\eta = \sum_\mu (\partial\chi_\mu/\partial\eta)c_{\mu,s}$. However, an exception is the matrix $\mathbf{U}^\eta$ of coupled perturbed coefficients which is defined by $\sum_\nu \chi_\nu \partial c_{\nu,s}/\partial\eta = \sum_r \psi_r U_{r,s}^\eta$. It follows that $U_{r,s}^\eta = \sum_{\mu,\nu} c_{\mu,r}^* S_{\mu,\nu}\partial c_{\nu,s}/\partial\eta$, where $S_{\mu,\nu} = \langle\chi_\mu|\chi_\nu\rangle$ is the AO overlap matrix. Taking the functional derivative of the MO orthonormality relation $\delta_{r,s} = \langle\psi_r|\psi_s\rangle = \sum_{\mu,\nu} c_{\mu,r}^* S_{\mu,\nu} c_{\nu,s}$ leads to what I call the "turnover rule," $U_{q,p}^{\eta,*} = -U_{p,q}^\eta - S_{p,q}^\eta$. For real coupled perturbed coefficients, $U_{p,p}^\eta = -S_{p,p}^\eta/2$.

It is now straightforward to take the derivative of the energy expression to obtain $\partial E/\partial\eta = E^\eta - \sum_{\mu,\nu} S_{\mu,\nu}^\eta W_{\nu,\mu}$, where $W_{\mu,\nu} = \sum_i c_{\mu,i}\epsilon_i n_i c_{\nu,i}^*$ is the energy-weighted density matrix. The first term is the Hellmann–Feynman force. In wavefunction terms, $E^\eta = \langle\Psi|(\partial\hat{H}/\partial\eta)|\Psi\rangle$. The second term is the Pulay force. It is there because the AOs move with the nuclei and it is necessary to ensure that the calculated forces are zero when the calculated energy is a minimum. Since the coupled perturbed coefficients do not enter into the calculation of the first analytic derivative, the calculation of this derivative is finally really relatively trivial.

Second analytical derivatives for the ground state (not discussed here) and first analytical derivatives of the TDA-TDDFT excitation energies require us to solve a coupled perturbed equation for $U_{p,q}^\eta$. Recognizing that

$$\frac{\partial P_{q,p}}{\partial\eta} = U_{q,p}^\eta n_p + U_{p,q}^{\eta,*} n_q = U_{q,p}^\eta(n_p - n_q) - S_{q,p}n_q, \tag{3.46}$$

we should anticipate an equation similar to the LR-TDDFT equations already found. This is indeed the case for, instead of Eq. (3.40), we find upon differentiating the MO eigencondition, $F_{p,q} = \delta_{p,q}\epsilon_q$, that

$$\sum\left(\delta_{p,p'}\delta_{q,q'}\frac{\epsilon_q - \epsilon_p}{n_q - n_p} + K_{pq,p'q'}\right)(n_{q'} - n_{p'})U_{p',q'}^\eta$$
$$= F_{p,q}^\eta - S_{p,q}^\eta\epsilon_q - \sum K_{pq,p'q'}n_{p'}S_{p',q'}^\eta. \tag{3.47}$$

However, it is clear from the turnover rule that there are many linear dependences among the coupled perturbed coefficients. By working a little harder, we arrive at the equation

$$\sum \mathcal{A}_{ai,bj}U_{bj}^\eta = \mathcal{B}_{ai}^0 \tag{3.48}$$

for the nonredundant coupled perturbed coefficients, where

$$\mathcal{A}_{ai,bj} = \delta_{i,j}\delta_{a,b}\frac{\epsilon_i - \epsilon_a}{n_i - n_a} - (K_{ai,jb} + K_{ai,bj})$$
$$\mathcal{B}^0_{ai} = F^\eta_{ai} - S^\eta_{ai}\epsilon_i - \sum K_{ai,kj}S^\eta_{jk}, \tag{3.49}$$

and I am using the MO index convention,

$$\underbrace{abc\cdots fgh}_{\text{unoccupied}}\;\underbrace{ijklmn}_{\text{occupied}}\;\underbrace{opq\cdots xyz}_{\text{free}}. \tag{3.50}$$

The redundant coupled perturbed coefficients may be calculated from the non-redundant coefficients by using the expression

$$U^\eta_{p,q} = \frac{1}{\epsilon_q - \epsilon_p}\left[F^\eta_{p,q} - S^\eta_{p,q}\epsilon_q - \sum K_{pq,jk}S^\eta_{jk} + \sum (K_{pq,bj} + K_{pq,jb})\,U^\eta_{bj}\right]. \tag{3.51}$$

It follows from the eigencondition (3.45) that the derivative of the TDA-TDDFT excitation energy is $\partial\omega/\partial\eta = \sum X^*_{ia}(\partial A_{ia,bj}/\partial\eta)X_{jb}$. This can be further developed as

$$\frac{\partial\omega}{\partial\eta} = \omega^\eta - \sum M_{kl}S^\eta_{kl} + \sum L_{ck}U^\eta_{ck}, \tag{3.52}$$

where

$$M_{kl} = \sum X^*_{ia}X_{jb}\left(K_{ab,kl} - K_{ji,kl} - G_{ia,bj,kl}\right)$$
$$L_{ck} = \sum X^*_{ia}X_{jb}\left[\left(K_{ab,kc} + K_{ab,ck}\right) - \left(K_{ji,kc} + K_{ji,ck}\right) + \left(G_{ia,jb,ck} + G_{ia,bj,kc}\right)\right]. \tag{3.53}$$

These are the same as for CIS except for the appearance of the term $G_{pq,p'q',pq} = (\partial K_{pq,p'q'}/\partial P_{pq})$, which is zero in CIS but involves a triple functional derivative of the xc-functional in TDDFT.

Direct implementation of Eq. (3.52) for calculation of geometric derivatives implies the solution of the coupled perturbed equation (3.48) for each geometric degree of freedom. As this would rapidly become prohibitively expensive, it is fortunate that the coupled perturbed coefficients can be replaced by a $Z$-vector defined implicitly by $\sum L_{ck}U^\eta_{ck} = \sum Z_{ck}\mathcal{B}^0_{ck}$, and explicitly by solving the new coupled perturbed equation$\sum \mathcal{A}_{ai,bj}Z_{bj} = L_{ai}$. This is a great saving because this new coupled perturbed equation is independent of the perturbation $\eta$ and so need to be solved only once for each geometry (see Table 3.2).

The difference between the reduced density matrix for the $I$th excited state and the ground state is given by

$$\gamma^I_{p,q} - \gamma^0_{p,q} = \frac{\partial\omega_I}{\partial h_{q,p}} = \begin{cases} -\sum X^*_{pa}X_{qa} & ; \quad p,q \text{ both occupied} \\ +\sum X^*_{iq}X_{ip} & ; \quad p,q \text{ both unoccupied} \\ Z_{p,q} & ; \quad \text{otherwise.} \end{cases} \tag{3.54}$$

**Table 3.2** Summary of formulae within the two-orbital model (Figure 3.3).[a]

| **ΔSCF Hartree–Fock** | **CIS (TDA-TDHF)** |
|---|---|
| $I_i = \epsilon_i$<br>$A_a = \epsilon_a$<br>$A_a(i^{-1}) = A_a - (aa\|f_H\|ii) + (ai\|f_H\|ia)$<br>$\omega_M = \epsilon_a - \epsilon_i - (aa\|f_H\|ii) + (ai\|f_H\|ia)$<br>$\omega_T = \omega_M - (ia\|f_H\|ai)$<br>$\omega_S = \omega_M + (ia\|f_H\|ai)$ | $I_i = \epsilon_i$<br>$A_a = \epsilon_a$<br>$A_a(i^{-1}) = A_a - (aa\|f_H\|ii) + (ai\|f_H\|ia)$<br>$\omega_M = \epsilon_a - \epsilon_i - (aa\|f_H\|ii) + (ai\|f_H\|ia)$<br>$\omega_T = \omega_M - (ia\|f_H\|ai)$<br>$\omega_S = \omega_M + (ia\|f_H\|ai)$ |
| **Linearized ΔSCF DFT** | **TDA-TDDFT** |
| $I_i = \epsilon_i - \frac{1}{2}(ii\|f_H + f_{xc}^{\uparrow,\uparrow}\|ii)$<br>$A_a = \epsilon_a + \frac{1}{2}(aa\|f_H + f_{xc}^{\uparrow,\uparrow}\|aa)$<br>$A_a(i^{-1}) = A_a - (aa\|f_H + f_{xc}^{\uparrow,\uparrow}\|ii)$<br>$\omega_M = \epsilon_a - \epsilon_i + \frac{1}{2}(aa - ii\|f_H + f_{xc}^{\uparrow,\uparrow}\|aa - ii)$<br>$\omega_T = \omega_M + (aa\|f_{xc}^{\uparrow,\uparrow} - f_{xc}^{\uparrow,\downarrow}\|ii)$<br>$\omega_S = \omega_M - (aa\|f_{xc}^{\uparrow,\uparrow} - f_{xc}^{\uparrow,\downarrow}\|ii)$ | $I_i = \epsilon_i$<br>$A_a = \epsilon_a + (aa\|f_H\|ii) + (ai\|f_{xc}^{\uparrow,\uparrow}\|ia)$<br>$A_a(i^{-1}) = A_a - (aa\|f_H\|ii) + (ai\|f_H\|ia)$<br>$\omega_M = \epsilon_a - \epsilon_i + (ai\|f_H + f_{xc}^{\uparrow,\uparrow}\|ia)$<br>$\omega_T = \omega_M - (ia\|f_H + f_{xc}^{\uparrow,\downarrow}\|ai)$<br>$\omega_S = \omega_M + (ia\|f_H + f_{xc}^{\uparrow,\downarrow}\|ai)$ |

[a] $I_i$ is minus the ionization potential of orbital $i$. $A_a$ is minus the electron affinity of orbital $a$. $A_a(i^{-1})$ is minus the electron affinity of orbital $a$ for the ion formed by removing an electron from orbital $i$. $\omega_T$, $\omega_S$, and $\omega_M$ are, respectively, $i \rightarrow a$ excitation energies to the triplet, singlet, and mixed symmetry states. ΔSCF quantities are obtained by the usual multiplet sum procedure [71] except that a truncated Taylor expansion of the xc-functional has been used in the DFT case. [61] The identification of $I_i$ and $A_a$ in the TDDFT case is based upon OEP theory (Section 3.2).

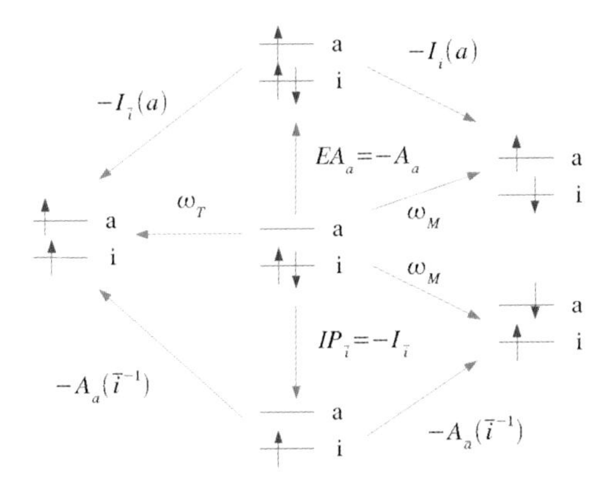


**Figure 3.3** Two-orbital model.

This allows the calculation of excited-state properties such as true dipole moments, rather than just transition dipole moments, showing that such properties are also accessible from TDDFT. In the past, some programs made an approximation when calculating excited-state properties by neglecting the $Z$-vector contribution to the excited-state reduced density matrix.

## 3.4
## Example: Oxirane

Our recent work [62, 72] aimed at making TDDFT a viable tool for photodynamics calculations is briefly reviewed here as an application of TDDFT. Some of the calculations are routine "safe" applications, meaning that we are avoiding the known main problems of present-day TDDFT, namely, (i) the underestimation of the ionization threshold [73, 74], (ii) the underestimation of charge transfer excitation energies [61, 75, 77], and (iii) the lack of explicit two- and higher electron excitations [2, 8–10]. In contrast, the study of photochemical pathways is a demanding application for TDDFT because of the apparent need for a coherent simultaneous description of several potential energy surfaces (PESs) over a wide range of geometries which typically involve either or both the formation of biradicaloid intermediates and charge transfer. Nevertheless TDDFT has a place in the photochemical modeler's toolbox –under the condition that other methods such as complete active space self-consistent field (CASSCF) calculations be used to validate and refine results from TDDFT for the more difficult parts of photochemical processes. Viewed this way, our objective is to reduce the need to fall back onto these more expensive traditional methods. The photochemical ring opening of oxirane has been chosen for troubleshooting TDDFT photodynamics calculations because it is a small-enough molecule that we can carry out very high quality comparison calculations and because it was felt to be a simple case where TDDFT "ought to work."

Figure 3.4 shows the generally high quality to be expected of geometries optimized by DFT, particularly for a "normal" organic molecule such as oxirane. In this case, HF underestimates CO and CH bond lengths, but overestimates the COC bond angle. DFT leads to longer CO and CH bond lengths and a smaller COC bond angle, giving results in better overall agreement with experiment.

Table 3.3 shows how the vertical stick spectrum of oxirane calculated using LR-TDHF, LR-TDLDA, and LR-TDB3LYP compares against experimental excitation

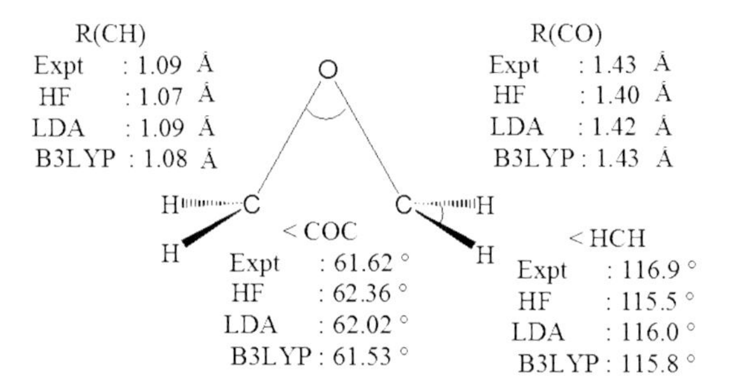


**Figure 3.4** Geometric parameters for oxirane obtained from Hartree–Fock (HF), pure DFT (LDA), and hybrid DFT (B3LYP) compared with experiment. Note that the structure has $C_{2v}$ symmetry. (Reused with permission from Ref. [62]. Copyright 2007, American Institute of Physics.)

**Table 3.3** Principal oxirane singlet excitation energies and oscillator strengths. (Reused with permission from Ref. [62]. Copyright 2007, American Institute of Physics.)

| Principal singlet excitation energies (eV) and oscillator strengths (in parentheses) | | | | |
|---|---|---|---|---|
| **TDHF** | **TDLDA** | **TDB3LYP** | **Experiment** | **Assignment[a]** |
| 9.14 (0.0007) | 6.01 (0.0309) | 6.69 (0.0266) | 7.24(s)[b,c,d] | $1^1B_1[2b_1(n) \rightarrow 7a_1(3s)]$ |
| 9.26 (0.0050) | 6.73 (0.0048) | 7.14 (0.0060) | 7.45(w)[c] | $2^1B_1[2b_1(n) \rightarrow 8a_1(3p_z)]$ |
| 9.36 (0.0635) | 6.78 (0.0252) | 7.36 (0.0218) | 7.88(s)[b], 7.89(s)[c] | $2^1A_1[2b_1(n) \rightarrow 3b_1(3p_x)]$ |
| 9.56 (0.0635) | 7.61 (0.0035) | 7.85 (0.0052) | | |
| 9.90 (0.0478) | 7.78 (0.0304) | 8.37 (0.0505) | | |
| 9.93 (0.0935) | 8.13 (0.0014) | 8.39 (0.0168) | | |
| 8.15 (0.0405) | 8.40 (0.0419) | | | |
| 12.27[e] | 6.40[e] | 7.68[e] | 10.57[f] | |

[a]TDB3LYP.
[b]Gas phase UV absorption spectrum [77].
[c]Obtained by a photoelectric technique [78].
[d]Gas phase UV absorption spectrum [79].
[e]Ionization threshold ($-\epsilon_{\text{HOMO}}$).
[f]Ionization potential [80].

energies. Neither LR-TDHF nor CIS (values not shown) is even remotely accurate enough to assign this spectrum. This is to be contrasted with the LR-TDDFT calculations, which are of comparable computational difficulty, but markedly better accuracy. Although some care must be taken not to overinterpret those features of the excitation spectrum which are close to or above the artificially low LR-TDDFT ionization threshold at $-\epsilon_{\text{HOMO}}$, the TDDFT calculations are in good enough agreement with experiment to allow us to make a credible assignment of the three principal UV absorption peaks. The assignment of the LR-TDB3LYP results is shown in Table 3.3 and is in agreement with the accepted interpretation of the experimental spectrum [81].

Figure 3.5 shows a comparison of the LR-TDLDA and TDA-TDLDA $C_{2v}$ ring opening potential energy curves of oxirane with those obtained from CASSCF and a high-quality diffusion quantum Monte Carlo (DMC) calculation. This is not the expected photochemical reaction path, but was chosen because the high symmetry facilitates analysis of the computed results. Before discussing the figure, it is worth pointing out that the manner in which the non-DFT models were constructed provides an excellent illustration of how TDDFT is often used in photochemical calculations. Analysis of TDDFT excitation energies significantly shortened the time it would otherwise take to choose the active space for CASSCF calculations. This same active space was then also used in the DMC calculations (see Ref. [62] and references therein for additional details). The very small differences between the LR-TDLDA and TDA-TDLDA calculations for most of the states in the figure can be explained by their Rydberg nature. In the two-level model, TDA excitation

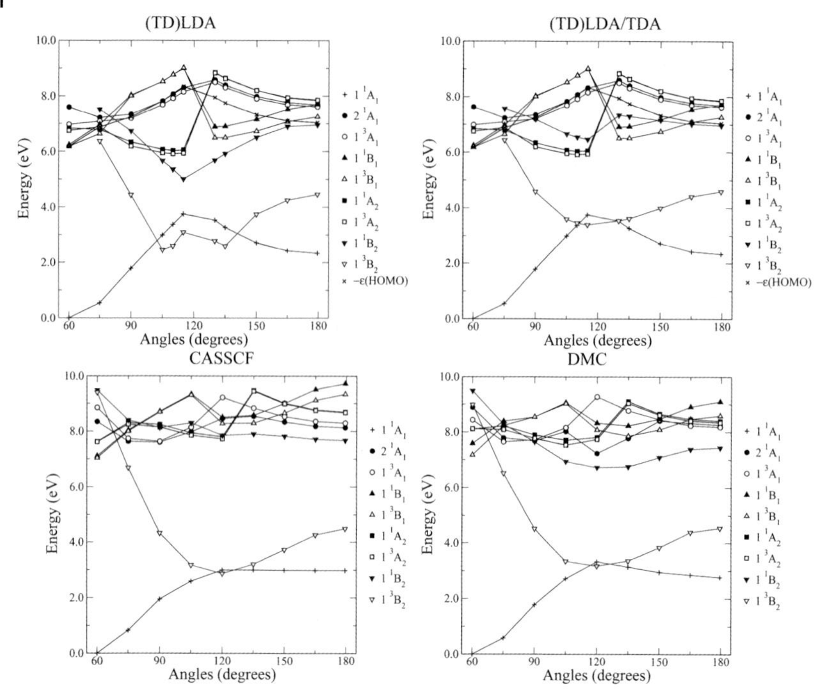


**Figure 3.5** $C_{2v}$ ring opening curves: upper left, LR-TDLDA; upper right, TDA-TDLDA; lower left, CASSCF; lower right, DMC. The energy zero has been chosen to be the ground-state energy for the 60° structure. Note that the "negative excitation energies" for the LR-TDLDA $1^3B_2$ state relative to the ground state are really imaginary excitation energies (triplet instability). On the other hand, the slight negative excitation energy for the TDA-TDLDA $1^3B_2$ state around 120° is real. Also shown is the TDLDA ionization threshold at $-\epsilon_{\text{HOMO}}$. (Reused with permission from Ref. [62]. Copyright 2007, American Institute of Physics.)

energies are always larger than full LR excitation energies,

$$\begin{aligned}\omega_{\text{TDA},S}^2 - \omega_{\text{LR},S}^2 &= (ai|2f_H + f_{\text{xc}}^{\uparrow,\uparrow} + f_{\text{xc}}^{\uparrow,\downarrow}|ia)^2 \geq 0 \\ \omega_{\text{TDA},T}^2 - \omega_{\text{LR},T}^2 &= (ai|f_{\text{xc}}^{\uparrow,\uparrow} - f_{\text{xc}}^{\uparrow,\downarrow}|ia)^2 \geq 0,\end{aligned} \tag{3.55}$$

but the two-electron integrals are small when orbital $a$ is diffuse. Thus the TDA and LR results are essentially identical for Rydberg states. Inspection of Table 3.2 shows that the singlet and triplet excitation energies just reduce to an orbital energy difference in this case.

Differences between the TDLDA and DMC curves are partly due to the proximity of the artificially low TDDFT ionization threshold and partly because the quality of the LDA ground state degrades around 120°. At this point, the $6a_1(\sigma)$ HOMO interchanges with the $4b_2(\sigma^*)$ LUMO. In wavefunction terms, the quasidegenerate $6a_1^2(\sigma)$ and $4b_2^2(\sigma^*)$ states are expected to undergo an avoided crossing here in what many will recognize as the signature of the breaking of the CC $\sigma$ bond to form a biradicaloid. This is exactly what happens in our CASSCF calculations where the avoided crossing is essentially described by taking a linear combination of the $6a_1^2(\sigma)$ and $4b_2^2(\sigma^*)$ states. To the extent that DFT is an exact single determinant theory, exact DFT should give us the exact ground-state curve without such artifices. However, practical DFT uses approximate functionals and so inherits some of the problems of the structurally similar HF model. The result is that the LDA ground-state curve has a cusp at around 120° (not shown because of convergence difficulties associated with a small HOMO-LUMO gap in the vicinity of this geometry). This suggests that there should be a lower energy broken symmetry solution that mimics the underlying physics of the biradicaloid by allowing each of the CC $\sigma$ electrons to localize on a different carbon. Our hypothesis is confirmed by the presence of an imaginary ${}^3B_2[a_1(\sigma), b_2(\sigma^*)]$ excitation energy in the cusp region. Note that the energy of the associated ${}^1B_2[a_1(\sigma), b_2(\sigma^*)]$ is also seriously underestimated.

Symmetry breaking is not typically something one would like to do in TDDFT. In addition to the fact that such symmetry breaking should not occur for the exact functional and that the amount of symmetry breaking can be extremely sensitive to the choice of functional (it is larger for hybrid functionals and less for the LDA and GGAs) and ignoring the fact that there may be more than one way to break symmetry in a polyatomic molecule, the main problem with symmetry breaking in TDDFT is that symmetry is very useful for assigning states. An alternative solution is simply to make the TDA. As shown in the figure, the TDA-TDLDA and DMC curves for the ${}^3B_2$ state are in remarkably good agreement. The agreement is not as good for the ${}^1B_2$ state, but at least the TDA-TDLDA curve is now in the right energy range, without the need to break symmetry.

Although symmetric CC ring-opening in oxirane is not infrequently used in advanced organic chemistry courses to illustrate the application of the Woodward–Hoffmann rules for photochemical reactivity, this photochemical process is not actually observed in unsubstituted oxirane. Our recent work [72] applying mixed TDA-TDDFT/classical trajectory surface hopping dynamics [83] to oxirane recovers the experimentally derived ring-opening mechanism of Gomer and Noyes (Figure 3.6). Analysis of ring-opening trajectories shows that the ${}^1[n, 3p_z]$ Rydberg excitation transforms easily into a ${}^1[n, \sigma^*_{CO}]$ valence-type excitation, leading to facile CO bond breaking. The excited-state trajectory hops to the ground state at a conical intersection which corresponds roughly to a mixed biradicaloid/carbonylide structure ((2) in Figure 3.6). The ground-state molecule is vibrationally hot and undergoes further hydrogen transfer and CC bond breaking reactions. This also is in line with the Gomer–Noyes mechanism.

It is a fundamental tenet of chemical kinetics that experiment can never prove a mechanism, only disprove hypotheses. In contrast, our photodynamics calculation produces a mechanism, provides state-specific information about alternative pathways, and gives lifetime information.

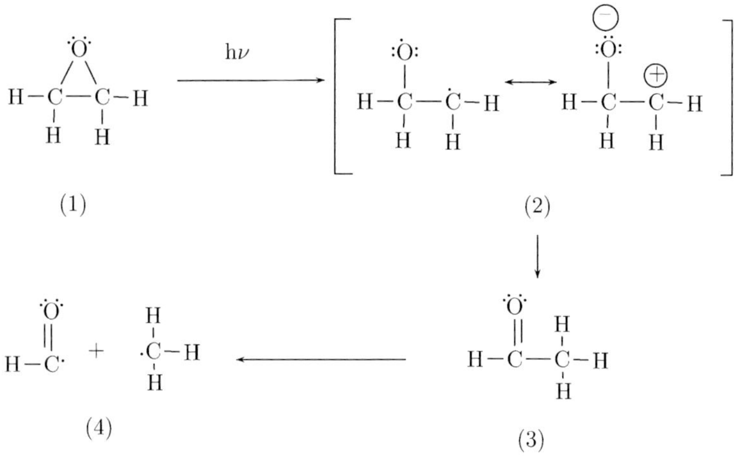

**Figure 3.6** Mechanism proposed by Gomer and Noyes [82].

## 3.5 The Future

This chapter began by noting that there has been a paradigm change in quantum chemistry where DFT has almost completely replaced HF for single determinant ground-state calculations and LR-TDDFT (or TDA-TDDFT) has largely replaced LR-TDHF (or CIS) for excited-state calculations. However, there has also been a paradigm change in DFT as should be evident from Section 3.3 on TDDFT technology–namely that DFT is taking on more and more of the aspects of traditional many-body theory. It is perhaps not too far fetched to expect that the quantum chemistry of the future will replace post-HF many-body theory calculations with post-DFT many-body theory calculations. The state-of-the-art *GW* and Bethe–Salpeter equation calculations of present day solid-state physics are certainly of this type [31]. Indeed quantum chemistry is currently expanding its frontiers to attack problems in nanoscience where reconciliation with current trends in solid-state physics theory seems both important and inevitable.

## Acknowledgments

Lars G.M. Pettersson is thanked for carefully reading this chapter, for his comments, and for his corrections. The assessment and development of TDDFT for photochemical modeling is the subject of two recent theses [84, 85]. Grenoble work on this subject continues as part of the *Réseau thématique de recherche avancée* (RTRA) project "Spectroscopy and Transport Properties in Nanomaterials: Applications and Research" and the Grenoble node of the European Theoretical Spectroscopy Facility.

## References


1 R. Singh and B. M. Deb, Developments in excited-state density-functional theory, *Phys. Rep.* **311**, 50 (**1999**).
2 M. E. Casida, Time-dependent density-functional response theory for molecules, in *Recent Advances in Density Functional Methods, Part I*, edited by D. P. Chong, p. 155, World Scientific, Singapore, **1995**.
3 E. Runge and E. K. U. Gross, Density functional theory for time-dependent systems, *Phys. Rev. Lett.* **52**, 997 (**1984**).
4 C. Jamorski, M. E. Casida, and D. R. Salahub, Dynamic polarizabilities and excitation spectra from a molecular implementation of time-dependent density-functional response theory: $N_2$ as a case study, *J. Chem. Phys.* **104**, 5134 (**1996**).
5 R. Bauernschmitt and R. Ahlrichs, Treatment of electronic excitations within the adiabatic approximation of time-dependent density functional theory, *Chem. Phys. Lett.* **256**, 454 (**1996**).
6 Z. Qian, A. Constantinescu, and G. Vignale, Solving the ultranonlocality problem in time-dependent spin-density-functional theory, *Phys. Rev. Lett.* **90**, 066402 (**2003**).
7 M. V. Faassen, Time-dependent current-density-functional theory applied to atoms and molecules, *Int. J. Mod. Phys. B* **20**, 3419 (**2006**).
8 R. J. Cave, F. Zhang, N. T. Maitra, and K. Burke, A dressed TDDFT treatment of the $2^1A_g$ states of butadiene and hexatriene, *Chem. Phys. Lett.* **389**, 39 (**2004**).
9 N. T. Maitra, F. Zhang, F. J. Cave, and K. Burke, Double excitations within time-dependent density functional theory linear response, *J. Chem. Phys.* **120**, 5932 (**2004**).
10 M. E. Casida, Propagator corrections to adiabatic time-dependent density-functional theory linear response theory, *J. Chem. Phys.* **122**, 054111 (**2005**).
11 Y. Shao, M. Head-Gordon, and A. I. Krylov, The spinflip approach within time-dependent density functional theory: Theory and applications to diradicals, *J. Chem. Phys.* **118**, 4807 (**2003**).
12 L. V. Slipchenko and A. I. Krylov, Electronic structure of the trimethylenemethane diradical in its ground and electronically excited states: Bonding, equilibrium geometries, and vibrational frequencies, *J. Chem. Phys.* **118**, 6874 (**2003**).
13 F. Wang and T. Ziegler, Time-dependent density functional theory based on a noncollinear formulation of the exchange-correlation potential, *J. Chem. Phys.* **121**, 12191 (**2004**).
14 Y. Tawada, T. Tsuneda, S. Yanagisawa, T. Yanai, and K. Hirao, A long-range-corrected time-dependent density functional theory, *J. Chem. Phys.* **120**, 8425 (**2004**).
15 S. Tokura, T. Tsuneda, and K. Hirao, Long-range-corrected time-dependent density functional study on electronic spectra of five-membered ring compounds and free-base porphyrin, *J. Theoret. Comput. Chem.* **5**, 925 (**2006**).
16 O. A. Vydrov and G. E. Scuseria, Assessment of a long-range corrected hybrid functional, *J. Chem. Phys.* **125**, 234109 (**2006**).
17 M. J. G. Peach, E. I. Tellgrent, P. Salek, T. Helgaker, and D. J. Tozer, Structural and electronic properties of polvacetylene and polyyne from hybrid and Coulomb-attenuated density functionals, *J. Phys. Chem. A* **111**, 11930 (**2007**).
18 M. E. Casida and T. A. Wesolowski, Generalization of the Kohn–Sham equations with constrained electron density (KSCED) formalism and its time-dependent response theory formulation, *Int. J. Quant. Chem.* **96**, 577 (**2004**).
19 T. A. Wesolowski, Hydrogen-bonding induced shifts of the excitation energies in nucleic acid bases: An interplay between electrostatic and

electron density overlap effects, *J. Am. Chem. Soc.* **126**, 11444 (**2004**).

20 J. Neugebauer, M. J. Louwerse, E. Baerends, and T. A. Wesolowski, The merits of the frozen-density embedding scheme to model solvatochromic shifts, *J. Chem. Phys.* **122**, 094115 (**2005**).

21 J. Neugebauer, Couplings between electronic transitions in a subsystem formulation of time-dependent density functional theory, *J. Chem. Phys.* **126**, 134116 (**2007**).

22 E. K. U. Gross and W. Kohn, Time-dependent density functional theory, *Adv. Quant. Chem.* **21**, 255 (**1990**).

23 E. K. U. Gross, C. A. Ullrich, and U. J. Gossmann, Density functional theory of time-dependent systems, in *Density Functional Theory*, edited by E. K. U. Gross and R. M. Dreizler, pp. 149–171, Plenum, New York, **1994**.

24 E. K. U. Gross, J. F. Dobson, and M. Petersilka, Density-functional theory of time-dependent phenomena, *Top. Curr. Chem.* **181**, 81 (**1996**).

25 M. E. Casida, Time-dependent density functional response theory of molecular systems: Theory, computational methods, and functionals, in *Recent Developments and Applications of Modern Density Functional Theory*, edited by J. M. Seminario, p. 391, Elsevier, Amsterdam, **1996**.

26 K. Burke and E. K. U. Gross, A guided tour of time-dependent density functional theory, in *Density Functionals: Theory and Applications*, edited by D. Joubert, volume 500 of Springer Lecture Notes in Physics, pp. 116–146, Springer, Berlin, **1998**.

27 R. van Leeuwen, Key concepts in time-dependent density-functional theory, *Int. J. Mod. Phys. B* **15**, 1969 (**2001**).

28 G. te Velde *et al.*, Chemistry with ADF, *J. Comput. Chem.* **22**, 931 (**2001**).

29 N. T. Maitra, K. Burke, H. Appel, E. K. U. Gross, and R. van Leeuwen, Ten topical questions in time-dependent density functional theory, in *Reviews in Modern Quantum Chemistry: A Celebration of the Contributions of R. G. Parr*, edited by K. D. Sen, pp. 1186–1225, World Scientific, Singapore, **2002**.

30 M. E. Casida, Jacob's ladder for time-dependent density-functional theory: Some rungs on the way to photochemical heaven, in *Accurate Description of Low-Lying Molecular States and Potential Energy Surfaces, ACS Symposium Series 828*, edited by M. R. Hoffmann and K. G. Dyall, pp. 199–220, Proceedings of ACS Symposium, San Diego, California, **2001**.

31 G. Onida, L. Reining, and A. Rubio, Electronic excitations: Density-functional versus many-body Green's-function approaches, *Rev. Mod. Phys.* **74**, 601 (**2002**).

32 C. Daniel, Electronic spectroscopy and photoreactivity in transition metal complexes, *Coordination Chem. Rev.* **238–239**, 141 (**2003**).

33 M. A. L. Marques and E. K. U. Gross, Time-dependent density functional theory, in *A Primer in Density Functional Theory*, edited by C. Fiolhais, F. Nogueira, and M. A. L Marques, volume 620 of Springer Lecture Notes in Physics, pp. 144–184, Springer, Berlin, **2003**.

34 M. A. L. Marques and E. K. U. Gross, Time-dependent density-functional theory, *Annu. Rev. Phys. Chem.* **55**, 427 (**2004**).

35 K. Burke, J. Werschnik, and E. K. U. Gross, Time-dependent density functional theory: Past, present and future, *J. Chem. Phys.* **123**, 062206 (**2005**).

36 A. Dreuw and M. Head-Gordon, Single-reference *ab initio* methods for the calculation of excited states of large molecules, *Chem. Rev.* **105**, 4009 (**2005**).

37 M. A. L. Marques, C. Ullrich, F. Nogueira, A. Rubio, and E. K. U. Gross, editors, *Time-Dependent Density-Functional Theory*, volume 706 of *Lecture Notes in Physics*, Springer, Berlin, **2006**.

38 A. Castro *et al.*, Octopus: A tool for the application of time-dependent

density functional theory, *Phys. Status Solidi* **243**, 2465 (**2006**).
39 P. Elliott, K. Burke, and F. Furche, Excited states from time-dependent density functional theory, Preprint cond-mat/0703590, March 22, **2007**.
40 P. Hohenberg and W. Kohn, Inhomogenous electron gas, *Phys. Rev.* **136**, B864 (**1964**).
41 M. Levy and J. P. Perdew, The constrained-search formulation of density functional theory, in *Density Functional Methods in Physics*, edited by R. Dreizler and J. da Providencia, p. 11, Plenum, New York, **1985**.
42 W. Kohn and L. J. Sham, Self-consistent equations including exchange and correlation effects, *Phys. Rev.* **140**, A1133 (**1965**).
43 J. P. Perdew and K. Schmidt, Jacob's ladder of density functional approximations for the exchange-correlation energy, in *Density Functional Theory and Its Applications to Materials*, edited by V. E. Van Doren, K. Van Alseoy and P. Geerlings, pp. 1–20, American Institute of Physics, **2001**.
44 W. Koch and M. C. Holthausen, *A Chemist's Guide to Density Functional Theory*, Wiley-VCH, New York, **2000**.
45 S. F. Sousa, P. A. Fernandes, and M. J. Ramos, General performance of density functionals, *J. Phys. Chem. A* **111**, 10439 (**2007**).
46 A. Becke, A new mixing of Hartree–Fock and local density functional theories, *J. Chem. Phys.* **93**, 1372 (**1993**).
47 S. Grimme and F. Neese, Double-hybrid density functional theory for excited electronic states of molecules, *J. Chem. Phys.* **127**, 154116 (**2007**).
48 R. T. Sharp and G. K. Horton, A variational approach to the unipotential many-electron problem, *Phys. Rev.* **90**, 317 (**1953**).
49 J. D. Talman and W. F. Shadwick, Optimized effective atomic central potential, *Phys. Rev. A* **14**, 36 (**1976**).
50 J. B. Krieger, T. Li, and G. J. Iafrate, Recent developments in Kohn–Sham theory for orbital-dependent exchange-correlation energy functionals, in *Density Functional Theory*, edited by E. K. U. Gross and R. M. Dreizler, p. 91, Plenum, New York, **1995**.
51 S. Hamel, P. Duffy, M. E. Casida, and D. R. Salahub, Kohn–Sham orbitals and orbital energies: Fictitious constructs but good approximations all the same, *J. Electr. Spectr. Relat. Phenom.* **123**, 345 (**2002**).
52 X. Gonze and M. Scheffler, Exchange and correlation kernels at the resonance frequency: Implications for excitation energies in density-functional theory, *Phys. Rev. Lett.* **82**, 4416 (**1999**).
53 R. van Leeuwen and E. J. Baerends, Exchange-correlation potential with correct asymptotic behavior, *Phys. Rev. A* **49**, 2421 (**1994**).
54 S. Hamel, M. E. Casida, and D. R. Salahub, Exchange-only optimized effective potential for molecules from resolution-of-the-identity techniques: Comparison with the local density approximation, with and without asymptotic correction, *J. Chem. Phys.* **116**, 8276 (**2002**).
55 S. Hirata, C. G. Zhan, E. Apra, T. L. Windus, and D. A. Dixon, A new, self-contained asymptotic correction scheme to exchange-correlation potentials for time-dependent density functional theory, *J. Phys. Chem. A* **107**, 10154 (**2003**).
56 C. A. Ullrich and I. V. Tokatly, Nonadiabatic electron dynamics in time-dependent density-functional theory, *Phys. Rev. B* **73**, 235102 (**2006**).
57 E. R. Davidson, The iterative calculation of a few of the lowest eigenvalues and corresponding eigenvectors of large real-symmetric matrices, *J. Comput. Phys.* **17**, 87 (**1975**).
58 C. W. Murray, S. C. Racine, and E. R. Davidson, Improved algorithms for the lowest few eigenvalues and eigenvectors of large matrices, *J. Comput. Phys.* **103**, 382 (**1991**).
59 J. Olsen, H. J. A Jensen, and P. Joergensen, Solution of the large matrix equations which occur in response theory, *J. Comput. Phys.* **74**, 265 (**1988**).

**60** R. E. Stratmann, G. E. Scuseria, and M. J. Frisch, An efficient implementation of time-dependent density-functional theory for the calculation of excitation energies of large molecules, *J. Chem. Phys.* **109**, 8218.

**61** M. E. Casida *et al.*, Charge-transfer correction for improved time-dependent local density approximation excited-state potential energy curves: Analysis within the two-level model with illustration for $H_2O$ and LiH, *J. Chem. Phys.* **113**, 7062 (**2000**).

**62** F. Cordova *et al.*, Troubleshooting time-dependent density-functional theory for photochemical applications: Oxirane, *J. Chem. Phys.* **127**, 164111 (**2007**).

**63** Y. Yamaguchi, Y. Osamura, J. D. Goddard, and H. F. Schaefer III, *A New Dimension to Quantum Chemistry: Analytic Derivative Methods in Ab Initio Molecular Electronic Structure Theory*, Oxford University Press, Oxford, **1994**.

**64** C. V. Caillie and R. D. Amos, Geometric derivatives of excitation energies using SCF and DFT, *Chem. Phys. Lett.* **308**, 249 (**1999**).

**65** C. V. Caillie and R. D. Amos, Geometric derivatives of density functional theory excitation energies using gradient-corrected functionals, *Chem. Phys. Lett.* **317**, 159 (**2000**).

**66** F. Furche and R. Ahlrichs, Adiabatic time-dependent density functional methods for excited state properties, *J. Chem. Phys.* **117**, 7433 (**2002**).

**67** D. Rappoport and F. Furche, Analytical time-dependent density functional derivative methods within the RI-J approximation, an approach to excited states of large molecules, *J. Chem. Phys.* **122**, 064105 (**2005**).

**68** J. Hutter, Excited state nuclear forces from the Tamm–Dancoff approximation to time-dependent density functional theory within the plane wave basis set framework, *J. Chem. Phys.* **118**, 3928 (**2003**).

**69** N. L. Doltsinis and D. S. Kosov, Plane wave/pseudopotential implementation of excited state gradients in density functional linear response theory: A new route via implicit differentiation, *J. Chem. Phys.* **122**, 144101 (**2005**).

**70** G. Scalmani *et al.*, Geometries and properties of excited states in the gas phase and in solution: Theory and application of a time-dependent density functional theory polarizable continuum model, *J. Chem. Phys.* **124**, 094107 (**2006**).

**71** T. Ziegler, A. Rauk, and E. J. Baerends, Calculation of multiplet energies by Hartree–Fock–Slater method, *Theor. Chim. Acta* **43**, 261 (**1977**).

**72** E. Tapavicza, I. Tavernelli, U. Röthlisberger, C. Filippi, and M. E. Casida, Mixed time-dependent density-functional theory/classical trajectory surface hopping study of oxirane photochemistry, *J. Chem. Phys.* **129**, 124108 (**2008**).

**73** M. E. Casida, C. Jamorski, K. C. Casida, and D. R. Salahub, Molecular excitation energies to high-lying bound states from time-dependent density-functional response theory: Characterization and correction of the time-dependent local density approximation ionization threshold, *J. Chem. Phys.* **108**, 4439 (**1998**).

**74** M. E. Casida, K. C. Casida, and D. R. Salahub, Excited-state potential energy curves from time-dependent density-functional theory: A cross-section of formaldehyde's $^1A_1$ manifold, *Int. J. Quant. Chem.* **70**, 933 (**1998**).

**75** D. J. Tozer, R. D. Amos, N. C. Handy, B. O. Roos, and L. Serrano-Andrés, Does density functional theory contribute to the understanding of excited states of unsaturated organic compounds? *Mol. Phys.* **97**, 859 (**1999**).

**76** A. Dreuw, J. L. Weisman, and M. Head-Gordon, Long-range charge-transfer excited states in time-dependent density functional theory require non-local exchange, *J. Chem. Phys.* **119**, 2943 (**2003**).

77 T.-K. Liu and A. B. F. Duncan, The absorption spectrum of ethylene oxide in the vacuum ultraviolet, *J. Chem. Phys.* **17**, 241 (**1949**).

78 A. Lowrey III and K. Watanabe, Absorption and ionization coefficients of ethylene oxide, *J. Chem. Phys.* **28**, 208 (**1958**).

79 G. Fleming, M. M. Anderson, A. J. Harrison, and L. W. Pickett, Effect of ring size on the far ultraviolet absorption and photolysis of cyclic ethers, *J. Chem. Phys.* **30**, 351.

80 W. von Niessen, L. S. Cederbaum, and W. P. Kraemer, The electronic structure of molecules by a many-body approach VIII. Ionization potentials of the three-membered ring molecules $C_3H_6$, $C_2H_4O$, $C_2H_5N$, *Theor. Chim. Acta* **44**, 85 (**1977**).

81 H. Basch, M. B. Robin, N. A. Kuebler, C. Baker, and D. W. Turner, Optical and photoelectron spectra of small rings. III. The saturated three-membered rings, *J. Chem. Phys.* **51**, 52 (**1969**).

82 E. Gomer and J. W. A. Noyes, Photochemical studies. XLII. Ethylene oxide, *J. Am. Chem. Soc.* **72**, 101 (**1950**).

83 E. Tapavicza, I. Tavernelli, and U. Röthlisberger, Trajectory surface hopping within linear response time-dependent density-functional theory, *Phys. Rev. Lett.* **98**, 023001 (**2007**).

84 F. Cordova, *Photochemistry from Density-Functional Theory*, PhD thesis, Université Joseph Fourier (Grenoble I), Grenoble, France, **2007**.

85 E. Tapavicza, *Nonadiabatic First Principles Molecular Dynamics: Development and Applications to Photochemistry*, PhD thesis, École Polytechnique Fédérale de Lausanne, Lausanne, Switzerland, **2008**.

# 4
# Periodic Systems, Plane Waves, the PAW Method, and Hybrid Functionals

*Martijn Marsman*

## 4.1
## Periodic Systems

In this chapter we consider systems made up of a unit cell periodically repeated throughout all space by rigid translation along the lattice vectors $\mathbf{a}_1$, $\mathbf{a}_2$, and $\mathbf{a}_3$ (see, for instance, the face-centered-cubic (fcc) lattice depicted in Figure 4.1(a)). The corresponding reciprocal lattice is spanned by the vectors

$$\mathbf{b}_1 = \frac{2\pi}{\Omega}\mathbf{a}_2 \times \mathbf{a}_3 \quad \mathbf{b}_2 = \frac{2\pi}{\Omega}\mathbf{a}_3 \times \mathbf{a}_1 \quad \mathbf{b}_3 = \frac{2\pi}{\Omega}\mathbf{a}_1 \times \mathbf{a}_2, \tag{4.1}$$

where $\Omega = \mathbf{a}_1 \cdot \mathbf{a}_2 \times \mathbf{a}_3$ is the volume of the unit cell. Note that with the above definition we have $\mathbf{a}_i \cdot \mathbf{b}_j = 2\pi\delta_{ij}$. The reciprocal lattice that corresponds to an fcc Bravais lattice is the body-centered-cubic (bcc) lattice depicted in Figure 4.1(b).

The Bloch theorem implies that under periodic boundary conditions (PBC) all eigenfunctions of the one-electron Kohn–Sham equations, $\psi_{n\mathbf{k}}(\mathbf{r})$, may be written as the product of a cell-periodic part $u_{n\mathbf{k}}(\mathbf{r})$ and a plane-wave-like modulation, i.e.,

$$\psi_{n\mathbf{k}}(\mathbf{r}) = u_{n\mathbf{k}}(\mathbf{r}) \exp[i\mathbf{k} \cdot \mathbf{r}], \tag{4.2}$$

with

$$u_{n\mathbf{k}}(\mathbf{r}) = u_{n\mathbf{k}}(\mathbf{r} + \mathbf{R}) \;\; \forall\, \mathbf{R} = \sum_{i=1}^{3} m_i \mathbf{a}_i \mid m_i \in \mathbb{Z}. \tag{4.3}$$

The corresponding eigenenergies are periodic in reciprocal space,

$$\epsilon_n(\mathbf{k}) = \epsilon_n(\mathbf{k} + \mathbf{G}) \;\; \forall\, \mathbf{G} = \sum_{i=1}^{3} m_i \mathbf{b}_i \mid m_i \in \mathbb{Z}. \tag{4.4}$$

The so-called *band* index $n$ labels the different solutions to the Schrödinger equation at fixed $\mathbf{k}$. The wave vector $\mathbf{k}$ is usually constrained to lie in the first Brillouin zone (BZ) of the reciprocal lattice.

The first BZ is a uniquely defined primitive (Wigner–Seitz) cell of the reciprocal lattice (see, for instance, Figure 4.1(c)). Its construction recipe will not be discussed

*Computational Methods in Catalysis and Materials Science.* Edited by Rutger A. van Santen and Philippe Sautet

ISBN: 978-3-527-32032-5

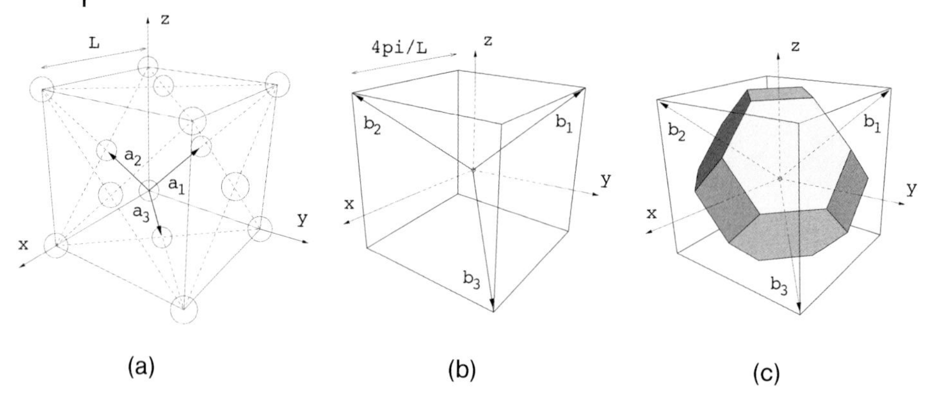


**Figure 4.1** The face-centered-cubic Bravais lattice (a), the corresponding body-centered-cubic reciprocal lattice (b), and the first Brillouin zone of the body-centered-cubic reciprocal lattice (c).

here; it is an important concept in crystallography and may be found in any good textbook on solid-state physics (see, for instance, Ref. [1]).

The evaluation of many key quantities (e.g., charge density, density of states, and total energy) requires integration over the first BZ. The charge density $\rho(\mathbf{r})$, for instance, is given by

$$\rho(\mathbf{r}) = \frac{1}{\Omega_{\mathrm{BZ}}} \sum_{n} \int_{\mathrm{BZ}} f_{n\mathbf{k}} |\psi_{n\mathbf{k}}(\mathbf{r})|^2 \, d\mathbf{k}, \tag{4.5}$$

where $\Omega_{\mathrm{BZ}}$ is the volume of the first BZ in reciprocal space, and $f_{n\mathbf{k}}$ are the occupation numbers, i.e., the number of electrons that occupy state $n\mathbf{k}$ (possibly fractional in the case of metallic systems: $0 \leq f_{n\mathbf{k}} \leq 1$). The direct evaluation of the charge density in accordance with Eq. (4.5) would involve the calculation of $n$ electronic wavefunctions at an infinite number of $\mathbf{k}$-points (an infinite system contains an infinite number of electrons), which is out of the question, of course.

However, exploiting the fact that the wavefunctions at $\mathbf{k}$-points that are close together will be almost identical, one may approximate the integration in Eq. (4.5) by a weighted sum over a discrete set of points

$$\rho(\mathbf{r}) = \sum_{n} \sum_{\mathbf{k}} w_{\mathbf{k}} f_{n\mathbf{k}} |\psi_{n\mathbf{k}}(\mathbf{r})|^2 \, d\mathbf{k}, \tag{4.6}$$

where the weights $w_{\mathbf{k}}$ sum up to 1. Like any other numerical integration, the accuracy of this approximation depends on the number of sampling points, and the particular choice of $\mathbf{k}$-points and corresponding weights $w_{\mathbf{k}}$. Several different sampling schemes that aim to provide an accurate integration over the first BZ, using as few $\mathbf{k}$-points as possible, may be found in the literature [2–5]. The most commonly used method is the straightforward sampling of the first BZ by means of Monkhorst–Pack [4] ($n_1 \times n_2 \times n_3$) regular grids in reciprocal space.

The minimal $\mathbf{k}$-point sampling density, required to reach a given accuracy, is system dependent. As a rule of thumb, however, one may state that the smaller the

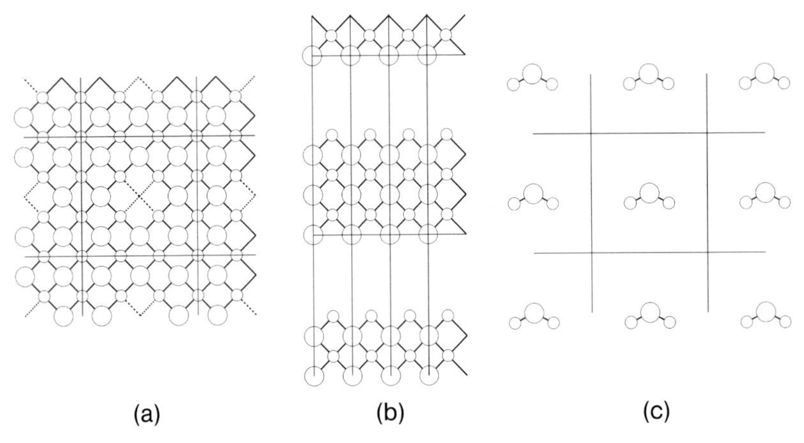


**Figure 4.2** Modeling aperiodic systems under periodic boundary conditions, by means of supercells: a vacancy in bulk (a), a surface slab (b), and an isolated molecule (c).

band gap of a material, the denser the first BZ has to be sampled. Metals require the densest **k**-point sampling (primarily in order to correctly represent the features of the Fermi surface). Furthermore, note that the larger $\Omega$, the volume of the unit cell in real space, the smaller the volume of the corresponding first Brillouin zone in reciprocal space [see Eq. (4.1)]. Consequently, the number of **k**-points needed to sample the first BZ decreases with increasing unit cell size. In the case of very large unit cells, the sampling of the first BZ is limited to a single **k**-point, usually the $\Gamma$-point ($\mathbf{k} = 0$).

The latter applies, for instance, to many so-called *supercell* calculations. Supercells are large unit cells, constructed to model aperiodic phenomena under periodic boundary conditions. An appreciable number of interesting systems do not obey periodic boundary conditions (along one or more spatial directions), e.g., isolated defects in bulk structures (like vacancies or dopants), surfaces (clean or covered with chemical reactants), nanowires or polymers, and isolated molecules. Figures 4.2(a)–(c) illustrate how the aforementioned aperiodic systems may be approximated by means of supercells: (a) a vacancy in bulk, (b) a slab geometry used to model surfaces, and (c) an isolated molecule in a box.

The essence of the supercell approach is to make sure that the unit cell is large enough, along the relevant spatial direction(s), to prevent the periodic images of the "aperiodic" entity (defect, surface, or molecule) to interact with each other. If that is the case, then the use of a supercell and periodic boundary conditions constitutes a good approximation to the original aperiodic system.

## 4.2
## Plane Waves, Pseudopotentials, and the PAW Method

The following presents a brief introduction to the representation of the one-electron wavefunctions and Kohn–Sham equations under periodic boundary conditions, using a plane-wave basis set. In addition we discuss the general concept of pseudopotential methods, and in particular the projector-augmented-wave (PAW) method. For a comprehensive introduction to *ab initio* total-energy calculations using plane-wave basis sets, we refer the reader to the excellent review article by Payne et al. [6].

### 4.2.1
### Plane Waves

Let us assume that the cell-periodic parts of the Bloch waves, $u_{n\mathbf{k}}(\mathbf{r})$, are defined by specifying their values at an $(N_1 \times N_2 \times N_3)$ regular grid in the unit cell

$$u_{n\mathbf{k}}(\mathbf{r}) = C_{\mathbf{r}n\mathbf{k}}, \quad \mathbf{r} = \sum_{i=1}^{3} \frac{n_i}{N_i}\mathbf{a}_i \mid n_i = 0, \ldots, N_i - 1. \tag{4.7}$$

In that case $u_{n\mathbf{k}}(\mathbf{r})$ can conveniently be expanded in a finite set of plane-wave basis functions,

$$u_{n\mathbf{k}}(\mathbf{r}) = \frac{1}{\sqrt{\Omega}} \sum_{\mathbf{G}} C_{\mathbf{G}n\mathbf{k}} \exp[i\mathbf{G} \cdot \mathbf{r}], \tag{4.8}$$

where $\mathbf{G} = \sum_{i=1}^{3} m_i \mathbf{b}_i \mid m_i = -N_i/2 + 1, \ldots, N_i/2$, are lattice vectors in reciprocal space. The Bloch waves [see Eq. (4.2)], therefore, can be written as

$$\psi_{n\mathbf{k}}(\mathbf{r}) = \frac{1}{\sqrt{\Omega}} \sum_{\mathbf{G}} C_{\mathbf{G}n\mathbf{k}} \exp[i(\mathbf{k} + \mathbf{G}) \cdot \mathbf{r}], \tag{4.9}$$

or in bracket notation

$$|\psi_{n\mathbf{k}}\rangle = \sum_{\mathbf{G}} C_{\mathbf{G}n\mathbf{k}} |\mathbf{k} + \mathbf{G}\rangle, \tag{4.10}$$

where

$$\langle \mathbf{r} | \mathbf{k} + \mathbf{G} \rangle = \frac{1}{\sqrt{\Omega}} \exp[i(\mathbf{k} + \mathbf{G}) \cdot \mathbf{r}] \tag{4.11}$$

and

$$C_{\mathbf{G}n\mathbf{k}} = \langle \mathbf{k} + \mathbf{G} | \psi_{n\mathbf{k}} \rangle. \tag{4.12}$$

The transformations to-and-fro between real and reciprocal space are easily accomplished by means of fast Fourier transform (FFT) algorithms, i.e.,

$$\{C_{\mathbf{r}n\mathbf{k}}\} \overset{\text{FFT}}{\longleftrightarrow} \{C_{\mathbf{G}n\mathbf{k}}\}, \tag{4.13}$$

with

$$C_{\mathbf{r}n\mathbf{k}} = \sum_{\mathbf{G}} C_{\mathbf{G}n\mathbf{k}} \exp[i\mathbf{G} \cdot \mathbf{r}], \tag{4.14}$$

$$C_{\mathbf{G}n\mathbf{k}} = \frac{1}{N_{\text{FFT}}} \sum_{\mathbf{r}} C_{\mathbf{r}n\mathbf{k}} \exp[-i\mathbf{G} \cdot \mathbf{r}], \tag{4.15}$$

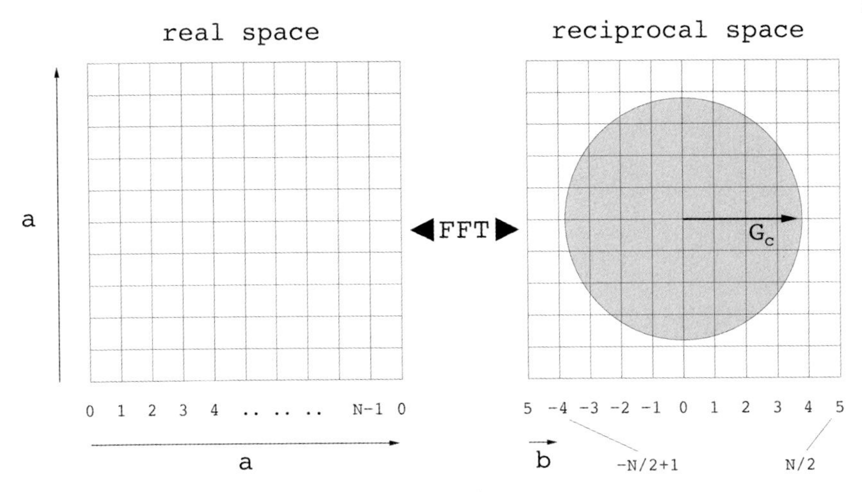


**Figure 4.3** A (10 × 10) regular grid in real space (left panel) and the reciprocal space grid related to it by FFT (right panel). The length of the vector $G_c$ is related to a hypothetical basis set kinetic energy cutoff of $E_c = G_c^2/2$; the shaded area contains all plane-wave components that are included in the basis set.

where $N_{\text{FFT}} = N_1 N_2 N_3$ is the total number of grid points in the unit cell. The computational cost of an FFT scales as $N_{\text{FFT}} \ln N_{\text{FFT}}$.

Usually the plane-wave basis set expansion of $u_{n\mathbf{k}}$ is truncated to include only those plane-wave components $|\mathbf{k}+\mathbf{G}\rangle$, with a kinetic energy $(1/2)|\mathbf{k}+\mathbf{G}|^2$ below a particular cutoff energy $E_c$. This means that the size of the $(N_1 \times N_2 \times N_3)$ regular grid in reciprocal space must be chosen large enough such that all plane waves, $|\mathbf{G}\rangle$, that are part of the basis are contained in the grid. This is illustrated in Figure 4.3.

Similarly, all other cell-periodic quantities such as the charge density and the potentials may be expanded in plane-wave basis functions as well,

$$\rho(\mathbf{r}) = \sum_{\mathbf{G}} \rho_{\mathbf{G}} \exp[i\mathbf{G}\cdot\mathbf{r}], \tag{4.16}$$

$$V(\mathbf{r}) = \sum_{\mathbf{G}} V_{\mathbf{G}} \exp[i\mathbf{G}\cdot\mathbf{r}]. \tag{4.17}$$

The plane-wave basis set expansion of the Kohn–Sham equations is quite straightforward:

$$\langle\mathbf{k}+\mathbf{G}| - \frac{1}{2}\Delta + V|\psi_{n\mathbf{k}}\rangle = \epsilon_{n\mathbf{k}}\langle\mathbf{k}+\mathbf{G}|\psi_{n\mathbf{k}}\rangle. \tag{4.18}$$

The kinetic energy contribution can be conveniently evaluated in reciprocal space,

$$\langle\mathbf{k}+\mathbf{G}| - \frac{1}{2}\Delta|\psi_{n\mathbf{k}}\rangle = -\frac{1}{2}|\mathbf{k}+\mathbf{G}|^2 C_{\mathbf{G}n\mathbf{k}}, \tag{4.19}$$

whereas the action of the local potential on the wavefunction is best computed in real space,

$$\langle \mathbf{k}+\mathbf{G}|V|\psi_{n\mathbf{k}}\rangle = \frac{1}{N_{\mathrm{FFT}}}\sum_{\mathbf{r}} V_{\mathbf{r}} C_{\mathbf{r}n\mathbf{k}} \exp[-i\mathbf{G}\cdot\mathbf{r}]. \tag{4.20}$$

Here, $V_{\mathbf{r}} = V(\mathbf{r})$, $\mathbf{r} = \sum_{i=1}^{3} n_i \mathbf{a}_i/N_i \mid n_i = 0, \ldots, N_i - 1$, is the local potential on the regular grid in the unit cell.

As discussed in Section 4.2 the local potential can be written as the sum of three contributions

$$V = V_{\mathrm{C}}[\rho] + V_{\mathrm{xc}}[\rho] + V_{\mathrm{ext}}, \tag{4.21}$$

where $V_{\mathrm{C}}$ and $V_{\mathrm{xc}}$ are the Coulomb and exchange-correlation potentials arising from the charge density $\rho$, and $V_{\mathrm{ext}}$ is the external potential. The easiest way to compute the Coulomb potential is to solve the Poisson equation in reciprocal space:

$$V_{\mathrm{C},\mathbf{G}} = \frac{4\pi}{|\mathbf{G}|^2}\rho_{\mathbf{G}}. \tag{4.22}$$

The exchange-correlation potential can be readily obtained from the charge density on the real space grid $V_{\mathrm{xc},\mathbf{r}} = V_{\mathrm{xc}}[\rho_{\mathbf{r}}]$, and by means of an FFT we obtain $V_{\mathrm{xc}}$ in reciprocal space $\{V_{\mathrm{xc},\mathbf{r}}\} \rightarrow \{V_{\mathrm{xc},\mathbf{G}}\}$. Let us assume for the moment that the external potential is known in reciprocal space as well (we will come back to this point as we discuss the frozen core approximation). The summation of Eq. (4.21) is then carried out in reciprocal space

$$V_{\mathbf{G}} = V_{\mathrm{C},\mathbf{G}} + V_{\mathrm{xc},\mathbf{G}} + V_{\mathrm{ext},\mathbf{G}}. \tag{4.23}$$

A subsequent FFT: $\{V_{\mathbf{G}}\} \rightarrow \{V_{\mathbf{r}}\}$ yields the local potential in real space. The simple shape of the Poisson equation in reciprocal space, coupled to the computational efficiency of the FFT, is a major advantage of using plane-wave basis sets.

With Eqs. (4.12), (4.19), and (4.20) we then obtain

$$-\frac{1}{2}|\mathbf{k}+\mathbf{G}|^2 C_{\mathbf{G}n\mathbf{k}} + \frac{1}{N_{\mathrm{FFT}}}\sum_{\mathbf{r}} V_{\mathbf{r}} C_{\mathbf{r}n\mathbf{k}} \exp[-i\mathbf{G}\cdot\mathbf{r}] = \epsilon_{n\mathbf{k}} C_{\mathbf{G}n\mathbf{k}}. \tag{4.24}$$

Note that the second term on the left-hand side of Eq. (4.24) corresponds to the expansion coefficient of the plane-wave component $|\mathbf{G}\rangle$ in the FFT of $\{V_{\mathbf{r}} C_{\mathbf{r}n\mathbf{k}}\}$ [see Eqs. (4.14) and (4.15)].

### 4.2.2 Pseudopotentials

The plane-wave basis set expansion of the electronic wavefunctions [see Eq. (4.8)] can quickly become prohibitively expensive from a computational point of view. The number of plane-wave components needed to (i) describe tightly bound (spatially strongly localized) states, and/or (ii) the rapid oscillations (nodal features) of the wavefunctions near the nucleus, exceeds any practical limit, except maybe for Li and H. To address these points one commonly introduces two interrelated approximations, the *frozen core* and the *pseudopotential* approximation.

The frozen core approximation recognizes the fact that most physical properties of compound materials are largely determined by the valence electrons of the constituent atoms. The tightly bound and spatially strongly localized inner electronic shells, the so-called core electrons, hardly adapt to the environment of the atom. Consequently, one may precalculate the electronic wavefunctions of the core electrons for each atomic constituent, and keep them fixed, or frozen, in the course of subsequent calculations, thus reducing the number of wavefunctions that have to be calculated.

The pseudopotential approximation replaces the ionic potential that arises from the nuclear charge and the frozen core charge density by an effective pseudopotential that is easily representable in a plane wave basis set. The pseudopotential is chosen in such a way that it gives rise to a set of nodeless pseudowavefunctions, and that its scattering properties for this set of wavefunctions are equal to the scattering properties of the true ionic potential acting on the true valence wave functions. A number of different pseudopotential construction schemes may be found in literature [7–14]. The removal of the nodal features in the wavefunctions strongly alleviates the costs of their representation in terms of plane waves. The pseudized ionic potential and wavefunctions match continuously onto their counterparts at a predetermined radius $r_c$. The above is illustrated in Figure 4.4, where $V$ ($\widetilde{V}$) and $\psi$ ($\widetilde{\psi}$) denote the true (pseudo) ionic potential and corresponding valence wavefunctions.

Within the framework of a pseudopotential approximation the external potential in reciprocal space [see Eq. (4.23)] is given by

$$V_{\text{ext},\mathbf{G}} = \sum_{\alpha} S_{\alpha,\mathbf{G}} \widetilde{V}_{\alpha,\mathbf{G}}, \tag{4.25}$$

where $\widetilde{V}_{\alpha,\mathbf{G}}$ is the local ionic pseudopotential in reciprocal space, and

$$S_{\alpha,\mathbf{G}} = \sum_{I} \exp[i\mathbf{G} \cdot \mathbf{R}_I] \tag{4.26}$$

is the structure factor of atomic species $\alpha$. The sum $I$ is taken over all lattice sites $\mathbf{R}_I$ in the unit cell that are occupied by atoms of species $\alpha$.

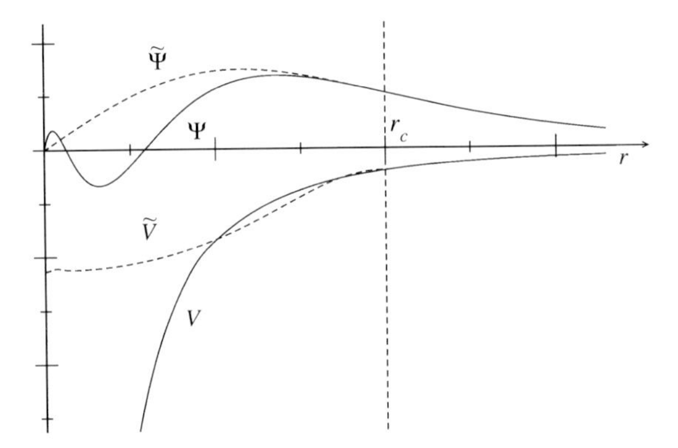


**Figure 4.4** The true (pseudo) ionic potential, $V$ ($\widetilde{V}$), and corresponding $3p$ valence wavefunctions $\psi$ ($\widetilde{\psi}$) for a Cl atom.

### 4.2.3
### The PAW Method

Amongst the currently available pseudopotential approximations, the PAW method, introduced by Blöchl [14], is probably the optimal tradeoff between accuracy and computational efficiency. In the PAW method, the true, or so-called *all-electron* (AE), wavefunctions $\psi_{n\mathbf{k}}$ are derived from pseudo (PS) wavefunctions $\widetilde{\psi}_{n\mathbf{k}}$ by means of a linear transformation:

$$|\psi_{n\mathbf{k}}\rangle = |\widetilde{\psi}_{n\mathbf{k}}\rangle + \sum_i (|\phi_i\rangle - |\widetilde{\phi}_i\rangle)\langle\widetilde{p}_i|\widetilde{\psi}_{n\mathbf{k}}\rangle. \tag{4.27}$$

The designation "all-electron" refers to the fact that these wave functions contain the required nodal features inside the ionic cores, which ensure orthogonality between the valence and core wavefunctions.

The PS wavefunctions $\widetilde{\psi}_{n\mathbf{k}}$ are the variational quantities of the PAW method and are expanded in plane waves or another appropriate basis set. The additional local basis functions, or *partial waves*, $\phi_i$ and $\widetilde{\phi}_i$, are nonzero only within nonoverlapping spheres centered at the atomic sites, the so-called PAW spheres. In the interstitial region between the PAW spheres, therefore, the PS wave functions $\widetilde{\psi}_{n\mathbf{k}}$ are identical to the AE wavefunctions $\psi_{n\mathbf{k}}$. Inside the spheres, however, the PS wavefunctions are only a computational tool and a bad approximation to the true wavefunctions, since even the norm of the AE wavefunction is not reproduced. Equation (4.27) is required to map the auxiliary quantities $\widetilde{\psi}_{n\mathbf{k}}$ onto the corresponding AE wavefunctions. The index $i$ is a shorthand for the atomic site $\mathbf{R}_i$, the angular momentum quantum numbers $(l_i, m_i)$, and an additional index $\varepsilon_i$ denoting the linearization energy.

In all practical implementations of the PAW method, the AE partial waves $\phi_i$ are chosen to be solutions of the spherical (scalar relativistic) Schrödinger equation for a *nonspinpolarized atom* at a specific energy $\varepsilon_i$ in the valence regime, and for a specific angular momentum $l_i$:

$$\left(-\frac{1}{2}\Delta + V\right)|\phi_i\rangle = \epsilon_i|\phi_i\rangle, \tag{4.28}$$

where $V$ is the (spherical) effective atomic AE potential. The, usually nodeless, PS partial waves $\widetilde{\phi}_i$ are equivalent to the AE partial waves outside a predetermined core radius $r_c$ and match continuously onto $\phi_i$ inside the core radius. As an example, Figure 4.5 shows the 3$d$ and 4$s$ AE- and PS partial waves for Mn.

The projector functions $\widetilde{p}_i$ must be constructed in such a way that they are dual to the PS partial waves: $\langle\widetilde{p}_i|\widetilde{\phi}_j\rangle = \delta_{ij}$. In practice, a two-step procedure is adopted to establish the projector functions. First we compute the intermediate functions $\chi_i$,

$$|\chi_i\rangle = \left(\epsilon_i + \frac{1}{2}\Delta - \widetilde{V}\right)|\widetilde{\phi}_i\rangle, \tag{4.29}$$

where $\widetilde{V}$ is the (spherical) effective atomic PS potential. The latter may be freely chosen inside of $r_c$, but must match its AE counterpart at $r_c$ [i.e., $\widetilde{V}(r) = V(r)$ for $r \geq r_c$]. Subsequently, the projectors are defined as a linear combination of the intermediate $\chi_i$, in accordance with [12]:

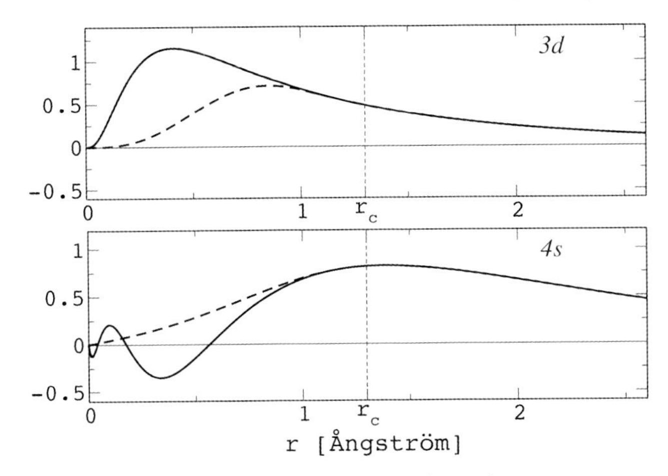


**Figure 4.5** The all-electron (drawn lines) and pseudo (dashed lines) Mn 3*d* and 4*s* radial partial waves.

$$|\widetilde{p}_i\rangle = \sum_j (B^{-1})_{ji}|\chi_j\rangle \quad B_{ij} = \langle\widetilde{\phi}_i|\chi_j\rangle. \tag{4.30}$$

The resulting projectors $\widetilde{p}_i$ are dual to the PS partial waves and $\langle r|\widetilde{p}_i\rangle = 0$ for $r > r_c$.

For this choice of projector functions one may show that the PS partial waves observe the generalized Kohn–Sham eigenvalue equation

$$\Big(-\frac{1}{2}\Delta + \widetilde{V} + \sum_{ij}|\widetilde{p}_i\rangle D_{ij}\langle\widetilde{p}_j|\Big)|\widetilde{\phi}_k\rangle = \epsilon_k\Big(1 + \sum_{ij}|\widetilde{p}_i\rangle Q_{ij}\langle\widetilde{p}_j|\Big)|\widetilde{\phi}_k\rangle \tag{4.31}$$

*exactly*, at the linearization energies $\epsilon_k$ [see Eq. (4.28)], provided the integrated compensation charges $Q_{ij}$ and the one-center strength parameters $D_{ij}$ are defined as

$$Q_{ij} = \langle\phi_i|\phi_j\rangle - \langle\widetilde{\phi}_i|\widetilde{\phi}_j\rangle \tag{4.32}$$

$$D_{ij} = \langle\phi_i| - \frac{1}{2}\Delta + V|\phi_j\rangle - \langle\widetilde{\phi}_i| - \frac{1}{2}\Delta + \widetilde{V}|\widetilde{\phi}_j\rangle. \tag{4.33}$$

Beware that in case the projector functions span a different space than that defined by Eq. (4.29), then Eq. (4.31) is necessarily never exactly fulfilled.

The construction of an AE wavefunction from its corresponding PS counterpart is illustrated in Fig 4.6.

The crucial point in the PAW method, however, is that the reconstruction of Eq. (4.27) is only carried out *implicitly*. The fact that the AE wavefunction is never constructed explicitly, by adding its separate contributions on a common grid, is what makes the PAW method a viable scheme. The following illustrates this principle. The expectation value of any quasi local operator $\hat{O}$, like for instance the kinetic energy operator $-\Delta/2$ or the density operator $|\mathbf{r}\rangle\langle\mathbf{r}|$, with respect to the AE wavefunction $|\psi\rangle$ is given by

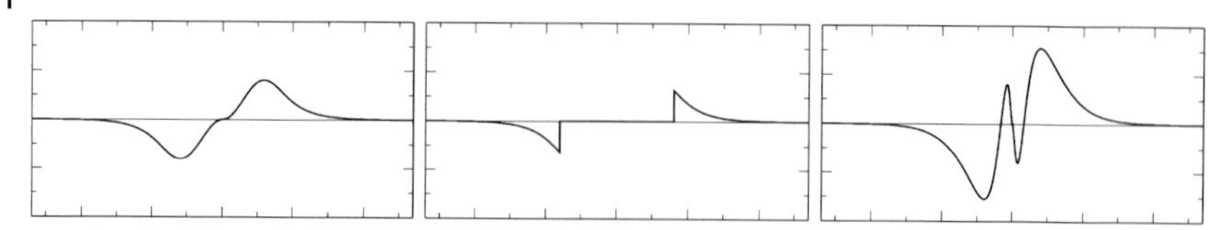

**Figure 4.6** The reconstruction of the all-electron PAW wavefunction $|\psi_{n\mathbf{k}}\rangle$ (rightmost panel), from the corresponding pseudized counterpart, by means of the linear transformation of Eq. (4.27).

$$\langle\psi|\hat{O}|\psi\rangle = \langle\widetilde{\psi}|\hat{O}|\widetilde{\psi}\rangle + \sum_{ij}\left(\langle\phi_i|\hat{O}|\phi_j\rangle - \langle\widetilde{\phi}_i|\hat{O}|\widetilde{\phi}_j\rangle\right)\langle\widetilde{\psi}|\widetilde{p}_i\rangle\langle\widetilde{p}_j|\widetilde{\psi}\rangle. \tag{4.34}$$

The kinetic energy, for instance, is given by

$$\sum_{n\mathbf{k}} w_{\mathbf{k}} f_{n\mathbf{k}}\langle\widetilde{\psi}_{n\mathbf{k}}| -\frac{1}{2}\Delta|\widetilde{\psi}_{n\mathbf{k}}\rangle - \sum_{ij}\rho_{ij}\langle\widetilde{\phi}_i| -\frac{1}{2}\Delta|\widetilde{\phi}_j\rangle + \sum_{ij}\rho_{ij}\langle\phi_i| -\frac{1}{2}\Delta|\phi_j\rangle, \tag{4.35}$$

where

$$\rho_{ij} = \sum_{n\mathbf{k}} w_{\mathbf{k}} f_{n\mathbf{k}}\langle\widetilde{\psi}_{n\mathbf{k}}|\widetilde{p}_i\rangle\langle\widetilde{p}_j|\widetilde{\psi}_{n\mathbf{k}}\rangle. \tag{4.36}$$

In other words with any quasi local operator $\hat{O}$ that acts on the AE wavefunctions, we can associate a pseudooperator $\widetilde{O}$,

$$\widetilde{O} = \hat{O} + \sum_{ij}|\widetilde{p}_i\rangle\left(\langle\phi_i|\hat{O}|\phi_j\rangle - \langle\widetilde{\phi}_i|\hat{O}|\widetilde{\phi}_j\rangle\right)\langle\widetilde{p}_j|, \tag{4.37}$$

that acts on the PS wavefunctions. In the case of the density operator $|\mathbf{r}\rangle\langle\mathbf{r}|$, the PS operator takes on the following form:

$$\widetilde{O} = |\mathbf{r}\rangle\langle\mathbf{r}| + \sum_{ij}|\widetilde{p}_i\rangle\left(\phi_i^*(r)\phi_j(r) - \widetilde{\phi}_i^*(r)\widetilde{\phi}_j(r)\right)\langle\widetilde{p}_j| \tag{4.38}$$

Note that Eqs. (4.34) and (4.37) only strictly hold when $\sum_i|\widetilde{\phi}_i\rangle\langle\widetilde{p}_i| = 1$ within the PAW spheres, that is, when the PS partial waves and the projectors provide a complete local basis set expansion of the PS wave function within the PAW spheres. The integrals in Eq. (4.34) that involve the AE and PS partial waves are evaluated on radial logarithmic support grids within the PAW spheres, and *not* on the regular (FFT) grid (see Section 4.2.1). Only quantities that involve the PS wavefunctions $\widetilde{\psi}_{n\mathbf{k}}$ are evaluated on the latter. In PAW language, the expectation value of an operator is said to be the sum of plane wave ($\langle\widetilde{\psi}_{n\mathbf{k}}|\hat{O}|\widetilde{\psi}_{n\mathbf{k}}\rangle$) and so-called *one-center* ($\sum_{ij}\ldots$) contributions.

An equivalent, though somewhat more involved, decomposition into a plane-wave part and one-center terms exists for truly nonlocal operators, such as the Coulomb operator or the Hartree–Fock nonlocal exchange operator, as well. The details of the

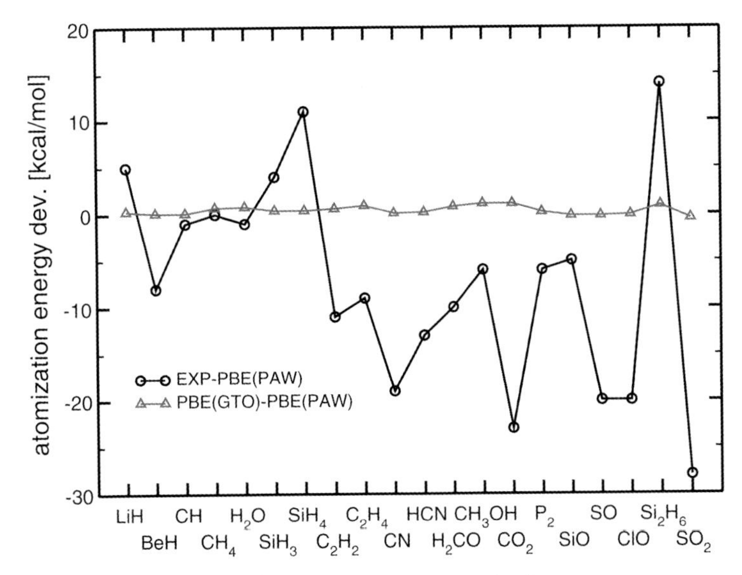


**Figure 4.7** Deviations in the PBE atomization energies from PAW calculations with respect to GTO calculations (triangles) and experiment (circles), for a subset of Pople's G2-1 test set of small molecules.

implementation of these operators within the PAW framework is beyond the scope of the present textbook, and may be found in Refs. [14,15] and [16], respectively.

In essence, this means the PAW method includes the effects of the nodal features of the AE wavefunctions on the quantum mechanical observables (total energy, forces, etc.), without having to represent this nodal behavior in terms of plane waves. The PAW method is highly accurate, and generally speaking the agreement between PAW and, for instance, full-potential linearized plane-wave (FLAPW) or Gaussian-type orbital (GTO) local basis set calculations is excellent (see, for instance, Refs. [15] and [16]). This is illustrated in Figure 4.7, which shows the deviations in the atomization energies from PAW calculations with respect to GTO calculations and experiment, for a subset of Pople's G2-1 test set of small molecules. The PAW and GTO results were obtained using the PBE semilocal exchange-correlation functional [17]. As can be seen, the agreement between the PAW and GTO calculations is excellent, especially compared to the deviation of the theoretical atomization energies with respect to experiment.

## 4.3
## Hybrid Functionals

Hybrid functionals are exchange-correlation functionals (see Section 4.2) that admix a certain amount of nonlocal Hartree–Fock exchange to a corresponding density functional counterpart. The nonlocal Hartree–Fock exchange energy $E_X^{HF}$ can be written as

$$E_X^{HF} = -\frac{1}{2}\sum_{\mathbf{k}n,\mathbf{q}m} 2w_{\mathbf{k}}f_{\mathbf{k}n}w_{\mathbf{q}}f_{\mathbf{q}m} \int\int \frac{\psi_{n\mathbf{k}}^*(\mathbf{r})\psi_{m\mathbf{q}}(\mathbf{r})\psi_{m\mathbf{q}}^*(\mathbf{r}')\psi_{n\mathbf{k}}(\mathbf{r}')}{|\mathbf{r}-\mathbf{r}'|} d\mathbf{r}d\mathbf{r}', \quad (4.39)$$

where $\{\psi_{n\mathbf{k}}\}$ is the set of one-electron Bloch states of the system, and $\{w_{\mathbf{k}}\}$ and $\{f_{n\mathbf{k}}\}$ are the **k**-point weights and occupational numbers (see Section 4.1). The sums over **k** and **q** run over all points chosen to sample the Brillouin zone, and the sums over $m$ and $n$ are performed over all bands at these **k**-points. The **k**-point weights $w_{\mathbf{k}}$ sum to 1, and the factor 2 accounts for the fact that we consider a spin-degenerated system.

The following presents an excerpt from a recently published review paper on the application of hybrid functionals to extended systems [18]. Sections 4.3.1–4.3.3 provide a brief sketch of three different well-known hybrid functional construction schemes: PBE0, HSE03, and B3LYP. A detailed description of the implementation of these functionals within the PAW method and of the computational aspects of the Hartree–Fock exchange interaction in reciprocal space, relevant to methods that use a plane-wave basis set and periodic boundary conditions, may be found in Refs. [16, 19]. These functionals were extensively benchmarked with respect to their description of the structural, electronic, and thermochemical properties of both molecular and solid state systems [16, 19–25]. These benchmarks are roughly summarized in Section 4.3.4. For a more substantive overview, we refer the reader to Refs. [18, 26] and references therein. Readers with an interest in the catalytic properties of metallic surfaces are recommended to read the work of Stroppa et al. [27] on the hybrid functional description of the adsorption of CO at $d$-metal surfaces.

### 4.3.1 PBE0

The PBE0 hybrid functional [28, 29] is a mix of 25% of the Hartree–Fock exchange with 75% of the well-known PBE exchange [17]. Electronic correlation is represented by the corresponding part of the PBE density functional. The resulting exchange-correlation energy expression takes the following simple form:

$$E_{XC}^{PBE0} = \frac{1}{4}E_X^{HF} + \frac{3}{4}E_X^{PBE} + E_c^{PBE}, \quad (4.40)$$

where $E_X^{PBE}$ and $E_c^{PBE}$ denote the PBE exchange and correlation energies, respectively. The choice of a 1:3 mix of Hartree–Fock to the semilocal density functional exchange stems from the theoretical work of Ernzerhof et al. [30–32]. Hybrid functionals that are based on their work are often called "parameter-free" or nonempirical, since the amount of Hartree–Fock exchange to be admixed is not determined by a fit to experimental data.

### 4.3.2 HSE03

The computational expense associated with the evaluation of $E_X^{HF}$ [Eq. (4.39)] depends on the decay of the Hartree–Fock interactions in real space. This decay is highly system dependent and may range from a few to up to hundreds of Ångströms. To avoid the calculation of expensive integrals over slowly decaying exchange interactions, Heyd et al. [20] introduced a hybrid functional, the so-called HSE03, that replaces the long-range part of the Hartree–Fock exchange in the PBE0 hybrid by a corresponding density functional counterpart. The resulting expression for the exchange-correlation energy is given by

$$E_{\text{xc}}^{\text{HSE03}} = \frac{1}{4} E_{\text{X}}^{\text{HF,sr},\mu} + \frac{3}{4} E_{\text{X}}^{\text{PBE,sr},\mu} + E_{\text{X}}^{\text{PBE,lr},\mu} + E_{\text{c}}^{\text{PBE}}, \tag{4.41}$$

where sr and lr denote the short- and long-range parts of the respective Hartree–Fock or PBE exchange interactions. Separating the exchange interactions into short- and long-range parts is realized through a decomposition of the Coulomb kernel

$$\frac{1}{r} = S_{\mu}(r) + L_{\mu}(r) = \frac{\operatorname{erfc}(\mu r)}{r} + \frac{\operatorname{erf}(\mu r)}{r}, \tag{4.42}$$

where $r = |\mathbf{r} - \mathbf{r}'|$, and $\mu$ is the parameter that defines the range separation. The short-range part of the Hartree–Fock exchange energy, $E_{\text{X}}^{\text{HF,sr},\mu}$ is given by Eq. (4.39), provided one replaces the Coulomb kernel by $S_{\mu}(r)$. The short- and long-range parts of the PBE exchange energy are consistently defined using the same decomposition [Eq. (4.42)] [20]. Empirically, it was shown that the optimal range-separation parameter $\mu$ is between 0.2 and 0.3 Å$^{-1}$ [20, 24]. Consequently, the HSE03 functional is a semiemperical functional. In general, one finds that the results using HSE03 are very similar to those obtained using the PBE0.

### 4.3.3 B3LYP

The construction scheme of the B3LYP hybrid functional [33, 34] adheres to the form proposed by Becke [35]. The exchange energy is given by

$$E_{\text{X}}^{\text{B3LYP}} = 0.8 E_{\text{X}}^{\text{LDA}} + 0.2 E_{\text{X}}^{\text{HF}} + 0.72 \Delta E_{\text{X}}^{\text{B88}}, \tag{4.43}$$

where $E_{\text{X}}^{\text{LDA}}$ and $\Delta E_{\text{X}}^{\text{B88}}$ are the LDA exchange, and the gradient corrections to the Becke88 exchange [36], respectively. The B3LYP correlation energy is defined as

$$E_{\text{C}}^{\text{B3LYP}} = 0.19 E_{\text{C}}^{\text{VWN3}} + 0.81 E_{\text{C}}^{\text{LYP}}, \tag{4.44}$$

where $E_{\text{C}}^{\text{VWN3}}$ and $E_{\text{C}}^{\text{LYP}}$ denote the correlation energy from the Vosko–Wilk–Nusair III [37] and the Lee–Yang–Parr [38] correlation functionals, respectively. The B3LYP functional is a semiempirical functional. The coefficients in Eqs. (4.43) and (4.44) are determined by a least-squares fit to atomization energies, electron and proton affinities, and the ionization potentials of the atomic species and molecules in Pople's G2 test set.

### 4.3.4 Summary

The following summarizes the broad characteristics of the hybrid functional description of the structural, thermochemical, and electronic properties of molecules and extended systems, in comparison to the PBE semilocal density functional:

1. *Structural properties:* Compared to the PBE semilocal density functional, the PBE0 and HSE03 hybrid functionals yield an improved description of the structural properties (lattice constants and bulk moduli) of extended systems: a reduction of the overestimation of the lattice constants and the corresponding underestimation of the bulk moduli. The overall performance of the B3LYP hybrid functional for extended systems is slightly *worse* than that of the PBE, mostly due to a poor description of *d* metals.

   The hybrid functional description of the structural properties of molecular systems presents a clear improvement with respect to the PBE: a reduction of the overestimation of the bond lengths with respect to experiment.

2. *Thermochemistry:* The description of the atomization energies of extended systems is best at the PBE level. The PBE functional yields an overall underestimation of the atomization energies, which is further enhanced by the hybrid functionals. The latter underperform for metallic systems, in the case of the B3LYP functional even dramatically so. Worse even, the B3LYP underbinds all systems that possess substantial itinerant character, i.e., small to medium gap systems too. This is probably due to the fact that the LYP correlation functional does not become exact in the homogeneous electron gas limit. Barring metallic systems, the PBE0 and HSE03 hybrid functionals show a similar overall agreement with experiment as the PBE does.

   For the calculation of heats of formation of extended systems, the use of hybrid functionals presents a substantial improvement over the PBE semilocal density functional. Where the PBE shows a pronounced underestimation of the heats of formation, the hybrid functionals yield markedly better results.

   The hybrid functional description of the thermochemistry of molecular systems presents a clear improvement with respect to the PBE: the underestimation of the standard enthalpy of formation, for instance, is strongly reduced.

3. *Electronic properties:* Conventional density functionals, like PBE, substantially underestimate the band gaps in extended systems. Admixture of Hartree–Fock exchange tends to widen the band gaps, and as such hybrid functionals often provide a significantly improved description of these band gap. Especially, the HSE03 band gaps for small to medium gap systems are in excellent agreement with experiment. For large gap systems they still remain underestimated, though considerably less so than with conventional density functionals. The PBE0 overestimates the band gap in semiconductors, and similar to the HSE03 underestimates the gap in large gap systems. Both the PBE0 and the HSE03 functional overestimate the band widths in metallic systems.

In the case of molecular systems the admixture of Hartree–Fock exchange leads to an increase in the energy gap between the highest occupied molecular orbital (HOMO) and the lowest unoccupied molecular orbital (LUMO).

In a broad generalization one may state that hybrid functionals perform well for molecules and insulators, but are lacking in the description of metallic systems.

## References


1 N. W. Ashcroft and N. D. Mermin, *Solid State Physics*, Chapter 5, HRW, Philadelphia, **1976**.

2 A. Baldereschi, "Mean-value point in the Brillouin zone", *Phys. Rev. B* **7**, 5212 (**1973**).

3 D. J. Chadi and M. L. Cohen, "Special points in the Brillouin zone", *Phys. Rev. B* **8**, 5747 (**1973**).

4 H. J. Monkhorst and J. D. Pack, "Special points for Brillouin-zone integrations", *Phys. Rev. B* **13**, 5188 (**1976**).

5 P. E. Blöchl, O. Jepsen, and O. K. Andersen, "Improved tetrahedron method for Brillouin-zone integrations", *Phys. Rev. B* **49**, 16233 (**1994**).

6 M. C. Payne, M. P. Teter, D. C. Allan, T. A. Arias, and J. D. Joannopoulos, "Iterative minimization techniques for *ab initio* total-energy calculations: molecular dynamics and conjugate gradients", *Rev. Mod. Phys.* **64**, 1045 (**1992**).

7 D. R. Hamann, M. Schlüter, and C. Chiang, "Norm-conserving pseudopotentials", *Phys. Rev. Lett.* **43**, 1494 (**1982**).

8 G. B. Bachelet, D. R. Hamann, and M. Schlüter, "Pseudopotentials that work: from H to Pu", *Phys. Rev. B* **26**, 4199 (**1982**).

9 L. Kleinman and D. M. Bylander, "Efficacious form for model pseudopotentials", *Phys. Rev. Lett.* **48**, 1425 (**1982**).

10 D. Vanderbilt, "Optimally smooth norm-conserving pseudopotentials", *Phys. Rev. B* **32**, 8412 (**1985**).

11 A. M. Rappe, K. M. Rabe, E. Kaxiras, and J. D. Joannopoulos, "Optimized pseudopotentials", *Phys. Rev. B* **41**, 1227 (**1990**).

12 D. Vanderbilt, "Soft self-consistent pseudopotentials in generalized eigenvalue formalism", *Phys. Rev. B* **41**, 7892 (**1990**).

13 N. Troulliers and J. L. Martins, "Efficient pseudopotentials for plane-wave calculations", *Phys. Rev. B* **43**, 1993 (**1991**).

14 P. E. Blöchl, "Projector augmented-wave method", *Phys. Rev. B* **50**, 17953 (**1994**).

15 G. Kresse and D. Joubert, "Form ultrasoft pseudopotentials to the projector augmented-wave method", *Phys. Rev. B* **59**, 1758 (**1999**).

16 J. Paier, R. Hirschl, M. Marsman, and G. Kresse, "The Perdew–Burke–Ernzerhof exchange-correlation functional applied to the G2-1 test set using a plane-wave basis set", *J. Chem. Phys.* **122**, 234102 (**2005**).

17 J. P. Perdew, K. Burke, and M. Ernzerhof, "Generalized gradient approximation made simple", *Phys. Rev. Lett.* **77**, 3865 (**1996**).

18 M. Marsman, J. Paier, A. Stroppa, and G. Kresse, "Hybrid functionals applied to extended systems", *J. Phys.: Condens. Matter* **20**, 064201 (**2008**).

19 J. Paier, M. Marsman, K. Hummer, and G. Kresse, "Screened hybrid density functionals applied to solids", *J. Chem. Phys.* **124**, 154709 (**2006**).

20 J. Heyd, G. E. Scuseria, and M. Ernzerhof, "Hybrid functionals based on a screened Coulomb potential", *J. Chem. Phys.* **118**, 8207 (**2003**).

21 V. N. Staroverov, G. E. Scuseria, J. Tao, and J. P. Perdew, "Comparative assessment of a new nonemperical density functional: molecules and hydrogen-bonded complexes", *J. Chem. Phys.* **119**, 12129 (**2003**).

**22** J. Heyd and G. E. Scuseria, "Assessment and validation of a screened Coulomb hybrid density functional", *J. Chem. Phys.* **120**, 7274 (**2004**).

**23** J. Heyd and G. E. Scuseria, "Efficient hybrid density functional calculations in solids: assessment of the Heyd–Scuseria–Ernzerhof screened Coulomb hybrid functional", *J. Chem. Phys.* **121**, 1187 (**2004**).

**24** A. V. Krukau, O. A. Vydrov, A. F. Izmaylov, and G. E. Scuseria, "Influence of the exchange screening parameter on the performance of screened hybrid functionals", *J. Chem. Phys.* **125**, 224106 (**2006**).

**25** J. Paier, M. Marsman, and G. Kresse, "Why does the B3LYP hybrid functional fail for metals?", *J. Chem. Phys.* **127**, 024103 (**2007**).

**26** S. Kümmel and L. Kronik, "Orbital-dependent density functionals: theory and applications", *Rev. Mod. Phys.* **80**, 3 (**2008**).

**27** A. Stroppa, K. Termentzidis, J. Paier, G. Kresse, and J. Hafner, "CO adsorption on metal surfaces: a hybrid functional study with plane-wave basis set", *Phys. Rev. B* **76**, 195440 (**2007**).

**28** M. Ernzerhof and G. E. Scuseria, "Assessment of the Perdew–Burke–Ernzerhof exchange-correlation functional", *J. Chem. Phys.* **110**, 5029 (**1999**).

**29** C. Adamo and V. Barone, "Towards reliable density functional methods without adjustable parameters: the PBE0 model", *J. Chem. Phys.* **110**, 6158 (**1999**).

**30** M. Ernzerhof, "Construction of the adiabatic connection", *Chem. Phys. Lett.* **263**, 499 (**1996**).

**31** M. Ernzerhof, J. P. Perdew, and K. Burke, "Coupling-constant dependence of atomization energies", *Int. J. Quantum Chem.* **64**, 285 (**1996**).

**32** J. P. Perdew, M. Ernzerhof, and K. Burke, "Rationale for mixing exact exchange with density functional approximations", *J. Chem. Phys.* **105**, 9982 (**1996**).

**33** GAUSSIAN Inc., *Gaussian News* **5**, 2 (**1994**).

**34** M. J. Frisch *et al.*, *GAUSSIAN03, Revision C.02.*, Wallingford, CT: Gaussian Inc., **1996**.

**35** A. D. Becke, "Density-functional thermochemistry: III. The role of exact exchange", *J. Chem. Phys.* **98**, 5648 (**1993**).

**36** A. D. Becke, "Density-functional exchange-energy approximation with correct asymptotic behavior", *Phys. Rev. A* **38**, 3098 (**1988**).

**37** S. H. Vosko, L. Wilk, and M. Nusair, "Accurate spin-dependent electron liquid correlation energies for local spin density calculations: a critical analysis", *Can. J. Phys.* **58**, 1200 (**1980**).

**38** C. Lee, W. Yang, and R. G. Parr, "Development of the Colle–Salvetti correlation-energy formula into a functional of the electron density", *Phys. Rev. B* **37**, 785 (**1988**).

# 5
# Periodic Linear Combination of Atomic Orbitals and Order-$N$ Methods

*Emilio Artacho*

## 5.1
## Introduction

Both in the study of heterogeneous catalysis and materials in general, extended solid systems become main players with their own peculiarities and challenges, in some aspects quite different from the ones faced when addressing isolated molecules and clusters. These systems are best approximated as infinite and are the traditional subject of study of solid-state physics. The main concepts of this field relevant to this book have been introduced in Chapter 4. In this chapter, we shall concentrate on two ideas: (a) using basis sets made of atomic orbitals in solid-state calculations (linear combination of atomic orbitals (LCAO)), and thus connect with previous chapters in this book, and (b) improving the efficiency and scaling of such calculations. We shall see how these two ideas are more related than they look.

### 5.1.1
### LCAO and Extended Systems

The study of extended systems with LCAO has a long history that started with the tight-binding approximation to the electronic structure of solids [1]. From its beginnings, assuming a minimal, orthogonal, and actually unspecified basis set (the matrix elements of the Hamiltonian were parameterized), it has come a long way. Nowadays, there are many different tight-binding methods [2] with different degree of empiricism and/or different approximations, all the way to first-principles schemes [3, 4]. This evolution has generated some confusion in what is nowadays meant by tight binding, especially on whether the term implies empirical character or not.

Tight-binding methods are still well differentiated from more general LCAO methods by the fact that the former tend to use minimal basis sets, that is, the atomic orbitals that would be occupied or partly occupied in the independent atoms. More general LCAO methods use atomic-like wavefunctions as basis sets, but aiming to approach the converged basis set limit, by extending the basis beyond the minimal, including radial and angular flexibility to respond to

*Computational Methods in Catalysis and Materials Science.* Edited by Rutger A. van Santen and Philippe Sautet

ISBN: 978-3-527-32032-5

the altered environment. First-principles methods using Gaussian basis sets for extended systems started in the mid-eighties from both the chemistry [5] and physics communities [6], but they did not become as popular as the plane-wave methods described in Section 5.1.2. They regained momentum with the advent of linear-scaling ideas in the mid-nineties.

### 5.1.2
### Linear Scaling

The computational complexity (the required computational effort) of electronic structure methods of a mean-field type (such as DFT) in the canonical form scales as $N^3$, $N$ being the number of atoms in the simulation cell. This means that an increase of an order of magnitude of computer power allows a mere doubling of the attainable system size. This complexity is intrinsic, not dependent on algorithmic choices. A simple way to see its origin is observing that the set of occupied Kohn–Sham (KS) solution functions $\{|\psi_n\rangle\}$ have to be orthonormal, that is,

$$\langle \psi_n \mid \psi_m \rangle = \int d^3r\ \psi_n^*(\vec{r})\psi_m(\vec{r}) = \delta_{nm}. \tag{5.1}$$

Since the number of occupied KS orbitals scales with $N$, the number of these integrals scales as $N^2$, and each integral involves the whole system volume, which scales as $N$, giving the total $N^3$.

The nineties witnessed the drive toward linear-scaling methods: it was realized that departing from the canonical solution, plus suitable approximations, can reduce the complexity to linear [7, 8]. The main (but not only) concept behind linear scaling is locality, understood as confinement of key quantities within regions of space smaller than the system size. The connection between locality and linear scaling is also understandable in simple terms. Take a large system and subdivide it in a set of subsystems of smaller size, as in Figure 5.1(a). If one can solve for a subsystem by only calculating on it and possibly on a wider environment, still smaller than the overall system, then the resolution of the whole system can be accomplished by calculating all of the subsystems, one at a time. Such a procedure is of linear complexity, since the overall effort is linear with the number of subsystems. Of course, this is a simplified illustration, and not the base of most linear-scaling methods (the "divide and conquer" method [7] would be the closest), but it transmits the basic idea. Going back to the orthonormality condition of Eq. (5.1), if the KS wavefunctions were confined in space, then wavefunctions in separate regions would not overlap, and thus only a set of wavefunctions would have to be considered around any given one. If that set does not scale with system size, the number of conditions that need to be considered would be of order $N$, and each integral would only need to be performed in a region independent of $N$, giving again the overall linear complexity. The KS functions themselves are in general not confined; we show next how this is actually done.

This chapter will confine itself to KS-DFT (Eq. (5.1) implies it). DFT does not require single-particle eigenfunctions but could obtain the energy directly

(a)

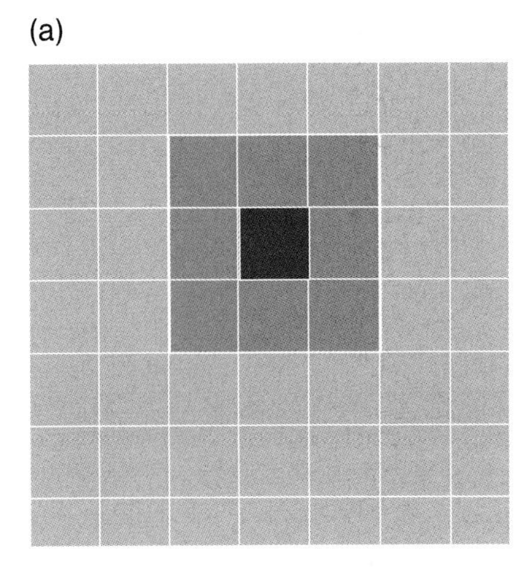

(b)

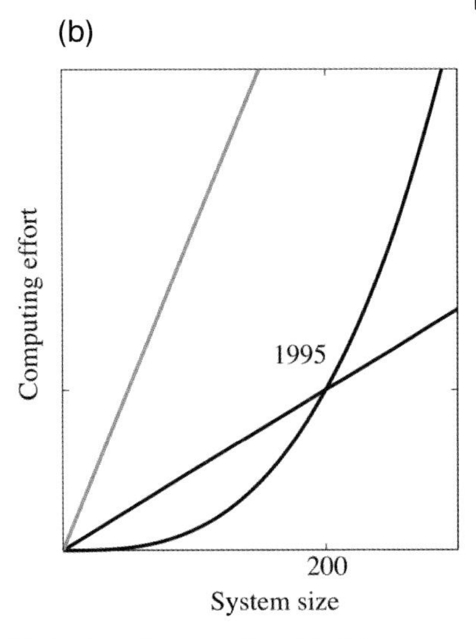


**Figure 5.1** (a) Illustration of the link between locality and linear scaling. Take a large system and divide it up in pieces such that the relevant equations can be solved for one piece (black) considering an environment (grey) which is smaller than the system. If this can be done to enough accuracy, then the solution of the whole can be achieved by following this procedure for all the pieces in the sample, and thus the effort depends on how many pieces the system contains, which leads to linear scaling. (b) Cube versus linear scaling. The black curves illustrate the situation for a crossover at around 200 atoms, which would imply a computational effort affordable since the mid nineties. If the prefactor to linear scaling is higher, the crossover can shift substantially pushing the interest for linear scaling very much into the future. The prefactor (*efficient* linear-scaling techniques) is thus very important for their being useful for a given computer setup.

as a functional of the electron density. The so-called orbital-free DFT proposes approximate functionals [9]. It is not only linear scaling but extremely efficient also, comparable to empirical force fields. The difficulty is to obtain good approximations for the kinetic energy functional. Progress so far allows the study of systems not too far off the homogeneous electron liquid. Interestingly, its comfort zone (metals) is precisely the unsolved problem for linear-scaling KS-DFT (see below).

Locality being a key to linear scaling, local or confined basis functions acquire new prominence, and so does LCAO in particular. It provides a very natural local language for linear scaling, and several of the linear-scaling methods available nowadays are based on LCAO. There are alternative methods that allow a more systematic route to basis convergence than LCAO. This is the case of methods in the line of finite differences on grids, or finite elements. These will not be described in detail here, but their main differences with respect to LCAO will be pointed out at different stages. At this time, it is useful to point out the importance of the prefactor over linear scaling for practical calculations: although eventually linear-scaling methods will clearly be of advantage over cube-scaling ones, the actual system sizes where the crossover occurs depend very much on the slope of the linear behavior, as illustrated in Figure 5.1(b). This prefactor depends on many technicalities, but a very important component is the number of basis functions per electron. Here LCAO becomes attractive again, being a rather efficient method in this respect. The fact that choosing a good atomic basis set for a particular problem demands more scientist intervention than is the case for a real-space grid is compensated by greater computational efficiency, and thus access to larger and more complex systems.

In the following, some general concepts of LCAO for extended systems are revised, setting the scene for the further exposition and discussion of main ideas for linear-scaling LCAO-DFT. The discussion will be centered on the SIESTA method [10], which pioneered the field and is nowadays widely used, either using the SIESTA program or other independent programs that have implemented the method. The main concepts, however, can be directly translated to other methods.

## 5.2 LCAO and Extended Systems

### 5.2.1 LCAO Basis Sets

Basis functions in LCAO have the general shape

$$\phi_\mu(\vec{r}) = R_{l(\mu)}(r)\, Y_{l(\mu)}^{m(\mu)}(\theta, \varphi), \tag{5.2}$$

where $R_l(r)$ describes the radial dependence, and $Y_l^m(\theta, \varphi)$, a spherical harmonic, describes the angular dependence, with the angular momentum quantum numbers $l$ and $m$ associated to $\phi_\mu$ ($r$, $\theta$, and $\varphi$ refer to the spherical coordinates of $\vec{r} - \vec{R}_\mu$). They are customarily centered on atom nuclei, but can also be put in arbitrary positions. The number of basis functions on a given center is quite arbitrary, both in terms of how many functions for a given angular momentum channel (value of $l$) and in terms of how many channels. A good starting point is the set of KS solutions of the isolated atom, including the occupied and partially occupied atomic states. The basis set generated by these states on all atoms of the condensed system constitutes the minimal basis, or single-$\zeta$ (SZ), which is used in many qualitative models, but does not provide good basis-set convergence in general. Additional basis functions are added to provide extra radial and angular flexibility. Although many options are there to enlarge the Hilbert space, the challenge is to generate a Hilbert space that includes as much of the occupied Hilbert space of the solid as possible with as few basis functions as possible.

In the quantum-chemistry tradition, Gaussian functions are used, $G_{\alpha l}(r) \propto r^l e^{-\alpha r^2}$, the atomic $R_l(r)$ being expanded in a set of a few Gaussians. Efficient radial flexibility is then achieved by taking the original Gaussian expansion of every single atomic orbital and separating it into two functions, one with the narrower Gaussians and another with the most extended one(s). This so-called split-valence procedure gives two basis function per original atomic orbital, and thus double-$\zeta$ (DZ). Triple-$\zeta$ is achieved by letting three pieces free, and so forth.

Angular flexibility is achieved by adding extra $l$-channels to the ones present in the free atom. They are said to allow the lower $l$-shells to "polarize." The radial shape for these extra polarization functions is normally chosen as a Gaussian, substantially overlapping with the radial shape of the orbitals it polarizes. With all these ingredients, a hierarchy of basis sets is built: SZ, DZP, TZDP, and so forth, meaning single-$\zeta$, double-$\zeta$ polarized, triple-$\zeta$ doubly polarized. Off-center functions can also be added, and so can diffuse ones, meaning functions with a slower decay than in the free atom.

### 5.2.2
### Non-orthogonal Bases

Basis functions centered on different atoms are not orthogonal to each other, which means that we shall be working with nonorthogonal basis sets. We then face the generalized eigenvalue problem, which can be written as [11,12]

$$H^{\mu}_{\nu}\psi^{\nu}_{n} = \varepsilon_n \psi^{\mu}_{n} \text{ or } H_{\mu\nu}\psi^{\nu}_{n} = \varepsilon_n S_{\mu\nu}\psi^{\nu}_{n}, \tag{5.3}$$

where tensor implicit summation over repeated indices is used. $S_{\mu\nu} = \langle\phi_\mu|\phi_\nu\rangle$ is the overlap matrix or the metric tensor, $H_{\mu\nu} = \langle\phi_\mu|H|\phi_\nu\rangle$ is the Hamiltonian matrix in the conventional form, and $H^{\mu}_{\nu} = \langle\phi^{\mu}|H|\phi_\nu\rangle = S^{\mu\eta}H_{\eta\nu} = \sum_{\eta}(\mathbf{S}^{-1})_{\mu\eta}(\mathbf{H})_{\eta\nu}$ is the natural representation of the Hamiltonian tensor [12] (bold symbols represent matrices). The unknown expansion coefficients of the sought eigenvectors $|\psi_n\rangle$ are $\psi^{\nu}_{n} = \langle\phi^{\nu} \mid \psi_n\rangle$, where $\langle\phi^{\nu}|$ refers to the dual basis of the original one [11]. The electronic-structure problem then implies computing the $\mathbf{H}$ and $\mathbf{S}$ matrices and then solving Eq. (5.3).

### 5.2.3
### Bloch Basis Functions

Given the unit cell of a crystalline system, the way to take advantage of the crystalline translational symmetries (Bloch's theorem) in a LCAO context is by transforming the basis set from the purely atomic one to the one made of Bloch states. A Bloch state is defined for each basis function in the unit cell as

$$|\phi_\mu, \vec{k}\rangle = \frac{1}{\sqrt{N_c}}\sum_{\vec{R}} e^{i\vec{k}\vec{R}}|\phi_\mu, \vec{R}\rangle, \tag{5.4}$$

where $\mu$ labels a particular basis function within a unit cell, $\vec{R}$ labels a particular unit cell in the crystal, $\vec{k}$ stands for Bloch's wavevector, and $N_c$ is the number of unit cells in the crystal. This transformation takes the atomic basis functions associated to the $N_c \times N$ atoms in the crystal to the Bloch functions associated to $N$ atoms in one cell for each one of the $N_c$ values of the wavevector $\vec{k}$ (periodic boundary conditions gives a direct correspondence between $N_c$ unit cells in the crystal and $N_c$ values of $\vec{k}$ in the first Brillouin zone). The Hamiltonian becomes block-diagonal in this basis, meaning that $\langle\phi_\mu, \vec{k}|H|\phi_\nu, \vec{k}'\rangle \propto \delta(\vec{k} - \vec{k}')$, and thus one needs to calculate $H_{\mu\nu}(\vec{k}) = \langle\phi_\mu, \vec{k}|H|\phi_\nu, \vec{k}\rangle$ and solve it for one wavevector at a time.

Integrals in $\vec{k}$ over the Brillouin zone are replaced by suitable $\vec{k}$-point samplings, normally homogeneous grids in $\vec{k}$-space, as fine as needed to converge the integrated quantities such as the total energy. This fineness is conveniently controlled by its reciprocal, a cut-off length [13]. For sufficiently large unit cells, the sampling can be reduced to the single point $\vec{k} = \vec{0}$, the so-called $\Gamma$-point approximation.

## 5.3
## Linear-Scaling DFT

In the LCAO electronic structure problem, two main stages are well differentiated, both representing sizeable effort and requiring separate treatment: the calculation of the Hamiltonian and overlap matrix elements, on one hand, and the solution of the eigenvector problem on the other.

### 5.3.1
### Basis Functions and Locality

In the line of the locality idea mentioned above, the key for the linear-scaling computation of matrix elements in all methods to date is the effective confinement of the basis functions (except for the Hartree term, as shown below), giving sparse **H** and **S** matrices, that is, matrix elements that are zero for basis functions centered in distant locations. This confinement is accomplished in two different ways by different methods available:

1. *Thresholding*. This is the method used for analytic basis functions, mainly Gaussians [14–16]. Although they are very rapidly decaying functions, they do not reach strict zero at any distance. Effective confinement is achieved by neglecting matrix elements of **H** and **S** whose value is expected to be below a given threshold.
2. *Finite support basis functions*. Basis functions are chosen that have nonzero value only within a finite region of space, its finite support, and hence matrix elements between basis functions whose support regions do not touch are strictly zero. This technique was proposed by Sankey and collaborators [3] and is the one used by the SIESTA [10] and other methods [17], which use basis functions with a numerically defined radial dependence.

Present implementations of the finite-support technique are based on spherical support for their basis functions, each of which is defined to be nonzero within a finite cut-off radius. Other support shapes are perfectly valid and could be useful, but have not been implemented so far, to the best of our knowledge. The cut-off radius may be different for each basis function, the key for linear scaling being that cut-off radii are smaller than, and do not scale with, system size. It is important to stress that the SIESTA method needs LCAO basis functions of finite spherical support, but the method itself does not prescribe anything on the actual cut-off radii, on the radial shape of its functions, on where they are centered, or on the overall size of the basis. The choice of good basis sets is a problem altogether independent of the method and the program, in the same way as most quantum chemistry methods need Gaussian LCAO basis sets, for which there are different published tabulations independent of different programs. SIESTA allows the user to introduce any radial shapes through numerical tables.

The most popular approach to build finite-support basis functions starts by building a minimal basis from the DFT solution of the free atom (or pseudoatom if using pseudopotentials) with an added spherical confining potential. It can be hard confinement, by using a potential that is zero within the sphere and infinite elsewhere [3, 18], giving a discontinuity in the derivative of the function at the cut-off radius, or soft confinement, for which the less abrupt divergence of the confining potential ensures continuity of value and derivatives of the basis function [19]. A double-$\zeta$ basis can then be built in the split-valence spirit described above. It is normally done by defining a second orbital that has the shape of the first one outside a given radius, and goes smoothly toward zero as $r^l(a + br^2)$, matching value and derivative at that radius [18]. Multiple-$\zeta$ functions are built by redoing this procedure. Polarization functions can be introduced by using confinement on the free-atom virtual orbitals of the needed angular momentum, by using tabulated Gaussians, or by obtaining them from a calculation of the free atom under a polarizing electric field [18]. Some guidelines on transferable basis sets can be found in [20].

### 5.3.2
### Calculation of H and S Matrix Elements

The KS Hamiltonian can be split into the following terms:

$$\hat{H}_{\text{KS}} = \hat{T} + \hat{V}_{\text{ps}} + \hat{V}_{\text{H}} + \hat{V}_{xc}, \tag{5.5}$$

corresponding to the kinetic energy, electron–core interactions (pseudopotentials), Hartree potential, and exchange-correlation potential, respectively. The electrostatic components (Hartree and electron–core attractions) are long range and obviously not amenable to localized treatments. It is its smoothness in the long range that allows its linear-scaling computation. This is accomplished in two different ways, coming from different traditions of quantum chemistry and solid-state physics. The former does it by explicitly calculating all the LCAO two-electron integrals in the near field (short range), and approximates the far field by a fast-multipole expansion [21, 22]. The latter expresses the electron density in a real-space grid (or equivalently in an expansion in plane waves), and solves the Poisson equation either using fast Fourier transforms (FFTs, not strictly linear scaling, rather scaling as $N \log N$, but extremely efficient) or by multigrid techniques [23].

The electron density can be decomposed separating the deformation density and the sum of free-atom densities,

$$\delta\rho \equiv \rho - \sum_{n}^{\text{atoms}} \rho_n^{\text{atomic}}. \tag{5.6}$$

The neutral-atom potential is then defined as

$$V_{\mathrm{NA}} \equiv \sum_{n}^{\mathrm{atoms}} \{V_{\mathrm{ps},n}^{\mathrm{loc}} - V_H(\rho_n^{\mathrm{atomic}})\}, \tag{5.7}$$

where $V_{\mathrm{ps},n}^{\mathrm{loc}}$ stands for the local part of the pseudopotential of each atom. $V_{\mathrm{NA}}$ is a short-range potential, each atomic term being of finite support again, equal to the largest basis support in the atom. The Hamiltonian then becomes

$$H_{\mathrm{KS}} = T + V_{\mathrm{ps}}^{\mathrm{nloc}} + V_{\mathrm{NA}} + V_{\mathrm{H}}(\delta\rho) + V_{xc}, \tag{5.8}$$

where $V_{\mathrm{ps}}^{\mathrm{nloc}}$ stands for the sum of short-ranged, nonlocal pseudopotential components over all the atoms, and $V_{\mathrm{H}}(\delta\rho)$ represents the Hartree potential on the deformation density. The matrix elements of the kinetic energy and the overlap matrices represent integrals that involve functions centered in one or two positions only, two-center integrals. These can be transformed into one-dimensional integrals, which depend only on the orbitals involved and the distance between the centers, after suitable manipulations of the spherical harmonics involved. These distance-dependent integrals are computed extremely efficiently, and are thus tabulated versus distance at the beginning of any calculation [3]. The subsequent evaluation of these matrix elements is extremely accurate and quick, whenever needed. This scheme is extended to $V_{\mathrm{ps}}^{\mathrm{nloc}}$, since, in the fully factorized form [24], this term looks like

$$V_{\mathrm{ps}}^{\mathrm{nloc}} = \sum_{l,m} |\chi_{lm}\rangle \varepsilon_{lm} \langle \chi_{lm}|, \tag{5.9}$$

and thus, its matrix elements

$$\langle \phi_\mu | V_{\mathrm{ps}}^{\mathrm{nloc}} | \phi_\nu \rangle = \sum_{l,m} \langle \phi_\mu \mid \chi_{lm}\rangle \varepsilon_{lm} \langle \chi_{lm} \mid \phi_\nu \rangle \tag{5.10}$$

are products of two-center integrals, analogous to overlap integrals between a basis function and a Kleinman–Bylander projector function [24]. These are tabulated and then retrieved as explained above for $S_{\mu\nu}$ and $T_{\mu\nu}$.

The remaining matrix elements, of $V_{\mathrm{NA}}$, $V_{\mathrm{H}}(\delta\rho)$, and $V_{xc}$, are then calculated by summation in a real 3D-space discretization in a grid. The Hartree potential associated to the deformation density is evaluated with either FFTs [9] or a multigrid solver [20, 23]. The three terms are then added in the points of the grid, and the matrix elements integrated. Figure 5.2 illustrates it in a two-dimensional diagram. The quality of this approximation is controlled by the grid fineness, equivalent to a plane-wave cutoff, or mesh cutoff. Once these elements are in place, the rest of the linear-scaling implementation of matrix calculation is just careful coding of algorithms for sparse matrix manipulation.

This exposition accounts for what is done in the SIESTA method. Quickstep [16] uses a similar hybrid method, which uses Gaussian basis sets (and thresholding) instead of finite-support numerical functions. There is an analogous partition between integrals done with plane waves as auxiliary basis (analogous

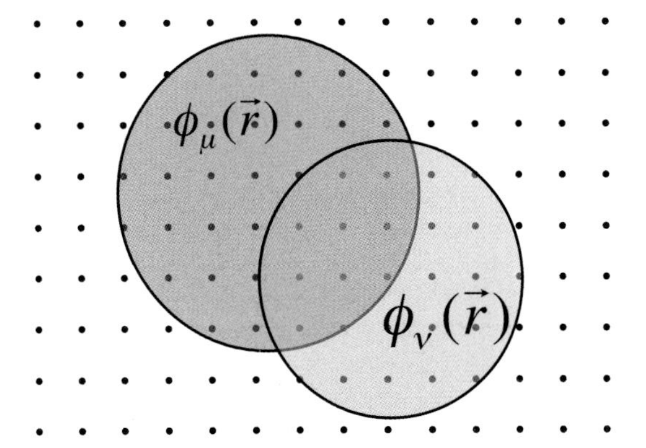


**Figure 5.2** Two-dimensional illustration of the integration of a matrix element on a grid. By knowing the value of, for instance, the exchange-correlation potential and the values of the two basis functions in the points at the intersection of both support regions, the integral is efficiently done.

to integrals in a grid) and integrals done within the quantum-chemistry artillery of analytic Gaussian integrals, instead of SIESTA's tabulated two-center integrals.

The approximate grid integrals break the homogeneity of space, and introduce a space rippling, in this context called egg-box effect [20], which of course reduces by increasing the mesh cutoff. It can, however, be quite disturbing in fine structure relaxations or in finite-difference calculations of second derivatives (like vibrations). The problem is very substantially reduced [25] by Fourier filtering the basis functions, the neutral-atom potential, and other components.

## 5.4 Linear-Scaling Solving of the Eigenvalue Problem

The KS eigenfunctions (or canonical orbitals in a mean-field theory) in extended systems are rarely localized. If confining approximations are to be used to attain linear scaling, it has to be done on other magnitudes. Linear-scaling solvers have been based on two alternatives: (a) the density matrix and (b) a localized basis of occupied space, the so-called localized wavefunctions (LWFs). The former is

$$\rho^{\mu\nu} = \langle\phi^\mu|\hat{\rho}|\phi^\nu\rangle = \sum_n^{\mathrm{occ}} \langle\phi^\mu \mid \psi_n\rangle\langle\psi_n \mid \phi^\nu\rangle, \tag{5.11}$$

where $\hat{\rho}$ is the density operator, $|\psi_n\rangle$ are the KS states, $\langle\phi^\mu \mid \psi_n\rangle$ are the expansion coefficients of these states in the basis, and the sum is over all occupied states.

The LWFs are functions obtained by a unitary transformation within the space spanned by the occupied KS states. It has been shown (see [26, 27] and references therein) that both are expected to display exponential confinement in quite general situations, with the notable exception of metals at zero temperature, for which the confinement is only algebraic. For LWFs, the exponential confinement means that there is a basis of occupied space whose functions are themselves exponentially decaying with the distance from their centers. For the density matrix, it means that

$$\rho^{\mu\nu} \sim e^{-|\vec{R}_\mu - \vec{R}_\nu|/L}, \tag{5.12}$$

where the characteristic decay length $L$ decreases with temperature and insulating character (measured by the band gap, for instance), as do the exponential sizes of the LWFs. These confinement characteristics of the one-particle solutions reflect more general confinement characteristics of the equilibrium state of a correlated electron liquid, as formalized by Walter Kohn [28], in his "nearsightedness principle."

The two keys for linear-scaling solvers are (a) finding either the LWFs or the density matrix directly, and (b) introducing a strict localization condition in the search. The first task can be achieved exactly, but an approximation is introduced in the second.

### 5.4.1 Density Matrix Approaches

The energy functional to be minimized when searching for the density matrix is in principle simple, since (forgetting for the moment double counting terms in DFT)

$$E = 2\mathrm{Tr}\{\hat{\rho}\hat{H}\} = 2\rho^{\mu\nu} H_{\nu\mu} \tag{5.13}$$

under the constrain that the density operator be idempotent (for zero temperature), that is, $\hat{\rho}^2 = \hat{\rho}$ (the factor of two accounts for spin in a nonpolarized state, assumed henceforth for simplicity). Instead, a general minimization was proposed [29] in which $\hat{\rho}$ is replaced by its "purified" version, $3\hat{\rho}^2 - 2\hat{\rho}^3$, together with an electronic chemical potential ($\eta$) term, which ensure idempotency and the right occupation, respectively, at the energy minimum. The matrix expression of this functional reflects the operator expression if the basis is orthogonal. For nonorthogonal bases, the translation is direct in the natural tensor representation

$$E = 2\,(3\rho^{\mu}_{\delta}\rho^{\delta}_{\nu} - 2\rho^{\mu}_{\delta}\rho^{\delta}_{\lambda}\rho^{\lambda}_{\nu})(H^{\nu}_{\mu} - \eta\delta^{\nu}_{\mu}) + \eta N_{\mathrm{el}}, \tag{5.14}$$

which in the usual matrix representation becomes

$$E = 2\,\mathrm{Tr}\{(3{-}2\rho\mathbf{S}\rho)\rho\mathbf{S}\rho(\mathbf{H} - \eta\mathbf{S})\} + \eta N_{\mathrm{el}}, \tag{5.15}$$

where bold font indicates the conventional matrices. There are different ways to do the minimization varying the chemical potential parameter such that the number of electrons remains constant [27, 30].

Some working linear-scaling DFT methods [30, 31] use this functional or variants of it. The ones based on finite-elements [30] or finite-differences [31] basis sets resort to a two-tier energy minimization: they look for an optimal density matrix expressed in an intermediate auxiliary (small) basis set of the so-called support functions, or generalized Wannier functions. The energy found with the functional is further minimized in an outer loop that re-expresses the intermediate basis in terms of the underlying basis set, until optimal support functions are found. It does not involve additional approximations (up to minimization tolerances).

### 5.4.2 Localized Wave-function Approaches

If looking for the LWFs, the energy (without double-counting terms) is simply the trace of the Hamiltonian operator in such basis, that is,

$$E = 2\sum_{i}^{\mathrm{occ}} \langle \chi_i | H | \chi_i \rangle \tag{5.16}$$

being the $|\chi_i\rangle$ the unknown LWFs, which are to be found by minimizing $E$ under the constrain that they be orthonormal. This can be transformed into an unconstrained minimization by using the modified functional [32]

$$E = 2\sum_{i,j}^{\mathrm{occ}} (S^{-1})_{ij} H_{ji}, \tag{5.17}$$

which requires the inversion of the overlap matrix of the LWFs (not of the basis) at every minimization step, or, alternatively, the functional [33, 34]

$$E = 2\sum_{i,j}^{\mathrm{occ}} (2\delta_{ij} - S_{ij}) H_{ji}, \tag{5.18}$$

which reaches the same minimum (for both energy and LWFs) as the constrained one (Eq. (5.16)), fulfilling the orthonormality condition at the minimum, without requiring to invert the overlap matrix.

After choosing an energy functional, the confining approximation is imposed. In the density-matrix case, this is done by establishing a sparsity pattern in the matrix *a priori*, that is, defining to be zero the matrix elements that link basis orbitals in atoms beyond some distance criterion. Analogously, LWFs are forced to have zero contribution from basis functions beyond some distance from their chosen centers. As before, confinement criteria do not need to be spherical [35], but they are customarily so for simplicity.

The confining approximation introduces small deviations in the energy and solutions that are controlled by the tolerance in the minimization procedure. It can also introduce pathologies, however. First, imposing (weak) confinement introduces a small curvature of the energy functional around the minimum, in the directions in the Hilbert space that would correspond to energy-invariant rotations if unconfined. This causes very ill-conditioned minimizations, demanding an unreasonably large number of steps (the condition number goes as the highest curvature divided by the lowest nonzero one; the number of steps in a conjugate-gradient relaxation scales with the condition number). This affects most linear-scaling solvers. Fortunately, once a solution has been found, further minimizations with slightly different atomic positions or effective potential converge very quickly if starting with the information of the previous step. The computational burden due to this pathology then becomes negligible in the overall effort of a long relaxation or dynamics run.

Another pathology affecting the functionals in Eqs. (5.17) and (5.18) is the appearance of unphysical local minima upon the introduction of confinement [35]. The problem is illustrated in Figure 5.3 for a simple two-dimensional square lattice, in which the atoms are joined by covalent bonds. Each atom has four valence electrons used in four bonds. The physical solution would have one LWF with its two electrons associated to each bond. The number of LWFs would be two per atom. If we centered the confinement region of two LWFs on each atom, the physical solution could be well described if the confinement radius was ample enough to allow each LWF to describe a bond nearby (note that the fact that the support region is centered on an atom does not mean that the weight and shape of the LWF are centered there). Figure 5.3(a) shows this solution schematically by indicating an LWF with a stick, which is sticking out of the atom on which it is centered. Figure 5.3(b) shows a pathological minimum, in which all LWFs have found the bond to describe, except one LWF (and its corresponding electron pair) which is missing for a bond, giving rise to an unphysical positive net charge there, in addition to the absence of the bond.

This problem has been partly addressed by recentering the confinement regions dynamically [37]: if the first minimization finds the physically meaningful minimum, then the centers of confinement for the next minimization (in the next self-consistency step or new atomic positions) are defined at the actual centers of the LWFs, meaning $\langle \chi_i | \vec{r} | \chi_i \rangle$. If the charge does not jump discontinuously from one minimization to the next, the solution will remain physical throughout the simulation, for short enough displacements.

A more general solution was proposed [36] in which more LWFs than needed are used (analogous to the support functions mentioned above), but the correct

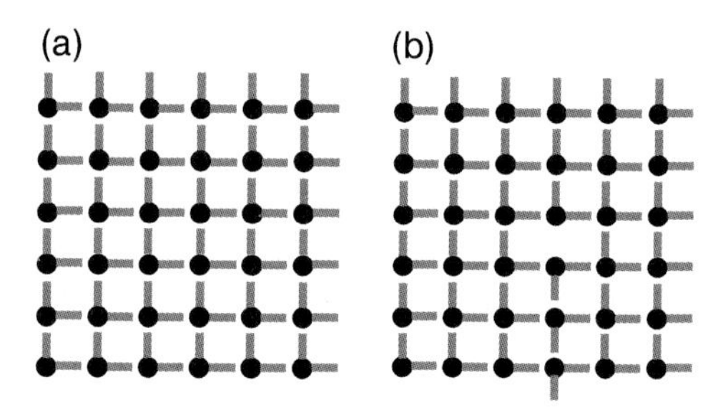


**Figure 5.3** Schematic illustration of physical and unphysical solutions of the linear-scaling functional of Eq. (5.18) by imposing confinement (see text).

number of electrons is obtained by direct minimization within the larger space if slightly modifying the functional in Eq. (5.18) to

$$E = 2\sum_{i,j}^{M} (2\delta_{ij} - S_{ij})(H_{ji} - \eta\delta_{ji}) + \eta N_{\text{el}}, \tag{5.19}$$

where $N_{\text{el}}$ is the number of electrons, $M > 2N_{\text{el}}$, and $\eta$ the electronic chemical potential of Eq. (5.14), that is, an energy value that should be above occupied states and below unoccupied. The minimum of this functional makes the norm of the states above $\eta$ to vanish, recovering the right number of electrons. The extra orbitals allow to put orbital weight where needed and permit tighter confinements, with confining radii of around 4–5 Å for wide band insulators as DNA.

This functional is the most popular within the SIESTA method, having been used for liquid water [38], biomolecules [39, 40], and carbon nanostructures [41]. A new problem arises with this scheme, however: the value of $\eta$ has to be guessed. If it does not fall within the band gap, the number of electrons is wrong and the functional becomes unstable, since for this functional, the physical minimum is a local one, separated by a barrier from a downward unbound fall. The crossing of one state across the value of $\eta$ puts LWFs on the barrier separating the physical minimum from the infinite cliff. This instability may, therefore, demand quite an initial effort before a long simulation is effectively launched. It is, however, worth the effort: the relaxation of a dry DNA double helix [40] took 20 times less computational effort than when using an efficient but conventional diagonalization solver (the calculation of **H** and **S** was done with the same linear-scaling procedure in this comparison).

### 5.4.3
### Other Linear-Scaling Solvers

A completely new method to approach the eigen-solving problem was proposed recently [42, 43], which has offered new ways of looking at linear-scaling solving. This so-called MOSAICO method introduces confinement through a Gilbert transformation of the KS states [42], which allows diagonalizations in blocks. Its coming from such a different direction (no functional minimization) brings new perspectives and possibilities to the field that should be explored.

Other linear-scaling solvers have been proposed that look attractive in principle, but we are not aware of their having been used in linear-scaling DFT production calculations, and cannot comment on them. Some of them can be found reviewed in [27], especially the ones using approximations to projectors onto occupied space. It is worth mentioning a proposal [44] that would supersede the functional of Eq. (5.19), in its more rigorous treatment of the singular LWFs' overlap matrix, leading to an absolute physical minimum.

### 5.4.4
### Exact Exchange for Linear-Scaling Methods

On a final note, let us go back to the computation of the Hamiltonian matrix. What exposed in that section is valid for canonical density functionals, but does not hold for hybrid ones (or for Hartree–Fock). These demand the explicit calculation of the nonlocal exchange as in the Hartree–Fock theory. In addition to its being much more computationally demanding, it is interesting to make a point here on its linear-scaling possibilities [21, 22]. Such a term breaks the clean division in tasks (calculation of matrix elements on one hand, eigensolving on the other) since the exchange operator depends on the density matrix,

$$\hat{K}\varphi(\vec{r}) = \int d^3\vec{r}' \frac{\rho(\vec{r},\vec{r}')}{\|\vec{r}-\vec{r}'\|}\varphi(\vec{r}') \tag{5.20}$$

and thus matrix elements of the exchange operator will decay exponentially only if the density matrix does, that is, for insulators and finite temperature in general. It is, however, not a cause of concern, since it is not clear that the Hartree–Fock approximation or hybrid functionals should be preferred over canonical functionals (as using the Generalized Gradient Approximation) for metals anyway. Interestingly, the problem of linear-scaling computation of the matrix elements of $\hat{K}$ displays some commonalities with the eigen-solving problem, which should be further explored.

## 5.5
## Conclusions and Outlook

We have concentrated on the linear-scaling solution of the eigenvalue problem and the computation of **S** and **H** for an arbitrary electron density. With these tools, reaching self-consistency and obtaining total energy, forces, stresses, and so forth are straightforward, and analogous to what is done for other methods. The relevant expressions can be found in Ref. [10]. Around a thousand groups worldwide use SIESTA to calculate on a large variety of systems, in fields including the physical-, geo-, bio-, and nanosciences. The method provides most of the capabilities of modern electronic structure methods, from energy, forces, geometrical relaxations, and DFT molecular dynamics to electronic structure analysis tools and vibrational modes, to name some representative capabilities. It is steadily being developed, incorporating new capabilities and improving the efficiency, flexibility, robustness, and accuracy of its algorithms. Other extended LCAO methods [16] are also growing in capability and user base. This versatility, dynamism, and demand show the vitality and good prospects of linear-scaling LCAO methods.

Most applications, however, use the linear-scaling calculation of **H** and **S**, but resort to the conventional cube-scaling solvers. Indeed, many extended LCAO methods do not offer linear-scaling solver options [5, 16, 17]. The main reason behind this is the configurational complexity of the nuclear degrees of freedom, be it long ill-conditioned relaxations, an exponentially growing number of local minima, or long time scales in molecular dynamics runs. This complexity currently limits the systems that can be addressed to sizes that rarely justify the extra hassle of present-day linear-scaling solvers. In addition to the foreseeable progress addressing nuclear complexity, much more efficient DFT methods are needed for complex systems. We thus face the challenge of substantially reducing the linear-scaling prefactor (see Figure 5.1(b)), while keeping standards of accuracy. In addition, the mentioned hassle in the use of linear-scaling solvers should be removed by introducing robustness into the algorithms. Notwithstanding the possibility of new ideas, we think there is a scope for substantial progress in exploration of the many ideas already proposed, and of variants and possible synergies among them. It should be kept in mind, however, that the linear-scaling computation of the Hamiltonian and overlap matrices, which is at the heart of methods like SIESTA, has been very successfully incorporated into the electronic structure community, and is being routinely used by hundreds of groups worldwide.

## References

1 W.A. Harrison, *Electronic Structure and the Properties of Solids: The Physics of the Chemical Bond*, Dover, Mineola, NY (**1989**).

2 C.M. Goringe, D.R. Bowler and E. Hernandez, *Rep. Prog. Phys.* **60**, 1447 (**1997**).

3 O.F. Sankey and D.J. Niklewski, *Phys. Rev. B* **40**, 3979 (**1989**).

4 D. Porezag, Th. Frauenheim, Th. Köhler, G. Seifert and R. Kaschner, *Phys. Rev. B* **51**, 12947 (**1995**).

5 C. Pisani, R. Dovesi, C. Roetti, M. Causa, R. Orlando, S. Casassa and V.R. Saunders, *Int. J. Quantum Chem.* **77**, 1032 (**2000**).

6 D. Vanderbilt and S.G. Louie, *Phys. Rev. B* **30**, 6118 (**1984**).

7 W. Yang, *Phys. Rev. Lett.* **66**, 1438 (**1991**).

8 G. Galli and M. Parrinello, *Phys. Rev. Lett.* **69**, 3547 (**1992**).

9 E. Smargiassi and P.A. Madden, *Phys. Rev. B* **49**, 5220 (**1994**).

10 J. Soler, E. Artacho, J.D. Gale, A. Garcia, J. Junquera, P. Ordejon and D. Sanchez-Portal, *J. Phys. Condens. Matter* **14**, 2745 (**2002**).

11 L.E. Ballentine and M. Kolár, *J. Phys. C* **19**, 981 (**1986**).

12 E. Artacho and L. Milans del Bosch, *Phys. Rev. A* **43**, 5770 (**1991**).

13 J. Moreno and J.M. Soler, *Phys. Rev. B* **45**, 13981 (**1992**).

14 J. Kong *et al.*, *J. Comp. Chem.* **21**, 1532 (**2000**).

15 M.J. Frisch *et al.*, Gaussian Inc., Wallingford CT (**2004**).

16 J. VandeVondele, M. Krack, F. Mohamed, M. Parrinello, T. Chassaing and J. Hutter, *Comp. Phys. Comm.* **167**, 103 (**2005**).

17 S.D. Kenny, A.P. Horsfield and H. Fujitani, *Phys. Rev. B* **62**, 4899 (**2000**).

18 E. Artacho, D. Sanchez-Portal, P. Ordejon, A. Garcia and J.M. Soler, *phys. status solidi (b)* **215**, 809 (**1999**).

19 J. Junquera, O. Paz, D. Sanchez-Portal and E. Artacho, *Phys. Rev. B* **64**, 235111 (**2001**).

20 E. Artacho, E. Anglada, O. Dieguez, J.D. Gale, A. Garcia, J. Junquera, R.M. Martin, P. Ordejon, J.M. Pruneda, D. Sanchez-Portal and J.M. Soler, *J. Phys. Condens. Matter* **20**, 064208 (**2008**).

21 C.A. White, B.G. Johnson, P.MW. Gill and M. Head-Gordon, *Chem. Phys. Lett.* **2130**, 8 (**1994**).

22 M.C. Strain, G.E. Scuseria and M.J. Frisch, *Science* **271**, 51 (**1996**).

23 W.H. Press, S.A. Teukosly, V.T. Vetterling and B.P. Flannery, *Numerical Recipes: The Art of Scientific Computing*, Third Edition, Cambridge University Press, Cambridge (**2007**).

24 L. Kleinmann and D.M. Bylander, *Phys. Rev. Lett.* **48**, 1425 (**1982**).

25 E. Anglada and J.M. Soler, *Phys. Rev. B* **73**, 115122 (**2006**).

26 P. Ordejon, *Comp. Mater. Sci.* **12**, 157 (**1998**).

27 S. Goedecker, *Rev. Mod. Phys.* **71**, 1085 (**1999**).

28 W. Kohn, *Phys. Rev. Lett.* **76**, 3168 (**1996**).

29 R.W. Nunes and D. Vanderbilt, *Phys. Rev. B* **50**, 17611 (**1994**).

30 D.R. Bowler, T. Miyazaki and M.J. Gillan, *J. Phys. Condens. Matter* **14**, 2781 (**2002**).

31 C.K. Skylaris, P.D. Haynes, A.A. Mostofi and M.C. Payne, *J. Chem. Phys.* **122**, 084119 (**2005**).

32 W. Hierse and E.B. Stechel, *Phys. Rev. B* **50**, 17811 (**1994**).

33 P. Ordejon, D.A. Drabold, M.P. Grumbach and R.M. Martin, *Phys. Rev. B* **48**, 14646 (**1993**).

34 F. Mauri, G. Galli and R. Car, *Phys. Rev. B* **47**, 9973 (**1993**).

35 U. Stephan, *Phys. Rev. B* **62**, 16412 (**2000**).

36 J. Kim, F. Mauri and G. Galli, *Phys. Rev. B* **52**, 1640 (**1995**).

37 J.L. Fattebert and F. Gygi, *Comp. Phys. Commun.* **162**, 24 (**2004**).

38 M.V. Fernandez-Serra and E. Artacho, *J. Chem. Phys.* **121**, 11136 (**2004**).

39 P.J. de Pablo, F. Moreno-Herrero, J. Colchero, J. Gómez Herrero, P. Herrero, A.M. Baró, P. Ordejón, J.M. Soler and E. Artacho, *Phys. Rev. Lett.* **85**, 4992 (**2000**).

40 E. Artacho, M. Machado, D. Sanchez-Portal, P. Ordejon and J.M. Soler, *Mol. Phys.* **101**, 1587 (**2003**).

41 P. Ordejon, E. Artacho and J.M. Soler, *Phys. Rev. B* **53**, R10441 (**1996**).

42 L. Seijo and Z. Barandiaran, *J. Chem. Phys.* **121**, 6698 (**2004**).

43 L. Seijo, Z. Barandiaran and J.M. Soler, *Theor. Chim. Acta* **118**, 541 (**2007**).

44 W. Yang, *Phys. Rev. B* **56**, 9294 (**1997**).

# 6
# *Ab Initio* Molecular Dynamics

*Marcella Iannuzzi*

## 6.1
## Introduction

The growing need to understand complex phenomena on an atomistic level, taking into account the interactions between atoms and electrons, has recently stimulated the extension of the scope of molecular dynamics (MD) simulations. Having translated the investigated problem into a suitable atomistic model containing a limited number of atoms, the simulation approach can be used to study specific aspects, thus explaining observed phenomena or predicting unforeseen events. Thanks to the constant improvement of the computational algorithms (toward linear-scaling procedures) and thanks to the rapidly increasing accessible computer power (massive parallel supercomputers), MD has become a powerful tool to investigate many-body condensed matter systems. The starting point of any MD simulation is the definition of the molecular systems by an initial set of $N$ particles (positions and momenta) within a volume $V$. The dynamics of the particles is then evolved by integrating numerically classical Newton's equations of motion, where the forces are computed as derivatives of a given interaction potential. The resulting deterministic trajectories explore the available phase space under the assigned thermodynamic conditions and should be able to properly sample all the relevant configurations, or macrostates. This means that long enough simulations are going to provide realistic descriptions of the thermodynamic equilibrium and of structural and dynamical properties at finite temperature [8–11]. On top of that, the analysis of an atomistic trajectory allows the identification of individual events and complex mechanisms that drive specific processes and structural transformations. These features are expected to have important impact on both fundamental and industrial research through their application to complex problems in materials science.

The affordable system size as well as the accuracy in the description of the interaction potential are determined by the choice of the model Hamiltonian. Empirical force fields, as described in the previous chapter of this book, allows to routinely run simulations for large systems (>100 K atoms) generating trajectories of several nanoseconds. However, they are usually characterized by limited transferability

*Computational Methods in Catalysis and Materials Science.* Edited by Rutger A. van Santen and Philippe Sautet

ISBN: 978-3-527-32032-5

and cannot account for strong variations in the structural and electronic properties, although more and more sophisticated models are being developed that include polarization effects, charge transfer, nonadditive many-body interactions, etc. (see the next chapter). In chemically complex systems, many types of atoms and molecules give rise to a myriad of interactions making the parameterization of reliable empirical potentials not always possible. Moreover, the electronic structure and the chemical bonding change qualitatively in time, during the course of rearrangements, reactions, or other physicochemical processes. Such transformations are normally not reproducible using parametric descriptions of the interaction, i.e., the microscopic mechanisms responsible for an important class of events cannot be resolved by molecular simulations based on empirical force fields.

These limitations are overcome by *ab initio* molecular dynamics (AIMD), where the forces acting on the nuclei are computed from electronic structure calculations performed on-the-fly as the trajectory is generated. The electronic structure variables are not integrated out beforehand and represented by fixed parameters, rather they are considered as active and explicit degrees of freedom. The chemical complexity of the system is accounted for and the rearrangements of the electronic structure associated with the dynamic evolution of the system of particles are properly described. Physical quantities that depend on the electronic properties can be computed and their expectation values can be obtained as ensemble averages. AIMD is able to describe the working state of a catalyst under realistic pressure and temperature conditions, to disclose possible reaction mechanisms, and to explain them in terms of the rearrangements of the electronic structure along the generated trajectory. Moreover, by characterizing the chemical process on an atomistic level, it is possible to highlight trends of reactivity thus having an impact on the development of new catalysts. Unforeseen phenomena can happen, if necessary, lending to AIMD a truly predictive power.

AIMD is a general approach that can be used with any electronic structure method, and the level of approximation is determined by the selected method to solve the Schrödinger equation for the many-electron problem [12–15]. The high computational costs of most of the quantum chemistry methodologies restrict the investigation to very small systems (few atoms), thus hampering their applicability to real-world-scale problems. Methods based on density functional theory (DFT) [16], in the self-consistent formalism of Kohn and Sham, [17] could afford larger time and size scales (a few hundred atoms and tens of picoseconds) and compare generally well with experimental results, thus providing a good compromise. The most popular scheme for AIMD simulations of condensed matter systems is, indeed, DFT in combination with plane wave (PW) basis set and pseudopotentials. The accessibility of the forces on nuclei and, as a consequence, the possibility to generate trajectories spanning relatively long-time windows have determined the success of this technique in addressing a large number of important problems in different areas of physics, chemistry, materials science, and more recently biology also. A clear breakthrough in this field has been the unified scheme combining molecular dynamics and density functional theory proposed by Car and Parrinello (CP) in 1985 [18]. Their revolutionary idea of propagating a fictitious electron dynamics coupled with the classical motion of the nuclei made it possible to compute simultaneously the evolution of the wave function and the forces acting on the atoms. The introduction of this method decisively fostered not only the application but also the further development of AIMD techniques. Today these methodologies have reached a level of maturity such to represent essential tools for the investigation of properties and behaviors at finite temperature for a wide variety of materials and molecular systems. By AIMD, it has been possible to reveal new physical phenomena, which could not have been uncovered using empirical models, often leading to new interpretations of experimental data and even suggesting new experiments to perform.

In the following sections, we will discuss some of the most important aspects of AIMD simulations, focusing in particular on the CP formulation. Closely related topics as DFT electronic structure calculations and general features of MD simulations have been discussed in previous chapters and will not be covered here.

## 6.2 Born–Oppenheimer Molecular Dynamics

In nonrelativistic quantum-mechanics, the dynamics of an atomic system constituted of nuclei ($\{\mathbf{R}_I\}$) and electrons ($\{\mathbf{r}_i\}$) should be derived from the solution of the time-dependent Schrödinger equation

$$i\hbar\frac{\partial}{\partial t}\Phi(\{\mathbf{r}_i\},\{\mathbf{R}_I\};t) = \mathcal{H}\Phi(\{\mathbf{r}_i\},\{\mathbf{R}_I\};t), \tag{6.1}$$

where the Hamiltonian can be written in terms of the kinetic energy of the nuclei and an electronic Hamiltonian including the kinetic energy of the electrons and all the interaction potentials (e–e, n–e, n–n)

$$\mathcal{H} = -\sum_I \frac{\hbar^2}{2M_I}\nabla_I^2 + \mathcal{H}_e(\{\mathbf{r}_i\},\{\mathbf{R}_I\}). \tag{6.2}$$

In order to simplify the problem, three levels of approximation are normally introduced, which are the adiabatic approximation, the Born–Oppenheimer (BO) approximation, and the semiclassical approximation. In practice, it is recognized that for the time-independent problem it is convenient to separate the light electrons from the heavy nuclei [19–21]. The time-independent Schrödinger equation is solved for each given atomic configuration, where the electronic Hamiltonian depends parametrically on the atomic positions

$$\mathcal{H}_e(\{\mathbf{r}_i\},\{\mathbf{R}_I(t)\})\Psi_k(\{\mathbf{r}_i\},\{\mathbf{R}_I(t)\}) = E_k(\{\mathbf{R}_I(t)\})\Psi_k(\{\mathbf{r}_i\},\{\mathbf{R}_I(t)\}). \tag{6.3}$$

If we know all the adiabatic eigenfunctions, $\Psi_l$, at all possible atomic configurations, the total wavefunction could be expanded as

$$\Phi(\{\mathbf{r}_i\},\{\mathbf{R}_I\};t) = \sum_{l=0}^{\infty}\Psi_l(\{\mathbf{r}_i\},\{\mathbf{R}_I\})\Xi_l(\{\mathbf{R}_I\};t), \tag{6.4}$$

where the nuclear wavefunctions $\Xi_l$ can be viewed as time-dependent expansion coefficients. The insertion of this expression in the time-dependent Schrödinger equation, followed by multiplication from the left by $\Psi_k^*$ and integration over the electronic coordinates, leads to a set of coupled differential equations

$$\left[-\sum_I \frac{\hbar^2}{2M_I}\nabla_I^2 + E_k(\{\mathbf{R}_I\})\right]\Xi_k + \sum_l C_{kl}\Xi_l = i\hbar\frac{\partial}{\partial t}\Xi_k. \tag{6.5}$$

The $C_{kl}$ coefficients are functions of the kinetic energy and of the momenta of the nuclei, and they represent the matrix elements of the operator that couples the adiabatic states, i.e., stationary solutions of the time-independent problem.

When nonadiabatic effects are taken into account and different potential energy surfaces are coupled, changes in the electronic state become possible as the propagation of the nuclear wavefunction proceeds. In the adiabatic approximation, instead, the off-diagonal matrix elements of the nonadiabatic coupling operator are neglected, thus implying that the quantum state $k$ does not change during the time evolution. In most of the cases the coupling effects can be neglected and the evolution is restricted on the ground state electronic surface. However, there are examples of interesting applications of various forms on nonadiabatic molecular dynamics, in particular for the investigation of reaction mechanisms that are accompained by electronic excitation and decay processes [22–24].

By neglecting also the diagonal coupling, we assume that the electrons respond instantaneously to the nuclear motion. According to the Born–Oppenheimer approximation, the electronic eigenvalues give rise to potential surfaces on which the nuclear dynamics evolves, as described by the time-dependent Schrödinger equation for the time-dependent nuclear wavefunction $\Xi_k(\{\mathbf{R}_I\}, t)$

$$\left[T_n + V_{nn}(\{\mathbf{R}_I\}) + E_k(\{\mathbf{R}_I\})\right]\Xi_k(\{\mathbf{R}_I\}, t) = i\hbar\frac{\partial}{\partial t}\Xi_k(\{\mathbf{R}_I\}, t). \tag{6.6}$$

Finally, if the nuclear quantum effects can be also neglected, the semiclassical approximation is obtained by rewriting the nuclear wavefunction in its polar representation, in terms of an amplitude $A$ and a phase $S$, $\Xi(\{\mathbf{R}_I\}, t) = A(\{\mathbf{R}_I\}, t)e^{iS(\{\mathbf{R}_I\},t)/\hbar}$. By disentangling the equations of motion for $A$ and $S$ and assuming the classical limit $\hbar \rightarrow 0$, i.e., neglecting all terms involving $\hbar$, it can be demonstrated that the approximate equation for $S(\{\mathbf{R}_I\}, t))$ is isomorphic to the Hamilton–Jacobi formulation [25, 26] of classical mechanics, where the classical Hamilton function is the total energy for the selected adiabatic state $k$ [27–29]. The corresponding equations of motion are

$$\frac{d\mathbf{P}_I}{dt} = -\nabla_I E_k = -\nabla_I V_k^{\mathrm{BO}}(\{\mathbf{R}_I(t)\}). \tag{6.7}$$

Thus, the nuclei move according to classical mechanics in an effective potential given by the BO potential energy surface, which has been obtained by solving the time-independent electronic Schrödinger equation for the k-th state at the given nuclear configuration. The electronic Hamiltonian is a function of all the nuclear coordinates, and through them it depends indirectly on the time $t$.

The *a priori* construction of the potential energy surface is a cumbersome task, becoming rapidly not feasible for chemically complex systems, with more than a few degrees of freedom. The alternative is the *on-the-fly* molecular dynamics, where the forces for the propagation of the atomic positions are obtained directly from the electronic structure calculated at each nuclear configuration,′ as $\mathbf{F}_I(t) = -\nabla_I\langle\Psi|\mathcal{H}_e|\Psi\rangle = \langle\mathcal{H}_e\rangle$.

The Ehrenfest molecular dynamics (E-MD) is the oldest *on-the-fly* formalism [30]. The Ehrenfest forces are computed by solving numerically the coupled set of quantum and classical equations simultaneously

$$M_I\ddot{\mathbf{R}}_I(t) = -\nabla_I\langle\mathcal{H}_e\rangle \qquad i\hbar\frac{\partial\Psi}{\partial t} = -\nabla_I\mathcal{H}_e\Psi. \tag{6.8}$$

The propagation of the electronic wavefunction starts from an initial wavefunction, obtained by a single self-consistent optimization. It proceeds by applying the electronic Hamiltonian to the wavefunction at each MD iteration, following the nuclear motion. If $\Psi$ is defined as an expansion in terms of several instantaneous adiabatic electronic states, this approach accounts rigorously for nonadiabatic transitions. Instead, by keeping only one adiabatic state in the expansion, typically the ground state, the wavefunction remains on the selected adiabatic surface as the nuclei propagate, without need of any further self-consistent optimization. When the electronic structure is described through the Kohn–Sham theory, the one-electron orbitals remain automatically orthonormal by the propagation through the time-dependent Schröedinger equation in (6.8). The Ehrenfest approach is a scheme to intrinsically propagate the electronic dynamics within the framework of classical nuclear motion and the mean field approximation. The integration of the fast electronic degrees of freedom, however, imposes MD time steps in the order of the attoseconds, thus limiting possible applications to small systems and/or very short trajectories. Nevertheless, this methodology has been successfully employed in the investigation of few-body collision- and scattering-type problems [31–34], and more recently also in the field of time-dependent DFT [35–44].

Another *on-the-fly* method, more frequently applied for simulations of larger systems, is the Born–Oppenheimer molecular dynamics (BO-MD) [45–47]. By this approach, the static electronic structure problem is solved through the time-independent Schrödinger equation, at each MD iteration, i.e., for fixed atomic configurations. The time evolution of the electronic structure is enforced only through its parametrically dependence on the nuclear positions, which, in turn, are propagated at each time step by using the forces derived from the self-consistently optimized electronic state. In contrast with E-MD, one electronic state is selected, which can be the ground state or any other excited state, and at each MD step the wavefunction is optimized onto the corresponding adiabatic surface, neglecting any interference with other surfaces. In the Kohn–Sham density functional theory, the ground state is determined as variational minimum of the expectation value of the effective one-particle Hamiltonian with respect to the one-particle orbitals $\phi_i$, subject to the orthonormality constraints, $\langle\phi_i|\phi_j\rangle = \delta_{ij}$. The molecular orbitals are expanded in terms of the selected basis set $\phi_i(\mathbf{r}) = C_{\mu i}\varphi_\mu(\mathbf{r})$ and the charge density can be written in terms of the density matrix $\mathbf{P}_{\mu\nu} = \sum_i C^*_{\mu i} C_{\nu i}$, as $\rho(\mathbf{r}) = \sum_i |\phi_i(\mathbf{r})|^2 = \sum_{\mu\nu} \mathbf{P}_{\mu\nu} \langle \varphi_\mu | \varphi_\nu \rangle$. The resulting BO equations of motion are

$$M_I \ddot{\mathbf{R}}_I(t) = -\nabla_I \min_{\rho} \left\{ \left\langle \Psi_0 \left| \mathcal{H}_{\mathrm{e}}^{\mathrm{KS}} \right| \Psi_0 \right\rangle \right\} \tag{6.9}$$

$$0 = -\mathcal{H}_{\mathrm{e}}^{\mathrm{KS}} \phi_i + \sum_j \Lambda_{ij} \phi_j, \tag{6.10}$$

where $\mathcal{H}_{\mathrm{e}}^{\mathrm{KS}}$ is the Kohn–Sham Hamiltonian and $\Lambda_{ij}$ are the Lagrange multipliers to enforce the orthonormality constraints. This approach implies the iterative solution of the Kohn–Sham equations at each MD step. In combination with DFT, BO-MD has been successfully employed for condensed matter and cluster simulations, [47–54] including MD in electronically excited states [55]. Since no dynamics of the electronic degrees of freedom is intrinsically involved, the BO equations of motion can be integrated on the nuclear time scales, which is much slower than the electronic one and allows time steps in the order of femtoseconds.

The extension of the scope of BO-MD simulations is conditional on the efficiency of the wavefunction optimization procedures. In this respect, important advancements have been achieved thanks to the development of improved minimization algorithms, to reduce the computational costs of the electronic structure calculations [47, 53, 54]. In particular, linear-scaling techniques either for the construction of the Hamiltonian and for the solution of the eigenvalues problem, have attracted large interest in the perspective of running samples of hundreds atom for trajectories of several picoseconds [56–61]. Another important ingredient to speed up the simulation is the reduction of the number of iterations needed by the self-consistent loop to converge. Obviously this cannot be achieved at the expense of accuracy, simply by requiring looser convergence criteria, since tight tolerance for the solution of the electronic problem is essential in order to yield the stability of atomic trajectories. One advantage of MD simulations is that atomic configurations are generated in a continuous fashion and one can hence predict an initial trial wave function for the SCF calculation by multilinear extrapolation using the previous wave functions [62]

$$C_{\mu i}(t_n) \approx \sum_{m=1}^{K} (-1)^{m+1} \binom{K}{m} C_{\mu i}(t_{n-m}). \tag{6.11}$$

Large fluctuations in the coefficients of the molecular orbitals, however, can introduce instabilities in the algorithm, in particular when higher order expansions are employed. The extrapolation of the density matrix is, instead, more robust and can be safely applied at higher orders. The coefficients of the trial orbitals at time $t_n$ are then obtained as the projection of the orbitals at time $t_{n-1}$ onto the occupied space defined by the extrapolated contra-covariant representation $PS(t_n)$, where the overlap matrix $S$ is the identity matrix only when orthogonal basis sets, like the PW, are employed. The effect on the number of iterations and on the error in the energy of the first guess depends on the order of the expansion, as discussed in Ref. [58].

Fundamental for any MD approach is the calculation of the forces on the nuclei, as derivatives of the total energy $\mathbf{F}_I = -\nabla_I \langle \Psi_0 | \mathcal{H}_e | \Psi_0 \rangle$. Numerical solutions given by finite differences are too expensive and not sufficiently accurate. Therefore, the gradients are typically computed analytically, by taking into account the derivatives of the Hamiltonian, i.e., the contributions known as the Hellman–Feynman forces, [63–65] but also the terms stemming from variations of the wavefunctions

$$\nabla_I \langle \Psi_0 | \mathcal{H}_e | \Psi_0 \rangle = \langle \Psi_0 | \nabla_I \mathcal{H}_e | \Psi_0 \rangle + \langle \nabla_I \Psi_0 | \mathcal{H}_e | \Psi_0 \rangle + \langle \Psi_0 | \mathcal{H}_e | \nabla_I \Psi_0 \rangle . \tag{6.12}$$

The Hellmann–Feynman theorem states that when the $\Psi_0$ is an exact eigenfunction of the Hamiltonian, only the first contribution remains, whereas the other two vanish exactly. However, this is never the case in numerical calculations, where we use representations of the molecular orbitals through finite (not complete) basis sets and the exact self-consistency is never reached. Considering only the first term in (6.12) leads to large errors in the propagation of the trajectory, since the errors in the Hellmann–Feynman forces are first order in the errors of $\rho$, whereas the errors in the total energy are only second order. The orbitals depend on the nuclear positions implicitly through the expansion coefficients, and may also carry an explicit dependence, in the case of atom-centered basis set functions

$$\nabla_I \phi_i = \sum_\mu \left( \nabla_I C_{\mu i} \right) \varphi_\mu(\mathbf{r}, \mathbf{R}_I) + \sum_\mu C_{\mu i} \left( \nabla_I \varphi_\mu(\mathbf{r}, \mathbf{R}_I) . \right) \tag{6.13}$$

From this expression, two types of analytical corrections to the forces can be derived to cancel the first-order errors [66,67]. The first contribution is the incomplete-basis-set (IBS) correction. It is given from the nuclear gradients of the basis functions and the effective, non-self-consistent (NSC) Hamiltonian, since the orbitals are not exact eigenstates of the Hamilton operator

$$\mathbf{F}_I^{\mathrm{IBS}} = -\sum_{i\nu\mu} C_{\nu i} C_{\mu i} \left( \left\langle \nabla_I \varphi_\nu \left| H_\mathrm{e}^{\mathrm{NSC}} - \epsilon_i \right| \varphi_\mu \right\rangle + \left\langle \varphi_\nu \left| H_\mathrm{e}^{\mathrm{NSC}} - \epsilon_i \right| \nabla_I \varphi_\mu \right\rangle \right) . \tag{6.14}$$

This term vanishes either for exact self-consistency, or for origin-less basis set functions. The second term is the NSC correction

$$\mathbf{F}_I^{\mathrm{NSC}} = -\int d\mathbf{r} \, (\nabla_I \rho) \left( V^{\mathrm{SCF}} - V^{\mathrm{NSC}} \right) \tag{6.15}$$

and is given by the difference between the exact potential and its approximation, associated to $H_\mathrm{e}^{\mathrm{NSC}}$. The NSC error can be made arbitrarily small by optimizing the effective Hamiltonian to very high accuracy, but it can never be suppressed completely within the numerical approach. The resulting drift in the constant of motion along the propagation can be considered as one criterion to judge the quality of the simulation.

We notice in passing that the forces computed in Ehrenfest MD are not affected by the NSC error, since full self-consistency is not required. The Ehrenfest forces are not derived from the minimized expectation value of the Hamiltonian, but instead from the Hamiltonian and the associated wavefunction at a certain time step.

## 6.3
## Car–Parrinello Molecular Dynamics

In 1985 Roberto Car and Michele Parrinello proposed an approach that somehow combines the advantages of E-MD and BO-MD [18]. The "Unified Approach for MD and DFT," as it was initially named, propagates the dynamics of the electronic degrees of freedom, thus generating smooth trajectories and avoiding the computationally demanding iterative solution of the electronic problem at each MD step. On the other hand, it takes advantage of the time scale separation between the fast electronic motion and the slow nuclear motion, in order to allow the use of reasonably large time steps for the integration of the equations of motion. This is achieved by mapping the two-component quantum/classical problem onto a two-component purely classical problem with two adiabatically separate energy scales. The novelty of this scheme is the replacement of the quantum dynamics of the electrons by a fictitious classical dynamics of the electronic orbitals, which are interpreted as classical fields. By maintaining the separation between the time scales of the two subsystems, it can be demonstrated that the fictitious dynamics lets the orbitals follow adiabatically the motion of the nuclei, while the electronic configuration is kept approximately minimized at each MD iteration, without performing any optimization. Over its characteristic time interval $\Delta t$, the nuclear motion is, instead, governed by the average effective potential generated by the faster fluctuations of the orbitals about the minimum energy surface. In the limit of very fast orbital dynamics, the fluctuations are small and the approximation to the correctly minimized surface is very good. The striking idea behind the CP approach is to borrow concepts from statistical mechanics to establish an acceptable compromise between proximity to the exact BO surface and length of the integration time step. The result is a rather simple and effective formalism for the propagation of the two-component dynamics, involving electronic and nuclear degrees of freedom simultaneously, for which integration time steps of intermediate length (fractions of femtosecond) between nuclear and electronic time scales can be safely used.

The definition of the CP dynamics starts from the observation that the energy of the electronic subsystem can be seen as a function of the atomic positions, but also as a functional of the electronic wavefunction, or of the Kohn–Sham orbitals in the DFT description. In classical mechanics the forces on the nuclei are obtained from the Lagrangian of the system as derivatives with respect to the nuclear positions. In analogy, providing a suitable Lagrangian that describes the extended dynamical system of nuclei and orbitals, its functional derivative with respect to the orbitals yield the force driving the propagation of the electronic structure. This is realized through the definition of orbital velocities $\{\dot{\phi}_i\}$ and a related fictitious kinetic energy (not to be confused with the quantum kinetic energy appearing in $\mathcal{H}_e$)

$$K_{\text{fict}} = \mu \sum_i \langle \dot{\phi}_i | \dot{\phi}_i \rangle, \tag{6.16}$$

where $\mu$ is the fictitious electronic mass, with units (energy $\times$ time$^2$). The time scale on which the orbitals evolve is controlled by their inertia in responding to the nuclear motion, which is determined by $\mu$. The resulting extended Lagrangian reads

$$\mathcal{L}_{\mathrm{CP}} = \sum_I \frac{1}{2} M_I \dot{\mathbf{R}}_I^2 + \sum_i \mu \langle \dot{\phi}_i | \dot{\phi}_i \rangle - E_{\mathrm{KS}}(\{\phi_i\}, \{\mathbf{R}_I\}) + \sum_{ij} \Lambda_{ij} (\langle \phi_i | \phi_j \rangle - \delta_{ij}), \tag{6.17}$$

where $\Lambda_{ij}$ are the Lagrange multipliers to impose the orbitals orthonormality as dynamic constraint. The Newtonian equations of motion are obtained from the Euler Lagrangian equations

$$\frac{d}{dt}\frac{\partial \mathcal{L}}{\partial \dot{\mathbf{R}}_I} = \frac{\partial \mathcal{L}}{\partial \mathbf{R}_I} \qquad \frac{d}{dt}\frac{\delta \mathcal{L}}{\delta \langle \dot{\phi}_i |} = \frac{\delta \mathcal{L}}{\delta \langle \phi_i |}, \tag{6.18}$$

where functional derivatives are taken with respect to the scalar fields representing the orbitals. The generic CP equations of motion can then be derived as

$$\begin{aligned} M_I \ddot{\mathbf{R}}_I(t) &= -\frac{\partial E_{\mathrm{KS}}}{\partial \mathbf{R}_I} + \sum_{ij} \Lambda_{ij} \frac{\partial}{\partial \mathbf{R}_I} \langle \phi_i | \phi_j \rangle \\ &= -\nabla_I \langle \Psi_0 | \mathcal{H}_{\mathrm{KS}} | \Psi_0 \rangle + \sum_{ij} \Lambda_{ij} \frac{\partial}{\partial \mathbf{R}_I} \langle \phi_i | \phi_j \rangle \end{aligned} \tag{6.19}$$

$$\mu \ddot{\phi}_i(t) = -\frac{\delta E_{\mathrm{KS}}}{\delta \langle \phi_i |} + \sum_j \Lambda_{ij} | \phi_j \rangle = -\mathcal{H}_{\mathrm{KS}} \phi_i + \sum_j \Lambda_{ij} \phi_j. \tag{6.20}$$

In fact, the equation for the orbitals is typically transformed into an equation of motion for the expansion coefficients. In the special case of origin-less and orthonormal basis set functions, like PW, the force on the coefficient becomes

$$\mu \ddot{C}_{gi} = -\frac{\partial E_{\mathrm{KS}}}{\partial C_{gi}^*} + \sum_{ij} \Lambda_{ij} C_{gj}, \tag{6.21}$$

where g is the index of the reciprocal vector in the PW expansion.

For Gaussian basis sets, the expression is somewhat complicated by the position dependence of the basis functions and by their nonorthonormality [58]. Moreover, it is important to remark that when the constraints are also the function of the nuclear positions, the nuclear forces include a constraint contribution that must be considered in order to generate proper energy-conserving dynamical evolution. This is always the case when ultrasoft pseudopotential and/or atomic-centered basis set functions are used [68–70].

Like E-MD, the CP scheme is based on the explicit time evolution of the electronic degrees of freedom and no Hamiltonian diagonalization is needed, except at the very first step. At difference with respect to E-MD, where the evolution occurs through a unitary propagation [71–73] and the orbitals remain automatically orthogonal, the orthogonality constraint has to be imposed when using the CP approach. The additional costs for the orthogonalization procedure required at each MD iteration are normally fully compensated by the possibility to employ longer time steps (at least one order of magnitude). The CP equations of motion can be interpreted as a mixture of BO and E-MD, as they include orthonormality constraint explicitly (like in BO), but they also propagate the wavefunction dynamically (like Ehrenfest).

### 6.3.1 The Extended CP Dynamics

Thanks to the definition of an intrinsic dynamics of the electronic degrees of freedom, the forces derived from the corresponding extended Lagrangian are exact, e.g., they consistently conserve the constant of motion

$$E_{\text{cons}} = \mu \sum_i \langle \dot{\phi}_i | \dot{\phi}_i \rangle + \sum_I \frac{1}{2} M_I \dot{\mathbf{R}}_I^2 + \langle \Psi_0 | \mathcal{H}_{\text{KS}} | \Psi_0 \rangle . \tag{6.22}$$

Since no self-consistent optimization of the electronic wavefunction is performed, the forces cannot be affected by any NSC error, as already pointed out in the case of E-MD.

By integrating the CP equations of motion, the evolution of the orbitals occur at the "fictitious temperature" $T_{\text{fict}} \propto \mu \sum_i \langle \dot{\phi}_i | \dot{\phi}_i \rangle$. The fundamental condition under which the CP-MD is going to work is the separation of nuclear and electronic time scales, i.e., $T_{\text{fict}} \ll T_{\text{n}}$, where $T_{\text{n}}$ is the physical temperature of the nuclear motion. Under this condition, the initial orbitals, which are the self-consistent solution of the electronic problem at the initial atomic configuration ($t = 0$), evolve at low temperature remaining close to the instantaneous minimum energy $min_{\{\phi_i\}} \langle \Psi_0 | \mathcal{H}_{\text{KS}} | \Psi_0 \rangle$, i.e., close to the BO surface. In practice, $T_{\text{fict}}$ can be interpreted as a measure of the distance of the electronic wave function from the BO surface. In order to keep it small, thermal exchanges between the "cold" electrons and the "hot" nuclei have to be avoided. This means that the two subsystems remain decoupled, while the electrons still follow adiabatically the nuclear motion. Such adiabatic separation is possible if the two power spectra, obtained as Fourier transform of the velocity autocorrelation functions, of nuclei and orbitals, respectively, do not have significant overlap and no efficient mechanism to exchange energy is available. The adiabaticity issue of CP-MD has been addressed in several works [74–77] since the early time of the method up to very recent publications. The large interest on this aspect is a clear sign of how crucial is the decoupling to establish the validity of the approach. The problem has been also considered from a rigorous mathematical point of view [78].

Pastore et al. [74] analyzed the conservation of the constants of motion for a trajectory generated with the CPMD program package [79] for a system of two silicon atoms in a periodic diamond lattice. Using a time step of 0.3 fs and a ficitious mass of 300 a.u., the dynamical separation is verified from the comparison between the vibrational density of states of the electronic orbitals and the highest-frequency phonon mode of the nuclei, as shown in Figure 6.1(a). The electronic energy, $E_{\text{KS}}$ in Figure 6.1(b), varies considerably as a function of time, and its oscillations have a mirror image in the oscillations of the fictitious kinetic energy, $K_{\text{fict}}$. These latter are then bound around a constant value, which

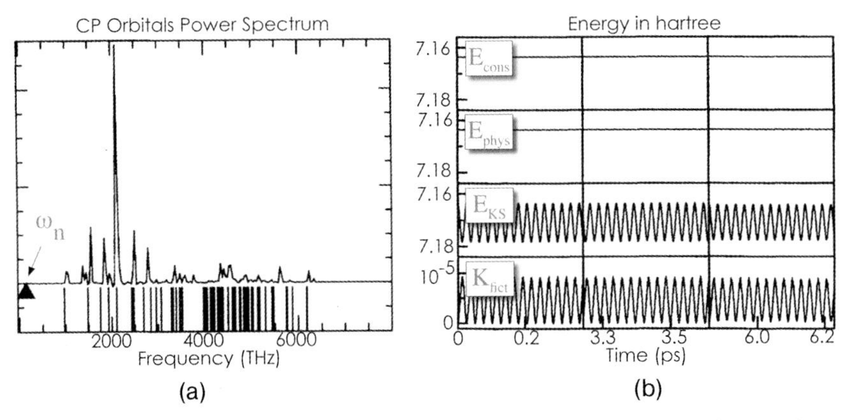


**Figure 6.1** (a) Vibrational density of states of the electronic degrees of freedom compared to the highest-frequency phonon mode $\omega_n^{max}$ (triangle) in bulk silicon. (b) Conserved energy $E_{cons}$, the physical energy $E_{phys}$, the electronic energy $E_{KS}$, and the fictitious kinetic energy $K_{fict}$ along the CP-MD trajectory of bulk silicon (see the discussion in text). The graphs are taken from Ref. [74].

means that the electrons do not heat up, and the deviation from the minimum electronic energy remains small. This oscillation represents the response of the electrons to the drag of the nuclear motion. Actually, the wavefunction is not instantaneously optimized on the BO surface, but follows the nuclei with a certain delay, which is determined by the inertia given by the fictitious mass. A closer inspection of $K_{fict}$ reveals the presence of a second time scale of oscillations, with much smaller amplitude and higher frequency. These latter are intrinsic of the dynamical evolution of the orbitals. In any case, even the larger fluctuations in $K_{fict}$ have amplitude three orders of magnitude smaller than the physically meaningful oscillations of $\langle \mathcal{H}_e \rangle$. As a consequence, $E_{phys} = E_{cons} - K_{fict}$ is essentially constant.

The smooth propagation of the orbitals and the consequent conservation of the total energy are possible only by choosing a proper time step $\delta t$, suitable to integrate the dynamics occurring on the orbital time scale, typically fractions of femtosecond. This guarantees that the force $|\mu\ddot{\phi}_i(t)|$ is small at any time, in comparison to the physically relevant forces, and the electronic state remains close to or at the minimum. When the electronic system is exactly on the BO surface, the expectation value $\langle |\mathcal{H}_e(t)| \rangle$ is a minimum for the instantaneous atomic configuration, and no forces act on the wavefunction, $\{|\mu\ddot{\phi}_i(t)| = 0\}$. By satisfying the condition of small forces, or equivalently of small fictitious kinetic energy, the CP equations are expected to produce asymptotically the correct dynamics of the nuclei.

This can be seen also considering the CP formulation from the viewpoint of the extended Lagrangian methodologies, commonly employed to generate proper sampling in the desired statistical ensemble through the propagation of the dynamics of an extended system [80, 81]. In practice, the dimensionality of the total phase space is increased by the definition of a set of auxiliary variables and the equations of motion derived from the resulting extended Lagrangian sample the microcanonical ensemble of the extended phase space. Therefore, the total energy of the original physical system is not strictly conserved, whereas the constant of motion is the total energy of the extended system. The coupling between the physical dynamic variables and the auxiliary variables determines the specific thermodynamic conditions and the effective environment in which the physical system evolves. In CP-MD, the orbitals $\phi_i$ are auxiliary variables with a precise physical meaning. Their momenta and masses, instead, are instrumental artificial variables to define the fictitious dynamics. The coupling to the physical system of the nuclei occurs through the electronic Hamiltonian $\mathcal{H}_e$. The fictitious degrees of freedom obviously affect the dynamics and thermodynamics; however, it is in general possible to obtain an essentially unperturbed time evolution of the physical system within the extended dynamics. This can be demonstrated also for the CP scheme, thanks to the separation of time scales. First, the fluctuations of the electronic system around the BO surface have maximal amplitude that is orders of magnitude smaller than the physical variations. Even more important, these high-frequency fluctuations, and with them the discrepancies from the BO surface, average out on the time scale of the nuclear motions. This means that the average forces governing the nuclear evolution can be considered as derived from an effective potential averaged over these fluctuations, and hence they are very close to the exact BO forces (without NSC error). This issue has been discussed also in Ref. [74], where the difference between the CP forces and the corresponding BO values is computed on same atomic configurations and using tight convergence criterion for the self-consistent optimization.

The fact that the fictitious kinetic energy is a measure of the deviation of the instantaneous electronic system from the BO surface suggests that the CP scheme can be used as a global optimization technique. Starting from an atomic configuration and an initial guess for the electronic wavefunction, the combined optimization of the geometry and of the electronic problem is achieved through an MD annealing driven by the CP equations of motions, where proper damping terms are added

$$\mu\ddot{\phi}_i(t) = -\frac{\delta}{\delta\langle\phi_i|}\langle\Psi_0|\mathcal{H}_{\mathrm{KS}}|\Psi_0\rangle + \frac{\delta}{\delta\langle\phi_i|}\{\text{constraints}\} - \gamma_e\mu\dot{\phi}_i. \tag{6.23}$$

$\gamma_e$ is a friction constant that governs the rate of energy dissipation in the electronic subsystem. At the same time, the nuclear velocities can be rescaled using a different friction coefficient, $\gamma_n$. Hence, geometries and charge densities are varied simultaneously, in the spirit of the CP method, leading the system toward the equilibrium without intermediate knowledge of the BO surface. If the excess energy is slowly pumped out, i.e., the dissipation coefficients are close to one, the extended system of nuclei and electrons should move progressively toward the global minimum. This technique has proved to be efficient for the optimization of complex non linear functions in high-dimensional parameter space, even including constraints [47, 82].

### 6.3.2
### Integration of the CP Equations of Motion

The integration of the CP equations of motion is typically carried out through the velocity Verlet algorithm [83]

$$\mathbf{R}(t+\delta t) = \mathbf{R}(t) + \mathbf{V}(t)\delta t + \frac{\mathbf{F}(t)}{2M}\delta t^2 \qquad \mathbf{V}(t+\delta t) = \mathbf{V}(t) + \frac{\mathbf{F}(t+\delta t)+\mathbf{F}(t)}{2M}\delta t. \tag{6.24}$$

Even if other algorithms can have better short time stability and allow longer time steps, this algorithm is considered one of the best integrators for MD simulations thanks to its simplicity and stability over long trajectories. It requires a limited storage space, since only positions, velocities and forces at time $t$ are needed (no higher derivatives or values at previous steps). Positions and velocities are always available at equal time. The introduced error is of the order of $(\delta t)^4$ and the algorithm conserves the volume in phase space, i.e., it is symplectic. Other interesting features are the fact that it is time reversible and is implemented in three steps, known as half kick, drift, half kick. This is realized by updating the velocities in two steps, before and after the update of positions and forces

```
V(:) := V(:) + dt/(2*M(:))*F(:)
R(:) := R(:) + dt*V(:)
Calculate new forces F(:)
V(:) := V(:) + dt/(2*M(:))*F(:)
```

This procedure turns out to be slightly complicated in the presence of constraints, as it is always the case for the CP equation of motions due to the required orthonormality of the orbitals. Starting from the orbitals $\{\phi_i(t)\}$ and the velocities $\{\dot{\phi}_i(t)\}$, the first step is a velocity update, immediately followed by the orbital update

$$|\dot{\phi}_i^{(1)}\rangle = |\dot{\phi}_i(t)\rangle + \frac{\delta t}{2\mu}|f_i(t)\rangle \qquad |\tilde{\phi}_i\rangle = |\phi_i(t)\rangle + \delta t|\dot{\phi}_i^{(1)}\rangle, \tag{6.25}$$

where $|f_i(t)\rangle = (\partial E_{KS}/\partial\langle\phi_i|)|_t$ is the force on the orbital at time $t$. The orthonormality condition

$$\langle\phi_i(t+\delta t)|\phi_j(t+\delta t)\rangle = \delta_{ij} \tag{6.26}$$

is enforced through the application of the constraint forces to orbitals and velocities resulting after the velocity Verlet update

$$|\phi_i(t+\delta t)\rangle = |\tilde{\phi}_i\rangle + \sum_j X_{ij}|\phi_j(t)\rangle \quad |\dot{\phi}_i^{(2)}\rangle = |\dot{\phi}_i^{(1)}\rangle + \frac{1}{\delta t}\sum_j X_{ij}|\phi_j(t)\rangle, \tag{6.27}$$

where $X_{ij} = (\delta t^2/2\mu)\Lambda_{ij}$. The matrix of Lagrange multipliers is determined from the matrix equation that results from the substitution of (6.27) into (6.25). The equation is solved iteratively starting from an initial guess, like $\{X_{ij} = \frac{1}{2}(1 - \langle\tilde{\phi}_i|\tilde{\phi}_j\rangle)\}$, until the orthonormality condition is satisfied within a given tolerance. The deviation from orthonormality of the initial orbitals is of order $\delta t^2$, and the typical convergence criterion for orthonormality is $10^{-6}$.

Once the new orbitals and the new nuclear positions are known, the forces at time $t + \delta t$ can be calculated and the second update of the velocities is performed

$$|\dot{\phi}_i^{(3)}\rangle = |\dot{\phi}_i^{(2)}\rangle + \frac{\delta t}{2\mu}|f_i(t + \delta t)\rangle. \tag{6.28}$$

Once more, these are not the final velocities, since also in this case the appropriate constraint forces need to be applied

$$|\dot{\phi}_i(t + \delta t)\rangle = |\dot{\phi}_i^{(3)}\rangle + \sum_j Y_{ij}|\phi_j(t + \delta t)\rangle. \tag{6.29}$$

The new set of Lagrange multipliers $Y_{ij}$ are directly determined by simply substituting the above expression into first-time derivative of the orthonormality condition

$$\langle \phi_i(t + \delta t)|\dot{\phi}_j(t + \delta t)\rangle + \langle \dot{\phi}_i(t + \delta t)|\phi_j(t + \delta t)\rangle = 0 \tag{6.30}$$

that also must be satisfied.

In the above description of the CP integration algorithm, we have assumed that the orthonormality constraint does not depend on the nuclear positions. This is not the case when atom-centered basis set functions are used, or with the ultrasoft pseudopotential scheme, [69, 84] due to the relaxation of the norm-conserving condition. In all these cases, the electronic and nuclear equations of motion are coupled through the constraint forces. Thereby, in order to satisfy the orthonormalitty condition, the constraint procedure has to be iterated through the nuclear update [69, 70]. The computational effort for these operations can be rather large, such that it can become convenient to reformulate the electronic structure problem in terms of nonorthogonal orbitals, employing the method known as the constrained nonorthogonal orbital approach [70].

### 6.3.3 How to Control Adiabaticity

The first necessary condition to perform correct CP-MD simulations is the separation of the electronic and nuclear time scales. This is satisfied when $\omega_e^{\min} \gg \omega_n^{\max}$, meaning that the phonon spectrum lays well below the characteristic frequencies of the orbitals dynamics. As already discussed, such adiabatic separation prevents significant energy flows between the two subsystems, thus keeping the "cold" electrons in the vicinity of the BO surface. Moreover, if the remaining intrinsic fluctuations of the electronic dynamics have small amplitude and high frequency, their effect is averaged out over the longer time scales of the nuclear motion. As expected, the parameter that controls the adiabaticity is the inertia assigned to the orbitals, i.e., the fictitious mass $\mu$. It can be shown that when the electronic system is close to the ground state, the frequencies of the orbitals are approximately given by

$$\omega_{ij} = \left( \frac{2(\varepsilon_i - \varepsilon_j)}{\mu} \right)^{1/2}, \tag{6.31}$$

where $\varepsilon_i$ and $\varepsilon_j$ are the Kohn–Sham eigenvalues of occupied and unoccupied orbitals, respectively. This harmonic approximation is valid in particular for the lowest frequency, hence

$$\omega_e^{\min} \propto \left( \frac{E_{\text{gap}}}{\mu} \right)^{1/2}, \tag{6.32}$$

where $E_{\text{gap}}$ is the energy difference between the last occupied and the first unoccupied Kohn–Sham states. Since the energy gap is a physical property of the system, the only parameter that can be tuned to control the adiabaticity is the mass $\mu$. On the other hand, $\mu$ cannot be decreased arbitrarily because the orbital dynamics would become too fast, thus requiring an excessively small $\delta t$, in order to be able to resolve the fastest modes. As a matter of fact, for the proper sampling of all relevant oscillations, the time step $\delta t$ should be chosen about 10 times smaller than the shortest period. The highest frequency of the orbitals dynamics can be estimated from the largest kinetic energy contribution in the basis set expansion. When PW basis sets are used, this value is the energy cutoff $E_{\text{cut}}$ determining the maximum reciprocal vector in the expansion, thus controlling the precision of the calculation. The maximum frequency is then $\omega_e^{\max} \propto \left( \frac{E_{\text{cut}}}{\mu} \right)^{1/2}$ and the time step should be lower than a maximum value $\delta t^{\max} \propto \sqrt{\frac{\mu}{E_{\text{cut}}}}$. The choice of $\mu$ is therefore system dependent and should result from a reasonable compromise between accuracy and efficiency. A too small mass gives "too light" electronic degrees of freedom, and the correct integration by the velocity Verlet algorithm would require a very small time step to avoid drifts in the constant of motion. On the other hand, too large $\mu$ makes the electrons "too heavy" and the adiabaticity is quickly lost. Typical values for $\mu$ are 400–1000 a.u. and for the time step 0.1–0.3 fs, depending on the mass of the nuclei.

A clear indication of loss of adiabaticity is the heating up of the electrons, accompanied by nuclear cooling. When two subsystems are not anymore decoupled, a steady energy flow tends to bring them to thermal equilibrium. As a consequence, the electronic system diverges from the BO surface and the CP forces become rapidly badly wrong, even if the dynamics might still conserve the constant of motion. Such a situation can occur when the nuclei are hot and the time scale of the nuclear motion gets closer to the electronic one. In Figure 6.2(a), we show such a situation, as obtained for a defective sample of bulk Si (63 atoms + one vacancy in a periodic box) simulated at rather high temperature by Pastore et al. [74]. One possible trick to avoid thermal exchanges with very fast nuclei is to increase their masses, for example replacing H with D. The cost is a renormalization of the dynamic quantities in terms of classical isotope effects.

In passing, it is important to remark that the loss of adiabaticity should not be "cured" by periodically quenching the wavefunction onto the BO surface. Such a procedure has no theoretical justification. The generated trajectory is discontinuous and does not represent any meaningful statistical-mechanics ensemble.

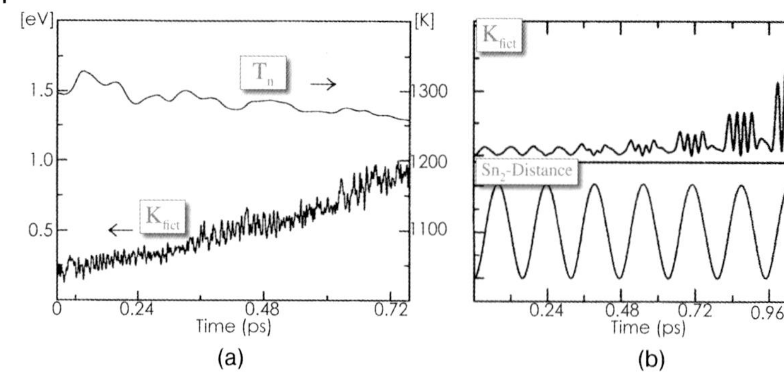


**Figure 6.2** (a) Steady variation of the nuclear temperature (Kelvin) and of the fictitious kinetic energy (eV) along the CP-MD of hot Si-crystal containing one vacant lattice site. The loss of adiabaticity is induced by the fast Si dynamics, whose time scale is not anymore separated from the electronic one. (b) Variations of the fictitious kinetic energy (eV) following $Sn_2$-dimer distance fluctuations (bohr). The loss in adiabaticity is induced by the periodically closure of the energy gap occurring below the dimer equilibrium distances. The graphs are taken from Ref. [74].

The adiabaticity can be easily controlled selecting the proper $\mu$ when the separation between the ground state and the first excited state is much larger than $KT$ for all the nuclear configurations and $\omega_e^{min} > 0$. This is not anymore the case for metallic systems or along certain reaction pathways, when surfaces crossing near the transition state are possible. Figure 6.2(b) displays the effect on the fictitious kinetic energy induced by the gap closure in bulk tin, as reported in Ref. [74]. The authors observe that the correlation between the phase of the ionic motion and the gain in the kinetic energy $K_{fict}$ is determined by the degeneracy of the lowest unoccupied level and the highest occupied one, occurring when the $Sn_2$ dimer distance is pushed below its equilibrium value ($\sim$5.0 au). The degeneracy is removed at larger distances, giving the characteristic periodic behavior. More details on the simulation can be found in the original paper. If this is the case, the electrons can be kept artificially cold through the coupling to a thermostat, for example using the Nosé–Hoover method [85, 86]. A target fictitious kinetic energy $K_{fict,0}$ and a thermostat mass have to be selected appropriately, so that the thermostat can absorb the excessive heat of the electronic subsystem. A too light thermostat would allow too large kinetic energy fluctuations and, as a consequence, the loss of the adiabaticity. If too heavy, instead, it might exert an excessive drag on the nuclear dynamics. In addition, also the nuclear subsystem needs to be thermostated in order to compensate the loss of heat induced by the coupling with the electronic dynamics. This technique has been used several times, in particular to allow the application of CP-MD to metallic systems [87]. Blöchl and Parrinello have shown that this technique, as applied to 64 atoms of molten aluminum [87], prevents the electrons from heating up, as shown in Figure 6.3(a), and, at the same time, is able to reproduce correctly the structural properties of the system, as demonstrated by the radial distribution function computed on the CP trajectory and reported in Figure 6.3.

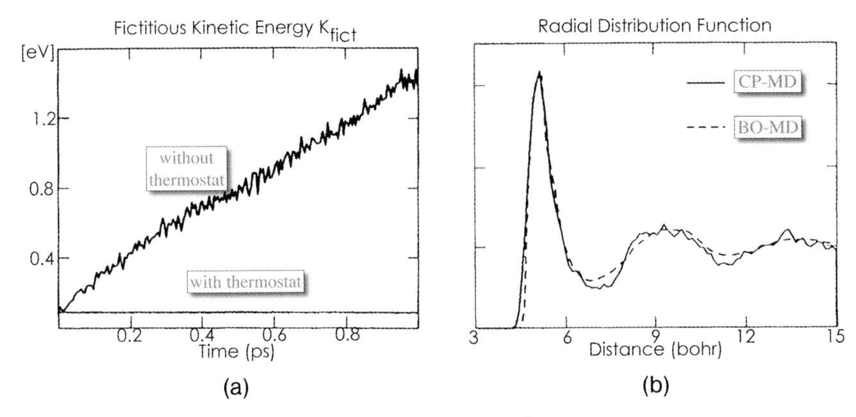


**Figure 6.3** (a) Behavior of the fictitious kinetic energy of 64 atoms of molten aluminum without and with the thermostat on the electronic dynamics, as computed by Blöchl and Parrinello [87]. (b) Aluminum radial distribution function obtained from the thermostated CP-MD run (solid line) compared to the result obtained by a BO-MD run.

As an alternative, the isokinetic ensemble method can be used by imposing a nonholonomic constraint to keep the fictitious kinetic energy at a fixed value [88]. To this purpose, an additional Lagrange multiplier $\alpha$ appears in the CP equations of motion, and the total force on the orthonormal orbitals reads

$$\mu|\ddot{\phi}_i\rangle = |f_i\rangle + \sum_j \Lambda_{ij}|\phi_j\rangle + \alpha|\dot{\phi}_i\rangle. \tag{6.33}$$

The orthogonality and the isokinetic conditions have to be satisfied simultaneously. However, the computational overhead is limited since the two constraints turn out to be uncoupled and the solution for $\alpha$ can be found analytically, retaining the numerical procedure for $\Lambda_{ij}$.

In spite of the fact that several techniques have been developed to extend the use of CP-MD to metallic systems, these solutions are not fully satisfactory from a theoretical point of view. For simulations on metals, the CP approach needs to be adapted, thereby loosing much of its simplicity. In general, we should recognize that the BO approach is better suited for small- or zero-gap systems, also because it allows the use of $k$-point sampling and fractional occupation numbers, and the charge sloshing problem can be handled more easily [47, 89, 90].

## 6.4 Error Estimate in CP-MD

Recently, there has been some concern in the literature about the impact of finite-$\mu$ effects on the results of the CP-MD simulations [91–97]. The validity of the CP approach has been questioned, claiming that the fictitious dynamics of the orbitals might influence in an uncontrollable and unknown way the evolution of the physical system. Similar fears are not justified when the method is applied in the appropriate manner, taking into serious consideration the issue of the adiabatic separation and selecting appropriate values for the fictitious mass and the integration time step. Under these conditions, the viability of the CP sampling has been confirmed by several numerical demonstrations [98–100].

As a matter of fact, already in 1998, it has been rigorously proven that the deviation of the CP trajectory from the exact BO surface is controlled by $\mu$ using a mathematical derivation of the absolute error [78]

$$\Delta_\mu = |\mathbf{R}^\mu(t) - \mathbf{R}^0(t)| + ||\phi^\mu; t\rangle - |\phi^0; t\rangle|. \tag{6.34}$$

The theorem demonstrated by Borenmann *et al.* says that if within a finite time interval $0 \leq t \leq t^{\max}$ a unique trajectory exists on the exact BO surface, $\{\mathbf{R}_I^0\}$, and if along the entire trajectory the energy gap $E_{\text{gap}}$ remains finite, it is always possible to generate a CP trajectory $\{\mathbf{R}_I^\mu\}$ which is arbitrarily close to $\{\mathbf{R}_I^0\}$. Starting exactly from the same initial conditions, the absolute error and the fictitious kinetic energy are bounded from above at every time

$$\Delta_\mu \leq C\mu^{1/2} \qquad K_{\text{e}} \leq C\mu, \tag{6.35}$$

where $C$ is a positive constant and $\mu$ is small enough to satisfy the adiabaticity condition. An important observation is that, given the definition of $\Delta_\mu$, not only the nuclear trajectory is close to the correct one, but also the orbitals are a good approximation of the fully converged ones.

As already extensively discussed, it is known that the CP forces deviate systematically from the BO forces and that this deviation is controlled by the fictitious mass $\mu$. In addition to that it has been observed that the finite mass $\mu$ exert an excessive inertia on the nuclear dynamics, and the nuclei are said to be "dressed nuclei." The resulting classical mass effect is very similar to typical isotope effects. Since the classical partition function depends only on the interaction potential, structural and thermodynamic properties are not affected by this rescaling. Mass-dependent dynamical quantities, instead, as for example the vibrational spectra, are affected by the $\mu$-renormalized nuclear masses. However, the systematic and well-controllable error can be easily corrected.

The finite-$\mu$ effects on dynamic observables have been investigated by a perturbative treatment of the CP single-particle wavefunctions with respect to the corresponding BO orbitals, thus writing the CP orbitals as $\phi_i(t) = \phi_i^0(t) + \delta\phi_i(t)$ [77,101]. By substituting this expression into the CP equations of motion, and retaining only the linear terms in the perturbation, the instantaneous CP forces turn out to be connected to the BO forces by

$$\mathbf{F}_I^{\text{CP}}(t) = \mathbf{F}_I^{\text{BO}}(t) + \sum_i \mu \left\{ \langle\ddot{\phi}_i| \frac{\partial|\phi_i^0\rangle}{\partial \mathbf{R}_I} + \frac{\partial\langle\phi_i^0|}{\partial \mathbf{R}_I}|\ddot{\phi}_i\rangle \right\} + \mathcal{O}(\delta\phi_i^2). \tag{6.36}$$

Thanks to the high-frequency fluctuations of the orbital dynamics, large part of the error in the instantaneous CP forces is averaged out on the time scale of the nuclear dynamics. Therefore, for a correct evaluation of the effective error, the above expression has to be averaged over a time interval that is intermediate between the characteristic electronic and nuclear time scales. It can be demonstrated that on this time scale the average acceleration $\overline{\ddot{\phi}_i(t)}$ vanishes independently from the value of the fictitious mass, whereas the remaining correction term is linear in $\mu$ and goes to zero only in the limit $\mu \rightarrow 0$.

$$\Delta F_{I\alpha} = 2\sum_i \mu \mathrm{Re}\Bigg\{\sum_{J,\beta} \ddot{\mathbf{R}}_{J,\beta} \frac{\partial < \phi_i^0|}{\partial \mathbf{R}_{I,\alpha}} \frac{\partial |\phi_i^0 >}{\partial \mathbf{R}_{J,\beta}} + \sum_{J,K,\beta,\gamma} \dot{\mathbf{R}}_{J,\beta}\dot{\mathbf{R}}_{K,\gamma} \frac{\partial < \phi_i^0|}{\partial \mathbf{R}_{I,\alpha}} \frac{\partial^2 |\phi_i^0 >}{\partial \mathbf{R}_{K,\gamma}\partial \mathbf{R}_{J,\beta}}\Bigg\} + \mathcal{O}\left(\delta\phi_i^2\right). \tag{6.37}$$

This correction varies on atomic time scales and therefore does not average out. However, its effect can be estimated by repeating the calculation with different masses, always reminding that smaller masses need smaller time steps ($\delta t \propto \mu^{1/2}$).

This partial averaging is equivalent to a sort of coarse graining of the CP equations. It shows that on the time scales relevant to the nuclear dynamics, the self-consistency problem is solved dynamically up to second order $\mathcal{O}(\delta\phi_i^2)$, whereas the forces on the nuclei feature an excess component linear in $\mu$. In order to have better control on the error, it is recommended to begin the dynamics with electrons and ions moving in consistent way, meaning that nuclear and orbitals dynamics should both start with zero velocities. If the initial nuclear velocity are different from zero, the electrons receive an initial kick, and the resulting high-frequency fluctuation causes a substantial increase of the error on the forces. This error may not affect too much the dynamics; however, the amplitudes are large and their influence is not easy to predict. Moreover, these oscillations survive a long time due to the adiabatic decoupling [77].

The largest contribution to the excess force on the nuclei is attributed to the mass renormalization. Actually, if the products $\dot{\mathbf{R}}_J\dot{\mathbf{R}}_K$ vanish and the tensor multiplying $\ddot{\mathbf{R}}$ is constant, the correction reduces to a rescaling of the atomic masses, which is known to leave thermodynamics intact. The inertia of the electrons, introduced with the finite fictitious mass, makes the nuclei heavier, such that the effective mass is $M_I + \Delta M_I^\mu$ and the correction term is proportional to the quantum kinetic energy of the orbitals located on the $I$th atom. In the extremely simplified case of a rigid ion (no distortion of the electronic cloud) moving in the field of all the other nuclei, the mass renormalization is estimated as a linear function in $\mu$

$$\Delta M_I^\mu = \frac{2}{3}\frac{m_e}{\hbar^2}\sum_j \mu\langle\varphi_{jI}| - \frac{\hbar^2}{2m_e}\nabla_I^2|\varphi_{jI}\rangle, \tag{6.38}$$

where $\varphi_{jI}$ are approximations to the atomic orbitals. As a consequence, systems where the electrons are strongly localized close to the nuclei will lead to more pronounced mass corrections.

The best procedure to estimate the finite $\mu$ effect on the dynamical quantities which are sensitive to the increased inertia of the nuclei is to repeat the same CP-MD simulation using different values for $\mu$ and derive the best guess of the values through the extrapolation to the $\mu \to 0$ limit. A computationally less demanding alternative is to run the simulation employing directly the renormalized nuclear masses, according to the estimate of $\Delta M_I^{\mu}$ given in (6.38) [77, 101, 102]. There are also quantities for which the renormalization can be evaluated *a posteriori*, as for example the vibrational spectra.

The characteristic finite-$\mu$ effect on power spectra is a general shift to lower frequencies, that can be observed also for small values of the mass, since the phonons distribution is sensitive to the instantaneous forces and to the reduced mass of the normal modes. A good approximation to the correct spectrum is obtained by rescaling *a posteriori* the frequencies computed over the CP trajectory through the proper mass correction, $\omega = \omega^{\mu}\sqrt{1 + \Delta M^{\mu}/M}$. While this is easily achieved in a homogeneous system, for heterogeneous systems the correction is not always trivial, since different mass rescaling factors must be applied to different atomic species, and a simple rigid shift of the spectrum is not anymore sufficient.

Tangney et al. [77] investigated the finite-$\mu$ effects for two very different systems: bulk Si and bulk MgO under external load. While the interactions in bulk Si have covalent character and the electronic valence states are rather delocalized, pressurized MgO is highly ionic where the electrons remain localized at the atomic centers. Moreover, since the ionic charge transfer of the two 3s electrons from Mg to O is almost complete, to a first approximation the mass renormalization effects affect only the O nuclei. This makes MgO an ideal system to study with the rigid-ion model, since only O masses will be rescaled. On the other hand, the low-quantum kinetic energy of the delocalized states makes silicon one of the most favorable cases for the CP approach, but the estimate of the error in terms of the rigid ion model turn out not to be very accurate. In Figure 6.4(a), we report the spectra calculated by Tangney et al. for Si and in Figure 6.4(b) for MgO using different values for $\mu$ (for details on the simulations see the original paper). As expected, for Silicon the mass effects turn out to be very small, while the correction via the rigid ion model may overestimate by a factor two the rescaling of the mass. Much larger differences are instead observed for MgO, when the fictitious mass varies from 100 to 400 au. However, by rescaling *a priori* the mass of the O nuclei, according to the estimate of the rigid ion model, the CP dynamics is corrected and the resulting spectra are greatly improved, showing only small differences in the position of the peaks. An accuracy of 4% in the position of the peaks can be obtained with $\mu = 50$ a.u., which would require a integration time step of 0.067 fs.

The renormalization of nuclei masses has also an effect on the calculation of the physical temperature. The measured $T$ is always too low, since the real temperature at which the system equilibrates in a microcanonical ensemble is

$$k_B T = \frac{1}{3N} \sum_{I\alpha} (M_I + \Delta M_I) \left\langle V_{I\alpha}^2 \right\rangle, \tag{6.39}$$

where the additional term is due to the inertia caused by the rigid ion dragging the electronic orbitals, as if they were moving in a viscous media. This shift in temperature may affect even structural properties that carry a temperature dependence, as it is the case for the radial distribution function. The actual effect can be quantified by calculating the same observable using different values of the fictitious mass. For example, Kuo et al. [103] could provide a rough estimate of the temperature shift in liquid water. From the analysis of the height of the first peak in the O–O radial distribution function, which is known to depend linearly on temperature, they found a temperature shift between 2 and 4 degrees for a change in the fictitious mass of 100 a.u.

Theoretical and numerical proofs have demonstrated by now that, with the appropriate choice of $\mu$, consistent structural properties can be obtained via CP-MD. The effect on the dynamical properties, instead, has been much less investigated. Even if, since long, it has been recognized that the vibrational frequencies are affected by the fictitious dynamics and several correction procedures have been proposed, the effects on other aspects of the dynamical evolution, as the transport properties, have been addressed only very recently. In the work of Kuo et al., the effect of $\mu$ on the dynamical properties has been examined by comparing five MD simulations of liquid water kept at the experimental density, and at 423 K. Three of them are CP-M runs performed with different fictitious masses (100, 200, and 300 au). The other two are BO-MD runs, both using $10^{-7}$ as self-consistent convergence criterion for the energy, but starting from different initial configurations. After equilibration of the system over 5 ps, the structural and dynamical properties have been computed over a trajectory of 30 ps. The structural properties have been analyzed through the comparison of the radial distribution functions. A part for a slight spread of the peak height of 0.23, which can be attributed to the temperature shift due to the mass renormalization, the five curves are in very good agreement. To compare transport properties, the self-diffusion coefficient can be computed from the Einstein relation $2tD = \frac{1}{3}\langle|\mathbf{R}(t) - \mathbf{R}(0)|\rangle$, where the slope is found from

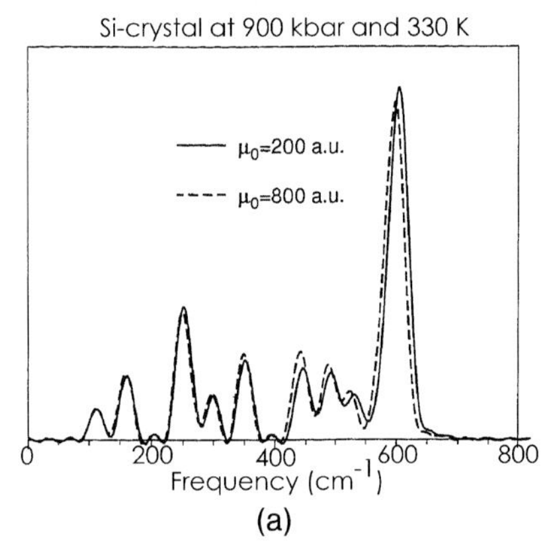


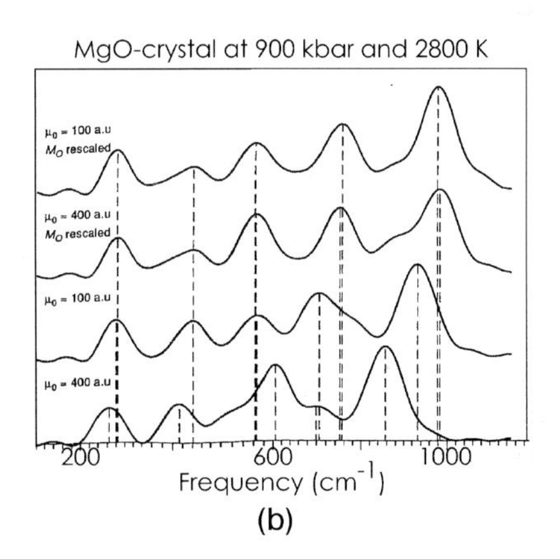


**Figure 6.4** (a) Power spectrum of Si at 900 kbar, as obtained with two different values of $\mu$. The CP forces error is estimated to be 0.94%. (b) Power spectrum of pressurized MgO obtained with $\mu = 400$ a.u and $\mu = 100$ a.u., first and second curve starting from the bottom, and with the same two values but by renormalizing the oxygen mass according to the rigid ion model, third and fourth curve. The graphs are taken from Ref. [77].

linear least-squares analysis. All the five simulations show large fluctuations and hence large uncertainties in the self-diffusion coefficient, with values ranging from 8.3 to 12.8 $\times 10^{-5}$ cm$^2$ s$^{-1}$, all of them at leasr a factor 3 lower than the experimental estimate. The DFT description used for these calculations provides a too structured model for the liquid, with sluggish diffusive properties. Anyway, on the explored time scale, there does not appear to be a statistically significant difference in transport properties between BO and CP, which leads to the conclusion that free energies and long-time processes are less affected by fictitious mass.

## 6.5 Conclusions

The experience accumulated from the large number of applications and technical investigations helped substantially in the identification of the validity range and the limits of the Car–Parrinello approach. It has been extensively demonstrated that, despite the forces always differ from the BO forces, properties that depend on the free energy or are governed by long-time processes involving free energy barriers (RDF, diffusive transport) should be significantly less affected, providing a reasonable choice of $\mu$. Indeed, the largest contribution to the instantaneous deviation of the CP electronic structure from the BO surface is given by bounded fluctuations, with small amplitudes and high frequencies, which average out on the time scale of the nuclear dynamics. The residual error of the averaged CP force is induced by the intrinsic fictitious dynamics of the orbitals and most of it can be attributed to the inertia that the fictitious electronic mass $\mu$ exerts on the nuclei, known as mass renormalization effect. Experience suggests that quantities extracted from CP simulations should be always checked against their possible dependence on $\mu$. This means that it is a good practice, before starting production runs, to assess the typical deviation of the CP wavefunction from the exact BO surface for the specific problem under investigation. This can be achieved by computing the BO forces for some configurations or performing preliminary simulations with different $\mu$ to check the scaling behavior. If errors are large, it is probably possible to correct the CP forces substantially by choosing better parameters and/or rescaling the masses. At any rate, for well-behaved systems, the error can be reduced arbitrarily by decreasing $\mu$. The drawback is the simultaneous reduction of the maximum time step that guarantees the correct integration of the equations of motion. Additional complications may arise if the dynamics lead to fundamental changes of the electronic structure: charge transfer, substantial rearrangements like during chemical reactions. Then the simple rescaling will not be sufficient and some time BO fails as well.

The selection of the optimal methodology to perform *ab initio* molecular dynamics simulations, among Ehrenfest, Born–Oppenheimer, and Car–Parrinello, remains a matter that has to be faced case by case, on the basis of the specific problem that one wants to investigate. Several works have been dedicated to the comparison among the different techniques from the point of view of accuracy, stability, and computational efficiency [1, 35, 37, 103–105]. However, such comparison is often delicate and cannot be easily generalized.

The first term of comparison is the maximum integration time step that still maintains the stability of the propagation of the dynamics for a sufficiently long-time window. Obviously, for a given simulation time, the longer is $\delta t$ the fewer are the necessary force evaluations, with significant savings in the computational time. For both E-MD and CP-MD, the choice of $\delta t$ has to cope with the fact that the faster electronic degrees of freedom are propagated explicitly, whereas in BO-MD the only relevant time scale is the nuclear one $\tau_n$, as determined by the highest phonon frequencies that need to be resolved. The fastest electronic motion is normally determined by the largest kinetic energy of the basis set functions, i.e., $\omega^{\mathrm{E}} \sim E_{\mathrm{cut}}$ for a simple PW basis set. In this respect, the advantage of CP-MD over E-MD is that the electronic motion is regulated by the fictitious electronic mass. The highest frequency is then $\omega^{\mathrm{CP}} \sim (E_{\mathrm{cut}}/\mu)^{1/2}$ and it increases more slowly with the basis set size. We could roughly estimate the relationship between the typical time scales involved in the three different representations as $\tau^{\mathrm{E}} \approx \tau^{\mathrm{CP}}/10$ and $\tau^{\mathrm{CP}} \approx \tau^{\mathrm{BO}}/10$. The time scale advantage of BO simulations is surely large when the nuclear dynamics is slow [89]. However, it has to be taken into account that by choosing too large time steps, certain information might be lost. For example a time step of about 3 fs allows the resolution of frequencies below 500 $\mathrm{cm}^{-1}$. Irrespective of the specific approach, multiple time step integration techniques can be applied. These tools have been ideated in the attempt to ameliorate the efficiency in simultaneously propagating dynamics that evolve on very different time scales [37, 106–108].

The comparison of the effective computational costs is complicated by the fact that the BO formalism requires the iterative optimization of the electronic structure at each MD step. The computer time needed for this task depends on various factors, like the quality of the minimization scheme, the selected convergence criterion, the system size (convergence might become more difficult), and the use of extrapolation procedures, which might greatly improve the initial trial function and thus reduce the number of iterations. Moreover, it should be considered that for a fixed simulation time, the computational time needed by CP-MD decreases linearly by increasing the time step. This is not the case in BO-MD, where larger time steps also imply a worse initial guess for the wavefunction optimization, and, as a consequence, more iterations. On the other hand, the choice of the time step, of the convergence criterion in BO-MD, and of the fictitious electronic mass in CP-MD, may affect significantly the conservation of the constant of motion. Therefore, performance and stability of the calculation are strictly related and both have to be considered in the evaluation of the computational setup. The total energy conservation law is a measure of the numerical quality of the simulation, and in most of the cases it is not a good idea to sacrifice the accuracy to gain something in cpu time.

These aspects have been carefully analyzed via several specific tests on toy systems, which are reported in previous review works on *ab initio* MD [1, 98, 109]. Test calculations on the well behaved case of bulk silicon show that CP-MD achieves perfect adiabaticity with $\mu = 400$ a.u. and $\delta t = 10$ a.u., conserving the energy to about $3 \times 10^{-7}$ a.u. $ps^{-1}$. Using the same time step and a convergence criterion for the maximum component of the orbital gradient equal to $10^{-6}$ a.u., BO-MD gives an energy conservation of about $1 \times 10^{-6}$ a.u. $ps^{-1}$, but requiring almost ten times more computational time on the same hardware (calculations performed with the CPMD program package [79]). An increase of $\delta t$ to 100 a.u. accelerates the simulation by a factor seven, at the expense of loosing in conservation of energy, $6 \times 10^{-6}$ a.u. $ps^{-1}$. Looser convergence criteria, like $10^{-5}$ a.u. or $10^{-4}$ a.u. speed up even more the calculation, but degrade consistently the conservation of energy, which decreases to $1 \times 10^{-5}$ a.u. $ps^{-1}$ and $1 \times 10^{-3}$ a.u. $ps^{-1}$, respectively.

For a more challenging system, like liquid water, selecting parameters that reasonably conserve the adiabaticity, the energy conservation provided by CP-MD is still better, but the difference with respect to BO-MD is reduced. Simulations with fictitious masses of 300, 400, and 700 a.u., and time steps of 3, 4, and 5 a.u., respectively, have been compared to BO runs carried out with $\delta t = 20$ a.u., using different convergence criteria ($10^{4}$–$10^{-7}$). While for the looser convergence criteria the BO conservation of energy is still several order of magnitude worse, the energy drift of the best converged BO simulation is analogous to the drift obtained by CP-MD with the largest time step, 0.033 (BO) Kelvin per degree of freedom against 0.019 (CP).

An important remark at the conclusion of this discussion is that the origin of the energy drift is also different for the two methods. In BO-MD, it is the numerical error in the forces, due to the incomplete minimization of the wavefunction, i.e., the non-self-consistency error. It can be systematically reduced by selecting tighter convergence criteria. In CP-MD simulations, instead, the drift is caused by the energy flow between electronic and nuclear subsystems, i.e., the deviation from adiabaticity. This error can also be systematically reduced by decreasing the fictitious electronic mass.

Going back to the question "which is the optimal method?", it really depends on the specific applications. For relatively well behaved systems, when plane waves are a suitable basis set, and in the presence of finite electronic gaps, the Car–Parrinello method appears to be superior. It is stable with respect to small deviations from the BO surface and it is more accurate, since the forces are exact. On the other hand, thanks to the development of very efficient and robust Hamiltonian minimization techniques, and thanks to the use of highly efficient extrapolation schemes, the speed of the two methods has become comparable [57, 58, 110]. Therefore, when the Car–Parrinello formalism needs to be readapted, as for example together with nonorthogonal basis set functions or in applications to metallic systems, then the Born–Oppenheimer molecular dynamics is often a better choice.

## Acknowledgments

I wish to thank Professor Jürg Hutter, who shared with me part of his incredible knowledge and experience in the field of *ab initio* molecular dynamics and always granted me his scientific support.

## References


1 D. Marx and J. Hutter, in *Modern Methods and Algorithms of Quantum Chemistry*, edited by J. Grotendorst (John von Neumann Institute for Computing, Forschungszentrum Jülich, Jülich, **2000**), pp. 301–449, see `http://www.theochem.rub.de/go/cprev.html`, URL http://www.theochem.rub.de/go/cprev.html.

2 M. E. Tuckerman, *Journal of Physics: Condensed Matter* **14**, R1297 (**2002**), URL http://dx.doi.org/10.1088/0953-8984/14/50/202.

3 J. S. Tse, *Annual Review Physical Chemistry* **53**, 249 (**2002**).

4 R. Iftimie, P. Minary, and M. E. Tuckerman, *Proceedings of the National Academy of Sciences of the United States of America* **102**, 6654 (**2005**).

5 J. Hutter and A. Curioni, *Chem Phys Chem* **6**, 1788 (**2005**), http://dx.doi.org/10.1002/cphc.200500059.

6 S. Raugei and P. Carloni, *Psi-K : Scientific Highlights of the month* (**2005**).

7 M. D. Peraro, S. Raugei, P. Ruggerone, F. L. Gervasio, and P. Carloni, *Current Opinion in Structural Biology* **17**, 149 (**2007**).

8 M. P. Allen and D. J. Tildesley, *Computer Simulation of Liquids* (Clarendon Press, Oxford, **1987** and **1990**).

9 R. Haberlandt, S. Fritzsche, G. Peinel, and K. Heinzinger, *Molekulardynamik– Grundlagen und Anwendungen* (Vieweg, Braunschweig, **1995**).

10 D. C. Rapaport, *The Art of Molecular Dynamics Simulation* (Cambridge University Press, Cambridge, **2001** and **2005**).

11 D. Frenkel and B. Smit, *Understanding Molecular Simulation–From Algorithms to Applications* (Academic Press, San Diego, **1996** and **2002**).

12 B. Martino, M. Celino, and V. Rosato, *Computional Physics Communications* **120**, 255 (**1999**).

13 Z. Liu, L. E. Carter, and E. A. Carter, *Journal of Physical Chemistry* **99**, 4355 (**1995**).

14 G. Lippert, J. Hutter, and M. Parrinello, *Molecular Physics* **92**, 477 (**1997**), URL http://dx.doi.org/10.1080/002689797170220.

15 G. Lippert, J. Hutter, and M. Parrinello, *Theoretical Chemistry Accounts* **103**, 124 (**1999**), URL http://dx.doi.org/10.1007/s002140050523.

16 P. Hohenberg and W. Kohn, *Physical Review* **136**, B864 (**1964**), URL http://dx.doi.org/10.1103/PhysRev.136.B864.

17 W. Kohn and L. J. Sham, *Physical Review* **140**, A1133 (**1965**), URL http://dx.doi.org/10.1103/PhysRev.140.A1133.

18 R. Car and M. Parrinello, *Physical Review Letters* **55**, 2471 (**1985**), see also Refs. [111], [112], URL http://dx.doi.org/10.1103/PhysRevLett.

19 M. Born and R. Oppenheimer, *Annalen der Physik (IV. Folge)* **84**, 457 (**1927**).

20 W. Kolos, *Advances in Quantum Chemistry* **5**, 99 (**1970**).

21 W. Kutzelnigg, *Molecular Physics* **90**, 909 (**1997**).

22 M. Böckmann, N. L. Doltsinis, and D. Marx, *unpublished* (**2005**).

23 M. Moret, E. Tapavicza, L. Guidon, U. Rohrig, M. Sulpizi, I. Tavernelli, and U. Röthlisberger, *Chimia* **59**, 493 (**2005**).

24 E. Tapavicza, I. Tavernelli, and U. Röthlisberger, *Physical Review Letters* **98**, 023001 (**2007**).

25 H. Goldstein, C. P. Poole, and J. L. Safko, *Classical Mechanics* (Addison-Wesley, San Fransisco, **2002**), 3rd ed., for errors see `http://astro.physics.sc.edu/goldstein/`.

26 F. Scheck, *Mechanics: From Newton's Laws to Deterministic Chaos* (Springer, Berlin, **2005**), 4th ed.

27 P. A. M. Dirac, *The Principles of Quantum Mechanics* (Oxford University Press, Oxford, **1947**), 3rd ed.

28 A. Messiah, *Quantum Mechanics* (North-Holland, Amsterdam, **1964**).

29 J. J. Sakurai, *Modern Quantum Mechanics* (Addison-Wesley, Redwood City, **1985**).

30 P. Ehrenfest, *Zeitschrift für Physik* **45**, 455 (**1927**).
31 J. B. Delos, W. R. Thorson, and S. K. Knudson, *Physical Review A* **6**, 709 (**1972**).
32 J. C. Tully, in *Modern Theoretical Chemistry: Dynamics of Molecular Collisions (Part B)*, edited by W. H. Miller (Plenum Press, New York, **1976**), p. 217.
33 H.-D. Meyer and W. H. Miller, *Journal of Chemical Physics* **70**, 3214 (**1979**).
34 J. B. Delos, *Reviews of Modern Physics* **53**, 287 (**1981**).
35 A. Selloni, P. Carnevali, R. Car, and M. Parrinello, *Physical Review Letters* **59**, 823 (**1987**).
36 R. N. Barnett, U. Landman, and A. Nitzan, *Journal of Chemical Physics* **89**, 2242 (**1988**).
37 E. S. Fois, A. Selloni, M. Parrinello, and R. Car, *Journal of Physical Chemistry* **92**, 3268 (**1988**).
38 J. Theilhaber, *Physical Review B* **46**, 12990 (**1992**), URL http://dx.doi.org/10.1103/PhysRevB.46.12990.
39 U. Saalmann and R. Schmidt, *Zeitschrift für Physik D* **38**, 153 (**1996**).
40 K. Yabana and G. F. Bertsch, *Physical Review B* **54**, 4484 (**1996**).
41 K. Yabana and G. F. Bertsch, *International Journal of Quantum Chemistry* **75**, 55 (**1999**).
42 W. Stier and O. V. Prezhdo, *Journal of Physical Chemistry B* **106**, 8047 (**2002**).
43 W. R. Duncan, W. M. Stier, and O. V. Prezhdo, *Journal of the American Chemical Society* **127**, 7941 (**2005**).
44 W. R. Duncan and O. V. Prezhdo, *Journal of Physical Chemistry B* **109**, 17998 (**2005**).
45 C. Leforestier, *Journal of Chemical Physics* **68**, 4406 (**1978**).
46 M. C. Payne, J. D. Joannopoulos, D. C. Allan, M. P. Teter, and D. Vanderbilt, *Physical Review Letters* **56**, 2656 (**1986**), URL http://dx.doi.org/10.1103/PhysRevLett.56.2656.
47 M. C. Payne, M. P. Teter, D. C. Allan, T. A. Arias, and J. D. Joannopoulos, *Reviews of Modern Physics* **64**, 1045 (**1992**), URL http://dx.doi.org/10.1103/RevModPhys.64.1045.
48 R. N. Barnett and U. Landman, *Physical Review B* **48**, 2081 (**1993**).
49 G. Kresse and J. Furthmüller, *Physical Review B* **54**, 11169 (**1996a**), URL http://dx.doi.org/10.1103/PhysRevB.54.11169.
50 G. Kresse and J. Furthmüller, *Computational Materials Science* **6**, 15 (**1996b**), URL http://dx.doi.org/10.1016/0927-0256(96)00008-0.
51 M. D. Segall, P. J. D. Lindan, M. J. Probert, C. J. Pickard, P. Hasnip, S. J. Clark, and M. C. Payne, *Journal of Physics: Condensed Matter* **14**, 2717 (**2002**).
52 G. Y. Sun, J. Kurti, P. Rajczy, M. Kertesz, J. Hafner, and G. Kresse, *Journal of Molecular Structure–THEOCHEM* **624**, 37 (**2003**).
53 CASTEP, CASTEP, Ref. [47], [51]; See http://www.tcm.phy.cam.ac.uk/castep/.
54 VASP, VASP: *Vienna ab-initio simulation package*, refs. [49], [50]; see http://cms.mpi.univie.ac.at/vasp/, URL http://cms.mpi.univie.ac.at/vasp/.
55 B. Hartke and E. A. Carter, *Chemical Physics Letters* **189**, 358 (**1992**).
56 CP2k, CP2k: *A General Program to Perform Molecular Dynamics Simulations*, distributed under the terms of the GNU General Public Licence; see http://cp2k.berlios.de/.
57 J. VandeVondele and J. Hutter, *Journal of Chemical Physics* **118**, 4365 (**2003**), URL http://dx.doi.org/10.1063/1.1543154.
58 J. VandeVondele, M. Krack, F. Mohamed, M. Parrinello, T. Chassaing, and J. Hutter, *Computer Physics Communications* **167**, 103 (**2005a**), URL http://dx.doi.org/10.1016/j.cpc.2004.12.014.
59 SIESTA, SIESTA: *Spanish Initiative for Electronic Simulations with Thousands of Atoms*, Ref. [113]; see http://www.uam.es/siesta/.
60 ONETEP, ONETEP *(Order-N Electronic Total Energy Package): Linear-Scaling Density-Functional Theory with Plane Waves*, Ref. [114]; see http://www.tcm.phy.cam.ac.uk/onetep/.

61 V. Weber and J. Hutter, *Journal of Chemical Physics* **128**, 064107 (**2008**).

62 T. A. Arias, M. C. Payne, and J. D. Joannopoulos, *Physical Review Letters* **69**, 1077 (**1992**), URL http://dx.doi.org/10.1103/PhysRevLett.69.1077.

63 H. Hellmann, *Zeitschrift für Physik* **85**, 180 (**1933**).

64 R. P. Feynman, *Physical Review* **56**, 340 (**1939**).

65 I. N. Levine, *Quantum Chemistry* (Allyn and Bacon, Boston, **1983**).

66 P. Bendt and A. Zunger, *Physical Review Letters* **50**, 1684 (**1983**).

67 G. P. Srivastava and D. Weaire, *Advances in Physics* **36**, 463 (**1987**).

68 D. Vanderbilt, *Physical Review B* **32**, 8412 (**1985**), URL http://dx.doi.org/10.1103/PhysRevB.32.8412.

69 K. Laasonen, A. Pasquarello, R. Car, C. Lee, and D. Vanderbilt, *Physical Review B* **47**, 10142 (**1993**), URL http://dx.doi.org/10.1103/PhysRevB.47.10142.

70 J. Hutter, M. E. Tuckerman, and M. Parrinello, *Journal of Chemical Physics* **102**, 859 (**1995**).

71 R. Kosloff, *Journal of Physical Chemistry* **92**, 2087 (**1988**).

72 C. Leforestier, R. H. Bisseling, C. Cerjan, M. D. Feit, R. Friesner, A. Guldberg, A. Hammerich, G. Jolicard, W. Karrlein, H.-D. Meyer, N. Lipkin, O. Roncero, and R. Kosloff, *Journal of Computational Physics* **94**, 59 (**1991**).

73 R. Kosloff, *Annual Review of Physical Chemistry* **45**, 145 (**1994**).

74 G. Pastore, E. Smargiassi, and F. Buda, *Physical Review A* **44**, 6334 (**1991**).

75 R. Car, M. Parrinello, and M. Payne, *Journal of Physics: Condensed Matter* **3**, 9539 (**1991**), comment to Ref. [115].

76 G. Pastore, in *Monte Carlo and Molecular Dynamics of Condensed Matter Systems*, edited by K. Binder and G. Ciccotti (Italian Physical Society SIF, Bologna, **1996**), chap. 24, pp. 635–647.

77 P. Tangney and S. Scandolo, *Journal of Chemical Physics* **116**, 14 (**2002**).

78 F. A. Bornemann and C. Schütte, *Numerische Mathematik* **78**, 359 (**1998**).

79 J. Hutter, A. Alavi, T. Deutsch, M. Bernasconi, S. Goedecker, D. Marx, M. Tuckerman, and M. Parrinello, An *Ab Initio* Electronic Structure and Molecular Dynamics Program; IBM Zurich Research Laboratory (1990–2007) and Max-Planck-Institut für Festkörperforschung Stuttgart (**1997–2001**); see http://www.cpmd.org/.

80 S. Nosé, *Journal of Chemical Physics* **81**, 511 (**1984a**).

81 M. Parrinello and A. Rahman, *Physical Review Letters* **45**, 1196 (**1980**), URL http://dx.doi.org/10.1103/PhysRevLett.45.1196.

82 F. Tassone, F. Mauri, and R. Car, *Physical Review B* **50**, 10561 (**1994**), URL http://dx.doi.org/10.1103/PhysRevB.50.10561.

83 L. Verlet, *Physical Review* **159**, 98 (**1967**), URL http://dx.doi.org/10.1103/PhysRev.159.98.

84 D. Vanderbilt, *Physical Review B* **41**, 7892 (**1990**), URL http://dx.doi.org/10.1103/PhysRevB.41.7892.

85 S. Nosé, *Molecular Physics* **52**, 255 (**1984b**).

86 W. G. Hoover, *Physical Review A* **31**, 1695 (**1985**).

87 P. E. Blöchl and M. Parrinello, *Physical Review B* **45**, 9413 (**1992**).

88 P. Minary, G. J. Martyna, and M. E. Tuckerman, *Journal of Chemical Physics* **118**, 2527 (**2003**).

89 G. Kresse and J. Hafner, *Physical Review B* **47**, 558 (**1993**), URL http://dx.doi.org/10.1103/PhysRevB.47.558.

90 G. Kresse, *Journal of Non-Crystalline Solids* **312**, 52 (**2002**).

91 D. Asthagiri, L. R. Pratt, and J. D. Kress, *Physical Review E* **68**, 041505 (**2003**).

92 D. Asthagiri, L. R. Pratt, J. D. Kress, and M. A. Gomez, *Proceedings of the National Academy of Sciences of the United States of America* **101**, 7229 (**2004**).

93 J. C. Grossman, E. Schwegler, E. W. Draeger, F. Gygi, and G. Galli, *Journal of Chemical Physics* **120**, 300 (**2004**).

94 E. Schwegler, J. C. Grossman, F. Gygi, and G. Galli, *Journal of Chemical Physics* **121**, 5400 (**2004**), URL http://dx.doi.org/10.1063/1.1782074.

95 H. Lapid, N. Agmon, M. K. Petersen, and G. A. Voth, *Journal of Chemical Physics* **122**, 014506 (**2005**).

96 S. Izvekov and G. A. Voth, *Journal of Chemical Physics* **123**, 044505 (**2005**), erratum: $Izvekov_2006_a$.

97 P. Tangney, *Journal of Chemical Physics* **124**, 044111 (**2006**).

98 I.-F. W. Kuo, C. J. Mundy, M. J. McGrath, J. I. Siepmann, J. VandeVondele, M. Sprik, J. Hutter, B. Chen, M. L. Klein, F. Mohamed, M. Krack, and M. Parrinello, *Journal of Physical Chemistry B* **108**, 12990 (**2004**), URL http://dx.doi.org/10.1021/jp047788i.

99 J. VandeVondele, F. Mohamed, M. Krack, J. Hutter, M. Sprik, and M. Parrinello, *Journal of Chemical Physics* **122**, 014515 (**2005b**), URL http://dx.doi.org/10.1063/1.1828433.

100 Y. A. Mantz, B. Chen, and G. J. Martyna, *Chemical Physics Letters* **405**, 294 (**2005**).

101 P. E. Blöchl, *Physical Review B* **65**, 104303 (**2002**).

102 P. E. Blöchl, *Physical Review B* **50**, 17953 (**1994**).

103 I.-F. W. Kuo, C. J. Mundy, M. J. McGrath, and J. I. Siepmann, *Journal of Chemical Theory and Computation* **2**, 1274 (**2006**).

104 S. S. Iyengar, H. B. Schlegel, and G. A. Voth, *Journal of Physical Chemistry A* **107**, 7269 (**2003**).

105 J. M. Herbert and M. Head-Gordon, *Journal of Chemical Physics* **121**, 11542 (**2004**).

106 M. Tuckerman, B. J. Berne, and G. J. Martyna, *Journal of Chemical Physics* **97**, 1990 (**1992**).

107 M. E. Tuckerman and M. Parrinello, *Journal of Chemical Physics* **101**, 1316 (**1994**), URL http://dx.doi.org/10.1063/1.467824.

108 M. Sprik, in *Monte Carlo and Molecular Dynamics of Condensed Matter Systems*, edited by K. Binder and G. Ciccotti (Italian Physical Society SIF, Bologna, **1996**), chap. 2, pp. 43–88.

109 M. E. Tuckerman, D. Marx, and M. Parrinello, *Nature* **417**, 925 (**2002**).

110 T. D. Kühne, M. Krack, F. R. Mohamed, and M. Parrinello, *Physical Review Letters* **98**, 066401 (**2007**).

111 J. Riordon, *APS News* **12**, 3 (**2003**).

112 N. D. Mermin, *Physics Today* **41**(4), 9 (**1988**).

113 J. M. Soler, E. Artacho, J. D. Gale, A. Garcia, J. Junquera, P. Ordejón, and D. Sanchez-Portal, *Journal of Physics: Condensed Matter* **14**, 2745 (**2002**).

114 P. D. Haynes, C.-K. Skylaris, A. A. Mostofi, and M. C. Payne, *physica status solidi (b)* **243**, 2489 (**2006**).

115 M. C. Payne, *Journal of Physics: Condensed Matter* **1**, 2199 (**1989**), comment: Ref. [75].

# Part II
# Force Fields, Classical Dynamics and Statistical Methods

# 7
# Molecular Simulation Techniques Using Classical Force Fields

*Thijs J.H. Vlugt, Kourosh Malek, and Berend Smit*

## 7.1
## Introduction

In this chapter, we present an overview of various simulation techniques used to study systems of particles that interact using classical force fields. In particular, we focus on the molecular dynamics (MD) technique (Section 7.2), simulation techniques to study rare events (Section 7.3), and the Monte Carlo (MC) technique (Section 7.4). For a more detailed review of these topics, we refer the reader to Refs. [1–8].

## 7.2
## Molecular Dynamics

### 7.2.1
### Introduction

In classical MD simulations, the system is treated as a set of $N$ interacting atoms (or molecules). The atoms are represented by spherical nuclei that attract and repel each other depending on the distance. Their electronic structure is not considered explicitly. After assigning point charges to each particle, the forces acting on the particles are usually obtained from a combination of bonded and nonbonded interactions. The motions of the atoms are calculated using the laws of classical mechanics. Before starting a simulation, a model system is built consisting all chemical components interacting in a simulation box. Just like any real experiment, this system needs to be carefully prepared. It should be a realistic representation of the system that is to be studied. The result of an MD simulation is a trajectory of the positions and velocities of all $N$ atoms in the system. If simulated with an appropriate time step and for a sufficiently long time, thermodynamic properties, time-dependent correlation functions, and transport properties can be reliably calculated. The required length of the trajectory depends on the length scale

*Computational Methods in Catalysis and Materials Science.* Edited by Rutger A. van Santen and Philippe Sautet

ISBN: 978-3-527-32032-5

of the system under study and the time scale needed for calculation of physical parameters. Nowadays, the typical trajectory time in atomistic MD simulations varies between few nanoseconds (ns) up to hundreds of nanoseconds.

The trajectory (positions and velocities as a function of time) is obtained from solving a system of second-order differential equations that follow from Newton's second law, and can be written as

$$m_i \frac{d^2\mathbf{r}_i}{dt^2} = \sum_j \mathbf{F}_{ij} + \sum \mathbf{F}_{\text{ex,i}} \quad i = 1, 2, \ldots, N. \tag{7.1}$$

Here $i$ denotes the current particle, and $m_i$ and $\mathbf{r}_i$ the mass and the position of this particle, respectively. The forces $\mathbf{F}_{ij}$ represent two-body interactions between atom $i$ and atom $j$. The forces $\mathbf{F}_{\text{ex,i}}$ due to external fields, for example, electric fields, are added as extra terms. After determining these forces as functions of atomic and molecular degrees of freedom, the equations of motion can be integrated. The combination of all forces in the model, including van der Waals interactions and electrostatic interactions, is called the force field. The choice of the force field is the key to accurate results that appropriately reproduce the true physical phenomena in the system. Popular force fields are, for example, TraPPE [9, 10], CHARMM [11], and AMBER [12]. Some force fields use a so-called united atom approach, in which carbon atoms with attached hydrogen atoms are treated as a single interaction site.

The forces acting on the atoms are derived from the gradients of the potential energy function,

$$\mathbf{F}_i = -\nabla U(\mathbf{r}^N), \tag{7.2}$$

where $U$ is the total potential energy of the system and the vector $\mathbf{r}^N$ represents the positions of all $N$ particles in the system. The force fields can be split up into two contributions: *nonbonded interactions* between all nuclei and *bonded interactions* between nuclei that are part of the same molecule. The nonbonded interactions consist of the following terms: electrostatic interactions, van der Waals interactions, and polarization effects. Polarization effects are the result of varying electron densities; they cannot be described explicitly using force-field methods that ignore electron dynamics. It is common practice to include them implicitly in the van der Waals interactions as an overall effect. This leaves two terms for the nonbonded interactions. The first type of nonbonded interactions corresponds to dispersion or van der Waals forces. These are the interactions between atoms that arise from (quantum) fluctuations of the electronic charge densities. Van der Waals interactions are usually modeled using the Lennard–Jones potential,

$$u(r_{ij}) = 4\epsilon \left[ \left( \frac{\sigma}{r_{ij}} \right)^{12} - \left( \frac{\sigma}{r_{ij}} \right)^{6} \right]. \tag{7.3}$$

In this equation, $\epsilon$ represents the depth of the potential at the minimum ($r_{\min} = 2^{1/6}\sigma$) and $r = \sigma$ is the point at which $u(r) = 0$ (see also Figure 7.1). The term $4\epsilon\sigma^{12}/r^{12}$ is due to the strong repulsion of atoms at short distances. The number of interactions that needs to be computed can be reduced by neglecting all interactions

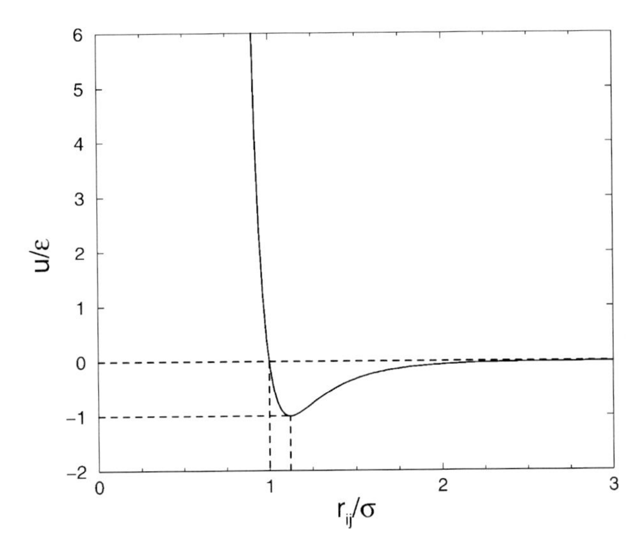


**Figure 7.1** The Lennard–Jones potential. Note that the axis labels $u/\epsilon$ and $r_{ij}/\sigma$ are dimensionless.

beyond a certain cut-off radius $r_{\text{cut}}$, which should not be too small (in practice, $r_{\text{cut}} = 2.5\sigma$ is often used). This results in the so-called truncated and shifted Lennard–Jones potential,

$$u\left(r_{ij}\right) = \begin{cases} 4\epsilon\left[\left(\frac{\sigma}{r_{ij}}\right)^{12} - \left(\frac{\sigma}{r_{ij}}\right)^{6}\right] - u_{\text{cut}} & r \leq r_{\text{cut}} \\ 0 & r > r_{\text{cut}} \end{cases}, \tag{7.4}$$

where

$$u_{\text{cut}} = 4\epsilon\left[\left(\frac{\sigma}{r_{\text{cut}}}\right)^{12} - \left(\frac{\sigma}{r_{\text{cut}}}\right)^{6}\right]. \tag{7.5}$$

It is important to note that the result of a computer simulation may depend on $r_{\text{cut}}$ and whether a truncated or truncated and shifted potential is used. Note that other truncation methods that remove discontinuities in higher order derivatives of $u(r_{ij})$ are also often used in simulations. The second type of nonbonded interactions are the Coulomb interactions

$$u(r_{ij}) = \frac{q_i q_j}{4\pi\epsilon_0 r_{ij}} \tag{7.6}$$

between two charged spheres at a distance $r_{ij}$ from each other. In this equation, $q_i$ represents the charge of particle $i$ and $\epsilon_0$ is the dielectric constant of vacuum. Particles can be assigned partial charges or integer values in the case of ions. Unlike the Lennard–Jones interactions, simple truncation of Coulombic interactions can lead to serious artifacts (e.g., an unphysical dipole moment at the cut-off radius). The generally accepted method to handle Coulombic interactions is the Ewald summation or equivalent method [4, 13, 14]. Recently, pairwise alternatives for the Ewald summation have been proposed [15, 16].

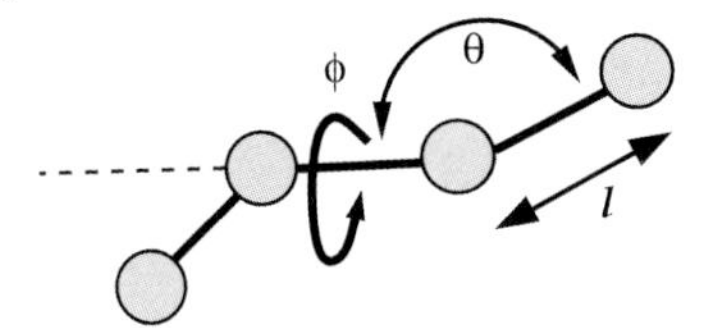


**Figure 7.2** Schematic representation of the bond-stretching (bond length $l$), bond-bending (angle $\theta$), and torsion interaction (torsion angle $\phi$).

For atoms that are part of molecules, the bonded interactions also have to be calculated. These bonded interactions are usually one of the following: bond-stretching (2-body), bond-bending (3-body), and dihedral angle (4-body) interactions. For clarity, the interactions are depicted in Figure 7.2. The first two interactions can be described using a harmonic potential. The same mechanism can also be used to describe the motion of a vibrating spring or a pendulum:

$$u_{\text{stretch}}(l) = \frac{k_l}{2}(l - l_0)^2, \tag{7.7}$$

$$u_{\text{bend}}(\theta) = \frac{k_\theta}{2}(\theta - \theta_0)^2, \tag{7.8}$$

where $l_0$ and $\theta_0$ represent the bond length and bond angle equilibrium values, and $k_l$ and $k_\theta$ are the force constants. The dihedral interaction cannot be described using a harmonic potential but rather a periodic function is used, because of the rotational symmetry,

$$u_{\text{torsion}}(\phi) = \sum_{i=0}^{n} c_i(\cos\phi)^i, \tag{7.9}$$

where $c_0, \ldots, c_n$ are constants. Together the bonded interactions provide flexibility to the molecular structure. They play a large role in, for example, simulations of a hydrated nafion membrane, the transport of proton and water through the membrane, simulations on protein folding, or the permeation event of an ion through an ion channel.

### 7.2.2
### Integrating the Equations of Motion

To integrate the equations of motion (Eq. (7.1)), special integration algorithms are used. The simplest algorithm is the so-called Verlet algorithm, which is based on a Taylor expansion of the coordinate of a particle at time $t + \Delta t$ and $t - \Delta t$ about time $t$:

$$r(t + \Delta t) = r(t) + v(t)\Delta t + \frac{f(t)}{2m}\Delta t^2 + \frac{\partial^3 r}{\partial t^3}\frac{\Delta t^3}{3!} + \mathcal{O}(\Delta t^4), \tag{7.10}$$

and

$$r(t - \Delta t) = r(t) - v(t)\Delta t + \frac{f(t)}{2m}\Delta t^2 - \frac{\partial^3 r}{\partial t^3}\frac{\Delta t^3}{3!} + \mathcal{O}(\Delta t^4). \tag{7.11}$$

Adding these two equations and subtracting $r(t - \Delta t)$ on both sides gives

$$r(t + \Delta t) = 2r(t) - r(t - \Delta t) + \frac{f(t)}{m}\Delta t^2 + \mathcal{O}(\Delta t^4). \tag{7.12}$$

Note that the new position is accurate to order $\Delta t^4$. The Verlet algorithm does not use the velocity to compute the new position, but the velocity can be derived from the trajectory, which is only accurate to order $\Delta t^2$:

$$v(t) = \frac{r(t + \Delta t) - r(t - \Delta t)}{2\Delta t} + \mathcal{O}(\Delta t^2). \tag{7.13}$$

The instantaneous temperature $T$ follows directly from the kinetic energy

$$E_{\text{kin}} = \sum_{i=1}^{N} \frac{mv_i^2}{2} = \frac{3Nk_B T}{2}, \tag{7.14}$$

where $\mathbf{v}_i^2 = v_{x,i}^2 + v_{y,i}^2 + v_{z,i}^2$ and $k_B$ is the Boltzmann constant. As $T \propto \sum_{i=1}^{N} \mathbf{v}_i^2$, the instantaneous temperature $T(t)$ can be adjusted to match the desired temperature $T$ by scaling all velocities with a factor $\sqrt{T/T(t)}$. It is important to note that the Verlet algorithm conserves:

- the total linear momentum of the system, i.e.,

$$\frac{d}{dt}\sum_{i=1}^{N} m_i \mathbf{v}_i(t) = 0 \tag{7.15}$$

- the *total* energy of the system, which is the sum of the potential energy $E_{\text{pot}}$ and the kinetic energy $E_{\text{kin}}$ (Eq. (7.14)).

If these quantities are not conserved in an actual simulation, there must be an error in the simulation program. Other well-known integration schemes are the leap-frog algorithm, the velocity–Verlet algorithm, and predictor–corrector algorithms [1,4]. The choice of the algorithm depends on several things. One consideration would be the accuracy, because this controls the size of the time step. The larger the time steps, the fewer evaluations of the forces are necessary, thus saving on CPU time. It is important to note that the goal of MD simulations is to compute time averages for a *representative* trajectory, and *not* to integrate the equations of motion as accurately as possible. In fact, no algorithm even exists that can predict the "real" trajectory over a large time scale. This is due to the fact that small integration errors and round-off errors will accumulate over time and the calculated trajectory will diverge from the "real" trajectory. In general, the choice for the algorithm will depend on how well the total energy of the system is conserved, since this property is more important for statistical predictions. It turns out that *time-reversible* integration algorithms like the Verlet algorithm conserve the total energy even on long time scales and that algorithms that are *not* time reversible show an energy drift for long time scales, even though they conserve the total energy very well for short time scales [4].

A program that performs an MD simulation consists of the following steps (see also Table 7.1):

**Table 7.1** Pseudocomputer code for a molecular dynamics simulation of `npart` particles that interact with a Lennard–Jones potential.[a]

|  |  |
|---|---|
| `program md` | Molecular Dynamics algorithm |
| `call init` | generate initial positions and velocities |
| `t=0.0` | start at time $t = 0$ |
| `do while (t.lt.tmax)` | for $t < t_{max}$ |
| `   call force` | calculate the forces on all particles |
| `   call integrate` | integrate equations of motion |
| `   t=t+deltat` | update the time with $\Delta t$ |
| `   call sample` | sample time averages |
| `enddo` | |
| `end` | |
| | |
| `subroutine force` | calculation of the forces |
| `en=0` | set total energy to zero |
| `do i=1,npart` | |
| `   f(i)=0` | set forces to zero |
| `enddo` | |
| `do i=1,npart-1` | consider all particle pairs |
| `   do j=i+1,npart` | |
| `      xr=x(i)-x(j)` | distance between $i$ and $j$ |
| `      xr=xr-box*nint(xr/box)` | apply periodic boundary conditions |
| `      r2=xr**2` | distance, note that $x{*}{*}2 = x^2$ |
| `      if(r2.lt.rc2) then` | $rc^2 = rc * rc$, where rc is the cut-off distance |
| `        r2i=1/r2` | |
| `        r6i=r2i**3` | |
| `        ff=48*r2i*r6i(r6i-0.5)` | Lennard–Jones forces |
| `        f(i)=f(i)+ff*xr` | update force for particle $i$ |
| `        f(j)=f(j)-ff*xr` | update force for particle $j$ |
| `        en=en+4*r6i*(r6i-1)-ecut` | update total potential energy |
| `      endif` | |
| `   enddo` | |
| `enddo` | |
| `return` | |
| `end` | |
| | |
| `subroutine integrate` | integration of equations of motion |
| `sumv=0` | |
| `sumv2=0` | |
| `do i=1,npart` | |

**Table 7.1** *(Continued)*.

| | |
|---|---|
| `    xx=2*x(i)-xm(i)+delt**2*f(i)` | Verlet algorithm (Eq. (7.12)) |
| `    vi=(xx-xm(i))/(2*delt)` | estimate for the velocity |
| `    sumv2=sumv2+vi**2` | compute $\sum v^2$ |
| `    xm(i)=x(i)` | update positions |
| `    x(i)=xx` | |
| `enddo` | |
| `temp=sumv2/(3*npart)` | compute temperature |
| `etot=(en+0.5*sumv2)/npart` | total energy per particle |
| `return` | |
| `end` | |

[a]The subroutine `init` generates initial positions and velocities. Counters to compute ensemble averages are updated in the subroutine `sample`. The function `nint` rounds off to the nearest integer.

1. Read the initial conditions: temperature, number of particles, integration time step, etc.
2. Initialize the positions and velocities of all atoms in the system.
3. Compute the forces on all particles. This is the most time-consuming step.
4. Integrate the equations of motion. Steps 3 and 4 form the core of the MD simulation. They are repeated until the desired number of time steps is reached.
5. Compute the average of the measured quantities and stop.

When the simulation has finished, the results can be analyzed. First the simulation results need to be validated. Do the pressure, temperature, and energy components remain constant during the simulation? Do large molecules such as proteins drift over time? These inspections can provide valuable clues to whether a simulation has been successful or not. After validation, macroscopic properties of the system can be estimated. Examples of such properties are the viscosity of liquids or the self-diffusion coefficient. The estimation usually takes place by averaging one or more quantities over the simulation time and over a large amount of molecules.

### 7.2.3
### Practical Issues

While evaluation of the molecular interactions and the integration of the equations of motion make up the largest part of the simulation, some practical problems also have to be accounted for. The simulated system is always finite, which means that a part of the molecules is located at the boundaries of the system. These molecules are not completely surrounded by other molecules. Such a situation would lead to serious boundary effects in systems with a relatively small number of atoms. Imagine a cubic system containing $n^3$ atoms arranged on a cubic lattice. A large fraction of them $(n^3 - (n-2)^3/n^3 \approx 6/n)$ is located at one of the surfaces of the box. This means that for 1000 particles $(n = 10)$, 60% of all particles belong to the

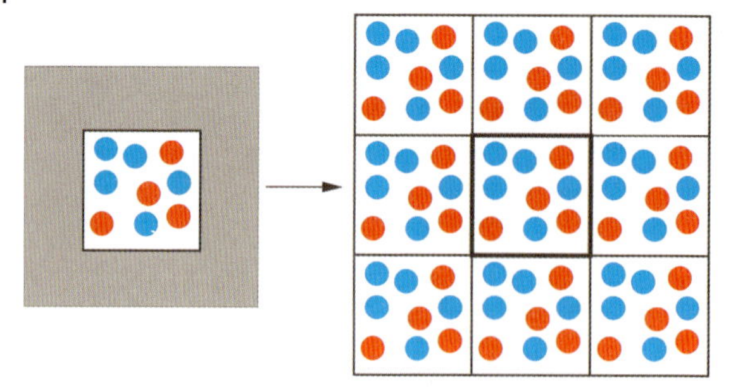

**Figure 7.3** Periodic boundary conditions. A central box is surrounded by copies of itself.

surface. Even for a system of $10^6$ particles ($n = 100$), still 6% of all particles belong to the surface. To minimize the influence of boundary effects, multiple images of the simulation box in all directions are used, in order to obtain a continuous system. This approach is known as periodic boundary conditions (see Figure 7.3). It is similar to treating the simulation system as a unit cell of a crystal. However, it is necessary to keep the amount of calculations manageable due to the limits in CPU time. Instead of calculating the interactions between all (original and periodic) atoms, only the interaction between one (original) atom and the closest image of the other atoms in one of the surrounding boxes could be calculated. This is called the nearest-image convention and it is satisfied automatically when the box length is at least twice the value for $r_{\text{cut}}$.

While molecular dynamics simulations have their obvious advantage in the amount of detail they provide, one should also be aware of the limitations:

1. The time span of current simulations is in the order of tens of nanoseconds for large systems. This is enough if the permeation of water molecules through channels is under investigation, but it will not be long enough to study the permeation of ions. It remains difficult to get information on ion-channel conduction and phenomena such as channel gating. Special techniques to study infrequent but important events will be discussed in Section 7.3.

2. The absence of atomic polarizability in the most force fields may present a flaw. Since the atoms are treated as point (partial) charges, a dynamic redistribution of electronic charge is not allowed. Instead, the polarizability is an average effect that is included by fitting the parameters of the Lennard–Jones potential to experimental data. This approach is sufficient to model bulk-like behavior of solvent molecules but it might not correctly represent individual water molecules inside a narrow channel pore.

3. The last limitation is caused by the classical treatment of the atoms. Using classical mechanics it is only possible to calculate the positions, accelerations, and forces between atoms. Due to the fact that protons and electrons are not explicitly modeled, proton conduction or reactions cannot be simulated. The blocking of protons would be worthwhile to investigate in this case, since it also occurs in the water channels.

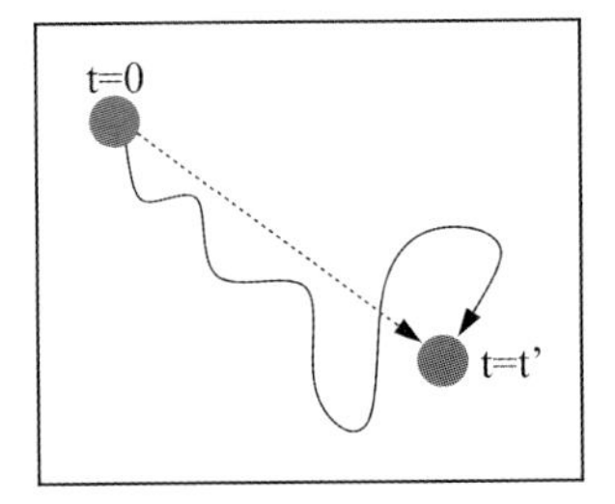


**Figure 7.4** The self-diffusion coefficient can be computed from the displacements of individual particles (Eq. (7.16)).

These issues are still under development. Force fields for atomistic simulations are constantly updated and improved. Addressing the third point, for large systems the use of quantum mechanical methods is still too detailed and is not expected to provide solutions in the next few years on the scale of molecular dynamics. However, with computational power doubling roughly every other year we might be able to apply quantum mechanics to these large systems in the near future.

### 7.2.4
### Diffusion

Transport properties like the diffusion coefficient can be computed directly from an MD simulation. The self-diffusion coefficient follows from the mean-squared displacements of individual particles:

$$D_{\alpha,\mathrm{self}} = \frac{1}{2N} \lim_{t\to\infty} \frac{d}{dt} \left\langle \sum_{i=1}^{N} \left(r_{i\alpha}(t) - r_{i\alpha}(0)\right)^2 \right\rangle, \tag{7.16}$$

where $r_{i\alpha}(t)$ is the position of particle $i$ at time $t$ ($\alpha = x, y, z$) (see also Figure 7.4).

To describe the *transport* of particles due to a gradient, the Maxwell–Stefan approach is often more useful than the traditional Fick formulation, especially for multicomponent systems [17]. In the Maxwell–Stefan approach, the driving force for transport is a gradient in chemical potential, which is balanced by frictional forces between compounds (and/or the host structure). The Maxwell–Stefan diffusion coefficients can be computed directly from MD simulations *at equilibrium* (i.e., without gradients in chemical potential or concentration) [18], or, alternatively, from nonequilibrium MD (NEMD) simulations [19,20]. For a review on diffusion in porous materials and liquids, we refer the reader to Refs. [17,21,22] and references therein.

## 7.3
## Rare Events

Diffusion coefficients of adsorbed molecules can vary as much as 10 orders of magnitude. If the diffusion coefficient is sufficiently high, straightforward MD can

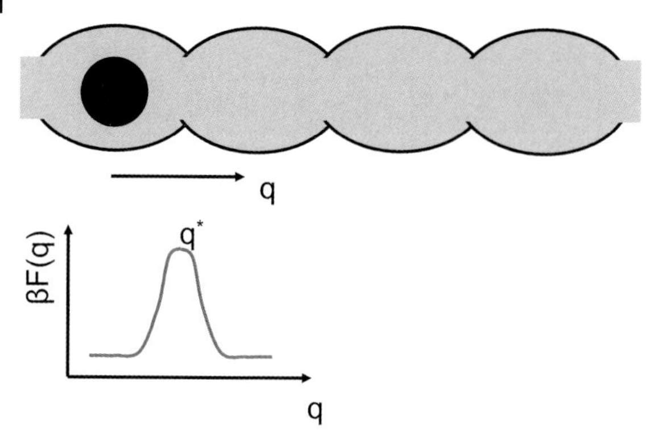


**Figure 7.5** Schematic representation of a single molecule diffusing in a one-dimensional pore (a) and the free energy of this molecule as a function of the position $q$ (b).

be used to simulate the system. If, however, the diffusion coefficient is very low, one often observes that molecules are trapped in low (free) energy sites and once in a while the molecule hops to another adsorption site. To compute a diffusion coefficient reliably, one has to observe a sufficient number of hops. Most of the CPU time is spent on molecules that "wait" at an adsorption site until a fluctuation gives them sufficient kinetic energy to take the barrier between adsorption site. The higher the barrier the longer the molecules remain trapped and – on the time scale of an MD simulation – such a hopping becomes a very rare event.

Special techniques have been developed to simulate such rare events [4]. The basic idea is to compute the hopping rate in two steps [23–25]. First, we compute the probability that a molecule can be found on top of the free-energy barrier followed by a separate simulation in which the average time is computed it takes for a molecule on top of the barrier to actually cross it.

Let us consider, as an example, the system shown in Figure 7.5. The adsorption sites are in the cavities and the windows are the diffusion barriers. In a rare-event simulation it is important to define an appropriate reaction coordinate which characterizes the progress of the "reaction." In case of a single atom, an obvious choice is the position along the tube and a typical free energy as a function of the reaction coordinate $q$ as shown in Figure 7.6. The probability to find a molecule on top of the barrier can be computed directly from the free-energy profile

$$P(q^*)\mathrm{d}q = \frac{\exp[-\beta F(q^*)]\mathrm{d}q}{\displaystyle\int_{-\infty}^{q^*} \mathrm{d}q \exp[-\beta F(q)]}, \tag{7.17}$$

where $q^*$ is defined as the top of the barrier. $F(q)$ is the free energy as a function of the order parameter. This free energy can be computed using the techniques described in Refs. [4, 26, 27].

The second step involves the average time taken by a molecule to cross the barrier. The simplest approach is to assume that transition state theory (TST) holds. A molecule that arrives at the top of the barrier is assumed to be in equilibrium

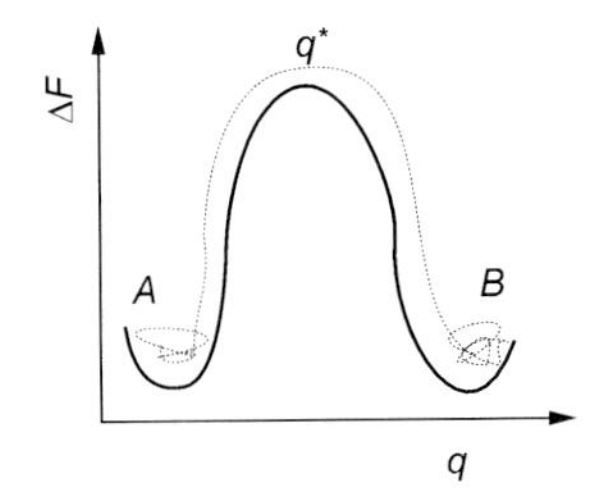


**Figure 7.6** Free energy as a function of the reaction coordinate $q$.

with its surrounding. As a consequence, the velocity distribution $\mathcal{P}(v)$ is given by the Maxwell distribution corresponding to the temperature of the system:

$$\mathcal{P}(v) \propto \exp[-\beta m v^2/2]. \tag{7.18}$$

TST assumes that half of the molecules that reach the barrier also cross the barrier, i.e, those with a positive velocity of the order parameter. The TST approximation of the hopping rate is

$$k^{\mathrm{TST}} = \frac{1}{2}|\dot{q}|P(q^*) = \sqrt{\frac{k_{\mathrm{B}}T}{2\pi m}} \frac{\exp[-\beta F(q^*)]}{\int_{-\infty}^{q^*} \mathrm{d}q \exp[-\beta F(q)]}. \tag{7.19}$$

The advantage of TST is that one has to compute only the free energy as a function of the reaction coordinate to compute the hopping rate. The disadvantage is that one does not know in advance whether the assumptions underlying TST hold. In addition, TST also assumes that the transition state is known exactly, i.e., the top of the free energy, $q^*$, exactly corresponds to the true transition state. In practice, we do not know the free energy exactly and we, therefore, can only approximate the transition state.

More importantly, in the system we consider in Figure 7.5, the choice of the reaction coordinate is straightforward. However, in practice one has to be very careful. Consider, for example, the zeolites shown in Figure 7.7. Both the zeolites are one-dimension channels of cages connected via narrow windows. In analogy with the system of Figure 7.5, one would take as order parameter the position of the atom projected on the axis of the channel (red-short dashed line), but we could have also taken a projection on a line through the window that has an angle with the channel axes (blue-long dashed line). Depending on the particular choice, the free energy of the transition state will be different. If we use these free energies and compute the hopping rate using Eq. (7.19), we would find different values of this hopping rate. In fact, TST gives an upper limit; the true hopping rate is always lower compared to the TST result. It is, therefore, important to select the reaction coordinate that gives the highest free energy of the transition state. Since the free energy appears in the exponential in Eq. (7.19), TST can give a large error in the hopping rate if the choice of reaction coordinate is far from optimal. It may look strange to use a reaction coordinate that does not correspond to the direction

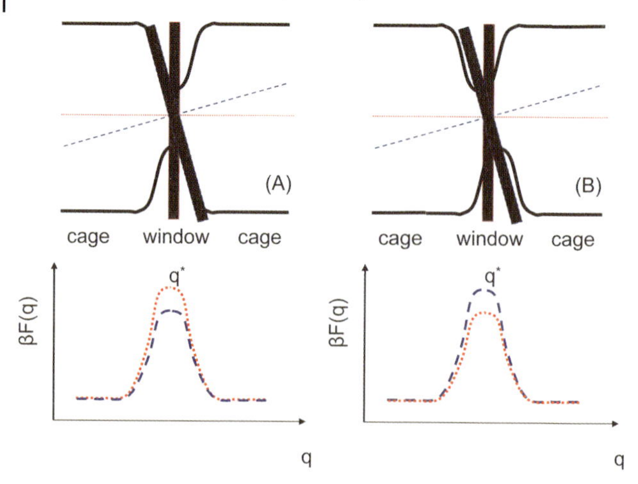


**Figure 7.7** Comparison of two different choices of the reaction coordinate $q$. The red (short dashed) choice is parallel to the axis of the zeolite, while the blue choice (long dashed) makes an angle with the horizontal axis. In the left figure the window is perpendicular to the axis of the zeolite while in the right the window takes an angle. The shaded areas show the part of the zeolite of which the free energy is projected on the transition state $q*$. The bottom figures show that depending on the choice of reaction coordinate, the free energy of the transition state, $F(q^*)$, has a different value.

of diffusion. Figure 7.7 shows that for the zeolite in which the window is not perpendicular to the channel axis this choice results in a higher free energy of the transition state. This example illustrates that even for diffusion of an atom the choice of reaction coordinate can be nontrivial. For molecules, the number of possible reaction coordinates increases dramatically and it will be impossible to compute the free energies for all possible reaction coordinates. Finally, even if one is able to select the optimal reaction coordinate, TST may not give the correct hopping rate since in Eq. (7.19) it is assumed that all particles that start in one cage and arrive on top of the barrier with a positive velocity of the order parameter arrive in the *product* cage. TST ignores the possibility that such a particle recrosses the barrier and returns in the cage it originates due to, for example, collisions with the zeolite atoms.

To compute the "true" hopping rate, one has to correct the TST to take into account the recrossing of the barrier. These recrossings can be intrinsic to the system or due to the nonoptimum choice of the reaction coordinate. This correction is obtained using, for example, the Bennet–Chandler [23–25] approach in which MD simulations are performed to compute the transmission coefficient $\kappa$. The computation of the transmission coefficient $\kappa$ involves many MD simulations, which all start on top of the barrier ($q^*$) and of which we determine the fraction that ends up in the product cage. The time-dependent transmission coefficient is defined as

$$\kappa(t) = \frac{k(t)}{k^{\mathrm{TST}}} = \frac{\langle \dot{q}(0)\delta(q(0) - q^*)\theta(q(t) - q^*)\rangle}{0.5\langle|\dot{q}(0)|\rangle}, \tag{7.20}$$

where $\theta(q)$ is the Heaviside function ($\theta(q) = 1$ if $q > 0$ and $\theta(q) = 0$ otherwise) and $\delta(q)$ the Dirac delta function. The delta function in Eq. (7.20) indicates that the trajectories are initiated on top of the barrier and the Heaviside function takes a value if the particle is on the product side of the barrier. Equation (7.20) shows that if all particles with a positive velocity of the order parameter stay in the product cage the transmission coefficient is one and TST gives the correct result. For those systems in which barrier recrossings are important, Eq. (7.20) gives a plateau value for intermediate times that can be used to correct the TST hopping rate. The important aspect of these MD simulations is that they are initiated on top of the barrier, which is a very unfavorable configuration for which the relaxation to equilibrium, one of the cages, is relatively fast. These simulations, therefore, do not require much CPU time for most systems.

In our discussion, we focus on one-dimensional order parameters and for some systems it can be desirable to use a multidimensional order parameter. TST can be generalized to higher dimensions and one has to locate the saddle point in such a multidimensional space. Special techniques have been developed for this [28]. For some systems, however, the dynamics on top of the barrier can be diffusive; because of collisions with the atoms of zeolite a particle may spend a relatively long time on top of the barrier before it falls in one of the cages. For such systems, it can be advantageous to compute the hopping rate using the approach of Ruiz–Montero et al. [29]. In the methods we have discussed so far we assume that a good estimate of the transition state can be obtained. Although Eq. (7.20) can be used to correct an unfortunate choice of reaction coordinate, if the transmission coefficient is very small it is expensive to compute it accurately. Special techniques like transition path sampling [7, 30–34] have been developed to compute hopping rates without prior knowledge of the reaction coordinate. This method can also be used to check whether the assumed transition state resembles to true transition state.

## 7.4 Monte Carlo

### 7.4.1 Introduction

In molecular dynamics, one integrates the equation of motion for a system of particles (atoms, molecules) that interact according to a certain force field, and from this trajectory one can calculate *time averages*. According to the ergodicity hypothesis [4], these *time averages* should, in principle, be equal to averages over *all possible configurations* of the particles in the system.

Consider a system where the positions of all $N$ particles in the system are denoted by the vector $\mathbf{r}^N$. From elementary statistical thermodynamics, it is well known that the statistical weight of each configuration $\mathbf{r}^N$ is not equal. At a fixed number of particles $N$, fixed volume $V$ and absolute temperature $T$, the statistical weight of a system with configuration $\mathbf{r}^N$ is proportional to $\exp[-\beta U(\mathbf{r}^N)]$, where $\beta = 1/(k_B T)$,

$k_B$ is the Boltzmann factor ($\approx 1.38066 \times 10^{-23}$ J K$^{-1}$), and $U(\mathbf{r}^N)$ is the potential energy that depends on the positions of all the particles in the system. Therefore, ensemble averages that only depend on the positions of the particles in the system (e.g., the average energy or average pressure) can be calculated in the following way:

$$\langle X \rangle = \frac{\int d\mathbf{r}^N X(\mathbf{r}^N) \exp[-\beta U(\mathbf{r}^N)]}{\int d\mathbf{r}^N \exp[-\beta U(\mathbf{r}^N)]}. \tag{7.21}$$

Naively, one might think that the ensemble average $\langle X \rangle$ can be evaluated by the conventional numerical integration techniques such as numerical quadrature. However, evaluating the integrand on a grid in the high-dimensional phase space is impossible as the number of gridpoints becomes more than astronomically large. For instance, $N = 100$ particles in $D = 3$ dimensions using a very rough grid of only $m = 5$ gridpoints already leads to $m^{DN} = 5^{300}$ gridpoints at which this integrand has to be evaluated. This is impossible within the lifetime of our universe. In addition, suppose that we would somehow be able to perform the integration of Eq. (7.21). Our claim is that the statistical error will then be so large that the result is not meaningful anyway. The reason is that when two particles overlap, the potential energy is extremely large and therefore the Boltzmann factor equals zero. In fact, it turns out that for a typical liquid this is the case for almost all configurations $\mathbf{r}^N$ and only an extremely small part of the phase space will have a nonzero contribution to the integrals of Eq. (7.21) (see also Figure 7.8).

We therefore have to resort to more suitable numerical techniques. One such a technique is the Monte Carlo method. The basic idea in the Metropolis Monte Carlo method is that phase points are generated in phase space according to the desired probability. In this way, one avoids the numerical evaluation of the integrand on a high-dimensional grid where most of the gridpoints result in configurations with an extremely low Boltzmann weight. Consider, for example, a sequence of configurations $\mathbf{r}_1^N, \mathbf{r}_2^N, \mathbf{r}_3^N, \ldots, \mathbf{r}_n^N$. If this sequence consists of *random* configurations, the ensemble average $\langle X \rangle$ is simply an unweighted average

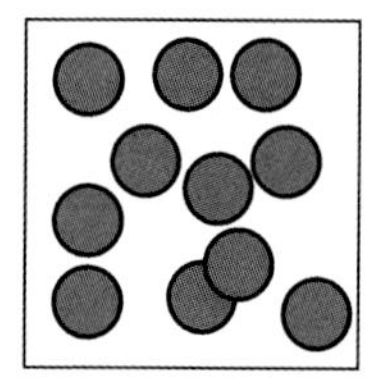

**Figure 7.8** If particles are inserted *randomly* in a simulation box, already at a moderate density it is very likely that there is at least a single particle overlap in the system, resulting in a Boltzmann weight $\exp[-\beta U]$ that is extremely low.

$$\langle X \rangle = \lim_{n \to \infty} \frac{\sum_{i=1}^{n} X(\mathbf{r}_i^N) \exp[-\beta U(\mathbf{r}_i^N)]}{\sum_{i=1}^{n} \exp[-\beta U(\mathbf{r}_i^N)]}. \tag{7.22}$$

Again, computing $\langle X \rangle$ in this way is often not meaningful as $\exp[-\beta U(\mathbf{r}_i^N)]$ is nearly always zero for random configurations. However, suppose that we are able to generate $\mathbf{r}_i^N$ in such a way that the probability of generating $\mathbf{r}_i^N$ is proportional to $\exp[-\beta U(\mathbf{r}_i^N)]$, then the ensemble average is simply

$$\langle X \rangle = \lim_{n \to \infty} \frac{\sum_{i=1}^{n} X(\mathbf{r}_i^N)}{n}. \tag{7.23}$$

This sampling is called Metropolis sampling [35]. As we cannot generate an infinitely long sequence of configurations on the computer, we estimate the ensemble average by taking a large value of $n$.

We will now present a method to generate $\mathbf{r}_i^N$ with a probability proportional to $\exp\left[-\beta U(\mathbf{r}_i^N)\right]$ (Metropolis sampling). For a system of $N$ particles in volume $V$, this works as follows: We first generate a configuration of $N$ particles at positions $o = \mathbf{r}_o^N$ with a nonvanishing Boltzmann weight $\exp[-\beta U(o)]$. Next we generate a random new trial configuration $n = \mathbf{r}_n^N$, for example, by picking randomly a particle and by displacing it randomly. The Boltzmann weight of this new trial configuration is $\exp[-\beta U(n)]$. We must now decide whether we accept or reject this trial configuration satisfying the constraint that on average the probability of finding the system in a configuration $\mathbf{r}^N$ is proportional to the probability distribution $f_c(\mathbf{r}^N) = \exp[-\beta U(\mathbf{r}^N)]$. If we reject this trial move, the next element in our sequence of configurations (Eq. (7.23)) will be $\mathbf{r}_o^N$, otherwise it will be $\mathbf{r}_n^N$. The rule to accept or reject $\mathbf{r}_n^N$ must satisfy three conditions: (1) the number of points in any configuration $o$ should be proportional to $f_c(o)$; (2) all configurations can, in principle, be visited, (3) in equilibrium, the average number of accepted trial moves leaving state $o$ should be equal to the average number of accepted trial moves from all other states to state $o$. This is called the balance condition [36]. It is however, convenient to impose a much stronger condition (detailed-balance condition); namely that in equilibrium the average number of accepted trial moves leaving state $o$ to state $n$ is equal to the average number of accepted trial moves leaving state $n$ to state $o$, which essentially means that all "fluxes of configurations" are equal:

$$f_c(o)\pi(o \to n) = f_c(n)\pi(n \to o), \tag{7.24}$$

where $\pi(o \to n)$ denotes the transition probability that can be split into two terms: (1) the probability $\alpha(o \to n)$ to perform a trial move from $o \to n$, and (2) the probability $\mathrm{acc}(o \to n)$ of accepting a trial move from $o \to n$:

$$\pi(o \to n) = \alpha(o \to n)\mathrm{acc}(o \to n). \tag{7.25}$$

In the original Metropolis scheme [35], $\alpha(o \rightarrow n)$ is chosen to be symmetric, i.e., $\alpha(o \rightarrow n) = \alpha(n \rightarrow o)$ and we arrive at

$$\frac{\text{acc}(o \rightarrow n)}{\text{acc}(n \rightarrow o)} = \frac{f_c(n)}{f_c(o)} = \exp[-\beta(U(n) - U(o))] = \exp[-\beta\Delta U]. \tag{7.26}$$

An example of such a symmetric transition is the random displacement of a randomly selected particle. Many choices for $\text{acc}(o \rightarrow n)$ satisfy this condition, but the commonly used choice is of Metropolis:

$$\begin{aligned} \text{acc}(o \rightarrow n) &= f_c(n)/f_c(o) \text{ if } \; f_c(n) < f_c(o) \\ &= \quad 1 \qquad \text{if } \; f_c(n) \geq f_c(o). \end{aligned} \tag{7.27}$$

More explicitly, to decide whether a trial move will be accepted or rejected we generate a random number, denoted by *ranf()*, from a uniform distribution in the interval [0,1]. If *ranf()*$<$ $\text{acc}(o \rightarrow n)$, we accept the trial move and reject it otherwise. This rule satisfies the Metropolis condition and can be written as

$$\text{acc}(o \rightarrow n) = \min(1, \exp[-\beta(U(n) - U(o))]) = \min(1, \exp[-\beta\Delta U]), \tag{7.28}$$

where $\min(a, b) = a$ when $a < b$, and $b$ otherwise. This means that new configurations that lower the energy are always accepted, and new configurations that increase the total energy are accepted with a certain probability that depends on their energy difference and the temperature. In summary, our Monte Carlo approach to compute ensemble averages of a system of $N$ particles in volume $V$ is as follows (see also Table 7.2):

1. Generate an initial configuration.
2. Start with a configuration $o$, and calculate its energy $U(o)$.
3. Select a particle at random.
4. Give the selected particle a random displacement $x(n) = x(o) + \Delta$, where $\Delta$ is a uniformly distributed random number from $[-\Delta x, \Delta x]$
5. Calculate the energy $U(n)$ of the new configuration $n$.
6. Accept the trial move with a probability

$$\text{acc}(o \rightarrow n) = \min(1, \exp[-\beta(U(n) - U(o))]) = \min(1, \exp[-\beta\Delta U]) \tag{7.29}$$

7. Update the calculation of ensemble averages, also after rejected trial moves.

A pseudocomputer code of this algorithm is listed in Table 7.2.

### 7.4.2 The Grand-Canonical Ensemble

In the previous section, we have discussed the Monte Carlo technique for a system with a constant number of particles $N$ and a constant volume $V$ at temperature $T$.

**Table 7.2** Pseudocomputer code of the Monte Carlo algorithm described in the text.[a]

```
 program mc                                basic Monte Carlo algorithm
 do icycle=1,ncycle                        number of MC cycles
    call move                              displace a randomly selected particle
    if(mod(icycle,nsample).eq.0)           collect ensemble averages
+      call sample                         each nsample MC cycles
 enddo
 end

 subroutine move                           perform a trial move
 i=int(ranf()*npart)+1                     select particle i at random
 call energy(x(i),eold,i)                  calculate energy of old configuration
 xnew=x(i)+(2*ranf()-1)*Δx                 random displacement of particle i
 call energy(xnew,enew,i)                  calculate energy of new configuration
 if(ranf().lt.exp(-beta*(enew-eold))       acceptance rule
+   x(i)=xnew                              accept new position of particle i
 return
 end

 subroutine energy(xi,e,i)                 subroutine to calculate the energy
 e=0.0
 do j=1,npart                              loop over all particles
    if (j.ne.i) then                       if particle j is not i
       dx=x(j)-xi                          calculate distance between particles i and j
       dx=dx-box*nint(dx/box)              apply periodic boundary conditions
       eij=4*(1.0/(dx**12)-1.0/(dx**6))    calculate Lennard–Jones potential energy
       e=e+eij
    endif
 enddo
 return
 end
```

[a]The program consists of three parts. The main program `mc` controls the simulation. The subroutine `move` displaces a randomly selected particle, and the subroutine `energy` computes the energy of a particle. The function `ranf()` generates a uniformly distributed random number between 0 and 1. The function `int` truncates a real number to its inter value, while the function `nint` converts a real number to its nearest integer value.

However, the choice of a constant number of particles $N$ is not always a convenient one.[1] Consider, for example, the case where one is interested in studying the adsorption of guest molecules inside a microporous host such as a zeolite or carbon nanotube. In this system, the gas phase (with pressure $P$) is in equilibrium with the host material in such a way that the chemical potential $\mu$ of the guest molecules in the gas phase and in the porous host is identical. A direct MD or

1) For a description of other ensembles that are useful in molecular simulation, we refer the reader to Ref. [4].

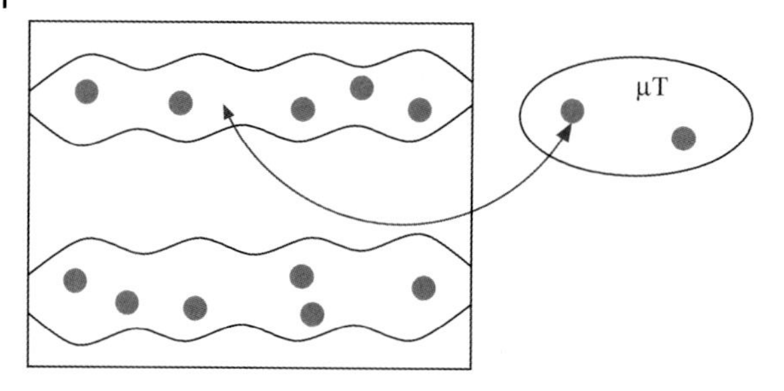


**Figure 7.9** Illustration of the grand-canonical ensemble. A porous host structure is coupled to an infinitely large reservoir of guest molecules at temperature $T$ and chemical potential $\mu$. There is no direct interaction between the molecules in the reservoir and in the host. Monte Carlo trial moves are used to insert/remove guest molecules in/from the host structure. In this way, one can compute the adsorption isotherm (average number of guest molecules $\langle N \rangle$ in the host structure as a function of the chemical potential $\mu$).

MC simulation of the gas phase and the porous solid in a single simulation box is not very convenient as this will create a large surface area and the zeolite cannot be periodic in all directions. In this case, the grand-canonical ($\mu VT$) ensemble is more useful. In this ensemble, the volume $V$ and chemical potential $\mu$ of the guest molecules are fixed, while the number of adsorbed guest molecules ($N$) is fluctuating. The host structure is coupled to an infinitely large particle reservoir at a certain chemical potential $\mu$ and temperature $T$ (see Figure 7.9). The reservoir and the host structure can exchange particles. For a given chemical potential $\mu$ and temperature $T$ one can compute average of the number of particles $\langle N \rangle$ adsorbed in the host structure.

From statistical mechanics, it is known that the statistical weight of the host structure with $N$ guest molecules each consisting of a single atom with coordinates $\mathbf{r}^N$ is proportional to

$$\mathcal{W}(N, \mathbf{r}^N) \propto \frac{V^N \exp[\beta\mu N - \beta U(\mathbf{r}^N)]}{\Lambda^{3N} N!}. \tag{7.30}$$

In this equation, $V$ is the volume of the host system, $\mu$ is the chemical potential of the guest molecules in the reservoir, and $\Lambda$ is the thermal wave vector

$$\Lambda = \frac{h}{\sqrt{2\pi m k_{\mathrm{B}} T}} = \frac{h}{\sqrt{2\pi m/\beta}}, \tag{7.31}$$

where $m$ is the mass of a guest molecule. The acceptance rules for the Monte Carlo method in the grand-canonical ensemble follow directly from Eq. (7.30), see also Refs. [4, 8]. The following trial moves are used:

1. *Particle displacement.* A randomly selected particle in the host is given a random displacement. This move is accepted according to Eq. (7.28)

2. *Particle insertion.* A guest molecule is inserted at a random position in the host structure. This move is accepted with a probability

$$\mathrm{acc}(N \rightarrow N+1) = \min\left(1, \frac{V}{\Lambda^3(N+1)} \exp[\beta(\mu - U(\mathbf{r}^{N+1}) + U(\mathbf{r}^N))]\right), \tag{7.32}$$

3. *Particle deletion.* A randomly selected guest molecule in the host structure is deleted. This move is accepted with a probability

$$\mathrm{acc}(N \rightarrow N-1) = \min\left(1, \frac{N\Lambda^3}{V} \exp[-\beta(\mu + U(\mathbf{r}^{N-1}) - U(\mathbf{r}^N))]\right). \tag{7.33}$$

It can be shown that Eqs. (7.32) and (7.33) obey detailed balance. The trial moves are schematically illustrated in Figure 7.10. It is selected at random which move is performed. For the simulation of mixtures, it can be very advantageous to include

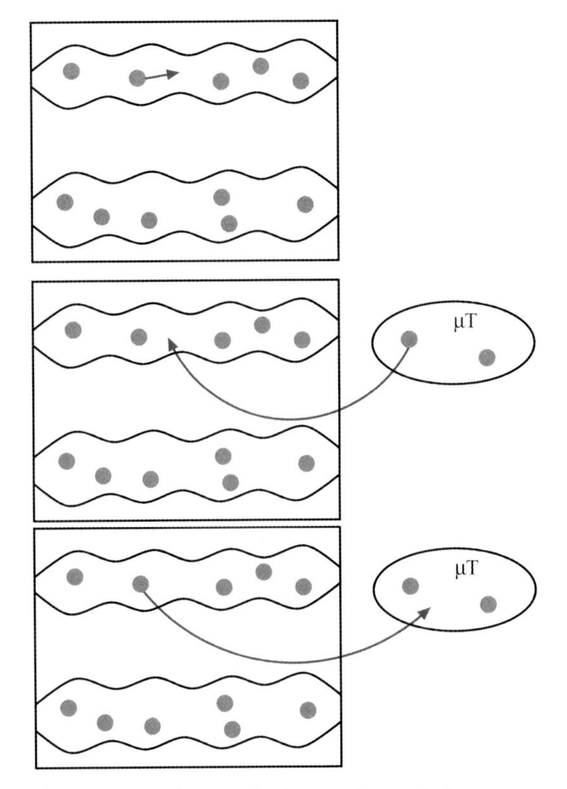


**Figure 7.10** Trial moves in the grand-canonical ensemble: particle displacement, particle insertion, and particle deletion. Here, it is convenient to select the trial moves at random [5].

trial moves that change the identity of a guest molecule in the host (semigrand ensemble). For details, we refer the reader to Refs. [37–40].

The chemical potential of the molecules in the reservoir can be converted to the pressure $P$ or fugacity $f$ of the reservoir. The fugacity of a system is defined as the pressure that the system would have if it would be an ideal gas, at exactly the same chemical potential. As for an ideal gas $\mu = k_B T \ln \rho\Lambda^3$ it follows directly that

$$f = \frac{\exp[\beta\mu]}{\beta\Lambda^3}. \tag{7.34}$$

It can be shown that the pressure $P$ and the fugacity $f$ are related according to [41]

$$\ln\frac{f}{P} = \int_0^P dP' \frac{Z(P') - 1}{P'}, \tag{7.35}$$

where $Z$ is the compressibility factor $Z = P\overline{V}/k_B T$ and $\overline{V} = V/N$ is the volume per molecule. The compressibility $Z$ can be taken from experiments or it can be computed in a separate simulation.

### 7.4.3 Chain Molecules

In the previous section, we have shown that the grand-canonical Monte Carlo technique critically relies on the insertion and deletion of guest molecules. For small guest molecules like methane or ethane, random insertion of guest molecules in the host structure does not pose any problem. However, for longer chain molecules, random insertion of a guest molecule becomes increasingly more difficult as nearly always there is an overlap with one of the zeolite atoms (see Figure 7.11).

The low acceptance probability can be overcome by using the configurational-bias Monte Carlo (CBMC) algorithm [4, 42–45]. This algorithm uses the Rosenbluth

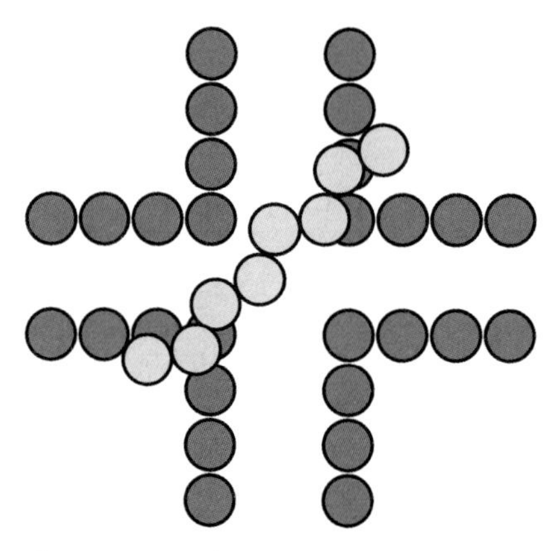

**Figure 7.11** Random insertion of long-chain molecules is difficult as nearly always there will be an overlap with the host structure.

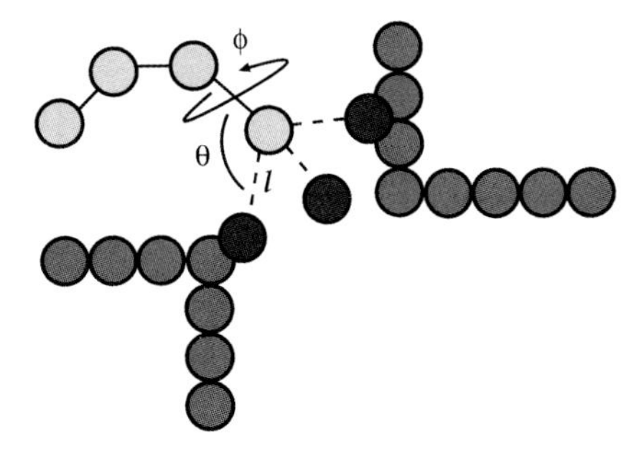


**Figure 7.12** Schematic representation of the CBMC technique. $k$ trial directions (each represented by values for $l$, $\theta$, $\phi$) are generated according to Eq. (7.36), and one of the trial directions is selected according to Eq. (7.37). Here, $k = 3$.

scheme [46] to generate chain configurations. This introduces a bias, which can be removed exactly in the acceptance rules. The CBMC algorithm significantly improves the acceptance probability of insertions and deletions in the grand-canonical ensemble, as well as the internal rearrangements of adsorbed chains.

The algorithm works as follows. Consider a linear chain consisting of $M$ monomers. The interactions of this chain can be split into *bonded* interactions (e.g., bond-stretching, bond-bending, torsion) and *nonbonded* interactions (Lennard–Jones, Coulombic interactions). The nonbonded interactions can be intramolecular (within the same molecule) or intermolecular (with other molecules or the host structure). For the insertion of a chain molecule, the following steps are executed (see also Figure 7.12):

1. The first monomer of a chain is placed at a random position in the system and its energy $u_1$ is recorded.
2. For the next monomer, $k$ trial positions are generated. The position of each trial direction is generated with a probability proportional to the Boltzmann factor of the *bonded* interactions. In general, the position of the next monomer can be characterized by the bond length $l$, bond-bending angle $\theta$, and a torsion angle $\phi$ with corresponding energies $u_{\text{stretch}}(l)$, $u_{\text{bend}}(\theta)$, and $u_{\text{torsion}}$. For each trial direction, the values of $l$, $\theta$, and $\phi$ are generated according to

$$
\begin{aligned}
P(l) &\propto dl\, l^2 \exp[-\beta u_{\text{stretch}}(l)] \\
P(\theta) &\propto d\theta\, \sin(\theta) \exp[-\beta u_{\text{bend}}(\theta)] \\
P(\phi) &\propto d\phi\, \exp[-\beta u_{\text{torsion}}(\phi)].
\end{aligned}
\tag{7.36}
$$

A method to draw random numbers from arbitrary distributions is presented in Ref. [47].

3. One of the trial directions (a) is selected with a probability proportional to the Boltzmann factor of the *nonbonded* energy of that monomer:

$$P_a = \frac{\exp[-\beta u_{ia}]}{\sum_{j=1}^{k} \exp[-\beta u_{ij}]}, \tag{7.37}$$

where $u_{ij}$ is the nonbonded energy of the $j$th trial direction for monomer $i$. In is way, it is very unlikely that unfavorable trial directions (with a low Boltzmann weight due to a large $u_{ij}$) are chosen and in this way overlaps with other particles are avoided.

4. This process is continued until the chain has been grown. The Rosenbluth weight of the generated chain equals

$$W = \frac{\exp[-\beta u_1] \prod_{i=2}^{M} \left[ \sum_{i=1}^{k} \exp[-\beta u_{ij}] \right]}{k^{M-1}}. \tag{7.38}$$

The insertion of a chain is accepted with a probability

$$\mathrm{acc}(N \to N+1) = \min\left(1, \frac{\beta V f W}{(N+1)\langle W_{IG}\rangle}\right), \tag{7.39}$$

where $f$ is the fugacity, $V$ the volume of the simulation box, $W$ is the Rosenbluth weight of the generated chain, and $\langle W_{IG}\rangle$ is the Rosenbluth weight of an isolated chain (ideal gas phase). In Ref. [4], it is shown that this scheme obeys detailed balance. Note that for the insertion of a chain consisting of a single monomer, this equation reduces to Eq. (7.32).

For the removal of a molecule, the same procedure is followed, but with the difference that the first trial direction is always the old chain, and this trial direction is always selected. One can show that the acceptance rule equals

$$\mathrm{acc}(N \to N-1) = \min\left(1, \frac{N\langle W_{IG}\rangle}{\beta V f W}\right). \tag{7.40}$$

The CBMC technique can also be applied to branched molecules [40,48–50] and to various ensembles [4]. Over the years, various improvements of the algorithm have been proposed, we refer the reader to Refs. [48–57].

### 7.4.4
### Calculating Adsorption Properties

In the previous sections, we have shown how to compute the adsorption isotherm (number of adsorbed guest molecules $\langle N\rangle$ as a function of the pressure, fugacity, or chemical potential). From the adsorption isotherm, two other properties that are directly accessible by experiments can be computed: the Henry coefficient and the heat of adsorption. For an overview on how to compute the adsorption entropy or free energy we refer the reader to Ref. [58].

### 7.4.5 Henry Coefficient

In the limit of a very low loading of guest molecules, the adsorption isotherm is a linear function:

$$\langle N \rangle = K_{\mathrm{H}} VP, \tag{7.41}$$

where $V$ is the volume of the host, $P$ the pressure of the gas phase, and $K_{\mathrm{H}}$ the Henry coefficient (in units of molecules per unit of volume per unit of pressure). In principle, the Henry coefficient follows from the result of grand-canonical simulations. However, the Henry coefficient can also be computed directly. In molecular simulations, the most convenient way to calculate the Henry coefficient is using Widom's test particle method (see Figure 7.13) [1, 59]:

$$K_{\mathrm{H}} = \beta \times \exp[-\beta \mu_{\mathrm{ex}}] = \beta \times \frac{\langle \exp[-\beta u^{+}] \rangle}{\langle \exp[-\beta u^{+}_{IG}] \rangle}, \tag{7.42}$$

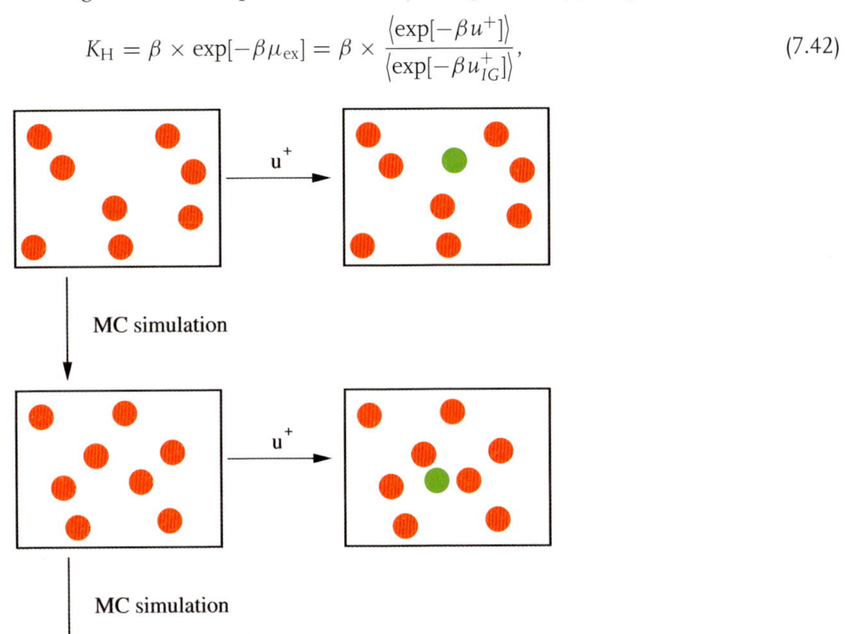


**Figure 7.13** Schematic illustration of Widom's test particle method. During a simulation in the *NVT* ensemble, a test particle is positioned at a random position and the energy change of the system due to this ($u^{+}$) is computed. Note that the insertion is actually never accepted. The excess chemical potential follows from $\mu_{\mathrm{ex}} = -k_{\mathrm{B}} T \ln \langle \exp[-\beta u^{+}] \rangle$, where the brackets $\langle \cdots \rangle$ denote an average over all positions of the test particle and over all configurations of the system. When CBMC is used, the excess chemical potential equals $\mu_{\mathrm{ex}} = -k_{\mathrm{B}} T \ln [\langle W \rangle / \langle W_{IG} \rangle]$.

where $\mu_{ex}$ is the excess chemical potential of the guest molecules, $\langle \exp[-\beta u^+] \rangle$ is the average Boltzmann factor of a test (guest) molecule inserted at a random position in the host, and $\langle \exp[-\beta u_{IG}^+] \rangle$ is the average Boltzmann factor of a test (guest) molecule inserted at a random position in an empty box without the presence of the host (often referred to as an isolated chain). For chain molecules like alkanes, it is well known that insertion of a test chain at a random position in the zeolite nearly always results in overlaps with zeolite atoms, and therefore the sampling statistics of the average $\langle \exp[-\beta u^+] \rangle$ will be extremely poor [60]. For chains that are not too long ($< 50$ monomers), it is convenient to use the CBMC method to insert test chains. In this case, the Henry coefficient is computed from [60]

$$K_{\mathrm{H}} = \beta \times \frac{\langle W \rangle}{\langle W_{IG} \rangle}, \tag{7.43}$$

where $\langle W \rangle$ is the average Rosenbluth weight of a test chain in the host and $\langle W_{IG} \rangle$ is the average Rosenbluth weight of an isolated test chain in an empty box.

### 7.4.6
### Heat of Adsorption

The heat of adsorption describes the change in enthalpy when a molecule is transferred from the gas phase into the pores of a zeolite. In experiments, the heat of adsorption is usually computed using the Clausius–Clapeyron equation [61] or direct calorimetry experiments [62]. In molecular simulations, there are several routes to compute this quantity [63]:

1. Directly from the Clausius–Clapeyron equation

$$-q = \Delta H = k_{\mathrm{B}} \left( \frac{\partial \ln [P/P_0]}{\partial T^{-1}} \right)_{\langle N \rangle = \mathrm{constant}} = \left( \frac{\partial \ln [P/P_0]}{\partial \beta} \right)_{\langle N \rangle = \mathrm{constant}}, \tag{7.44}$$

   where $P_0$ is an arbitrary reference pressure and $R = k_{\mathrm{B}}/N_{\mathrm{av}}$ is the gas constant. Note that the differentiation is carried out at a constant number of adsorbed guest molecules ($\langle N \rangle$). At low loading, this equation becomes

$$-q = \Delta H = -\frac{\partial \ln [K_{\mathrm{H}}/K_{H0}]}{\partial \beta}, \tag{7.45}$$

   where $K_{H0}$ is an arbitrary constant (that has the same units as the Henry coefficient $K_{\mathrm{H}}$). Note that this method requires more than one simulation as one needs to compute temperature derivatives.

2. From energy differences computed in the canonical ($NVT$) ensemble [60, 61]

$$-q = \Delta H = \langle U_1 \rangle_1 - \langle U_0 \rangle_0 - \langle U_{\mathrm{g}} \rangle - \frac{1}{\beta}, \tag{7.46}$$

where $U_N$ is the total energy of a host with $N$ guest molecules present, $\langle \cdots \rangle_X$ refers to an ensemble average at constant $V$,$T$, and $X$ guest molecules, and $\langle U_g \rangle$ is the average energy of an isolated guest molecule (without the host present). The average $\langle U_g \rangle$ for a certain guest molecule only depends on temperature and needs to be calculated only once. Note that Eq. (7.46) only applies at low loading. It turns out that this method has severe difficulties when applied to zeolites with strongly interaction nonframework cations [63].

3. From energy/particle fluctuations in the grand-canonical ($\mu VT$) ensemble [64]. We can approximate the change in potential energy upon adsorption of a single guest molecule:

$$\langle U_{N+1} \rangle_{N+1} - \langle U_N \rangle_N \approx \left( \frac{\partial \langle U \rangle_\mu}{\partial \langle N \rangle_\mu} \right)_\beta = \frac{\left( \frac{\partial \langle U \rangle_\mu}{\partial \mu} \right)_\beta}{\left( \frac{\partial \langle N \rangle_\mu}{\partial \mu} \right)_\beta}$$

$$= \frac{\langle U \times N \rangle_\mu - \langle U \rangle_\mu \langle N \rangle_\mu}{\langle N^2 \rangle_\mu - \langle N \rangle_\mu \langle N \rangle_\mu}, \qquad (7.47)$$

where the brackets $\langle \cdots \rangle_\mu$ denote an average in the grand-canonical ensemble, $N$ is the number of guest molecules, and $\mu$ is the chemical potential of the guest molecules. This leads to [64]

$$-q = \Delta H = \frac{\langle U \times N \rangle_\mu - \langle U \rangle_\mu \langle N \rangle_\mu}{\langle N^2 \rangle_\mu - \langle N \rangle_\mu \langle N \rangle_\mu} - \langle U_g \rangle - \frac{1}{\beta}, \qquad (7.48)$$

where $\Delta H$ in Eq. (7.48) is usually defined as the *isosteric heat of adsorption* and it is often applied at nonzero loading. It is assumed here that the gas phase is ideal.

4. From a direct application Widom's test particle method using the Rosenbluth method [63]

$$-q = \Delta H = \frac{\langle (U_N + u^+) \times W \rangle_N}{\langle W \rangle_N} - \langle U_N \rangle_N - \langle U_g \rangle - \frac{1}{\beta}, \qquad (7.49)$$

where $W$ is the Rosenbluth weight of a test chain and $u^+$ its energy. $\langle \cdots \rangle_X$ refers to an ensemble average at constant $V$,$T$, and $X$ guest molecules. This method is equivalent to Eq. (7.48) and it requires only a single simulation of the host structure.

## Acknowledgments

TJHV acknowledges financial support from the Netherlands Organization for Scientific Research (NWO-CW) through a VIDI grant.

## References


1. Allen, M.P., Tildesley, D.J. *Computer Simulation of Liquids*, Clarendon Press; Oxford, **1987**.
2. Landau, D.P., Binder, K. *A Guide to Monte Carlo Simulations in Statistical Physics*, Cambridge University Press; Cambridge, **2000**.
3. Bolhuis, P.G., Chandler, D., Dellago, C., Geissler, P.G. *Annu. Rev. Phys. Chem.* **2002**, 53, 291–318.
4. Frenkel, D., Smit, B. *Understanding Molecular Simulation: from Algorithms to Applications*, 2nd edn., Academic Press; San Diego, **2002**.
5. Frenkel, D. *Pnas* **2004**, 101, 17571–17575.
6. Rapaport, D.C. *The Art of Molecular Dynamics Simulation*, 2nd edn., Cambridge University Press; Cambridge, **2004**.
7. van Erp, T.S., Bolhuis, P.G. *J. Comput. Phys.* **2005**, 205, 157–181.
8. Vlugt, T.J.H., Van der Eerden, J.P.J.M., Dijkstra, M., Smit, B., Frenkel, D., *Introduction to Molecular Simulation and Statistical Thermodynamics*, http://www.phys.uu.nl/~vlugt, **2008**.
9. Martin, M.G., Siepmann, J.I. *J. Phys. Chem. B* **1998**, 102, 2569–2577.
10. http://siepmann6.chem.umn.edu.
11. MacKerell, A.D., Banavali, N., Foloppe, N. *Biopolymers* **2001**, 56, 257–265.
12. Case, D.A., Cheatham, T.E., Darden, T., Gohlke, H., Luo, R., Merz, K.M.M., Onufriev, A., Simmerling, C., Wang, B., Woods, R.J. *J. Comput. Chem.* **2005**, 26, 1667–1803.
13. Ewald, P.P. *Ann. Phys.* **1921**, 64, 253–287.
14. Fincham, D. *Mol. Sim.* **1994**, 13, 1–9.
15. Wolf, D., Keblinski, P., Phillpot, S.R., Eggebrecht, J. *J. Chem. Phys.* **1999**, 110, 8254–8282.
16. Fennell, C.J., Gezelter, J.D. *J. Chem. Phys.* **2006**, 124, 234104.
17. Krishna, R., Wesselingh, J.A. *Chem. Eng. Sci.* **1997**, 52, 861–911.
18. Wheeler, D.R., Newman, J. *J. Phys. Chem. B* **2004**, 108, 18353–18361.
19. Chempath, S., Krishna, R., Snurr, R.Q. *J. Phys. Chem. B* **2004**, 108, 13481–13491.
20. Wheeler, D.R., Newman, J. *J. Phys. Chem. B* **2004**, 108, 18362–18367.
21. Arya, G., Chang, H.C., Maginn, E.J. *J. Chem. Phys.* **2001**, 115, 8112–8124.
22. Dubbeldam, D., Snurr, R.Q. *Mol. Sim.* **2007**, 33, 305–325.
23. Bennett, C.H. *J. Comp. Phys.* **1976**, 22, 245–268.
24. Bennett, C.H., in *Algorithms for Chemical Computations*, ACS Symposium Series, edited by Christoffersen, R.E. American Chemical Society; Washington, D.C., **1977**, pp. 63–97.
25. Chandler, D. *J. Chem. Phys.* **1978**, 68, 2959–2970.
26. Beerdsen, E., Dubbeldam, D., Smit, B. *Phys. Rev. Lett.* **2004**, 93, 0248301.
27. Dubbeldam, D., Beerdsen, E., Vlugt, T.J.H., Smit, B. *J. Chem. Phys.* **2004**, 122, 224712.
28. Sevick, E.M., Bell, A.T., Theodorou, D.N. *J. Chem. Phys.* **1992**, 98, 3196–3212.
29. Ruiz-Montero, M.J., Frenkel, D., Brey, J.J. *Mol. Phys.* **1997**, 90, 925–941.
30. Bolhuis, P.G., Dellago, C., Chandler, D. *Faraday Discuss.* **1998**, 110, 421–436.
31. Dellago, C., Bolhuis, P.G., Csajka, F.S., Chandler, D. *J. Chem. Phys.* **1998**, 108, 1964–1977.
32. Dellago, C., Bolhuis, P.G., Chandler, D. *J. Chem. Phys.* **1999**, 110, 6617–6625.
33. Allen, R.J., Warren, P.B., ten Wolde, P.R. *Phys. Rev. Lett.* **2005**, 94, 018104.
34. van Erp, T.S. *Phys. Rev. Lett.* **2007**, 98, 268301.
35. Metropolis, N., Rosenbluth, A.W., Rosenbluth, M.N., Teller, A.N., Teller, E. *J. Chem. Phys.* **1953**, 21, 1087–1092.
36. Manousiouthakis, V.I., Deem, M.W. *J. Chem. Phys.* **1999**, 110, 2753–2756.
37. Kofke, D.A., Glandt, E.D. *Mol. Phys.* **1988**, 64, 1105–1131.

38 Panagiotopoulos, A.Z. *Int. J. Thermophys.* **1989**, 10, 447.

39 Martin, M.G., Siepmann, J.I. *J. Am. Chem. Soc.* **1997**, 119, 8921–8924.

40 Vlugt, T.J.H., Krishna, R., Smit, B. *J. Phys. Chem. B* **1999**, 103, 1102–1118.

41 McQuarrie, D.A., Simon, J.D., *Physical Chemistry: A Molecular Approach*, 1st edn., University Science Books; Sausalito, **1997**.

42 Siepmann, J.I., Frenkel, D. *Mol. Phys.* **1992**, 75, 59–70.

43 Frenkel, D., Mooij, G.C.A.M., Smit, B. *J. Phys.: Condens. Matter* **1992**, 4, 3053–3076.

44 De Pablo, J.J., Laso, M., Suter, U.W. *J. Chem. Phys.* **1992**, 96, 6157–6162.

45 Siepmann, J.I., in *Computer Simulation of Biomolecular Systems: Theoretical and Experimental Applications*, edited by van Gunsteren, W.F., Weiner, P.K., Wilkinson, A.J. Escom Science Publisher, Leiden, **1993**, pp. 249–264.

46 Rosenbluth, M.N., Rosenbluth, A.W. *J. Chem. Phys.* **1955**, 23, 356–359.

47 Press, W.H., Flannery, B.P., Teukolsky, S.A., Vetterling, W.T. *Numerical Recipes: The Art of Scientific Computing*, Cambridge University Press; Cambridge, **1986**.

48 Martin, M.G., Siepmann, J.I. *J. Phys. Chem. B* **1999**, 103, 4508–4517.

49 Macedonia, M.D., Maginn, E.J. *Mol. Phys.* **1999**, 96, 1375–1390.

50 Martin, M.G., Frischknecht, A.L. *Mol. Phys.* **2006**, 104, 2439–2456.

51 Esselink, K., Loyens, L.D.J.C., Smit, B. *Phys. Rev. E* **1995**, 51, 1560–1568.

52 Vlugt, T.J.H., Martin, M.G., Smit, B., Siepmann, J.I., Krishna, R. *Mol. Phys.* **1998**, 94, 727–733.

53 Vlugt, T.J.H. *Mol. Sim.* **1999**, 23, 63–78.

54 Consta, S., Wilding, N.B., Frenkel, D., Alexandrowicz, Z. *J. Chem. Phys.* **1999**, 110, 3220–3228.

55 Consta, S., Vlugt, T.J.H., Wichers Hoeth, J., Smit, B., Frenkel, D. *Mol. Phys.* **1999**, 97, 1243–1254.

56 Houdayer, J. *J. Chem. Phys.* **2002**, 116, 1783–1787.

57 Combe, N., Vlugt, T.J.H., ten Wolde, P.R., Frenkel, D. *Mol. Phys.* **2003**, 101, 1675–1682.

58 Dubbeldam, D., Calero, S., Vlugt, T.J.H., Krishna, R., Maesen, T.L.M., Smit, B. *J. Phys. Chem. B* **2004**, 108, 12301–12313.

59 Widom, B. *J. Chem. Phys.* **1963**, 39, 2802–2812.

60 Smit, B., Siepmann, J.I. *J. Phys. Chem.* **1994**, 98, 8442–8452.

61 Wood, G.B., Panagiotopoulos, A.Z., Rowlinson, J.S. *Mol. Phys.* **1988**, 63, 49–63.

62 Dunne, J. A., Mariwala, R., Rao, M., Sircar, S., Gorte, R. J., Myers, A. L. *Langmuir* **1996**, 12, 5888–5895.

63 Vlugt, T.J.H., García-Pérez, E., Dubbeldam, D., Ban, S., Calero, S. *J. Chem. Theory Comp.* **2008**, 4, 1107–1118.

64 Karavias, F., Myers, A.L. *Langmuir* **1991**, 7, 3118–3126.

# 8
# Coarse-Grained Molecular Dynamics

*Albert Jan Markvoort*

## 8.1
## Introduction

In the previous chapter, molecular dynamics and classical force fields have been discussed. Although compared to *ab initio* calculations, the use of classical force fields increases the system size (both in time and length) that can be treated, typically system sizes remain too small to study many interesting systems and phenomena. To simulate larger systems, so-called coarse-grained molecular dynamics (CGMD) simulations can be performed. The basic idea of such coarse-grained simulations is to represent a system with a reduced number of degrees of freedom compared to its all-atom description. Due to this reduction in degrees of freedom and elimination of fine interaction details, the simulation of a coarse-grained system requires fewer resources (both memory and computation time) than the simulation of the same system in its all-atom description. As a result, an increase of orders of magnitude in simulated time and length scales can be achieved. In this chapter, we discuss coarse-graining approaches in molecular dynamics in general, several techniques to attain a coarse-grained model, and, as an example, our application of CGMD on lipid membranes. This example shows that with such simulations, we can bridge the time scale and length scale gap between atomistic simulations and experimental and continuum methods. Of course, coarse graining is not restricted to such lipid systems. Other fields where CGMD has been used already range from polymers and dendrimers to DNA, peptides, and food materials [1–4].

The idea of coarse graining is best seen in the typical example of coarse-graining butane, which is illustrated in Figure 8.1. One butane molecule consists of 4 carbon and 10 hydrogen atoms, and thus 14 atoms in total. In a fully atomistic model, each butane molecule thus has 42 degrees of freedom (3 degrees of freedom per atom times 14 atoms) and for every 2 molecules there are 196 intermolecular interactions (14 atoms interacting with 14 other atoms). A first step in coarse graining would be incorporating the hydrogen atoms into the heavy atoms, creating so-called united atoms. In this representation, a butane molecule is thus described by 4 particles, thus having 12 degrees of freedom and 16 intermolecular interactions. Fewer particles need to be stored meaning that less memory is needed. However, as

*Computational Methods in Catalysis and Materials Science.* Edited by Rutger A. van Santen and Philippe Sautet

ISBN: 978-3-527-32032-5

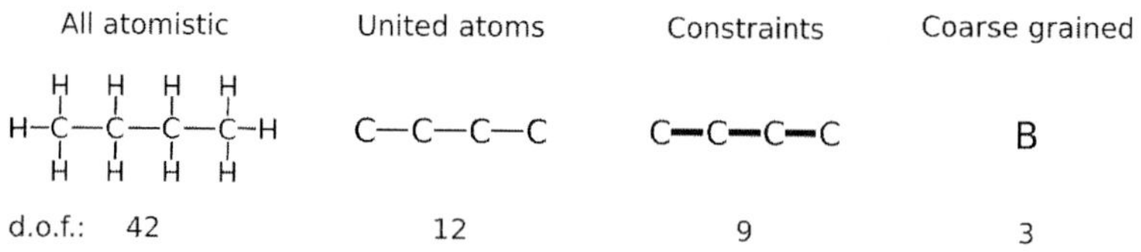

| | All atomistic | United atoms | Constraints | Coarse grained |
|---|---|---|---|---|
| | H–C–C–C–C–H (with H above and below each C) | C—C—C—C | C━C━C━C | B |
| d.o.f.: | 42 | 12 | 9 | 3 |

**Figure 8.1** Various degrees of coarse graining illustrated for a butane molecule.

memory is nowadays not the limiting factor for molecular dynamics simulations, the major advantage is the computational speedup. This advantage is twofold: not only there are less interactions to be calculated, but the highest frequency motions (those of bonded hydrogen atoms) are removed from the system as well. The latter allows larger time steps in integrating the equations of motion, such that with the same number of iterations (time steps) a larger physical time interval is simulated. In a second coarse-graining step, three more degrees of freedom can be removed by adding constraints on the bonds between the remaining heavy atoms. This removes the next highest frequency motion from the system, allowing again for larger time steps. However, in a coarse-grained model, we go even one step further and describe the average behavior of all four heavy atoms with a single particle, leaving only three degrees of freedom and only a single intermolecular interaction between two butane molecules. Thus, coarse graining decreases the computational cost by a reduction of the number of particles and interparticle interactions, but also because larger time steps can be made in integrating the equations of motion as the highest frequency motions are removed from the system. Simultaneously, the dynamics in the coarse-grained system is likely to be further increased compared to the real system, as additional friction and noise from the eliminated degrees of freedom are absent.

Obviously, for the study of many phenomena, the loss of detail is a drawback. It must be kept in mind, however, that this is not limited to CGMD as all steps we have seen in previous chapters (i.e., from Hartree to DFT, from DFT to classical atomistic molecular dynamics, and all possible approximations in between) are in a sense a matter of coarse graining as well. For every problem, the appropriate method should be chosen that still incorporates the necessary physics and chemistry, while it ignores unnecessary details to keep the calculation feasible.

## 8.2 The Coarse-Graining Approach

From the example of coarse-graining butane, we see that a typical way of achieving a simpler description through coarse graining is to reduce the structural details of a complicated system by grouping atoms into fewer interaction sites. In the case of butane, the whole molecule was replaced by a single coarse-grained particle. For larger and more complex molecules, not necessarily all atoms are grouped together into a single coarse-grained particle. Instead, the molecule can be represented by several coarse-grained particles connected to each other with bonds. Except for the choice of which atoms to group together into coarse-grained particles, a major point in coarse graining is to determine the effective interactions between these coarse-grained particles. Coarse graining thus involves two distinct steps: grouping atoms into coarse-grained sites and determining effective interaction potentials.

### 8.2.1
### Grouping Atoms into Coarse-Grained Sites

The first step is to choose which degrees of freedom to group into a new structural unit. In the first place, this includes the choice of the number of atoms to be grouped together, something that we will call *degree of coarse graining*. Subsequently, this includes the question which individual atoms are combined and finally where the coarse-grained particles are positioned relative to each other. There are no fixed rules for this grouping of atoms; thus, this step typically relies on the chemical intuition of the modeler, but some general remarks can be made.

The appropriate degree of coarse graining will depend on the objective of the coarse-grained simulations. Just as for choosing the appropriate method for your problem, the degree of coarse graining should be chosen such that the necessary physics and chemistry is still incorporated while unnecessary details are ignored in order to keep the calculation feasible. For example, to study phase behavior in lipid bilayers, it may be convenient to have a high degree of coarse graining, the coarse-grained particles being whole lipids or sterols. On the other hand, in studies of vesicle fusion real three-dimensional amphiphiles with headgroups and tails are clearly needed. A common choice in such models is to group three or four heavy atoms together into a single coarse-grained particle. Coalescing fewer atoms obviously reduces the desired speedup, whereas grouping more atoms reduces the suitability of a spherical particle to represent these atoms, especially for chainlike molecules.

After it has been decided how many atoms are going to be grouped together, the question remains which individual atoms to group together. At this stage, one has two options. In the first place, one can have a one-to-one coupling between the atoms and coarse-grained particles. This way, for every coarse-grained particle it is known exactly which atoms it represents, and high chemical specificity can be incorporated in the model as for every chemical group in the molecule there will be a different coarse-grained particle type. A drawback is that it typically results in a large number of coarse-grained particle types with even more (pair) interactions that all need to be parametrized. The other option is more phenomenological. A limited number of different coarse-grained particle types can be chosen, and by using these building blocks a molecule is built that closely resembles the target molecule.

After it has been decided which coarse-grained particles to use, the position of the coarse-grained particles has to be determined. For coarse-grained models with a one-to-one relation between the atoms and the coarse-grained particles, the positioning of the coarse-grained center relative to its constituting atoms can still be chosen in various ways. Common choices are the geometric center or the center

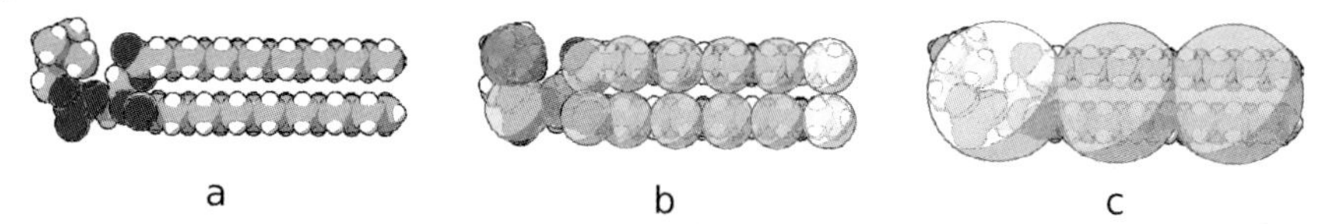


**Figure 8.2** Grouping of the atoms into coarse-grained particles illustrated for DPPC. (a) The all-atom model; (b) a low-degree coarse-grained model where every coarse-grained particle represents 3–4 heavy atoms; and (c) a high-degree coarse-grained model where only three coarse-grained particles are used for the whole molecule.

of mass of either all atoms or the heavy atoms only. For coarse-grained models without a one-to-one relation between atoms and coarse-grained particles, the coarse-grained particles can also be placed such that a good steric representation of the atomistic molecule is obtained.

In Figure 8.2, the grouping of atoms into coarse-grained particles is illustrated for a dipalmitoylphosphatidylcholine (DPPC) phospholipid. Figure 8.2(a) shows the all-atom model, whereas in Figures 8.2(b) and (c) two different degrees of coarse graining are considered. In the first case, each coarse-grained particle contains three or four heavy atoms, resulting in 15 coarse-grained centers: two different types of coarse-grained particles for the tails (differing only by a single hydrogen atom), one type for the ester groups, one type for the glycerol backbone, one for the phosphate, and a sixth type for the choline. Because every coarse-grained particle represents a specific chemical group, the coarse-grained sites are positioned at the center of mass of that groups heavy atoms. In the second case, a much higher degree of coarse graining is applied such that the whole phospholipid is represented by only three particles of two different types: a single particle for the headgroup and two particles of a second type for the tails. In this case, the particles are positioned such that the shape of the coarse-grained molecule resembles that of the atomistic molecule best.

### 8.2.2
### Determining Effective Force Fields

Once the coarse-grained particles have been chosen, an effective force field has to be devised to describe the interactions between these coarse-grained particles. In the example of Figure 8.2(b), there are six different types of coarse-grained particles, meaning that $(6 + 5 + 4 + 3 + 2 + 1)$ 21 different nonbonded interactions have to be determined. Within a molecule, the particles are bonded, so also for these bonds, as well as for other possible bonded interactions like bond-bending potentials and torsion potentials, parameters are needed. As the interactions between the coarse-grained centers are not known *a priori*, determination of these effective interactions is usually the most difficult phase in developing a proper coarse-grained model.

A large diversity of approaches to determine these interactions has been suggested over the course of time. These different approaches can be categorized in a variety of ways. One division could be the level of chemical detail as these approaches range from phenomenological to models including high chemical specificity. A second division criterion could be the kind of data used to parametrize the force fields, for example, whether primarily structural or thermodynamic data are used. And yet another division could be made based on the kind of potentials that the approaches allow. Namely, a first category of approaches only uses potentials of preselected analytical form, whereas a second category uses completely free (tabulated) forms. In case preselected analytical forms are used, parametrization of the force field boils down to determining typically two or three parameters per interaction type. In case completely free forms are used, for every interaction type and for every possible distance (or angle) a value is needed. To be able to determine all these values for the free-form potentials, coupling to atomistic simulations is needed, whereas the parameters for the preselected analytical potentials can also be tuned based on other (e.g., thermodynamic) data. The above-mentioned divisions are thus partly overlapping and not always unambiguous. Therefore, we will divide different strategies into three groups, namely phenomenological models, thermodynamic models, and multiscale models.

The basic idea of the phenomenological methods is to distinguish only between hydrophobic and hydrophilic particles. Hydrophilic particles will attract each other as will hydrophobic particles, but the attraction between hydrophilic and hydrophobic particles is reduced or replaced by a purely repulsive interaction. In these models, often few particles are used such that the geometry is only roughly described and there is not necessarily a one-to-one relation between atoms and coarse-grained particles. Although simple, these models often qualitatively predict the behavior of many amphiphilic systems well. As such, these phenomenological models were the first methods used to simulate aggregation of amphiphiles [5].

One way to add more chemical specificity to such models is to parametrize them on thermodynamic data. The basic idea of parametrizing models based on thermodynamic quantities is that if the local thermodynamic properties are correct, the dynamics on long time scales will also be correct. By extensive calibration based on a diversity of thermodynamic quantities, building blocks are parametrized. Once parametrized, they can be used to build various molecules. When a new building block is needed, the force field is extended with this new building block by calibrating it extensively against the already present ones. This approach is in fact analogous to the way many classic atomistic force fields are parametrized.

The basic idea of multiscale models is that high chemical specificity is added by basing the coarse-grained model on underlying atomistic simulations. All atoms in an atomistic simulation are coupled one-to-one to a coarse-grained site, and all interactions between these coarse-grained sites are directly derived from simulations of the atomistic system, either from the structure via radial distribution functions or directly from the forces in the atomistic simulation. Typical for these multiscale methods is that the model is reparametrized for every new coarse-grained simulation. The philosophy behind this is that as the effective interaction between coarse-grained sites is defined by the average structures of the underlying atoms (which can vary in different environments), coarse-grained potentials are expected to be less transferable than atomistic ones. Thus, for every new molecular mixture and even for every state point, the coarse-grained interactions are redetermined from a new atomistic simulation at the same conditions at which the coarse-grained system is intended to be simulated.

## 8.3 Methods to Obtain Effective Coarse-Grained Interactions

In the following, we go into some more detail on three different coarse-graining approaches in particular, namely, Boltzmann inversion, force matching, and matching thermodynamic properties.

### 8.3.1 Boltzmann Inversion

The main idea of the Boltzmann inversion method is to obtain an accurate reproduction of structural details [1,6,7]. The interaction potentials for the coarse-grained model are chosen such that structural properties of the coarse-grained simulations match those of atomistic simulations or known data. It is based on the idea that for particles that interact with each other via central forces (thus forces that only depend on the scalar distance between the particles), a one-to-one correspondence exists between the potential and the radial distribution function.

The radial distribution function $g(r)$ describes how for a particle the density of surrounding particles $n(r)$ varies as a function of the distance $r$ from it. For a gas with average density $n_0$, $g(r) = n(r)/n_0$. The radial distribution function thus shows the correlations in the distribution of particles arising from the forces they exert on each other due to their pair potential $V(r)$. For a dilute gas,

$$g(r) = \exp(-V(r)/k_B T), \tag{8.1}$$

where $k_B$ is the Boltzmann constant and $T$ the temperature. For a higher density gas, the distribution will also be influenced as surrounding particles attract and repel each other as well, such that some correction terms with various powers of the average density $n_0$ are needed, but, as we will see, Eq. (8.1) can still be used to obtain the pair potential $V(r)$.

When the coarse-grained interactions are derived from atomistic simulations, this Boltzmann inversion method is thus an example of a multiscale method. The procedure is to perform a fully atomistic molecular dynamics simulation and extract from this simulation the radial distribution function for the chosen coarse-grained centers. This radial distribution function is indicated here as $g_A$, where A stands for atomistic. The coarse-grained interaction potentials are then obtained (following Eq. (8.1)) by taking the logarithm of this radial distribution function, corrected by the temperature, that is,

$$V(r) = -k_B T \, ln(g_A(r)). \tag{8.2}$$

This potential can now be used directly to perform coarse-grained simulations. However, because the correction terms for higher densities are not included in this approach, because often multiple potentials for different pairs of particle types are needed that mutually influence each other, and because the atomistic system may be inhomogeneous, this potential can serve as an initial guess only and must be improved upon in an iterative scheme. In such a scheme, a coarse-grained simulation is run using the potential obtained from Eq. (8.2). The radial distribution function of this coarse-grained simulation is then compared to the ones from the atomistic simulation, which serve as the reference radial distribution function. The coarse-grained potential is then repeatedly improved upon by changing it in the following way:

$$V^{i+1}(r) = V^{i}(r) - k_B T \, ln(g_A(r)/g^{i}(r)). \tag{8.3}$$

After a number of iterations, a potential is obtained that reproduces the desired radial distribution function closely. An example of this can be seen in Figure 8.3 where this procedure has been applied to butane in gas–liquid coexistence.

The above is an example of an approach using free-form potentials. The radial distribution functions are not obtained in analytical form but as tabulated data, thus so are the resulting potentials. However, the parameters for preselected analytical potentials can be tuned using radial distribution functions as well. Instead of taking the logarithm of the radial distribution function directly, the parameters for the best fit are used and these parameters can then be improved upon with an iterative scheme as well. The final potential and radial distribution function for such an approach for butane are shown in Figure 8.3 as well. As the Lennard–Jones potential that is used here as the preselected analytical form allows for much less fluctuations in the potential, the radial distribution function is reproduced less accurately.

So far the approach has been illustrated only for one type of nonbonded interaction; however, the same scheme can be used to determine multiple nonbonded interactions simultaneously as well as to determine bond, bond angle bending, and torsion potentials.

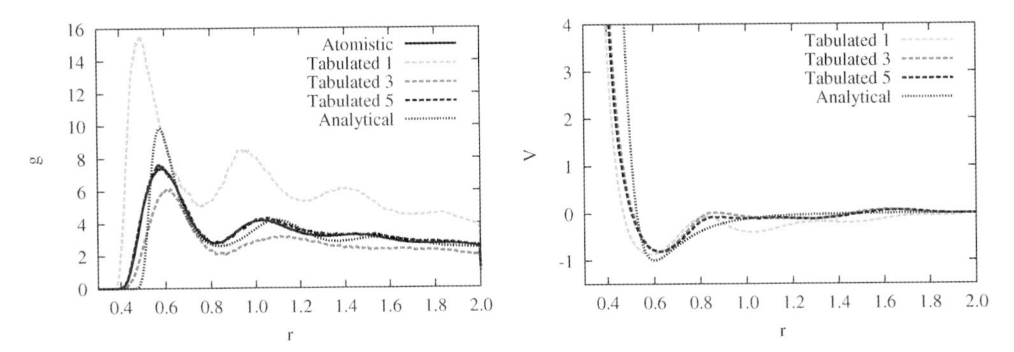


**Figure 8.3** Example of the Boltzmann inversion method applied to butane both for tabulated and analytical potentials. The left part shows the radial distribution functions and the right part the potentials.

### 8.3.2
### Force Matching

The second approach that we will discuss in more detail is the force-matching method [8, 9]. This method is also an example of a multiscale method, meaning that the interaction potentials are again determined from atomistic simulations. Contrary to the Boltzmann inversion method, no structural information is used, but, as the name suggests, forces.

The basic idea of the force-matching method is to run an atomistic simulation and obtain from this atomistic simulation the positions $\vec{R}_{i,m}^{\mathrm{CG}}$ of and the forces $\vec{F}_{i,m}^{\mathrm{ref}}$ on the selected coarse-grained centers for all $N$ coarse-grained particles at $M$ (independent) configurations. As for each of these configurations the positions of all coarse-grained sites are known, the force due to a coarse-grained force field on each of these sites can be calculated as well. When the force between two coarse-grained sites at a distance $r$ apart is given by $f^{\mathrm{CG}}(r)$, the total force on particle $i$ is given by

$$\vec{F}_{i,m}^{\mathrm{CG}} = \sum_{j \neq i} f^{\mathrm{CG}}\left(\left|\vec{R}_{ij,m}^{\mathrm{CG}}\right|\right) \frac{\vec{R}_{ij,m}^{\mathrm{CG}}}{\left|\vec{R}_{ij,m}^{\mathrm{CG}}\right|} . \tag{8.4}$$

A measure for the error $E$ of the coarse-grained force field can be defined as the average squared difference between the reference force on a coarse-grained particle from the atomistic simulation and the force on that particle due to the coarse-grained force field, that is,

$$E = \frac{1}{3MN} \sum_{i=1}^{N} \sum_{m=1}^{M} \left|\vec{F}_{i,m}^{\mathrm{ref}} - \vec{F}_{i,m}^{\mathrm{CG}}\right|^2 . \tag{8.5}$$

The idea of the force-matching method is now to choose the coarse-grained interaction $f^{\mathrm{CG}}(r)$ such that Eq. (8.5) is minimized. Thus, the coarse-grained force field is chosen such that the average error in the forces over the whole set of configuration data in an atomistic simulation is minimized.

Forces with preselected analytical form could be used, as is done in the classical force-matching method, but when the number of interaction types (and thus parameters) grows, the fitting rapidly becomes intractable. However, when tabulated potentials are used with spline interpolation, the forces depend linearly on the fitting parameters such that the least-squares problem can be written in the form of an overdetermined system of linear equations, which can be solved with, for example, matrix triangulation. In this way, a coarse-grained force field, which can be very fluctuant due to the tabulated form, can be derived for every new molecular mixture and for every new state point from a new atomistic simulation at the same conditions at which the coarse-grained system is intended to be simulated.

### 8.3.3
### Based on Thermodynamic Properties

The third approach that we will consider in more detail is one based on thermodynamic properties [10, 11]. Instead of focusing on an accurate reproduction of structural details at a particular state point for a specific system, the aim of methods based on thermodynamic properties is for a broader range of applications without the need to reparametrize the model each time. The idea of these approaches is that if the local thermodynamic properties are correct, the dynamics on long time scales will also be correct.

In these approaches, usually analytical form potentials are chosen. Common choices are Lennard–Jones potentials for nonbonded interactions and harmonic bond stretching and bond bending potentials just like in atomistic simulations. However, where atomistic simulations often use the 12-6 Lennard–Jones potential, sometimes softer versions (like the 9-6) are used as groups of atoms may overlap each other more than individual atoms. The parameters for these predefined analytical potentials are then tuned such that the thermodynamic properties of the coarse-grained system compare as good as possible to known thermodynamic data, like density, surface tension, isothermal compressibility, mutual solubilities, free energies of vaporization/hydration, etc.

A nice example of a force field thoroughly parametrized on thermodynamic data is the Martini force field [11], which focuses on lipids. Because processes such as lipid self-assembly, peptide membrane binding, or protein–protein recognition depend critically on the degree to which the constituents partition between polar and nonpolar environments, parametrization is done by extensive calibration of the building blocks of the coarse-grained force field against thermodynamic data, in particular oil/water partitioning coefficients. A typical calibration simulation based on partitioning coefficients is shown in Figure 8.4. Solute molecules are added to a system with both water and organic solvent. The partitioning free energies are directly obtained from the equilibrium densities of the solute particles in the water and organic solvent.

## 8.4
## Application of Coarse Graining to Lipid Membranes

After having demonstrated how coarse-grained models can be derived, we will show an example of coarse-grained simulations of phospholipid membranes. Such

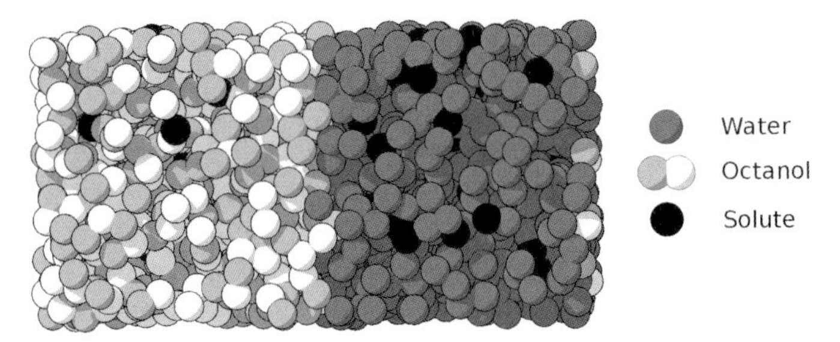


**Figure 8.4** Simulation setup to calibrate based on partitioning coefficients between polar and apolar environment.

membranes exhibit interesting properties at a large variety of length scales. In principle, one would like to simulate everything with the highest degree of detail, thus using electronic structure calculations. However, system sizes that can be treated that way in reasonable time are rather limited compared to the size of these membranes. A typical application would be the study of the interaction of a ligand with the interaction site of a protein in much detail. However, to be able to study a whole protein or a protein in a small piece of membrane, one has to resort to classical force fields, thus using fully atomistic molecular dynamics simulations where the interactions between the atoms are described by empirical potentials. But still, such a system is limited to say 10 nm and tens of nanoseconds. On the other hand, to describe phenomena on much longer length and/or time scales, the membrane could be described as a continuum using elasticity models. Experiments on membranes usually provide a picture at this higher length and time scale as well. However, mechanisms at the molecular level are then no longer visible. With the examples below, we demonstrate that with CGMD the gap between fully atomistic simulations on the one hand and continuum elastic models and experiment on the other can be bridged by performing simulations on longer time and length scales with reasonable molecular detail. As such, lipid membranes form an interesting example for demonstrating CGMD, although coarse graining is, as mentioned before, not restricted to lipids.

### 8.4.1
### The Coarse-Grained Lipid Model

The coarse-grained lipid model that we use is shown in Figure 8.5. To start with, we chose to have only three types of coarse-grained particles: one for the tails, one for the headgroups, and one for the solvent (water). One tail particle represents four methylene groups ($-(CH_2)_4-$) and one water particle represents four water molecules, but our headgroup particles are not directly related to underlying atoms. In order to keep the model as general as possible, the headgroup is represented by four identical particles instead of having different particle types for different chemical groups in the headgroup. This choice not only keeps the lipid model

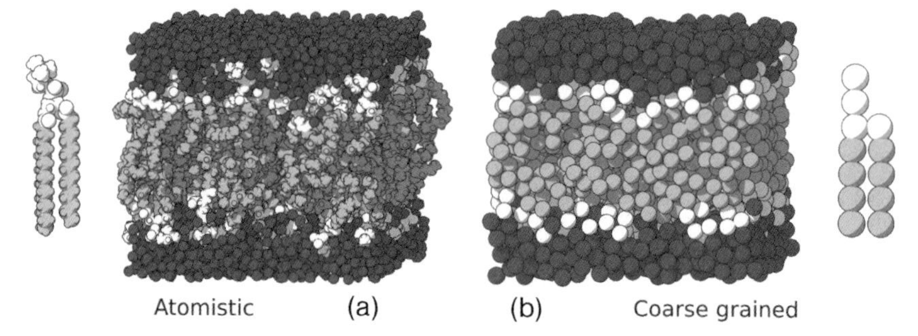


**Figure 8.5** Comparison of the atomistic lipid and bilayer (a) with their coarse-grained counterparts (b).

as general as possible, it also reduces the number of interaction potentials to be parameterized and allows in a later stage for a systematic study of the influence of different headgroup–water interactions. In this sense, the approach of building the coarse-grained lipid molecule is thus mostly phenomenological. However, this example also shows immediately that the division in different strategies that were discussed before is not black and white because to derive the interaction parameters between our coarse-grained particles, we use both atomistic simulations and thermodynamic quantities as well [12]. For example, the sizes and interaction parameters for the coarse-grained water and tail particles have been derived from comparison between coarse-grained simulation with atomistic simulations and known thermodynamic data like densities and melting temperatures of water and various alkanes. The mass and diameter of the headgroup particles were subsequently chosen equal to those of the tail particles, whereas the interaction parameters for these headgroup particles were chosen equal to those of the other hydrophilic particles, that is, the water. Furthermore, the bonded interactions were derived from atomistic simulations. By following the coarse-grained centers in atomistic simulations of eicosane, the equilibrium bond length and the force constant for the coarse-grained bonds were obtained from the frequency spectrum. From comparison with the same simulations, the force constant for the bond-bending potential was derived, which introduced the proper stiffness in the alkyl chains.

The first test of the model comprises its ability to form proper bilayers. When 128 lipids are randomly placed in a simulation box with the remaining space filled with water, during the simulation indeed spontaneous bilayer formation occurs. Although the parameters were not tuned on a bilayer, but on pure water and alkanes instead, many properties of the membranes compare well with those of atomistically simulated bilayers and experimentally known data. However, the applied coarse-graining increases the simulation speed by up to three orders of magnitude compared to fully atomistic simulations such that larger systems can be studied during longer periods of time.

### 8.4.2 Vesicle Formation

An important advantage of being able to study larger systems is that the influence of boundary conditions becomes less prominent. In the small bilayer of Figure 8.5, every particle is never far from a boundary of the simulation box, and the periodic boundary conditions that are used in these simulations highly favor the formation of the (periodic) bilayer. For larger systems, for example when 512 lipids are dispersed in a larger water box, the boundary conditions are already less prominent as no periodic structures are formed anymore. Instead, as can be seen from Figure 8.6, the lipids that are initially randomly dispersed in water (a) first form micelles and bicelles (b). These micelles and bicelles aggregate further into larger bilayers (c). Once sufficiently large, such a bilayer can start to curl (d) and once it closes, it envelopes part of the solvent thus forming a vesicle (e) [12].

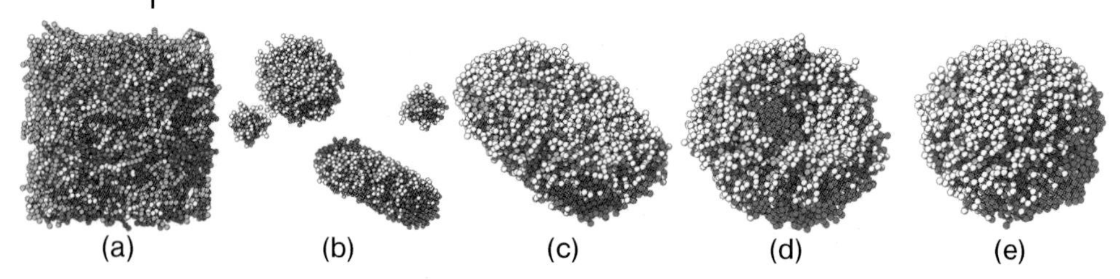


**Figure 8.6** Spontaneous vesicle formation: the lipids that are initially randomly dispersed in water (a) now first form micelles and bicelles (b), which aggregate into a larger bilayer (c) that, once sufficiently large, starts to curl (d) and finally closes to a vesicle (e).

This is a first example of a simulation, which is still out of reach of fully atomistic simulations and cannot be compared with such atomistic simulations. Yet, the vesicle formation process follows exactly the pathway as suggested by experimental studies. However, this simulation shows the process in more detail and allows all kinds of properties, like energies but, for example, also solvent accessible surfaces of the bilayer, to be followed as a function of time. The latter again allows for a comparison with bilayer–vesicle transition as described by continuum elasticity models. As such, this simulation is a first example of bridging the gap between fully atomistic simulations on the one hand and continuum elastic models and experiment on the other.

### 8.4.3 Vesicle Fusion

A second example of how these simulations can bridge the gap is the study of vesicle fusion. The fusion can be visualized experimentally with, for example, fluorescence microscopy. In such studies, both fusing vesicles can be labeled with a different fluorescent probe such that the formation of a hemifusion diaphragm and mixing of the lipids can be seen. However, these experiments give no insight in the mechanisms at the molecular level, such that still many different hypotheses on the molecular mechanisms of vesicle fusion exist. Whereas these mechanisms cannot be readily asserted experimentally, our simulations can show the fusion at a molecular level. When two spherical vesicles are placed next to each other, the vesicles may, by random Brownian motion, move away from each other or come together. In the latter case, the fusion process starts with fusion of the outer monolayers. This is still something that can be observed from a side view using transparently colored water as in Figure 8.7(a). One advantage of these simulations is, however, that we can also make intersections at arbitrary positions. With the intersection plane through the hearts of the two fusing vesicles, one can see how the two inner monolayers of the original vesicles subsequently form the hemifusion diaphragm while the outer monolayers are already completely fused (Figure 8.7(b)). Full fusion is subsequently obtained once this hemifusion

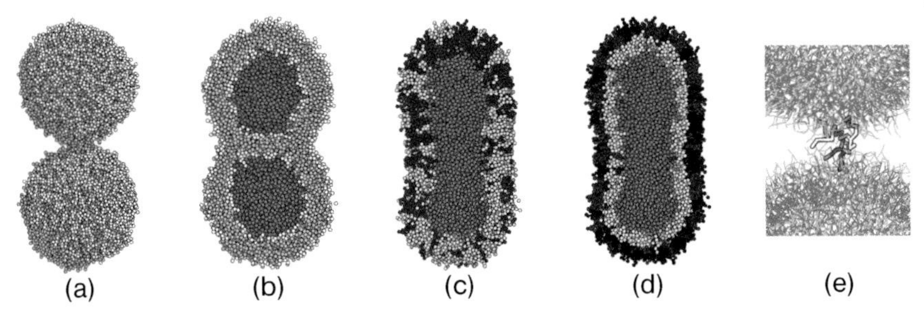


**Figure 8.7** Different stages in the vesicle fusion process (initial contact (a and e), hemifusion diaphragm (b), and fused (c and d)) in different perspectives (side view (a), intersection (b–d), and detail (e)) and color schemes (colored per particle type (a, b, and e) or per molecule (c and d)).

diaphragm breaks resulting in the fusion of the two inner monolayers as well as of the two original interiors as shown in Figure 8.7(c). This also shows an additional advantage of the simulations, namely that we can follow individual or groups of lipids. For example, we can color the lipids of one vesicle in a different color from those of the other vesicle (Figure 8.7(c)) such that we can observe the lateral mixing of the two vesicles, which was also observed in the experiments, in detail. Moreover, by coloring the lipids initially in the inner monolayer again in a different shade from those initially in the outer monolayer (Figure 8.7(d)), we can also see that during the fusion process there is hardly any exchange of lipids between the inner and outer monolayers.

The major advantage of the simulations is that we can also zoom in both in time and space and as such study important stages in the fusion process at a molecular level. For example, the initial contact is shown in detail in Figure 8.7(e). Other important stages, namely stalk formation, stalk expansion, hemifusion diaphragm formation, and hemifusion diaphragm breakage are discussed in detail in Ref. [13].

### 8.4.4 Vesicle Deformation and Fission

So far, we have seen a spherical vesicle and an elongated vesicle as a result of two fused spherical vesicles. Many more vesicle shapes have been observed experimentally and predicted theoretically using continuum elasticity models like the spontaneous curvature or bilayer-coupling model. Shape determining factors are assumed to be the volume ratio (i.e., the ratio between the volume of the interior and the interior of a spherical vesicle with the same membrane area) and the preferred (spontaneous) curvature of the membrane arising from an asymmetry in the composition of the two monolayers of the membrane, an asymmetry that can be either in type or in number of lipids.

To study the influence of such asymmetries, we introduce a second type of headgroup particle, which has a slightly different interaction strength with water. Whereas the effective shape of the original lipids is rather cylindrical, an increase or

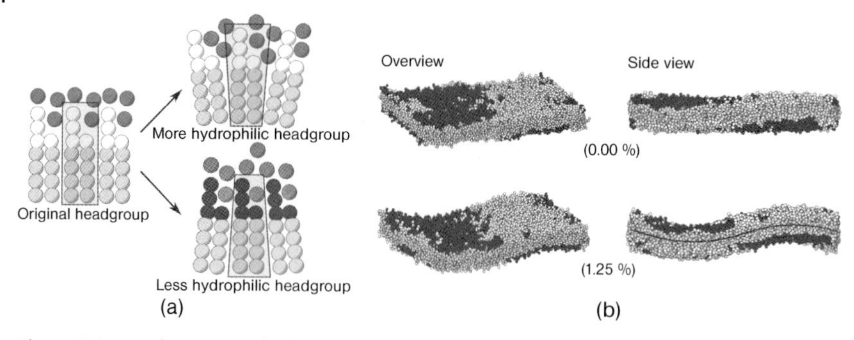


**Figure 8.8** (a) Changes in the headgroup–water interaction change the lipids effective shape. (b) Domains of lipids with slightly differing headgroup–water interactions can already result in highly curved membranes. In the side view of the bilayer with spontaneous curvature and also the membrane shape as calculated with the spontaneous curvature model is shown that allows to couple the microscopic change in headgroup–water interaction with the macroscopic spontaneous curvature.

decrease in headgroup–water interaction will make its effective shape more cone-shaped as illustrated in Figure 8.8(a). As shown in Figure 8.8(b), domains of lipids with small differences in the water–headgroup interaction strength (change given as a percentage) already result in bilayers with a notable spontaneous curvature [14].

Such a spontaneous curvature can be applied to the membrane of a vesicle as well by having one type of lipid in the inner monolayer and a second type in the outer monolayer. However, the headgroup–water interaction difference between the inner and outer monolayers does not require different lipid types *per se*; it could also result from identical lipids with a slightly different solvent, like a pH or ion concentration gradient over the membrane.

Figure 8.9 shows that such a spontaneous curvature in the membrane of a vesicle has a large effect on the vesicles shape. By only slightly changing the

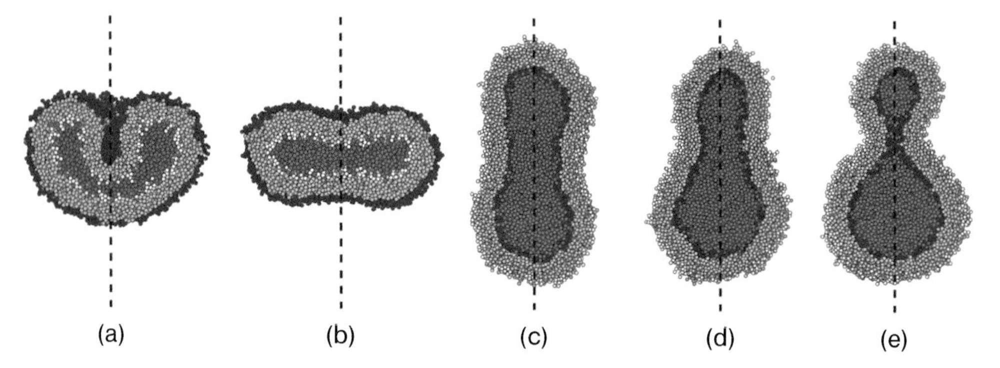


**Figure 8.9** Different vesicle shapes as a result of different spontaneous membrane curvatures. From negative (left) to positive (right) spontaneous curvature, the vesicle transforms from stomatocyte via discoid, elongated, and pear-shaped to a budded shape.

water–headgroup interaction strengths, a vesicle can adopt a wide variety of vesicle shapes. When the water–headgroup interaction strength of the lipids in the outer monolayer is reduced, the originally elongated ellipsoid vesicle (c) transforms into discoid (b) and stomatocyte (a) vesicles. Contrarily, when this headgroup–water interaction is increased, pear-shaped (d) and budded vesicles (e) are formed.

The beauty of this latter example is that by means of our simulations, we can couple (as demonstrated in Figure 8.8(b)) a difference in microscopic interaction parameters with the macroscopic spontaneous curvature. As such, we can compare the vesicle shapes from our simulations with those from the elasticity models, again bridging the gap while providing insight in how microscopic interactions between the molecules result in such a macroscopic spontaneous curvature.

Especially interesting is the vesicle shape on the right of Figure 8.9 as this could be a prestage for vesicle fission. In Ref. 15, we showed that under certain conditions a positive spontaneous curvature can indeed drive a vesicle to split into two smaller vesicles.

## 8.5 Conclusion

The basic idea of coarse-grained molecular dynamics (CGMD) simulations is to represent a system with a reduced number of degrees of freedom compared to its all-atom representation while still incorporating the necessary physics and chemistry. The major advantage of this is that by reducing the number of degrees of freedom, larger simulations can be performed, thus allowing to increasingly bridge the gap between atomistic simulations on one hand and continuum models and experiments on the other.

There is not a single strategy to obtain a coarse-grained model. Instead, many different methods have been suggested over course of time, each having their own advantages, disadvantages, and possible applications. A number of divisions can be made between these approaches, for example, between phenomenological approaches and methods that maintain high chemical specificity, between methods focusing either on structural or thermodynamic data, between methods using preselected analytical form potentials or free-form tabulated potentials, or between methods that are either or not directly coupled with underlying atomistic simulations (multiscale).

A possible application of CGMD was demonstrated by simulations of lipid membranes. Although the model used is still rather phenomenological, the simulations already show important insights in, for example, bilayer and vesicle formation, vesicle fusion, and vesicle deformation. Moreover, with the help of one of the coarse-graining strategies described in Section 8.3, higher chemical specificity could be added to such a lipid model providing insight in for which specific lipids and under which exact circumstances the observed phenomena will occur. With the ability of the latest coarse-graining strategies to incorporate quite some chemical detail into coarse-grained models, these methods become increasingly interesting for chemists.

## References


1 R. Faller, "Automatic coarse graining of polymers", *Polymer*, **45**, 3869–3876, (**2004**).

2 J. Zhou, I.F. Thorpe, S. Izvekov, G.A. Voth, "Coarse-grained peptide modeling using a systematic multiscale approach", *Biophys. J.*, **92**, 4289–4303, (**2007**).

3 J.S. Chen, H. Teng, A. Nakano, "Wavelet-based multi-scale coarse graining approach for DNA molecules", *Finite Elements in Analysis and Design*, **43**, 346–360, (**2007**).

4 H.J. Limbach, K. Kremer, "Multiscale modeling of polymers: Perspectives for food materials", *Trends Food Sci. Technol.*, **17**, 215–219, (**2006**).

5 B. Smit, P.A.J. Hilbers, K. Esselink, L.A.M. Rupert, N.M. van Os, A.G. Schlijper, "Computer simulations of a water/oil interface in the presence of micelles", *Nature*, **348**, 624–625, (**1990**).

6 D. Reith, M. Pütz, F. Müller -Plathe, "Deriving effective mesoscale potentials from atomistic simulations", *J. Comput. Chem.*, **24**, 1624–1636, (**2003**).

7 F. Müller -Plathe, "Coarse-graining in polymer simulation: From the atomistic to the mesoscopic scale and back", *Chem. Phys. Chem.*, **3**, 754–769, (**2002**).

8 S. Izvekov and G.A. Voth, "Multiscale coarse graining of liquid-state systems", *J. Chem. Phys.*, **123**, 134105, (**2005**).

9 J.W. Chu, S. Izveko, G.A. Voth, "The multiscale challenge for biomolecular systems: Coarse-grained modeling", *Mol. Sim.*, **32**, 211–218, (**2006**).

10 S.O. Nielsen, C.F. Lopez, G. Srinivas, M.L. Klein, "A coarse grain model for *n* -alkanes parametrized from surface tension data", *J. Chem. Phys*, **119**, 7043–7049, (**2003**).

11 S.J. Marrink, H.J. Risselada, S. Yefimov, D.P. Tieleman, A.H. de Vries, "The Martini force field: Coarse-grained model for biomolecular simulations", *J. Phys. Chem. B*, **111**, 7812–7824, (**2007**).

12 A.J. Markvoort, K. Pieterse, M.N. Steijaert, P. Spijker, P.A.J. Hilbers, "The bilayer-vesicle transition is entropy driven", *J. Phys. Chem. B*, **109**, 22649–22654, (**2005**).

13 A.F. Smeijers, A.J. Markvoort, K. Pieterse, P.A.J. Hilbers, "A detailed look at vesicle fusion", *J. Phys. Chem. B*, **110**, 13212–13219, (**2006**).

14 A.J. Markvoort, R.A. v. Santen, P.A.J. Hilbers, "Vesicle shapes from molecular dynamics simulations", *J. Phys. Chem. B*, **110**, 22780–22785, (**2006**).

15 A.J. Markvoort, A.F. Smeijers, K. Pieterse, R.A. v. Santen, P.A.J. Hilbers, "Lipid based mechanisms for vesicle fission", *J. Phys. Chem. B*, **111**, 5719–5725, (**2007**).

# 9 Reactive Force Fields: Concepts of ReaxFF

*Adri van Duin*

## 9.1 Introduction

Many of the challenges in the chemical and materials technologies require information regarding the dynamical properties of atoms, covering both the physics (diffusion, phase behavior) and chemistry (reactions) of atom–atom interactions. Experimental techniques can provide information on these properties, but usually do not have the spatial or time-resolution required to capture the atomistic-scale events. As such, we require computational methods to fully understand these events and obtain knowledge that allows us to improve existing or design new materials. The foundation for such a computational-based understanding is quantum mechanics (QM). There has been enormous progress in the last decade in using these QM methods to predict and analyze many important systems in materials science. Despite this progress, the time and length scales of accurate QM (100 to 1000 atoms) falls short of the scales required for designing and understanding experiments, (for example, a 20 nm cube has 1 million atoms). To solve this problem, allowing materials design based on QM principles, one can adopt the multiscale strategy illustrated in Figure 9.1. The idea here is to use the results of QM on small systems (~100 atoms) to determine an energy expression (force field, FF) that can calculate the forces directly for use in molecular dynamics (MD) calculations (solving coupled Newton's equations). This speeds up the calculation by a factor of a million, allowing simulations of the properties of systems with up to ~1 billion atoms in massively parallel simulations. One can go one step further in the multiscale strategy by using the results of such FF-based MD simulations to determine the parameters of some coarse-grain or mesoscale description in which the elements might represent 100s or 1000s of atom to permit a scale suitable for $10^{12}$ atoms, which begins to approach the size and timescales required for real materials design.

The success or failure of this multiscale strategy depends on whether it is possible to develop paradigm-bridging strategies that link the versatile, but more computationally expensive (e.g., QM, Figure 9.1) to the more limited, but less

*Computational Methods in Catalysis and Materials Science.* Edited by Rutger A. van Santen and Philippe Sautet

ISBN: 978-3-527-32032-5

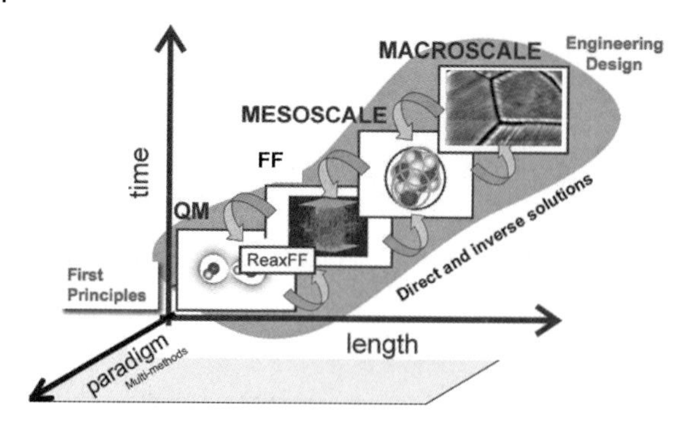


**Figure 9.1** Hierarchy of simulation methods.

expensive (e.g., FF, Figure 9.1) methods. On the QM/FF interface, the crucial challenge for the paradigm-bridging method is to properly describe the making and breaking of the chemical bond. Here we discuss the development of one of these paradigm-bridging methods, the ReaxFF reactive force field method [1, see the Application section for other ReaxFF references], which aims to bridge between traditional QM methods on one side and classical nonreactive force fields on the other side by enabling dynamical simulations on large (≫1000 atoms) reactive systems. To this end, we will start by introducing the concepts of traditional nonreactive force fields and describe how these concepts can be extended to enable the simulation of bond dissociation and formation. These concepts include bond distance/bond-order methods, coupling of bond orders to multibody interactions, and charge delocalization.

The ReaxFF reactive force field is certainly not the only method that is active at the QM/FF interface. ReaxFF is an empirical method, with an energy description that is more connected with FF tradition than with QM concepts. Other approaches for extending the time- and system size limits of QM methods include pseudo-*ab initio* [2] or tight-binding approaches [3], which are more directly linked with QM concepts. Empiricism has advantages in method development, as one is not tied to the shape of the potential functions dictated by first principles. Furthermore, empiricism makes it easier to "cherry-pick" between different approaches (like the bond order, multibody terms, and polarization methods that are included in the ReaxFF method). However, by abandoning direct links with the first-principle theory it becomes much harder to define the application area of a method; for a first-principle-based method the application range can, in principle, be directly defined by the approximations made in its development. Without the intrinsic justification of first-principles roots, empirical methods can only be validated by extrapolation; if the method shows good similarity to QM data or experimental results, for a particular material under certain conditions, one may assume that the method is also valid for a modified material or for the same material under different conditions.

## 9.2 Force Field Methods

Force field (FF) methods are a widely used tool for simulating the energetics and dynamics of large (≫1000 atoms) systems [4]. In general terms, FF methods abandon the electronic degrees of freedom described explicitly in QM methods and instead aim to describe the nuclear interactions by a set of empirical equations. Focusing on FF methods for covalent materials, we can identify a set of common elements between FF methods:

- The system energy is divided into contributions from bonded and nonbonded interactions (Scheme 9.1).
- The bonded interactions are linked to the internal coordinates of the molecule, most commonly including bonds, angles, and torsion angles (Figure 9.2, Scheme 9.1).
- For each of these internal coordinates an equilibrium state is defined (equilibrium bond length $r_o$, equilibrium angle $\theta_o$, Scheme 9.1)
- An energy function is linked to the internal coordinate, describing the energy and force required to displace the internal coordinate from its equilibrium state. In the simplest case, these energy functions can be harmonic (Scheme 9.1); however, the simple harmonic description fails even at small deviations from equilibrium. As such, more advanced FF methods employ cubic- or higher order terms and coupling between the internal coordinates (e.g., bond/angle cross-terms) to extend the application range of the force field to higher distortions.
- Nonbonded interactions are commonly divided into van der Waals- and Coulomb contributions (Scheme 9.1). Scheme 9.1 shows a commonly used Morse potential for the van der Waals, other popular choices include power functions

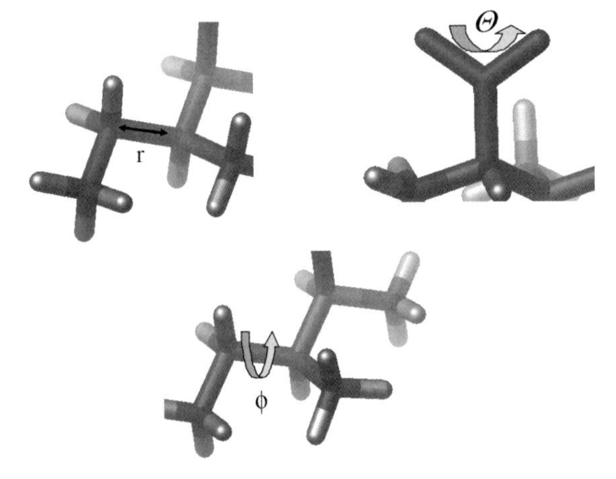


**Figure 9.2** Definition of bond ($r$), valency angle ($\Theta$), and torsion angle ($\phi$) molecular internal coordinates.

$$E_{\text{system}} = E_{\text{bond}} + E_{\text{angle}} + E_{\text{torsion}} + E_{\text{vdWaals}} + E_{\text{Coulomb}}$$

$$E_{\text{bond}} = k_b (r - r_o)^2$$

$$E_{\text{angle}} = k_v (\Theta - \Theta_o)^2$$

$$E_{\text{torsion}} = V_2 \cdot (1 - \cos 2\varphi) + V_3 \cdot (1 + \cos 3\varphi)$$

$$E_{\text{vdWaals}} = D_{ij} \left\{ \exp\left[ \alpha_{ij} \cdot \left( 1 - \frac{r_{ij}}{r_{\text{vdW}}} \right) \right] - 2 \cdot \exp\left[ \frac{1}{2} \alpha_{ij} \cdot \left( 1 - \frac{r_{ij}}{r_{\text{vdW}}} \right) \right] \right\}$$

$$E_{\text{Coulumb}} = C \cdot \frac{q_i \cdot q_j}{r_{ij}}$$

**Scheme 9.1** System energy definition for a simple harmonic nonreactive force field.

(Lennard–Jones potentials) and mixed power/exponent functions. The charges used to calculate the Coulomb interactions can come from various sources; in protein force fields like CHARMM [5] and AMBER [6] charges are derived for each peptide segment from QM calculations; these charges are kept fixed during the simulation. In most cases, nonbonded interactions are excluded between atoms sharing a bond, a valence angle or even a dihedral angle. These nonbonded exclusions completely separate the length scale of the bonded (short length scales) and nonbonded interactions (long range).

Once the set of empirical equations linking the molecular geometry to energy and forces is defined one obtains a set of parameters (like the force constants $k_b$ and $k_v$ in Scheme 9.1). Provided that a direct link between an experimental observable or QM-based property exists (like for example the equilibrium bond length $r_o$, which can be linked to the bond length in a strainless molecule, or the force constant $k_b$ which can be related to the vibrational stretch frequency) one can try to formally assign these parameter values. Quite often, however, especially for more complicated force fields, one finds that significant parameter correlation exists, which makes it complicated to link a single parameter to an experimental observable or QM-based property. In this case, one can obtain the parameter values by fitting them to a database of experimental and/or QM data (training set). The contents of the training set define the transferability of the force field; a force field that successfully reproduces data for a wide range of structure diversity can be expected to have a broader range of applications. When using a force-field-based method it is vital to familiarize oneself with the approach employed in its parameterization, as this defines the material and condition range within which the force field can be expected to give reliable results.

Currently, force fields that match the description above are available for a wide range of materials, including hydrocarbons [7], proteins [5,6], water [8], and covalent materials like silicon and silicon dioxide [9,10]. These force fields generally give good descriptions of the molecular or materials equilibrium properties (Figure 9.3), thus enabling them to be applied to energy minimization, molecular dynamics, and Monte Carlo applications to low-strain states. However, the equations described in

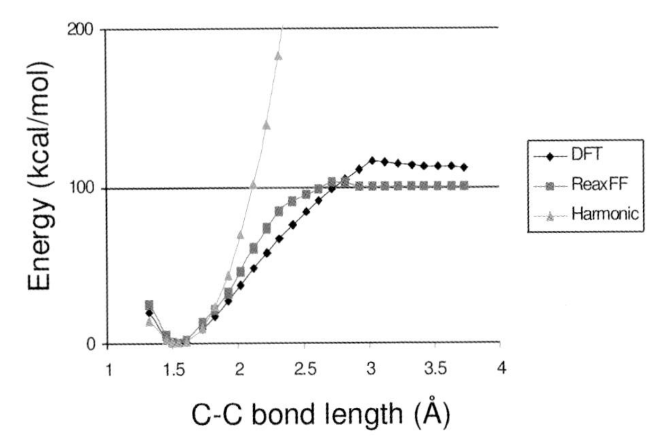


**Figure 9.3** Bond length/bond-energy relationship for the C–C bond in ethane according to a harmonic, nonreactive FF (harmonic), a reactive force field (ReaxFF), and a QM method (DFT).

Scheme 9.1, even when augmented with nonharmonic terms, cannot be expected to describe bond dissociation, as the harmonic description leads to the wrong asymptotic behavior as extended bond lengths (see Figure 9.3 for an example for the C–C bond in ethane). As such, their application is restricted to nonreactive simulations.

## 9.3 Making a Force Field Reactive

This section describes how the FF concepts described in the previous section can be extended to obtain a proper energetic and force description of bond formation and dissociation, thus providing a simulation tool that may not be as universal as QM methods but that can be trained to simulate selected reactive processes at a computational expense that is magnitudes faster than QM.

### 9.3.1 The Bond Order Concept

First and foremost, the harmonic approximation employed in nonreactive FF methods to describe bond stretching needs to be replaced by a description that converges to the bond dissociation energy at infinite atom separation, rather than to infinite energy as in the harmonic description (Figure 9.3). Furthermore, this bond distance/energy relation has to be continuous and should, preferably, be rooted in physical/chemical theory. One approach that fits these criteria is the bond order/bond-energy concept, as proposed by Pauling [11]. In this approach, the bond order, which effectively describes the number of shared electrons in the

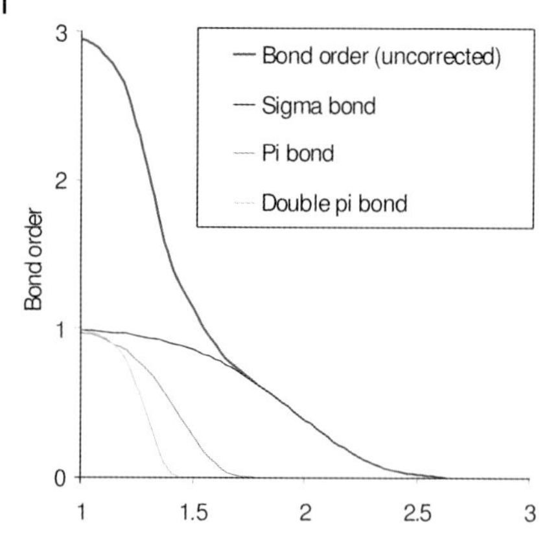


**Figure 9.4** ReaxFF total bond order and $\sigma$, $\pi$, and double-$\pi$ bond carbon–carbon bond order as a function of interatomic distance.

bond between two atoms, is linked by means of a continuous function to the bond length. At short distances, this bond order/bond-distance relation approaches a maximum (e.g., a triple bond for a C–C pair) while at infinite distance the bond order goes to zero (Figure 9.4). For noncovalent materials other approaches can be employed to allow FF methods to simulate coordination changes. For ionic materials one can replace the harmonic bond description with a purely Coulomb-based description [10], while metals can be successfully described by replacing the bond concept with a density term [12]. Both these Coulomb-based and density approaches satisfy the condition that the system energy goes to zero at infinite atom separation, but they lack the localized character required to describe covalent materials. The bond order/bond-distance concept does provide the necessary localization and the FF methods developed by Tersoff for silicon [13] and by Brenner for hydrocarbons [14] demonstrated that this approach could be used to formulate a FF method that can simulate reactions in covalent materials.

### 9.3.2
### Capturing Transition States with a Bond-Order Approach

The bond order/bond-distance approach adds the ability to FF methods to simulate connectivity changes and converges to the right dissociative bond limit. However, a simple bond order/bond-distance equation, like that depicted in Figure 9.4, does not capture all aspects of chemical reactivity. The range of the attractive covalent interactions related to bond orders is strongly coupled to the coordination of the atoms involved in the bond order; for fully coordinated atoms these attractive interactions decay quickly with bond distance, while undercoordinated, radical atoms have more delocalized valence electrons, resulting in a much longer attractive range. Both of these aspects need to be described properly by a reactive force field to both capture the stability of ground state molecules as well as the reactivity of radicals. In the ReaxFF reactive force field, this is achieved by employing a

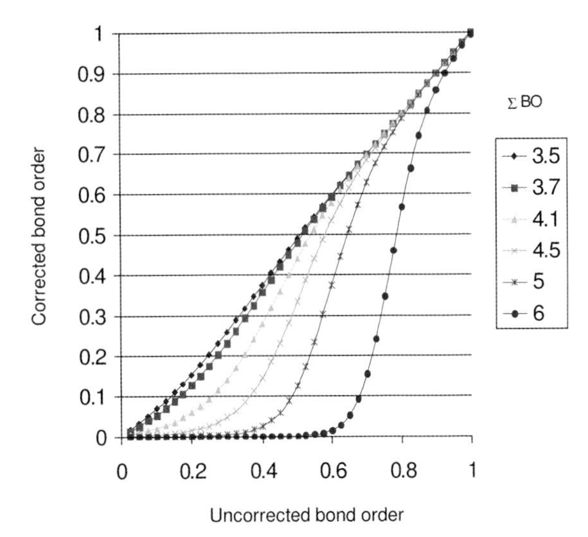


**Figure 9.5** Bond-order correction scheme in ReaxFF. Σ BO is the sum of the uncorrected bond orders around the atom. For carbon Σ BO = 4 describes a normal coordinates (4 strongly bound neighbors) while Σ BO = 3.5 describes a carbon with a partial radical character.

set of analytical functions that translate in the bond-order correction scheme in Figure 9.5. This corrects the initial bond orders, obtained from the interatomic distances, based on the level of undercoordination of the atoms, which is obtained by summing up all bond orders the atom is involved in. If the total bond-order approaches or exceeds the number of valence electrons for the element (4 for carbon, 1 for hydrogen), all weak bond orders, often related to 1–3 interactions, are significantly reduced in magnitude, while the strong bond orders remain virtually uncorrected (Figure 9.6). For undercoordinated radical systems, however,

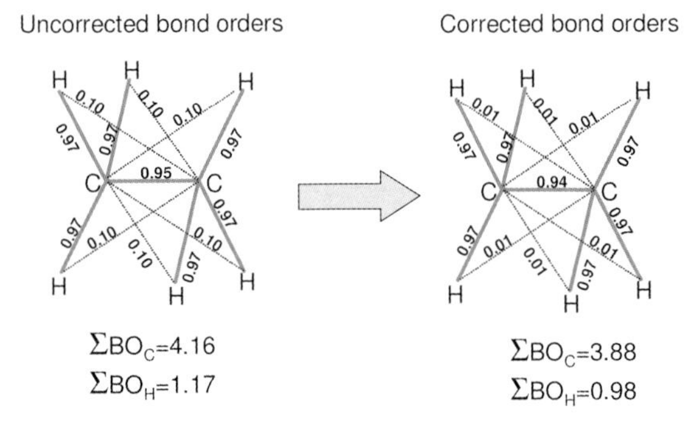


**Figure 9.6** Effect of the ReaxFF bond-order correction scheme on the bond orders in ethane. The weak bond orders are reduced by a magnitude while the strong bond orders remain almost unaffected.

these weak bond orders remain uncorrected, thus allowing a radical carbon sites to initiate bonding at distances as far as 3 Å. This bond-order correction scheme enables ReaxFF to properly capture QM-transition state energies, enabling the method to not only predict reaction products but also describe reaction kinetics.

This bond-order correction scheme effectively turns the bond orders from a 2-body interaction (depending only on the positions of the atoms sharing the bond) into a multibody interaction, where every atom in the coordination sphere affects every bond order that atom is involved in. While this is certainly realistic, this does come at a significant computational expense, as all bond-order related forces need to be determined over multiple atoms.

### 9.3.3 Coupling Bond Orders to 3-body Terms

In traditional nonreactive FF methods for first row elements the effects of molecular orbital hybridization on molecular geometry are captured by 3-body, valence angle, and 4-body, torsion (a.k.a. dihedral) terms. These terms ensured that all H–C–H angles around a $sp^3$ hybridized carbon in methane would have values of 109.47° and enforced planarity on the H–C=C–H torsion angles. These nonreactive FF methods usually employed different atom types for different atom hybridizations (e.g., $sp^1$, $sp^2$, and $sp^3$ hybridization for carbon); by simply assigning different equilibrium angles (180° for $sp^1$, 120° for $sp^2$, and 109.47° for $sp^3$ carbon) they ensure that the FF method gives a correct description of molecular geometry. This approach works fine in a nonreactive environment, since atoms will retain their initial hybridization. However, a reactive force field needs to provide a continuous transition between transition states; changing atom type based on local geometrical considerations is not viable as this will likely result in a discontinuity in energies and forces, which severely implicates molecular dynamics applications. As such, the reactive force field needs to have the capability to identify the hybridization state by itself and needs to have rules on how this state affects the valence- and torsion angle forces around the atoms.

Using the bond order/bond-distance scheme introduced in the previous sections, identification of hybridization state is relatively straightforward. In ReaxFF, separate bond distance/bond-order relationships are used for the single ($\sigma$), double ($\pi$i), and triple (double-$\pi$) bond order (Figure 9.4) and this information can be directly used to derive the hybridization effects on equilibrium angles and torsion angle force constants. Scheme 9.2 and Figure 9.7 describe how information regarding the amount of pi-character in a bond can be used to derive a valence angle equilibrium angle that smoothly goes from tetrahedral (109.47°) for $sp^3$-carbons to linear (180°) for $sp^1$-carbons. After deriving the bond-order dependent equilibrium angle, the valence angle energy can be calculated according to Scheme 9.2. An important difference between the valence angle/energy relationship in Scheme 9.2 compared to the nonreactive FF description in Scheme 9.1 is that the valence angle energy is multiplied by a function of the bond orders involved in the angle, thus ensuring that the angle energy smoothly goes to zero upon bond dissociation. This latter

$$E_{\text{val}} = f(\text{BO}_{ij}) \cdot f(\text{BO}_{jk}) \cdot \left\{ p_{\text{val1}} - p_{\text{val1}} \exp\left[ - p_{\text{val2}} \left( \Theta_o \left( \text{BO}^{\pi} \right) - \Theta_{ijk} \right)^2 \right] \right\}$$

$$\Theta_0 \left( \text{BO}^{\pi} \right) = 180^o - \Theta_{0,0} \cdot \{ 1 - \exp[ - p_{\text{val10}} \cdot (2 - \text{SBO}') ] \}$$

$$f(\text{BO}_{ij}) = 1 - \exp\left( - p_{\text{val3}} \cdot \text{BO}_{ij}^{p_{\text{val4}}} \right)$$

**Scheme 9.2** Bond order incorporation in the ReaxFF valence angle function. $\Theta_o(\text{BO}^{\pi})$ is the equilibrium angle, SBO′ is the sum of the $\pi$- and double-$\pi$ bond orders around the central atom $j$ in the valence angle, $p_{\text{val1}}, p_{\text{val2}}, p_{\text{val3}}, p_{\text{val10}}$, and $\Theta_{o,o}$ are force field parameters; $\Theta_{ijk}$ is the value of valence angle $ijk$.

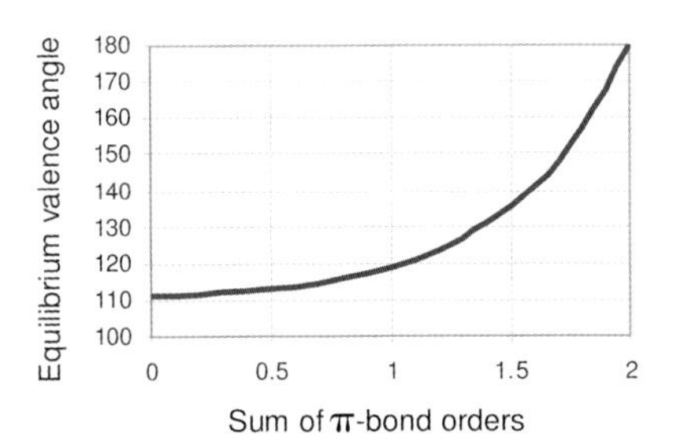


**Figure 9.7** Relationship between $\pi$-bond order (SBO′, Scheme 9.2) and equilibrium valence angle ($\Theta_o$(SBO′), Scheme 9.2) in ReaxFF.

condition is vital to retain the energy and force continuity required for molecular dynamics simulations.

These two concepts of calculating equilibrium properties from the bond orders and including the bond orders in the energy/force calculations can be extended to all covalent interactions; in ReaxFF similar concepts are used to describe torsion angles, 3-body, and 4-body conjugation and hydrogen bonds enabling description of rotational barriers, aromatic and partially conjugated systems and hydrophilic functional groups.

### 9.3.4 Charge Equilibration and Nonbonded Interactions

The previous sections describe how a bond order/bond-distance scheme can be used to describe coordination changes and how these bond orders can be employed to derive the atom hybridization and the related multibody interactions. This, by itself, already provides a platform for describing chemistry in nonpolar environments, as demonstrated by the widely used Brenner potential. However, to properly capture the physics of intermolecular interactions and to extend reactive methods to more polar materials the reactive force field should also describe the nonbonded van der Waals and Coulomb interactions. As mentioned previously, nonreactive FF methods for first-row elements usually exclude these interactions between atoms that share a bond, valence angle or sometimes even a torsion angle and commonly use fixed point charges. Both the exclusion and fixed-charge model cannot be transferred into a reactive environment; the exclusion concept fails since the definition of bonds, valence, and torsion angles continuously changes in a reactive environment, while the charge of a atom needs to be updated during reactive events to reflect modifications in the electronegativity of its neighbors.

Starting with the van der Waals interactions; an attractive approach for handling these interactions in a reactive environment is to introduce a switching function, which effectively is a continuous equivalent of the exclusion principle. Basically, when the distance between two atoms approaches their covalent range their repulsive interactions are smoothly scaled down to zero; the weakly long-range attractive sections of the van der Waals forces are completely retained. This approach, which has been implemented in the AIREBO potential [15] and in LCBOPII [16], keeps the length scales of the van der Waals and covalent interactions separate, which has distinct advantages in the force field development. An alternative approach, implemented in ReaxFF, is to fully abandon the exclusion principle and include the full van der Waals, both the highly repulsive short-range part as well as the weak long-range attractive part, for the entire spectrum of bond distances. In this approach, the van der Waals repulsion strongly contributes to the bond energy; as such the attractive, bond-order-based energies have to compensate for this so that the sum of attractive and repulsive interactions equals the bond dissociation energy (Figure 9.8). This approach strongly correlates the nonbonded and bond-order-based interactions, which complicates force field development and clouds the physical meaning of the force field parameters, but has the advantage that van der Waals interactions directly contribute to transition state energies. Furthermore, the van der Waals repulsion effectively turns into a density-based term at short

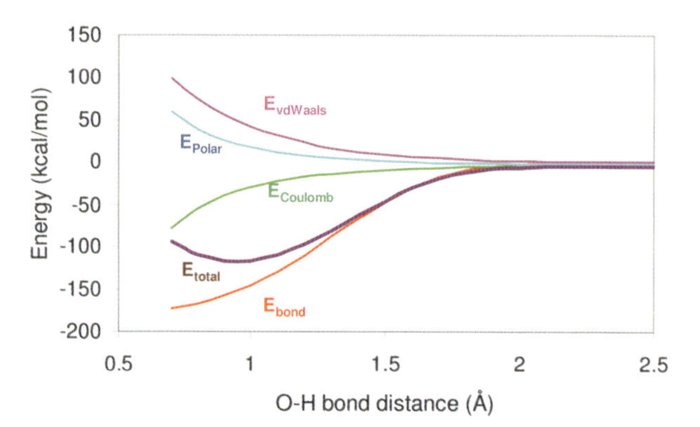


**Figure 9.8** Bond, van der Waals, Coulomb, and polarization energy contributions to the ReaxFF total bond energy for a OH radical. The polarization energy describes the energy required to delocalize the charges from the EEM scheme; this is calculated using $E_{\text{polar}} = \Sigma \chi_i q_i + \eta_i q_i^2$ (see Scheme 9.3).

distances, as commonly present in FF methods for metals [12], thus facilitating transfer to metallic materials.

To describe Coulomb interactions in a reactive FF, a drastic deviation from the nonreactive concept of fixed point charges and the exclusion principle is required. What we need here is a method that updates the charge distribution based on local electronegativity changes. Such methodology is available in the EEM [17] and QEq [18] approaches (Scheme 9.3). Both these approaches assign a electronegativity (the first derivative of the atom energy to its number of electrons) and a hardness (the second derivative of the atom energy to its number of electrons) to each element. This describes the energy required to polarize the atoms; when a Coulomb term is added to describe the energy obtained from the atom polarization, a set of equations is obtained that can be solved using the total system charge as a restraint. While the EEM- and QEq methods are also useful in nonreactive FFs, their concepts are much more crucial for reactive FF methods and greatly expand their application range. An even greater transferability can be obtained by abandoning the exclusion principle, allowing Coulomb interactions to directly affect the bond dissociation energies. By this means, the atom polarization directly affects the local chemistry, which directly integrates electrophilic and nucleophilic reactive concepts. To abandon the exclusion principles we need to shield the short-range Coulomb energy, since an undamped Coulomb repulsion or attraction would completely overwhelm the covalent, bond-order derived interactions (Figure 9.8). Such a shielding makes physical sense, as electron cloud overlap would indeed

$$\frac{\partial E}{\partial q_1} = \chi_1 + 2q_1\eta_1 + C \cdot \sum_{j=1}^{n} \frac{q_j}{\left(r_{1,j}^3 + \left(\frac{1}{\gamma_{1,j}}\right)^3\right)^{\frac{1}{3}}}$$

$$\frac{\partial E}{\partial q_2} = \chi_2 + 2q_2\eta_2 + C \cdot \sum_{j=1}^{n} \frac{q_j}{\left(r_{2,j}^3 + \left(\frac{1}{\gamma_{2,j}}\right)^3\right)^{\frac{1}{3}}}$$

........

........

$$\frac{\partial E}{\partial q_n} = \chi_n + 2q_n\eta_n + C \cdot \sum_{j=1}^{n} \frac{q_j}{\left(r_{n,j}^3 + \left(\frac{1}{\gamma_{n,j}}\right)^3\right)^{\frac{1}{3}}}$$

$$\sum_{i=1}^{n} q_i = 0$$

**Scheme 9.3** Set of equations used to derive atomic charges ($q$) using the shielded EEM approach. $\chi$ is the atomic electronegativity, $\eta$ is the atomic hardness, and $\gamma$ is the EEM shielding parameter and $C$ is a conversion constant.

diminish Coulomb effects at short ranges. In the QEq method, this overlap effect is calculated from Slater-orbital functions, while in the EEM method, which is integrated into ReaxFF, this shielding effect is included in a more empirical fashion by a straightforward modification of the Coulomb-energy function, adding a shielding constant $\gamma$ to the Coulomb equation (Scheme 9.3).

While the EEM/QEq methods do not describe all aspects of charge polarization correctly [19], they certainly extend the transferability of the reactive FF methods. They are, however, computationally the most demanding part of the reactive FF approach; for larger systems (>1000 atoms) the calculation time will be dominated by these methods. But even with these polarization methods the reactive FF methods remain magnitudes faster, and scale significantly better with the number of atoms, than QM methods.

## 9.4 Transferability, Training, and Applications of ReaxFF

The previous section describes how the concepts of nonreactive FF methods were modified to obtain the ReaxFF reactive force field model, which is capable of simulating chemical reactions at computational expense magnitudes lower than QM methods. These modifications include:

- Introduction of a bond order/bond-distance method, giving a continuous description of covalent bond-order dissociation.
- Integration of a bond-order correction scheme, which allows the bond order to distinguish between radical (delocalized) and ground-state (localized) bonds.
- Integration of these bond orders in all bonded interactions (including interactions based on valence angles and torsion angles).

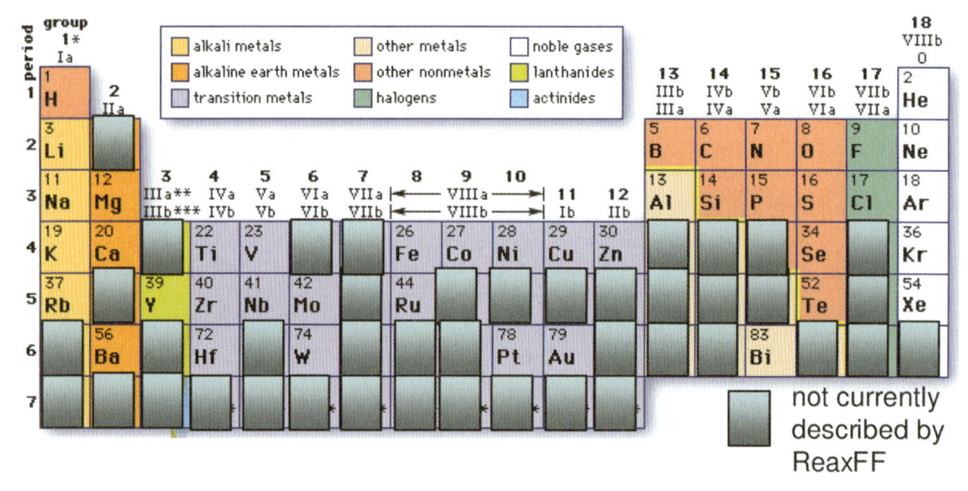


**Figure 9.9** Elements in the periodic table described by ReaxFF.

- Calculation of nonbonded (van der Waals and Coulomb) interactions between *all* atoms, thus abandoning the exclusion principle.
- Integration of a polarizable charge model.

These modifications yield a reactive model that can describe covalent [1,20–24], metallic [25–28], and (partionally) ionic materials [25,26,29] and has been successfully parameterized to simulate the physical and chemical behavior of materials all across the periodic table (Figure 9.9). To allow ReaxFF to describe a material its parameters are optimized against a mainly QM-derived database (a.k.a. training set) describing molecular and/or material properties. These properties include atomic charges, bond dissociation energies, angle strain, heats of formation (here

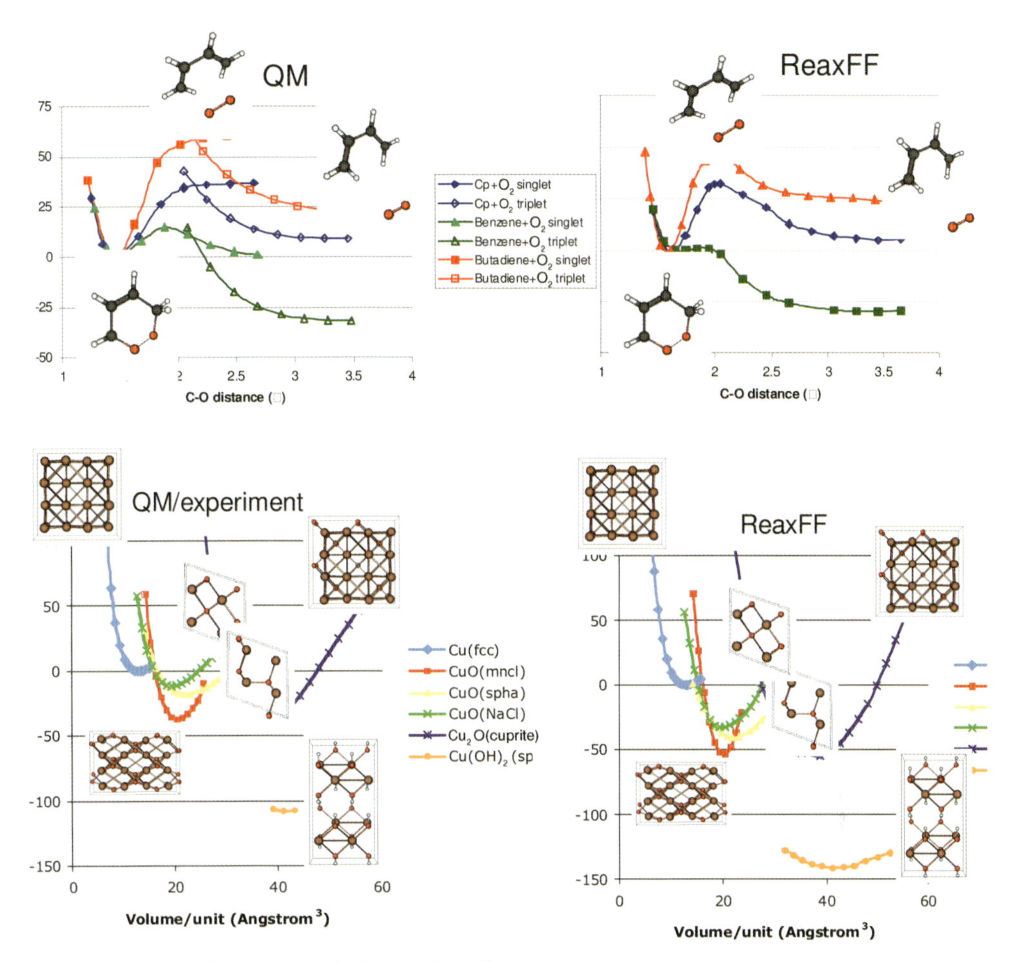


**Figure 9.10** QM- and ReaxFF results for covalent ($O_2$ reactions with hydrocarbons, up) metallic (Cu-fcc metal equation of state, bottom) and partially ionic (CuO, $Cu_2O$, $Cu(OH)_2$ condensed phases equations of state, bottom) materials.

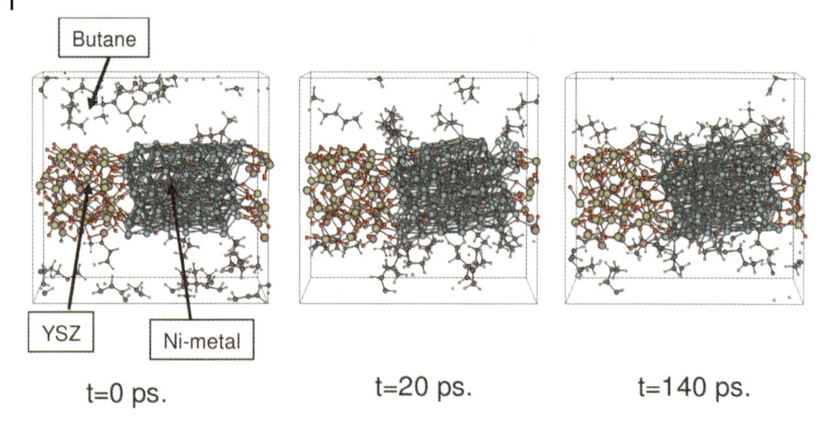


**Figure 9.11** Example of a ReaxFF simulation of the reactive dynamics at a butane/nickel-metal/ytrium stabilized zirconium dioxide (YSZ) interface. This simulation was performed using an NVT ensemble at $T = 750$ K. During the simulation we observe hydrocarbon conversion on the Ni surface, hydrogen transfer from the Ni- to the YSZ-surface and oxygen transfer from the YSZ to the Ni bulk.

usually experimental data is used instead of QM), vibrational frequencies, and transition-state energies. Figure 9.10 gives examples of ReaxFF and QM results for covalent, metallic, and (partially) ionic materials. All these materials are described with exactly the same set of potential functions, enabling ReaxFF to describe material interfaces. Figure 9.11 shows an example of a ReaxFF application to a mixed ionic/metallic/covalent system; in this simulation the reactions of *n*-butane on a nickel surface, interfaced to an yttrium-stabilized zirconium (YZr) dioxide phase, is simulated at a temperature of 750 K. During this simulation we observe many aspects of cross-interface chemistry; butane is dissociated on the Ni surface generating hydrogen, which migrate to the YSZ surface, while oxygen from the YSZ surface migrate into the Ni bulk. It is in these multimaterial dynamic environments, which require QM-based accuracy for barriers and reaction energies, combined with dynamics on large (>1000 atoms) systems, that reactive force fields like ReaxFF have their most obvious applications.

## References

1 A.C.T. van Duin, S. Dasgupta, F. Lorant, and W.A. Goddard (**2001**) *J. Phys. Chem. A* **105**, 9396–9409.

2 J.J.P Stewart (**1989**) *J. Comput. Chem.* **10**, 209.

3 A.K. McMahan and J.E. Klepeis (**1997**) *Phys. Rev. B* **56**, 12250.

4 M.P. Allen and D.J. Tildesley (**1987**) *Computer Simulation of Liquids*, Oxford University Press, New York.

5 A.D. MacKerell, D. Bashford, M. Bellott, R.L. Dunbrack, J.D. Evanseck, M.J. Field, S. Fischer, J. Gao, H. Guo, S. Ha, D. Joseph-McCarthy, L. Kuchnir, K. Kuczera, F.T.K. Lau, C. Mattos, S. Michnick, T. Ngo, D.T. Nguyen, B. Pro hom, W.E. Reiher, B. Roux, M. Schlenkrich, J.C. Smith, R. Stote, J. Straub, M. Watanabe,

J. Wiorkiewicz-Kuczera, D. Yin, and M. Karplus (**1998**) All-atom empirical potential for molecular modeling and dynamics studies of proteins. *J. Phys. Chem., B* **102**, 3586–3617.

6 W.D. Cornell, P. Cieplak, C.I. Bayly, I.R. Gould, K.M. Merz, D.M. Ferguson, D.C. Spellmeyer, T. Fox, J.W. Caldwell, and P.A. Kollman (**1995**) *J. Am. Chem. Soc.* **117**, 5179–5197.

7 N.L. Allinger, Y.H. Yuh, and J.-H. Lii (**1989**) *J. Am. Chem. Soc.* **111**, 8551.

8 W.L. Jorgensen, J. Chandrasekhar, J.D. Madura, R.W. Impey, and M.L. Klein (**1983**) *J. Chem. Phys.* **79**, 926.

9 G.J. Kramer, N.P. Farragher, B.W.H. van Beest, and R.A. van Santen (**1991**) *Phys. Rev. B* **43**, 5068.

10 T.S. Bush, J.D. Gale, C.R.A. Catlow, and D. Battle (**1994**) *J. Mater. Chem.* **4**, 831.

11 L. Pauling (**1947**) *J. Am. Chem. Soc.* **69**, 542.

12 M.S. Daw and M.I. Baskes (**1984**) *Phys. Rev. B*, **29(12)**, 6443–6453.

13 J. Tersoff (**1988**) *Phys. Rev. Lett.* **61**, 2879.

14 D.W. Brenner (**1990**) *Phys. Rev. B* **42**, 9458.

15 S.J. Stuart, A.B. Tutein, and J.A. Harrison (**2000**) *J. Chem. Phys.* **112**, 6472.

16 J.H. Los, L.M. Ghiringhelli, E.J. Meijer, and A. Fasolino (**2005**) *Phys. Rev. B* **72**, 214102.

17 W.J. Mortier, S.K. Ghosh, and S.J. Shankar (**1986**) *J. Am. Chem. Soc.* **108**, 4315.

18 A.K. Rappe' and W.A. Goddard (**1991**) *J. Phys. Chem.* **95**, 3358.

19 J. Chen and T.J. Martinez (**2007**) *Chem. Phys. Lett.* **438**, 315.

20 A. Strachan, A.C.T. van Duin, D. Chakraborty, S. Dasgupta, and W.A. Goddard III (**2003**) *Phys. Rev. Lett.* **91**, 098301.

21 K. Chenoweth, S. Cheung, A.C.T. van Duin, W.A. Goddard, and E.M. Kober (**2005**) *J. Am. Chem. Soc.* **127**, 7192–7202.

22 A.C.T. van Duin, Y. Zeiri, F. Dubnilova, R. Kosloff, and W.A. Goddard (**2005**) *J. Am. Chem. Soc.* **127**, 11053–11062.

23 M.J. Buehler, A.C.T. van Duin, and W.A. Goddard (**2006**) *Phys. Rev. Lett.* **96**, 095505.

24 K.D. Nielson, A.C.T. van Duin, J. Oxgaard, W. Deng, and W.A. Goddard (**2005**). *J. Phys. Chem. A*. **109**, 493.

25 Q. Zhang, T. Cagin, A.C.T. van Duin, W.A. Goddard, Y. Qi, and L.G. Hector (**2004**) *Phys. Rev. B* **69**, 045423.

26 S. Cheung, W. Deng, A.C.T. van Duin, and W.A. Goddard. (**2005**) *J. Phys. Chem. A*. **109**, 851.

27 J. Ludwig, D.G. Vlachos, A.C.T. van Duin, and W.A. Goddard (**2006**) *J. Phys. Chem. B* **110**, 4274–4282.

28 W.A. Goddard, A.C.T. van Duin, K. Chenoweth, M. Cheng, S. Pudar, J. Oxgaard, B. Merinov, Y.H. Jang, and P. Persson (**2006**) *Topics in Catalysis* **38(1–3)** 93–103.

29 A.C.T. van Duin, B. Merinov, and W.A. Goddard (**2008**) *J. Phys. Chem. A* **112**, 3133–3140.

# 10 Kinetic Monte Carlo

*Tonek Jansen*

## 10.1 Introduction

If one looks at chemical processes on surfaces from an atomic point of view, then the field of chemical kinetics is very complicated. Atomic scales are of the order of Ångstrøms and femtoseconds. Typical length scales in laboratory experiments vary between micrometers to centimeters, and typical time scales are often of the order of seconds or longer. This means that there are many orders of difference in length and time between the individual reactions and the resulting kinetics.

The length gap is not always a problem. In most cases an area with a length scale of a few dozen nanometers is representative for the whole catalyst's surface (see Section 10.4 for an exception). More of a problem is the time gap. The typical atomic time scale is given by the period of a molecular vibration. The fastest vibrations have a reciprocal wavelength of up to 4000 $cm^{-1}$, and a period of about 8 fs. Reactions in catalysis take place in seconds or more. It is important to be aware of the origin of these 15 orders of magnitude difference. A reaction can be regarded as a movement of the system from one local minimum on a potential-energy surface to another. In such a move a so-called activation barrier has to be overcome. Most of the time the system moves around one local minimum. Every time that the system moves in the direction of the activation barrier can be regarded as an attempt to react. The probability that the reaction actually succeeds can be estimated by $\exp[-E_{\mathrm{bar}}/k_{\mathrm{B}}T]$, where $E_{\mathrm{bar}}$ is the height of the barrier, $k_{\mathrm{B}}$ is the Boltzmann constant, and $T$ is the temperature. A barrier of $E_{\mathrm{bar}} = 100$ kJ $\mathrm{mol}^{-1}$ at room temperature gives a Boltzmann factor of about $10^{-18}$. Hence we see that the very large difference in time scales is due to the large number of times that the system needs to attempt to overcome activations barriers.

In molecular dynamics a reaction with a high activation barrier is called a rare event, and various techniques have been developed to simulate them. These techniques, however, work for one reacting molecule or two molecules that react together, but not when one is interested in the combination of the many reacting molecules that one has when studying kinetics. It turns out that if one uses a

*Computational Methods in Catalysis and Materials Science.* Edited by Rutger A. van Santen and Philippe Sautet

ISBN: 978-3-527-32032-5

slightly less accurate description of the positions of the atoms and molecules on a surface, one can disregard vibrations and use the reactions themselves as elementary events in a simulation. This leads to kinetic Monte Carlo (kMC). This method has atomic resolution, but can work on the same time scales as experiments.

The purpose of this chapter is to give an overview of the theoretical background of kMC. Details can be found on the kMC website [1]. Sections 10.2 and 10.3 contain established theory. Section 10.4 has an example showing what can be accomplished with kMC. Finally, Section 10.5 describes the areas where new kMC theory is being developed.

## 10.2 The Lattice-Gas Model and the Master Equation

Kinetic Monte Carlo of surface reactions is based on the concept that atoms or molecules adsorb onto well-defined positions called sites. If the surface of the catalyst has two-dimensional translational symmetry, or when it can be modeled as such, the sites form a regular grid or a lattice. We then get a so-called *lattice-gas* model.

We can describe how the occupation of the sites changes in time with a master equation

$$\frac{dP_\alpha}{dt} = \sum_\beta \left[W_{\alpha\beta}P_\beta - W_{\beta\alpha}P_\alpha\right]. \tag{10.1}$$

In this equation $t$ is time, $\alpha$ and $\beta$ are configurations of the adlayer, $P_\alpha$ and $P_\beta$ are their probabilities, and $W_{\alpha\beta}$ and $W_{\beta\alpha}$ are so-called transition probabilities per unit time that specify the rate with which the adlayer changes due to reactions and diffusion. A configuration is nothing but a specification of the occupation of each site. Instead of transition probability we will also use the term rate constant.

The master equation can be derived from first principles, and hence forms a solid basis for all subsequent work. There are other advantages as well. It has links with all other theoretical methods that are important for kinetics. First, the derivation of the master equation yields expressions for the rate constants that can be computed with quantum chemical methods. This makes *ab initio* kinetics for catalytic processes possible (see Section 10.2.2). Second, the master equation can be used to derive the normal macroscopic rate equation (see Section 10.2.3). In general, it forms a good basis to compare different theories of kinetic quantitatively, and also to compare these theories with simulations. Third, the master equation forms the basis for the actual kMC simulations that are the topic of this chapter (see Section 10.3). Solving the master equation with kMC simulations involves no approximations, and the results therefore can be used as a benchmark for other theories of kinetics.

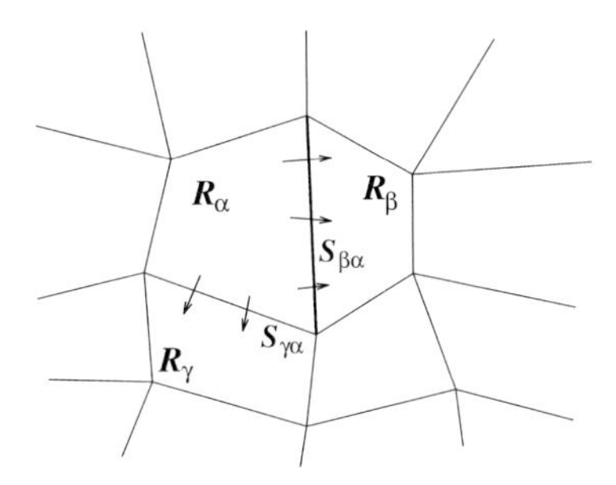


**Figure 10.1** Schematic drawing of the partitioning of phase space into regions $R$, each of which corresponds to some particular configuration of the adlayer. The reaction that changes $\alpha$ into $\beta$ corresponds to a flow from $R_\alpha$ to $R_\beta$. The transition probability $W_{\beta\alpha}$ for this reaction equals the flux through the surface $S_{\beta\alpha}$, separating $R_\alpha$ from $R_\beta$, divided by the probability to find the system in $R_\alpha$.

### 10.2.1 The Derivation of the Master Equation

In the derivation of the master equation one regards the evolution of a system in phase space, which is partitioned into various regions (see Figure 10.1) [1]. Each of these regions corresponds to a particular configuration of the adlayers. Reactions and diffusional hops of adsorbates from one site to a neighboring site are then nothing but motion of the system in phase space from one region to another.

### 10.2.2 The Master Equation and Quantum Chemistry

The derivation of the master equation also gives an expression for the rate constants. In most cases this can be cast in the following familiar form:

$$W_{\alpha\beta} = \frac{k_B T}{h} \frac{Q^\ddagger}{Q} \exp\left[-\frac{E_{\text{bar}}}{k_B T}\right], \tag{10.2}$$

with $T$ the temperature, $k_B$ the Boltzmann constant, $E_{\text{bar}}$ the height of the activation barrier for the process that changes the configuration $\beta$ to $\alpha$, $Q$ the partition function of the reactants, and $Q^\ddagger$ the partition function of the transition state.

The above expression for the rate constant can be computed using quantum chemistry. The main properties that should be determined in a quantum chemical calculation is the barrier height $E_{\text{bar}}$, and the vibrational frequencies of the reactants and the transition state to compute the partition functions.

### 10.2.3 The Master Equation and the Macroscopic Equation

It is possible to derive rate equations for macroscopic properties from the master equation. If a system is in a well-defined configuration then a macroscopic property can generally be computed easily. For example, the number of molecules of a particular type in the adlayer can be obtained simply by counting. If the property that we are interested in is denoted by $X$, then its value when the system is in configuration $\alpha$ is given by $X_\alpha$. The expectation value of $X$ is given by

$$\langle X \rangle = \sum_\alpha P_\alpha X_\alpha. \tag{10.3}$$

Kinetic experiments measure changes, so we have to look at $d\langle X \rangle / dt$. This is given by

$$\frac{d\langle X \rangle}{dt} = \sum_\alpha \frac{dP_\alpha}{dt} X_\alpha, \tag{10.4}$$

because $X_\alpha$ is a property of a fixed configuration. We can remove the derivative of the probability using the master equation. This gives us

$$\begin{aligned} \frac{d\langle X \rangle}{dt} &= \sum_{\alpha\beta} \left[ W_{\alpha\beta} P_\beta - W_{\beta\alpha} P_\alpha \right] X_\alpha, \\ &= \sum_{\alpha\beta} W_{\alpha\beta} P_\beta \left[ X_\alpha - X_\beta \right]. \end{aligned} \tag{10.5}$$

The second step is obtained by swapping the summation indices. The final result can be regarded as the expectation value of the change of $X$ in the reaction $\beta \to \alpha$ times the rate constant of that reaction. This equation is called the macroscopic or phenomenological equation [2]. It forms the basis for deriving relations between macroscopic properties and rate constants.

Suppose we have atoms or molecules that adsorb onto one particular type of site, and that the only processes are diffusion and simple first-order desorption. If we take for $X$ in the macroscopic equation the number of adsorbates, then Eq. (10.5) becomes after dividing by the number of sites

$$\frac{d\theta}{dt} = -W_{\text{des}}\theta. \tag{10.6}$$

Here $\theta$ is the coverage and $W_{\text{des}}$ is the rate constant for the desorption. This macroscopic *rate* equation holds exactly. This generally holds for processes in which only one site is involved [1]. In particular, the rate constant in the master equation and the one in the macroscopic rate equation are the same.

For bimolecular reactions it is not possible to derive macroscopic rate equations from the master equation exactly. If we look at the reaction A + B we get [1]

$$\frac{d\theta_\text{A}}{dt} = -ZW_{\text{rx}}\theta_\text{A}\theta_\text{B}, \tag{10.7}$$

where $W_{\text{rx}}$ is the rate constant of the reaction in the master equation, and $Z$ the number of nearest-neighbor sites of each site. In the derivation of this expression there are two approximations. The first is that we assume that the adsorbates are randomly distributed over the surface. The second is that fluctuations in the number of adsorbates are negligible. The latter approximation is usually valid, although fluctuations can be important in small systems, and for highly reactive intermediates of which there are very few molecules. The former approximation however often does not hold. This may be because the substrate is heterogeneous (defects, bimetals), there are interactions between adsorbates, and also a bimolecular reaction itself tends to cause segregation of the reactants (see Section 10.5.3). Note that even if these approximations hold, there is still a geometric factor $Z$ that leads to a difference in the rate constant in the master equation and in the macroscopic rate equation.

Deriving macroscopic equations often involves approximations. In cases where it does not, or when the approximations are very good, then these macroscopic equations provide a way to determine the rate constants in the master equation from experiments. This way to determine rate constants may be preferable, because it may be more accurate than using quantum chemistry.

## 10.3 Kinetic Monte Carlo Algorithms

Deriving analytical results from the master equation is not possible for most systems of interest. Approximations can of course be used, but they may not be satisfactory. In such cases one can resort to kMC simulations. Such simulation yields an ordered set of configurations and reaction times that can be written as

$$(\alpha_0, t_0) \xrightarrow{t_1} \alpha_1 \xrightarrow{t_2} \alpha_2 \xrightarrow{t_3} \alpha_3 \xrightarrow{t_4} \cdots. \tag{10.8}$$

Here $\alpha_0$ is the initial configuration of a simulation and $t_0$ is the time at the beginning of the simulations. The changes $\alpha_{n-1} \to \alpha_n$ are caused by reactions taking place at time $t_n$. There are many algorithms that can be used to determine the reaction times $t_n$ and the reactions $\alpha_{n-1} \to \alpha_n$, but they are all equivalent because they all give at time $t$ a configuration $\alpha$ with probability $P_\alpha(t)$ that is the solution of the master equation with boundary condition $P_\alpha(t_0) = \delta_{\alpha\alpha_0}$. It is in this sense that kMC simulations solve the master equation. The sequence of reactions and reaction times that are actually generated during a kMC simulation is called a realization of the master equation.

### 10.3.1 The Variable Step Size Method

If the system is in initial configuration $\alpha$ at time $t = 0$, then the probability that the master equation tells us that the system is still in $\alpha$ at a later time $t$ is given by $\exp[-t\sum_\gamma W_{\gamma\alpha}]$. We can then generate a time $t'$ when the first reaction actually occurs by solving

$$\exp[-t'\sum_\gamma W_{\gamma\alpha}] = r, \tag{10.9}$$

where $r$ is a random number in the interval $[0, 1]$. The different reactions that transform configuration $\alpha$ to another configuration $\beta$ have transition probabilities $W_{\beta\alpha}$. This means that the probability that the system will become configuration $\beta$ at time $t'$ is then proportional to $W_{\beta\alpha}$. We therefore generate a new configuration $\alpha'$ by picking it out of all possible new configurations $\beta$ with a probability proportional to $W_{\alpha'\alpha}$. This gives us a new configuration $\alpha'$ at time $t'$. At this point we are in the same situation as when we started the simulation, and we can proceed by repeating the previous steps. We call this whole procedure the variable step size method (VSSM), as it is equivalent to the algorithm with the same name developed by Gillespie for solving rate equation [3,4]. It has also been discovered independently by Bortz et al. and called the $n$-fold way [5]. Actually some people use the name kMC for this algorithm only [6]. It is a method that is conceptually simple and can be made very efficient; it is possible to get the computer time needed per reaction independent of system size [7,8].

### 10.3.2 The Random Selection Method

Efficient VSSM involves quite a bit of bookkeeping. This can be avoided by using a technique called *oversampling*. Replace $\sum_{\gamma} W_{\gamma\alpha}$ in VSSM by the fixed total rate constant $k = SW_{\max}$, where $W_{\max}$ is the maximum of the rate constants $W^{(i)}$'s of reaction type $i$ and $S$ is the number of sites in the system. At each step in the algorithm, time is increased by

$$\Delta t = -\frac{1}{k} \ln r, \tag{10.10}$$

where $r$ is a random number in the interval $[0, 1]$. The type of reaction and the site where the reaction is to take place are chosen randomly. If the type of reaction is possible at the chosen site, then it is accepted with probability $W^{(i)}/W_{\max}$ and the configuration is then changed accordingly.

This algorithm is called the random selection method (RSM) [7,8]. Note that the reaction time, the type of the reaction, and the location of the reaction can be done in any order. Only time and the configuration of the system need to be updated. The method can therefore be extremely efficient, provided that reactions are accepted often. (As for VSSM the computer time needed per reaction is independent of system size.) This is the case when each reaction type has a rate constant of the same order of magnitude, and each reaction type can occur at a substantial fraction of all sites. Unfortunately, these conditions are in practice seldom met.

### 10.3.3 The First Reaction Method

Instead of splitting the determination of the time, the type, and the location of a reaction as in RSM, it is also possible to combine them. This is done in the first reaction method (FRM) [7,8]. In this method a reaction time is determined for each reaction using

$$t_{\beta\alpha} = t - \frac{1}{W_{\beta\alpha}} \ln r, \tag{10.11}$$

where the reaction changes configuration $\alpha$ into $\beta$ and $t$ is the current time. The next reaction that actually occurs is then the one with the smallest reaction time. This algorithm is called discrete event simulation in computer science.

The disadvantage of FRM is the determination of the reaction with the smallest reaction time. The optimal way to do this scales logarithmically with system size, and consequently also the computer time needed per reaction scales in this way with system size [7, 8]. This is not as good as constant time, but it is not particularly bad either. In fact, for normal system sizes FRM is faster than VSSM if the rate constants depend on time (e.g., in a TPD experiment), or when there are interactions between the adsorbates that affect the rate constants.

### 10.3.4
### Practical Considerations

The efficiency of the methods described above depends very much on details of the algorithm that we have not discussed, and we refer the interested reader to Refs. [7, 8] for a more extensive analysis if he or she wants to implement one of the methods. The important difference between FRM on the one, and VSSM and RSM on the other hand is the dependence on the system size. Computer time per reaction in VSSM and RSM does not depend on the size of the system. In FRM the computer time per reaction depends logarithmically on the system size. So for large systems VSSM and RSM are generally to be preferred, and FRM should only be used if really necessary.

There are, however, a number of cases that occur quite frequently in which VSSM and RSM are not efficient. This is when there are many reaction types and when the rate constants depend on time. Many reaction types arise, for example, when there are lateral interactions. In that case FRM is the method of choice. (In general, one should realize that simulations of systems with lateral interactions are always costly.) The same holds if the rate constants depend on time (see Section 10.3.5.)

If VSSM and RSM can be used, then the choice between them depends on how many sites in the system the reactions can occur. RSM is efficient for reactions that occur on many sites, because the probability that a reaction is possible on the randomly chosen location is then high. If this is not the case then VSSM should be used.

The choice between FRM, VSSM, and RSM need not be made for all reactions in a system together, but can be made per reaction type, because it is easy to combine different methods. Suppose that the reaction type 1 is best treated by VSSM, but the reaction type 2 best by RSM. We then determine the first reaction of type 1 using VSSM, and the first of type 2 by RSM. The first reaction to actually occur is then simply the first reaction to occur of these two.

### 10.3.5
### Time-Dependent Transition Probabilities

In some experiments (e.g., TPD or voltammetry) the rate constants themselves are time dependent. The equations to determine the reaction times, Eqs. (10.9) and (10.11), are then no longer valid.

The drawback of VSSM for time-dependent transition probabilities is that the equation for the time of the reactions becomes often too difficult to be solved efficiently. In FRM we compute a time for each reaction. So if we are currently at time $t$ and in configuration $\alpha$, then we compute for each reaction $\alpha \to \beta$ a time $t_{\beta\alpha}$ using

$$\exp\left[-\int_t^{t_{\beta\alpha}} dt'\, W_{\beta\alpha}(t')\right] = r, \tag{10.12}$$

where $r$ is as usual a random number in the interval $[0, 1]$. The first reaction to occur is then the one with the smallest $t_{\beta\alpha}$. This method is FRM for time-dependent reaction rate constants [9]. Equation (10.12) can sometimes be solved analytically (e.g., for voltammetry [10]), but may need to be solved numerically (e.g., for TPD). It can have an interesting property, which is that it may have no solution. This is the case with some reactions in cyclic voltammetric experiments, which then will never occur [10].

## 10.4
## An Example: Oscillations in the CO Oxidation on Pt Surfaces

One problem for which extensive kMC simulations have been done by various groups is the problem of CO oscillations on Pt(100) and Pt(110). A crucial role in these oscillations is played by the reconstruction of the surface, and the effect of this reconstruction on the adsorption of oxygen. The explanation of the oscillations is as follows. A bare Pt surface reconstructs into a structure with a low sticking coefficient for oxygen. This means that predominantly CO adsorbs on bare Pt. However, CO lifts the reconstruction. The normal structure has a high sticking coefficient for oxygen. So after CO has adsorbed in a sufficient amount to lift the reconstruction oxygen can also adsorb. The CO and the oxygen react, and form $CO_2$. This $CO_2$ rapidly desorbs leaving bare Pt which reconstructs again. An important aspect of this process, and also other oscillatory reactions on surfaces, is the problem of synchronization. The cycle described above can easily take place on the whole surface, but oscillations on different parts on the surface are not necessarily in phase, and the overall reactivity of surface is then constant. To get the whole surface oscillating in phase there has to be a synchronization mechanism.

The most successful model to describe oscillations on Pt surfaces is the one by Kortlüke, Kuzovkov, and von Niessen [11]. This model has CO adsorption and desorption, oxygen adsorption, $CO_2$ formation, CO diffusion, and surface reconstruction. The surface is modeled by a square grid. Each site in the model is either in the state $\alpha$ or in the state $\beta$. The $\alpha$ state is the reconstructed state which has a reduced sticking coefficient for oxygen. The $\beta$ state is the unreconstructed state with a high sticking coefficient for oxygen. An $\alpha$ site will convert a neighboring $\beta$ site into an $\alpha$ state if neither site is occupied by CO. A $\beta$ site will convert a neighboring $\alpha$ site into $\beta$ if at least one of them is occupied by CO.

The model shows a large number of phenomena depending on the rate constants. Figure 10.2 shows snapshots obtained from some large simulations in which the diffusion is just about fast enough to lead to global oscillations provided the initial conditions are favorable [12]. However, it is also possible to choose the initial conditions so that the oscillations are not synchronized properly. In that case one can see the formation of patterns as the right half of the figure shows.

Synchronization is obtained when the diffusion rate is fast enough. The minimal value is related to the so-called Turing-like structures that are formed in the substrate. These labyrinthine structures can best be seen in the lower two pictures on the left and all pictures on the right of Figure 10.2. If diffusion is so fast that within one oscillatory period CO can move from one phase ($\alpha$ or $\beta$) to a neighboring island of the other phase, then the oscillations are well synchronized. If the diffusion rate is slower, then we get pattern formation. Note that the system has two length scales. The characteristic length scale of the adlayer is much larger than the characteristic length scale of the Turing-like structures as can be seen in the right half of the figure.

## 10.5 New Developments

### 10.5.1 Diffusion

Some reaction systems contain processes that have a high rate constant and that lead to more of the same type of processes. A simulation of such a system spends almost all of its time on such processes. If these processes are not of interest in themselves, then this makes the simulation very inefficient or even useless. This can occur, in particular, with diffusion. Therefore, there has been quite some effort in how to handle fast diffusion.

The simplest approach is to reduce the rate constants of all diffusional hops by some constant factor. The idea is that all diffusion does is to bring the adlayer to equilibrium, which is also achieved with (much) slower diffusion. A more sophisticated approach using the same equilibration idea is to remove the diffusion completely. Let $C(\alpha)$ be the set of all configurations that can be reached from configuration $\alpha$ by diffusional hops only. We can partition all configurations in such sets. Let $P_{C(\alpha)}$ be the probability that the system is in one of the configurations of $C(\alpha)$. We then have

$$\frac{dP_{C(\alpha)}}{dt} = \sum_{C(\beta)} \left[\Omega_{C(\alpha)C(\beta)} P_{C(\beta)} - \Omega_{C(\beta)C(\alpha)} P_{C(\alpha)}\right] \tag{10.13}$$

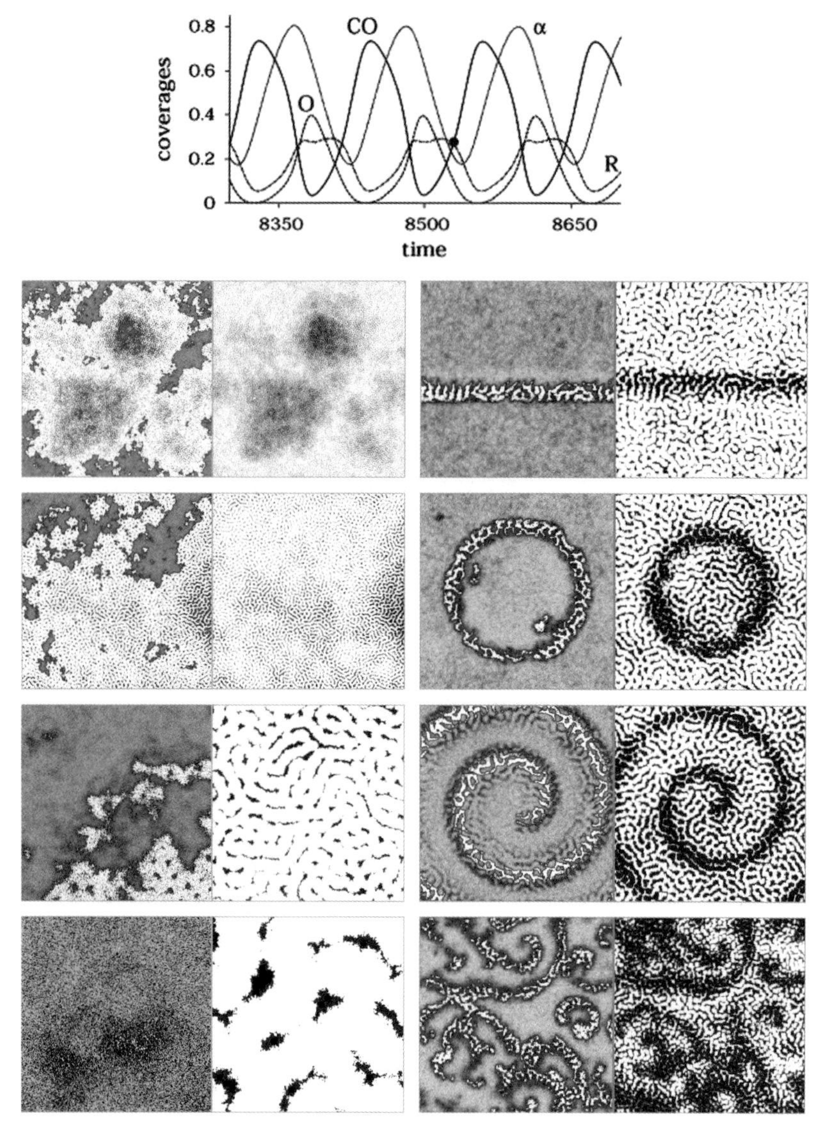


**Figure 10.2** Global oscillations and pattern formation for CO oxidation on Pt(110). The top shows temporal variations of the coverages, the fraction of the substrate in the $\alpha$ phase, and the $CO_2$ production $R$. Each picture has two parts. In the left part we plot the chemical species; CO particles are grey and O particles are white, and empty sites are black. The right part shows the structure of the surface; $\alpha$-phase sites are black, and $\beta$-phase sites are white. Sections of the upper-left corner with grid sizes $L = 8192$, 4096, 1024, and 256 are shown on the left half of the figure. The sections correspond to the dot in the temporal plot at the top. On the right half of the figure we have a wavefront, a target, a spiral, and turbulence ($L = 2048$), which can be obtained with different initial conditions [12].

with

$$\Omega_{C(\alpha)C(\beta)} = \sum_{\alpha' \in C(\alpha)} \sum_{\beta' \in C(\beta)} W_{\alpha'\beta'} \frac{P_{\beta'}}{P_{C(\beta)}}. \tag{10.14}$$

This equation is again a master equation provided that $P_{\beta'}/P_{C(\beta)}$ is a constant, which it is for infinitely fast diffusion. The reason is that such diffusion brings all $\beta' \in C(\beta)$ at equilibrium instantaneously and the ratio is nothing but a conditional probability that the system in $\beta'$ given that it is in one of the configurations of $C(\beta)$. This is determined by the diffusion only.

The applicability of Eq. (10.13) depends a how easy $P_{\beta'}/P_{C(\beta)}$ can be determined. For diffusion in a one-dimensional zeolite this proved to be easy, and instead of a master equation with configurations a master equation with the numbers of molecules could be used [13]. For higher dimensional systems things might not be so easy. Note that the partitioning of the configurations can also be used for other problems. The only requirement is that $P_{\beta'}/P_{C(\beta)}$ is constant.

Another approach is to simulate various diffusional hops in one step. This was done by Mason et al. in their work on flicker processes [14]. It has the advantage that it can also be used for processes that are not very fast, although the extra computational costs of the more complicated algorithm may not be worthwhile if the processes become too slow. A flicker process is change $\alpha \to \beta$ of configurations immediately followed by the reverse change $\beta \to \alpha$. In their work Mason et al. grouped all changes $\alpha \to \beta \to \alpha$ for a fixed $\alpha$ and all possible $\beta$s, and computed a time when the chain of flicker processes is broken. They also determined which configuration the system then goes to. Consequently, the flicker processes need no longer be simulated explicitly which can speed up a simulation substantially.

### 10.5.2
### Longer Time Scales

In many simulations, the period that one has to simulate a system is determined by how long it takes the system to reach the state of interest (e.g., a steady state), and the time one needs to simulate the system to get good statistics for the results. Neither times are generally very long. There are exceptions however. Relaxation times sometimes are long, or one is interested in processes with intrinsically long time scales; e.g., time-dependent experiments like TPD or oscillations. It is not always possible to get to the time scale of interest with a conventional kMC simulation.

As for diffusion it may be possible to remove processes from the simulation that take lot of computer time. An alternative approach has been used by Chatterjee and Vlachos [15]. Their approach uses ordinary reaction rate equations for the large time scale and kMC simulations to get accurate values for the rates with which coverages change. The procedure starts with a kMC simulation to determine rates. These are then put in the reaction rate equations, which are integrated over some time period. (Here the large time scales are obtained.) At the end of the period a new kMC simulation is done for new values of the rates, and the whole procedure is repeated a number of times.

The main problem of this procedure is the configuration at the start of each kMC simulation. After the integration of the rate equations one only knows coverages, whereas the kMC simulations also need to now how the adsorbates are distributed over the sites. For the procedure to work it is necessary that one can reconstruct a configuration from the coverages accurately. Whether this is possible seems to depend on the system.

The treatment of flicker processes, as described above, can be regarded as a method to get to longer time scales by doing multiple reactions in one step [14]. An extension that combines whole chains of reactions has recently been given by Sun and by Trygubenko and Wales [16, 17].

### 10.5.3
### Longer Length Scales

Very often the characteristic length scale of a system is small; e.g., correlation between occupation of sites is generally only a few times the distance between neighboring sites. If the substrate has, however, a long characteristic length scale or when there is pattern formation one may want to simulate a larger system than is possible with conventional kMC.

A possible approach is to use coarse graining. The idea is to replace the lattice of sites by a lattice with a larger lattice spacing. Instead of a lattice point representing a single site, points of the new lattice represent a block of sites. Such a point does not specify an occupation of a single site, but for each type of adsorbates the number of such adsorbates in the block.

This approach was used by Katsoulakis et al. and by Mastny et al. [18, 19]. The problem with the approach however is that the kinetics within the blocks is not completely determined by the number of adsorbates. One therefore has to introduce an approximation. Mastny et al. used the mean field and the quasichemical approximation, but that may not be good enough, as can be shown for the simple A + B model [20]. In this model there is only one reaction; A and B next to each other can react to form a AB that immediately desorbs. The As and Bs also diffuse by hopping to neighboring sites. The model shows anomalous kinetics. Because the adsorbates can only react to form a product that desorbs, the coverages decrease in time. If we start with equal numbers of A and B, macroscopic rate equations predict that at large times $t$ the coverage decreases as $t^{-1}$. Scaling arguments show, however, that it should be as $t^{-1/2}$ [20]. The reason is that the adsorbates segregate and the reaction only occurs where areas with only As and those with only Bs meet. KMC simulations confirm this behavior.

One needs large grids to study low coverages, and therefore one might want to do coarse graining. If one uses a mean-field approximation for the blocks, then one gets the following results (see Figure 10.3). Initially, the As and Bs are randomly distributed and the rate equations and both simulations (normal and

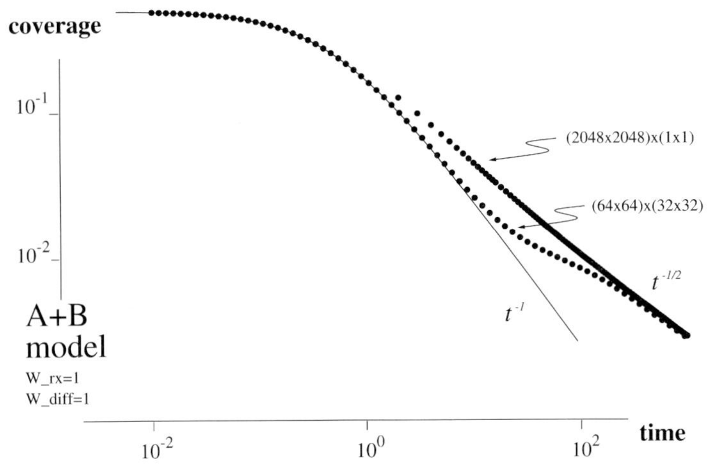


**Figure 10.3** The coverage as a function of time in the A + B model. The solid line is the result obtained with a macroscopic rate equation. The dots marked (2048 × 2048) × (1 × 1) are from a normal kMC simulation with a 2048 × 2048 grid. The dots marked (64 × 64) × (32 × 32) are from a coarse-grained kMC simulation with a 64 × 64 grid with each grid point representing a block of 32 × 32 sites. The kinetics in a block was approximated with mean field.

coarse-grained kMC) give identical results. When the areas with only As or only Bs are larger than the size of the blocks, then the coarse-grained simulation gives the same results as the normal kMC simulations, because the kinetics is determined by the boundaries of the areas, which are well approximated by the blocks. After the segregation has started, but the areas are still smaller than the blocks, then the coarse-grained simulation follows the results of the macroscopic rate equation instead of the normal kMC. The reason is that the adsorbates in a block are not randomly distributed and mean field does not hold.

A solution to this problem is to use a two-step procedure. First, normal kMC simulations are done for the blocks. These simulations are small, so this does not add too much computational effort to the coarse-grained simulation. These kMC simulations are used to determine the kinetics within the blocks. Second, the kinetics thus obtained are used in the coarse-grained simulation. This two-step procedure works well for the A + B model; the result is the same as for a normal kMC simulations with the full lattice. For more complex reaction systems this has not been used yet.

### 10.5.4
### Working Without a Lattice

Although the use of a lattice is very important in the theory above, one should realize that it is really not needed from a theoretical point of view. No reference is made to a lattice in the derivation of the master equation, and indeed one can also use the master equation for reactive systems that do no have translational or any other kind of symmetry.

The idea is to look at the potential-energy surface (PES) of a system [21], and associate each "configuration" $\alpha$ with a minimum of the PES. The region $R_\alpha$ consists of the points in phase space around the minimum. The derivation otherwise does not change, and we get a master equation describing processes/reactions corresponding to transitions between the minima of the PES.

This approach has been used, even in combination with density functional theory calculations, for the diffusion of a cluster of Al atoms on Al surface [22], the formation of defects during crystallization [23], and for the segregation of atoms [24]. The computational efforts are much larger than when a lattice is used, because the rate constants have to be computed during the simulation on the fly. Another problem is to find all transition states [25]. Apart from finding all transition states, it is also not clear how one knows if all transition states have been found. On the other hand, the range of systems that kMC can be applied to increases enormously if no lattice has to be used.

## References


1 The kinetic Monte Carlo website. http://www.catalysis.nl/chembond/kMC/.

2 N. G. van Kampen. *Stochastic Processes in Physics and Chemistry* (North-Holland, Amsterdam, **1981**).

3 D. T. Gillespie. *J. Comput. Phys.*, **22**, 403, **1976**.

4 D. T. Gillespie. *J. Phys. Chem.*, **81**, 2340, **1977**.

5 A. B. Bortz, M. H. Kalos, and J. L. Lebowitz. *J. Comput. Phys.*, **17**, 10, **1975**.

6 Wikipedia. http://www.wikipedia.org/.

7 J. J. Lukkien, J. P. L. Segers, P. A. J. Hilbers, R. J. Gelten, and A. P. J. Jansen. *Phys. Rev. E*, **58**, 2598, **1998**.

8 J. P. L. Segers. *Algorithms for The Simulation of Surface Processes*. Ph.D. thesis, Eindhoven University of Technology, Eindhoven, **1999**.

9 A. P. J. Jansen. *Comput. Phys. Comm.*, **86**, 1, **1995**.

10 M. T. M. Koper, J. J. Lukkien, A. P. J. Jansen, P. A. J. Hilbers, and R. A. van Santen. *J. Chem. Phys.*, **109**, 6051, **1998**.

11 V. N. Kuzovkov, O. Kortlüke, and W. von Niessen. *Phys. Rev. Lett.*, **83**, 1636, **1999**.

12 R. Salazar, A. P. J. Jansen, and V. N. Kuzovkov. *Phys. Rev. E*, **69**, 031604, **2004**.

13 S. V. Nedea, A. P. J. Jansen, J. J. Lukkien, and P. A. J. Hilbers. *Phys. Rev. E*, **65**, 066701, **2002**.

14 D. R. Mason, R. E. Rudd, and A. P. Sutton. *Comput. Phys. Comm.*, **160**, 140, **2004**.

15 A. Chatterjee and D. G. Vlachos. *J. Comput. Phys.*, **211**, 596, **2006**.

16 S. X. Sun. *Phys. Rev. Lett.*, **96**, 210602, **2006**.

17 S. A. Trygubenko and D. J. Wales. http://arXiv.org/, paperno. cond-mat/0603830, **2006**.

18 M. A. Katsoulakis, A. J. Majda, and D. G. Vlachos. *J. Comput. Phys.*, **186**, 250, **2003**.

19 E. A. Mastny, E. L. Haseltine, and J. B. Rawlings. *J. Chem. Phys.*, **125**, 194715, **2006**.

20 S. Redner. In V. Privman, ed., *Nonequilibrium Statistical Mechanics in one Dimension*, pp. 3–27 (Cambridge University Press, Cambridge, **1996**).

21 P. G. Mezey. *Potential Energy Hypersurfaces* (Elsevier, Amsterdam, **1987**).

22 G. Henkelman and H. Jónsson. *Phys. Rev. Lett.*, **90**, 116101, **2003**.

23 F. Much, M. Ahr, M. Biehl, and W. Kinzel. *Comp. Phys. Comm.*, **147**, 226, **2002**.
24 T. F. Middleton and D. J. Wales. *J. Chem. Phys.*, **120**, 8134, **2004**.
25 R. A. Olsen, G. J. Kroes, G. Henkelman, A. Arnaldsson, and H. Jónsson. *J. Chem. Phys.*, **121**, 9776, **2004**.

# Part III
# Properties

# 11
# Theory of Elastic and Inelastic Electron Tunneling

*Marie-Laure Bocquet, Hervé Lesnard, Serge Monturet, and Nicolás Lorente*

## 11.1
## Introduction

The scanning tunneling microscope (STM) is an exceptional device for the study of the electronic structure of molecules and adsorbates in general on surfaces [1]. The reason of the extraordinary capabilities of the STM lies in the tunneling character of the established electronic structure. Indeed, tunneling electrons exponentially decay with distance. Hence, electrons tunneling from virtually one atom, will decay in the almost 3D surrounding space, leading to a very sharp length scale. The exponential dependence of the tunneling current on distance offers the opportunity of great experimental control in the tip–substrate distances. Hence, the final contours of the tip position keeping the tip–substrate current constant are relatively easily attained via a feedback electronic device.

In 1998, Stipe, Rezaei, and Ho succeeded in measuring changes in conductance at the vibrational threshold of an acetylene molecule on a Cu(100) surface [2]. It is possible to obtain the vibrational structure of a single molecule based on a traditional vibrational spectroscopy technique called inelastic electron tunneling spectroscopy (IETS). IETS was developed in 1966, when Jacklevic and Lambe observed that tunneling electrons were able to excite vibrational modes of a thin layer of molecules buried between two metallic electrodes and an oxide layer [3]. The oxide represents a tunneling barrier for electronic transport between the metallic electrodes. The excitation of vibrations was a consequence of inelastic scattering processes taking place in the tunnel junction. Such inelastic processes induced a slight increase in the tunnel conductance due to the opening of those additional transport channels mediated by the inelastic scattering processes. Thus, the energy threshold to excite a vibrational mode was detected to investigate the vibrational structure of the buried molecular layer. The STM offers a tunneling junction constituted by the vacuum gap between the STM tip and the sample. Thus, it has many common properties with the traditional IETS. However, there are also some differences; in particular in the STM configuration, the majority of the electron current tunnels through the adsorbate itself, therefore having a strong component of molecular derived electronic structure.

*Computational Methods in Catalysis and Materials Science.* Edited by Rutger A. van Santen and Philippe Sautet

ISBN: 978-3-527-32032-5

In this chapter, we briefly present techniques to simulate (i) constant current STM images and (ii) inelastic STM currents or IETS-STM signals. We exemplify the theory with some realistic simulations that are readily comparable with experimental data on the same systems.

## 11.2 Simulations of Constant Current STM Images

The most successful approximation for the treatment of tunneling currents is the one developed by Bardeen [4]. The idea behind this approach is that the electronic structures of both tip and sample are largely unperturbed by the presence of the other electrode. Hence, one can neglect all effects of the potentials of the other electrode in the wavefunctions and compute the probability that an electron is transferred from one electrode state to the other under the condition that all common potentials are zero. Instead of using potentials, one can use gradients by inverting Schrödinger's equation, and after using several vectorial calculus identities, one reaches an intuitive compact expression. The main advantage of this approximation is that it can use the result of very accurate quantum chemistry calculations for the electronic structure of the electrodes, thus being very accurate. The approach breaks down as soon as the mutual interaction between the electrodes is not negligible.

Part of the achievement of recent approaches [5, 6] lies on the simplicity and accuracy of Bardeen's approximations as compared to more exact but complex treatments. For this reason, in the following sections, we sketch a brief derivation of Bardeen's approximation in order to facilitate access to the approximations and physics contained in it.

### 11.2.1 The Bardeen Approximation

We divide the space into two parts. Each is described by a Hamiltonian and its states. The left-hand side is given by

$$\left(-\frac{\hbar^2\nabla^2}{2m} + V_L\right)\psi_{\mu,L} = \epsilon_{\mu,L}\psi_{\mu,L}$$

and the right-hand side by

$$\left(-\frac{\hbar^2\nabla^2}{2m} + V_R\right)\psi_{\nu,R} = \epsilon_{\nu,R}\psi_{\nu,R}.$$

The total Hamiltonian takes into account the perturbation that the two contacts exert on each other; hence

$$\hat{H} = -\frac{\hbar^2 \nabla^2}{2m} + V_L + V_R + \Delta V.$$

The Bardeen approach only needs three approximations:

1. we can neglect the overlap between the left and right states,
2. the left potential cannot cause transitions in the right states and vice versa, and
3. we can neglect all mutual interactions in the transmission matrix elements $\Delta V$.

The rest of the approach develops keeping the first contribution to tunneling. A number of authors have included more conditions to reach Bardeen's results. In what follows, we will reproduce Bardeen's results just by using only the above conditions.

The first step is to show that we can write a Fermi's Golden rule-like expression for the transition rate of left states into right states. This is conceptually different from Fermi's Golden rule where the initial and final states diagonalize the same unperturbed Hamiltonian.

We proceed with a Fermi's Golden rule demonstration (see any quantum mechanics textbook, for example, C. Cohen-Tannoudji, B. Diu and F. Laloë, Méchanique quantique, Hermann, Paris, 1977, p. 1293). We expand the wavepacket in left and right states, and use Schrödinger's equation under the full Hamiltonian:

$$\psi(\mathbf{r}, t) = \sum_{\nu} a_\nu(t) \psi_{\nu,R}(\mathbf{r}) e^{-i\epsilon_\nu t/\hbar} + \sum_{\mu} a_\mu(t) \psi_{\mu,L}(\mathbf{r}) e^{-i\epsilon_\mu t/\hbar} \tag{11.1}$$

and

$$\sum_{\nu} i\hbar \frac{da_\nu}{dt} \psi_{\nu,R}(\mathbf{r}) e^{-i\epsilon_\nu t/\hbar} + \sum_{\mu} i\hbar \frac{da_\mu}{dt} \psi_{\mu,L}(\mathbf{r}) e^{-i\epsilon_\mu t/\hbar}$$
$$= \sum_{\mu} (V_L + \Delta V) a_\nu(t) \psi_{\nu,R} e^{-i\epsilon_\nu t/\hbar} + \sum_{\mu} (V_R + \Delta V) a_\mu(t) \psi_{\mu,L} e^{-i\epsilon_\mu t/\hbar}. \tag{11.2}$$

The boundary conditions are that initially the wavepacket is prepared on a left state:

$$\psi(\mathbf{r}, t = 0) = a_0(t) \psi_{0,L}(\mathbf{r}) e^{-i\epsilon_0 t/\hbar}.$$

A right state $\psi_{\alpha,R}$ is the final one. So we will project Schrödinger's equation on this final state:

$$i\hbar \frac{da_\alpha}{dt} e^{-i\epsilon_\alpha t/\hbar} + \sum_{\mu} i\hbar \frac{da_\mu}{dt} \langle \alpha, R|\mu, L\rangle e^{-i\epsilon_\mu t/\hbar}$$
$$= \sum_{\mu} \langle \alpha, R| V_L + \Delta V|\nu, R\rangle a_\nu e^{-i\epsilon_\nu t/\hbar} + \sum_{\mu} \langle \alpha, R| V_R + \Delta V|\mu, L\rangle a_\mu e^{-i\epsilon_\mu t/\hbar}. \tag{11.3}$$

The above approximations lead us to neglect the terms on $\alpha, R|\mu, L\rangle$ and $\langle\alpha, R|V_L + \Delta V|\nu, R\rangle$.

The probability of finding the wavepacket on the final state $\alpha$ is

$$P_{0\to\alpha} = |\langle\psi(\mathbf{r}, t)|\alpha, R\rangle|^2 = |a_\alpha(t)|^2.$$

By trivially solving the previous differential equation we find

$$P_{0\to\alpha} = \left|\frac{e^{i(\epsilon_\alpha-\epsilon_0)t/\hbar} - 1}{\epsilon_\alpha - \epsilon_0}\right|^2 |\langle\alpha, R|V_R + \Delta V|0, L\rangle|^2.$$

The transition rate is the speed of variation of the transmission probability:

$$\frac{1}{\tau} = \sum_\alpha \frac{dP_{0\to\alpha}}{dt}.$$

This leads to a Fermi's Golden rule-like expression in the long time and continuulm limits (see C. Cohen-Tannoudji, B. Diu and F. Laloë, Méchanique quantique, Hermann, Paris, 1977, p. 1293):

$$\frac{1}{\tau} = \frac{2\pi}{\hbar}\sum_\alpha |\langle\alpha, R|V_R + \Delta V|0, L\rangle|^2\delta(\epsilon_\alpha - \epsilon_0).$$

We need to evaluate the matrix element. In order to achieve this, we restrict the integration to the right hemispace. This is justified by the short range of both the potential and the wavefunctions, and we neglect the mutual action of both electrodes:

$$\langle\alpha, R|V_R + \Delta V|0, L\rangle \approx \langle\alpha, R|V_R|0, L\rangle_R.$$

If we assume that

$$\langle\alpha, R|V_L|0, L\rangle_R \approx 0,$$

then we can write

$$\langle\alpha, R|V_R + \Delta V|0, L\rangle \approx \langle\alpha, R|V_R|0, L\rangle_R + \langle 0, L|V_L|\alpha, R\rangle_R.$$

A common expression is to use the kinetic operator instead of the potential one:

$$V_R|\alpha, R\rangle = \left(-\frac{\hbar^2\nabla^2}{2m} - \epsilon_\alpha\right)|\alpha, R\rangle$$

and

$$V_L|0, L\rangle = \left(-\frac{\hbar^2\nabla^2}{2m} - \epsilon_0\right)|0, L\rangle.$$

So finally we end up having

$$\langle \alpha, R|V_R + \Delta V|0, L\rangle \approx \langle \alpha, R| - \frac{\hbar^2\nabla^2}{2m} - \epsilon_\alpha|0, L\rangle_R + \langle 0, L| - \frac{\hbar^2\nabla^2}{2m} - \epsilon_0|\alpha, R\rangle_R$$
$$= \int_R -\frac{\hbar^2}{2m}\vec{\nabla}\cdot[\psi_{\alpha,R}\vec{\nabla}\psi^*_{0,L} - \psi_{0,L}\vec{\nabla}\psi^*_{\alpha,R}]d^3r$$
$$= \hbar i \int_S \vec{J}_{\alpha,0}\cdot d\vec{S}.$$

The current (Eq. (11.4)) in Bardeen's approximation is then calculated by replacing the matrix element in Eq. (11.1) by the above relation. Now, the initial state $0, L$ is extended to all the left-electrode occupied states $\nu, L$, so we consider only the occupied state and hence multiply the matrix element by the Fermi occupation factor for the left electrode, $f_L(\epsilon_\nu)$, where $\epsilon_\nu$ is the eigenenergy of the occupied state $\nu$. In the same way, the electron has to tunnel into an unoccupied state $\alpha, R$; hence the current is given by

$$I_1 = -e\frac{1}{\tau} = -\frac{2\pi e}{\hbar}\sum_{\nu,\alpha}(f_L(\epsilon_\nu))(1 - f_R(\epsilon_\mu))\left|\hbar\int_S \vec{J}_{\alpha,\nu}\cdot d\vec{S}\right|^2 \delta(\epsilon_\alpha - \epsilon_\nu). \quad (11.4)$$

At finite temperature, we can expect to have some current also flowing from the right to the left electrode (hole injection, due to the smearing of the Fermi occupation factors at finite temperature); hence we have a term

$$I_2 = -\frac{2\pi e}{\hbar}\sum_{\nu,\alpha}(1 - f_L(\epsilon_\nu))f_R(\epsilon_\mu)\left|\hbar\int_S \vec{J}_{\alpha,\nu}\cdot d\vec{S}\right|^2 \delta(\epsilon_\alpha - \epsilon_\nu). \quad (11.5)$$

Therefore, the total current is

$$I = -\frac{2\pi e}{\hbar}\sum_{\nu,\alpha}(f_L(\epsilon_\nu) - f_R(\epsilon_\mu))\left|\hbar\int_S \vec{J}_{\alpha,\nu}\cdot d\vec{S}\right|^2 \delta(\epsilon_\alpha - \epsilon_\nu) \quad (11.6)$$

At zero temperature $I_2$ is zero as can be seen by realizing that the product of Fermi factors is identically zero.

### 11.2.2 Practical Implementation in Plane Wave Codes

In order to simplify the notation let us label electronic wavefunctions as $\psi_t$ for the tip functions and $\psi_s$ for the substrate ones. The first step is to approximate the $\delta$-function by a finite-broadening function, using the mathematical identity

$$\delta(x) = \lim_{\sigma\to 0}\frac{1}{\sqrt{\pi}\sigma}e^{-x^2/\sigma^2}.$$

For numerical applications, $\sigma$ is given a finite numerical value. For usual constant current STM images of molecular adsorbates, $\sigma$ in the range of 0.1 eV seems to be a good compromise. On one hand, 0.1 eV is much smaller than typical electronic features associated with molecular levels broadened by a strong chemisorption interaction with the substrate (hence, in the case of physisorbed molecules, one needs to pay extra attention to this parameter). On the other hand, it is large enough to include several electronic states contributing to the electronic current. The number of states will finally depend on the k-point sampling used.

Indeed, in a plane wave implementation of an electronic structure calculation, there are two good quantum numbers to label the electronic states: the band index and the k-point or k-vector. This is just a simple way of stating Bloch's theorem for periodic systems. In a plane-wave code, there is a unit cell that is periodically repeated. This unit cell contains the physically interesting problem. This periodicity imposes a band structure. Numerically, a minimum set of k-vectors is used, in order to describe as accurately as possible the electronic properties at the minimum numerical cost.

A plane-wave code uses a plane-wave basis set. This means that every wavefunction can be expanded in a linear combination of plane waves. Numerically, this is efficiently achieved by using fast Fourier transform techniques. Hence, for a wavefunction determined by its band index $s$ and k-vector $\mathbf{k}$, we can write

$$\psi_{s,\mathbf{k}}(\mathbf{r}) = \frac{1}{vol} \sum_{\mathbf{G}} C_{s,\mathbf{k}} e^{i(\mathbf{G}+\mathbf{k})\cdot\mathbf{r}}, \tag{11.7}$$

where $\mathbf{G}$ are the reciprocal lattice vectors giving rise to the plane-wave basis set, and $vol$ is the unit-cell volume. $C_{s,\mathbf{k}}$ is the coefficient of the expansion in plane waves for this particular wavefunction. These last coefficients together with the energy eigenvalues $\epsilon_{s,\mathbf{k}}$ are the results of a plane-wave code run.

We can now compute Bardeen's expression, Eq. (11.6), by using these quantities

$$I = -\frac{2\pi e}{\hbar} \sum_{s,\mathbf{k}_s,t,\mathbf{k}_t} (f(\epsilon_{t,\mathbf{k}_t}) - f(\epsilon_{s,\mathbf{k}_s})) \left| \hbar \int_S \vec{J}_{s,t} \cdot d\vec{S} \right|^2 \delta(\epsilon_{t,\mathbf{k}_t} - \epsilon_{s,\mathbf{k}_s} + eV). \tag{11.8}$$

Here, we have assumed that the tip is grounded, and hence, positive voltages correspond to probing empty states of the sample. Now, the matrix element, Eq. (11.4), can be computed using an integration surface $S$ that is a plane in the vacuum region between the sample and the tip. In Bardeen's implementation made by Hofer [5], the sample and tip do not have the same unit cell. This is the most general case and in principle poses no serious problem. Following their implementation, the integration is over the $x$- and $y$-axes of the surface, and the gradients of Eq. (11.4) are partial derivatives in the $z$-direction, which are computed numerically by finite differences in bSKAN [5]. Hence, the Bardeen matrix element, Eq. (11.4), becomes

$$\int_S \vec{J}_{s,t} \cdot d\vec{S} = \int_{x,y} -\frac{\hbar^2}{2m}[\psi_s \vec{\nabla}\psi_t^* - \psi_t \vec{\nabla}\psi_s^*]d^2r$$

$$= \frac{-\hbar^2}{2m\ vol_1\ vol_2} \int_{-L_x}^{L_x} dx \int_{-L_y}^{L_y} dy[\psi_s \frac{\partial \psi_t^*}{\partial z} - \psi_t \frac{\partial \psi_s^*}{\partial z}], \tag{11.9}$$

where $\psi_s$ (and $\psi_t$) can be replaced by Eq. (11.7). In the bSKAN implementation, the sum over **G** is divided into two parts : first the summation along the $z$ component is performed using a 1D fast Fourier transform, and second the $x, y$ in-plane summation is performed in Eq. (11.9), in this way the $z$ derivative is performed numerically. It is also possible to perform the derivative analytically, and then Fourier transforming the full 6D quantity. The six dimensions stem from the assumption that the tip's and sample's unit cells are different and one cannot exploit any simplification by integrating over $x, y$ in the unit cell. We see that using two different unit cells then brings about one more complication, which should be considered as the lateral dimensions $L_x$ and $L_y$ of the integration. In principle, this should be consistent with the volume $vol$, but there will be two different volumes, $vol_1$ and $vol_2$, for the two unit cells, and the wavefunctions will be uncorrectly normalized if we use only one of these. The solution of this problem is to take for $L_x$ and $L_y$ the smaller unit cell dimensions.

One more complication of using different unit cells is that the k-point set is different for each of the two cells. Hence, one cannot make use of the nice properties of the Fourier transform of an exponent in a periodic lattice (yielding Krönecker $\delta$'s) because the nonperiodic part (given by the k-vector) does not factor out in the product of the wavefunction of the tip by the complex conjugate of the substrate.

The first advantage is that instead of two summations over k-points, we are left with one summation. This can be a substantial time improvement. The second is that the $x, y$ integration becomes analytical, yielding Krönecker $\delta$'s that will trivially remove one of the **G** summations, being left with only one, and thus with a faster algorithm.

One has to be careful with finite-size effects. The tip belongs to a periodic system, and instead of simulating the case of a single molecule under an STM tip, we rather have a surface coverage of molecules under an infinite number of tips. The first situation can nevertheless be approximated if the distance between the tip is large enough so that no interference effect among the tips takes place. Typically, this means that the geometrical corrugation of the tip (the tip's dimensions) must be much smaller than the unit cell's dimensions. In this way, the tunneling current will be negligibly small between the cells and the departure from the single-tip case will not be measurable. Tunneling currents are very confined, and large lateral vacuum regions lead to good isolation of the tunneling current to unit cell of interest.

Planewave calculations need to be correctly converged in the vacuum region. This means that the summation over **G** has to be very complete in order to correctly reproduce the exponential decay of wavefunctions in the vacuum region. This leads to very time-consuming calculations. The implementation by Monturet *et al.* [7] goes around this problem. Without loss of generality and accuracy, except for the unique unit cell of the calculation that forces commensurability of the tip and substrate, each of the Fourier components of expression (11.7) are matched to its asymptotic expression in the vacuum region. In order for this approximation to work, one has to make sure that the vacuum region is attained (surface potential must be negligible). But once this is achieved, the accuracy and numerical efficiency of this approximation lead to very reliable and improved calculations. Typically, the STM constant current simulations are a factor of 1000 faster than other Bardeen approaches.

The new wavefunctions for the sample now look like (for the sake of simplicity, we use atomic units $\hbar = m = e$ unless otherwise specified)

$$\psi_s(\mathbf{r}) = \frac{1}{\text{vol}} \sum_{\mathbf{G}} A_{\mathbf{G},s}(z_s) e^{-\sqrt{2\phi_s + (\mathbf{k}_s+\mathbf{G})^2}(z-z_s)} e^{i(\mathbf{k}_s+\mathbf{G})\cdot\mathbf{r}}, \tag{11.10}$$

and for the tip

$$\psi_t(\mathbf{r}) = \frac{1}{\text{vol}} \sum_{\mathbf{G}} C_{\mathbf{G},t}(z_t) e^{-\sqrt{2\phi_t + (\mathbf{k}_t+\mathbf{G})^2}(z_t-z)} e^{i(\mathbf{k}_t+\mathbf{G})\cdot(\mathbf{r}-\mathbf{R}_t)}, \tag{11.11}$$

where now $\mathbf{k}_t = \mathbf{k}_s$ as we described above. The matching distances from the surface and the tip are, respectively, $z_s$ and $z_t$. We have taken $\phi_s$ and $\phi_t$ as the work functions. The sample is assumed to be centered at the origin of coordinates, but the tip is centered at $\mathbf{R}_t$. The Fourier components of the wavefunctions exponentially decrease with $z$, as can be seen in the above expressions; here we have adopted the convention that the tip is further away in positive $z$ from the surface, thus $z_t > z_s$. The Fourier coefficients $A_{\mathbf{G},s}(z_s)$ and $C_{\mathbf{G},t}(z_t)$ are obtained by Fourier transforming the real-space wavefunction in 2D and finding the coefficient at a height $z_s$ and $z_t$ from the sample and tip, respectively.

These expression can be fed into Eq. (11.9). In order to simplify the discussion let us assume that the tip and substrate work functions are the same, $\phi$. Hence,

$$\int_S \vec{J}_{s,t} \cdot d\vec{S} = \frac{\text{surf}}{\text{vol}} \sum_{\mathbf{G}} \sqrt{2\phi + (\mathbf{k}+\mathbf{G})^2} C^*_{\mathbf{G},t}(z_t) A_{\mathbf{G},s}(z_s) \times e^{-\sqrt{2\phi+(\mathbf{k}+\mathbf{G})^2}(z_t-z_s)} e^{i\mathbf{G}\cdot\mathbf{R}_t)} \tag{11.12}$$

The factor *surf* is the lateral area of the unit cell that appears from obtaining the Krönecker $\delta$ functions in 2D. The distance $z_t - z_s$ can be brought back to the surface–tip distance and then Eq. (11.12) becomes very transparent, because in order to perform the Bardeen matrix elements and to perform a 2D Fourier transform over $\mathbf{G}$, one needs to vary the surface–tip distance $z_t - z_s$. The sets of all points $\mathbf{R}_t, z_t - z_s$ define the 3D grid where the STM image is performed.

Finally, the numerical implementation of the Bardeen current is

$$I = \frac{2\sqrt{\pi}}{\sigma} \sum_{\mathbf{k},s,t} (f(\epsilon_t) - f(\epsilon_s)) \left| \int_S \vec{J}_{s,t} \cdot d\vec{S} \right|^2 e^{-(\epsilon_t - \epsilon_s + V)^2/\sigma^2}, \qquad (11.13)$$

where $V$ is the sample–tip bias voltage for a grounded tip.

### 11.2.3
### The Tersoff–Hamann Approximation

Starting with expression (11.6), Tersoff and Hamann [8] reduced the tunneling current calculation to a simplified expression that has had a lot of echo in the STM simulation community. Tersoff and Hamann assumed that the tip's wavefunction could be expanded in spherical harmonics about some curvature center. They retained the first term, the s-wave of the spherical harmonic expansion. Chen [13] later on improved on this assumption by giving a hierarchy of successive improvements by including higher spherical harmonics. The Tersoff and Hamann approximation is equivalent to assuming a spherical tip.

Next, Tersoff and Hamann considered linear response theory. In this case, the current is assumed to be proportional to the bias voltage, the conductance being the proportionality constant. In the case where the current is indeed proportional, the conductance is solely determined by the electronic structure at the Fermi level. And this is an exact result.

With these two approximations (s-wave tip wavefunction and linear response), the conductance is then proportional to the local density of states. In other words, $I = GV$ and

$$G \propto \sum_{n\mathbf{k}} |\psi_{n\mathbf{k}}|^2 \delta(E_F - \epsilon_{n\mathbf{k}}). \qquad (11.14)$$

The local density of states is a density of states as seen by the $\delta$-function (number of states per unit energy) weighed by the spatial distribution of each state with weight at the Fermi energy. This interpretation is very attractive: the constant current STM images are then maps of the constant local density of the states lying at the Fermi energy. A constant current STM image gives direct information about the spatial distribution of states at the Fermi energy. It is thus a very powerful electronic analysis tool.

The two underlying assumptions with this interpretation are the linear behavior of current with voltage and the spherical symmetry of the electronic structure of the tip's wavefunction. In sofar as the current is basically linear with voltage (the electronic structure should not be rapidly changing with energy: in the case of sharp resonances at the Fermi energy, this theory will not be valid. This is rarely the case in chemisorbed molecules on metal surfaces) and the tip electronic structure is rather featureless (correctly approximated by some s-function; this means that each individual atomic component can be a d-wave, as far as the sum of all atomic wavefunctions to constitute the electronic structure of the tip is an s-wave, the theory will be correct), the Tersoff–Hamann theory is an excellent approach to constant current STM imaging.

## 11.3
## Example of Constant Current STM Simulation: Acetylene on Cu(100)

Acetylene on Cu(100) is the first system where IETS was realized [2]. Preliminary studies of STM constant current images were realized both by the experimental group and the theory groups working on it [9, 14]. It is then interesting to study this system here, in order to analyze later on the IET spectra.

Tersoff–Hamann images of $C_2H_2$/Cu(100) could reproduce the experimental images, which were characterized by a deep depression perpendicular to the C–C axis of this molecule. Soon, this depression was recognized as originating from the nodal plane of the $\pi^*$ LUMO orbital of the acetylene molecule [2]. Indeed, chemisorption is strong and large charge transfer takes place; hence the LUMO is the dominating molecular electronic structure at the Fermi energy on Cu(100) [12]. Two small protrusions were found and attributed to the H atoms. However, they also originate in the $\pi^*$ LUMO, since the nodal plane is between two protrusions in the LUMO spatial distribution. The picture is somewhat more complex because due to chemisorption the molecule distorts to incorporate charge from the substrates. This leads to a substantial rehybridization of the C–C triple bond and changes from sp hybridization to basically $sp^2$ hybridization. Hence, the H atoms do indeed stick out of the surface plane, leading to some hyperconjugation: the initial experimental interpretation cannot be taken as wrong.

Figure 11.1 shows a Tersoff–Hamann constant current image of an acetylene molecule. The calculation was performed with VASP in a $3\times3$ supercell, with 64 k-points in the surface Brillouin zone. The results are very similar to the ones obtained by another plane-wave code [9, 15]. The wavefunctions had to be extended

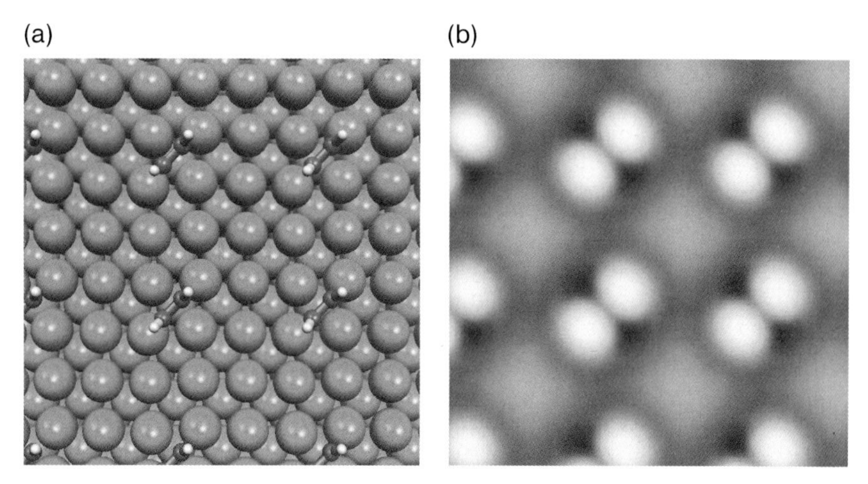


**Figure 11.1** (a) Acetylene molecules in a 3 × 3 supercell of the Cu(100) surface. (b) Tersoff–Hamann constant current image of the structure on the left panel. The calculations have been performed with 64 k-points in the surface Brillouin zone, and a plane wave cutoff of 270 eV.

**Figure 11.2** Bardeen constant current image corresponding to the previous figure, Figure 11.1. The tip has been taken as a 3×3 Cu(100) slab with a single adsorbate Cu atom on a hollow site.

asymptotically into vacuum as in Eq. (11.10), for the same reasons as above: the plane-wave calculations in supercell geometry are either poorly converged because of the lack of plane waves in the expansion or because the next slab is very close, and hence the vacuum wavefunctions are artificially perturbed by the next slab of the next supercell.

This calculation shows that the 3×3 supercell is small, and a substantial corrugation of the metallic density between molecules is artificially obtained: these are finite-size artifacts. Nevertheless, the agreement with the experimental images is impressive; refer to Ref. [15] for a detailed Tersoff–Hamann and experiment comparison.

Figure 11.2 presents Bardeen's constant current image obtained under the same conditions as above. The tip is modeled by a single Cu atom adsorbed on the hollow site of a 3×3 Cu slab. It is the simplest tip one can think of, but it contains the basic ingredients of an STM tip yielding atomic resolution; it has a rather flat curvature (in our case infinite, because it is a planar slab) and an atomic ending yielding its atomic resolution capabilities. In our implementation of the Bardeen equations, the tip geometrical corrugation must be negligible with respect to the supercell dimensions; otherwise the periodic structure induces multiple-tip effects. This is quickly recognizable because the Tersoff–Hamann symmetry has no relation with the Bardeen one. In order to avoid this, we keep the fastest varying dimensions of the tip, much smaller than the lateral dimensions of the tip. In the present 3×3 case, this means that an atomic tip is probably at the limit of a single-tip modelization.

The main difference between Figures 11.1 and 11.2 is the enlargening of the depression in the acetylene molecule. This is due to the blurring effect of the tip: Tersoff–Hamann pictures are too sharp, and the tip induces some smoothening of the image features. As a consequence, the Bardeen picture is noticeably in better agreement than the Tersoff–Hamann one. However, the fictitious increase of the intermolecular metal density is enhanced. This is due to finite-size effects and it is worsened by the presence of the tip, because the size of the tip is not negligible compared to the lateral size of the simulation cell. In order to approach the experimental image of a single acetylene molecule, a larger simulation cell is needed [15]. The agreement with the experimental images is substantially improved in the Bardeen image, Figure 11.2 in particular with respect to the absolute calibration of the tunneling current versus distance and in the overall shape of the image. The calibration leads to realistic values while the Tersoff–Hamann theory systematically underestimates the tunneling current, leading to tip–surface distances that are too short [16]. But the features of the STM image are also systematically improved in Bardeen's treatment. The depression over the acetylene molecule is basically indistinguishable between theory and experiment in Figure 11.2; however, the Tersoff–Hamann undervalues the size of the depression, Figure 11.1. A thorough comparison between the Tersoff–Hamann theory and the experimental images can be found in Ref. [15].

## 11.4
## Extension of the Tersoff–Hamman Theory to IETS-STM

Tersoff and Hamann [8] used the wavefunction of a spherical tip and obtained that the tunneling conductance, $\sigma$, is proportional to the local density of states (LDOS), $\rho$, evaluated at the Fermi level, as we just expounded in the previous section:

$$\sigma \propto \sum_{\nu} |\psi_{\mu}(\mathbf{r}_0)|^2 \delta(E_F - \epsilon_{\mu}) = \rho(\mathbf{r}_0, \epsilon_F). \qquad (11.15)$$

The LDOS is a density of states ($\delta(E_F - \epsilon_{\mu})$) weighted by the spatial information of each state contributing ($|\psi_{\mu}(\mathbf{r}_0)|^2$). This quantity is evaluated at the tip's center of curvature ($\mathbf{r}_0$) and at the Fermi level ($E_F$). A first success of such a result is that the STM is interpreted as a probe to read the electronic structure of the substrate at certain distance from the surface ($\mathbf{r}_0$) and at the substrate's Fermi level. This last result is a consequence of linear-response theory which is justified in the limit of low bias voltage. In the case of molecules adsorbed on metal surfaces, this approach yields a very good description of the STM constant current image when the substrate's electronic structure (eigenvalue $\epsilon_{\mu}$ and wavefunction $\psi_{\mu}(\mathbf{r})$) is evaluated within DFT and a plane-wave code, see the previous section.

In Refs. [9, 11] the many-body extension of the Tersoff–Hamann theory to the treatment of IETS-STM is presented. Briefly, the inelastic contribution to the change in conductance, $\Delta\sigma$, will be caused by the change in the LDOS due to the vibration. Now, the problem is complicated by the many-body aspects of the theory. There is a first term that can be traced back to a transfer of a quantum of vibration by the impinging electron. This is called the inelastic contribution to the change in conductance [10, 18]. The relative change in differential conductance, $\eta_{\text{ine}}$, is then given by

$$\eta_{\text{ine}}(\mathbf{r}_0) = \frac{\Delta\sigma}{\sigma}. \qquad (11.16)$$

To leading order in electron–vibration coupling and using Eq. (11.15) this term is given by

$$\eta_{\text{ine}}(\mathbf{r}_0) = \frac{1}{\rho(\mathbf{r}_0, \epsilon_F)} \sum_{\mu} \left| \sum_{\lambda} \frac{\langle \psi_\lambda | \delta H | \psi_\mu \rangle \psi_\lambda(\mathbf{r}_0)}{\epsilon_\mu - \epsilon_\lambda + i0^+} \right|^2 \delta(\epsilon_F - \epsilon_\mu). \tag{11.17}$$

Here we have taken the quasistatic limit, $\hbar\omega_i \to 0$. $|\psi_\lambda\rangle$ are the unperturbed one-electron states appearing naturally in the lowest order perturbation theory (LOPT) and $\epsilon_\lambda$ the corresponding eigenenergies. For a detailed account of the numerical implementation of these equations, read Refs. [10, 11, 18].

This equation says that there is an increase in conductance due to the modulation of the wavefunction by the vibration, because the squared term is just the square of the perturbed electronic wavefunction. The spatial resolution of the wavefunction carries the information of the exponential decay in vacuum of the tunneling probability. Hence, during the vibration this tunneling probability will be modulated, in a way given by the change of the wavefunction.

The second contribution to the change in conductance at the same order in the electron–vibration coupling is termed the elastic contribution. The name originates from the fact that the initial and final electron states are at the same energy; they do not differ in a quantum of vibration. The physical origin of this term is the many-body character of electron transport in the presence of vibrations. In the absence of vibrations one can approximate the many-body wavefunctions in terms of one-electron wavefunctions, which are solutions of an effective one-body Hamiltonian. When the electron–vibration coupling is included, the one-electron wavefunctions are no longer eigenstates of the Hamiltonian. The vibration mixes them up. The complexity appears because the full wavefunction is antisymmetric under the electron exchange, i.e., two electrons cannot be in the same quantum state. electrons mediated by the electron–vibration interaction. This exchange term gives a negative contribution to the change in conductance due to the antisymmetric character of the many-body wavefunction under the exchange of two electrons. This elastic contribution is given by

$$\eta_{\text{ela}}(\mathbf{r}_0) = \frac{-2\pi^2}{\rho(\mathbf{r}_0, \epsilon_F)} \sum_{\mu} \left| \sum_{\lambda} \langle \psi_\lambda | \delta H | \psi_\mu \rangle \psi_\lambda(\mathbf{r}_0) \delta(\epsilon_\mu - \epsilon_\lambda) \right|^2 \delta(\epsilon_F - \epsilon_\mu). \tag{11.18}$$

The notation is the same as in Eq. (11.17). There are two fundamental differences between Eqs. (11.17) and (11.18). The first one is the sign: the elastic contribution, Eq. (11.18), is negative. It is the term responsible for the decrease in conductance as we announced. The second difference is the range of evaluation of the inner summation over electronic states: in Eq. (11.18) this summation is restricted to states at the Fermi level, while in Eq. (11.17) it extends over all energies. Hence, the elastic contribution, Eq. (11.18), will become particularly important when the density of states is very high at the Fermi level: namely, in the case of a sharp resonance at the Fermi level.

The results obtained to date are in good agreement with the experimental data and allow us to discuss and analyze the molecular modes excited with the STM.

## 11.5
## Applications of the IETS Theory to Realistic Systems

### 11.5.1
### Acetylene Molecules on Cu(100)

The first case where this theory showed predictive capabilities was the IETS of acetylene molecules on Cu(100) [9]. Further calculations permitted us to determine the actual symmetry of the excited modes, and in Ref. [12] it was predicted that the mode leading to the largest change in conductance was the antisymmetric stretch C–H mode. This prediction was enhanced by the analysis that the Tersoff–Hamman theory permitted us to perform. In the Tersoff–Hamman theory the tip's electronic structure is completely symmetrical (s-wave, see above), and our calculations predicted that the LUMO dominated the molecular electronic structure at the Fermi energy. Since the LUMO is antisymmetric with respect to the nodal plane perpendicular to the C–C axis, the matrix element present in Eq. (11.17) can only be different from zero if the electronic component of the electron–vibration coupling is antisymmetric. The electron–vibration coupling has an electronic component and a nuclear one, and the product of both is the perturbing potential which is part of the Hamiltonian. The Hamiltonian has to be symmetric [17]. Hence, the nuclear component has to be antisymmetric, and thus the mode is antisymmetric [12, 18].

Figure 11.3 shows the spatial distribution of the change in conductance over an acetylene molecule on Cu(100) when the C–H stretch mode is excited. The experiment gives a maximum value of $8 \pm 1\%$ [12] and the theoretical change in conductance is 9.6% at the center of the molecule. Both the spatial distribution and the values of the change in conductance are in excellent agreement between theory and experiment. Thus, this analysis of the theoretical images of the IETS signal for DCCD allows us to conclude that the detected IETS signal originates from the antisymmetric mode.

### 11.5.2
### Benzene, Phenyl, and Benzyne Molecules on Cu(100)

In a series of experiments performed by the Ho group [19] and the Kawai group [20], molecules of benzene are subjected to bias voltage pulses under an STM tip. While the IETS show no signal for benzene molecules in the expected voltage range of the C–H stretch modes [19, 20], the molecular fragments found after the pulses present peaks at the C–H stretch frequencies. After a thorough analysis of the constant current images of both the benzene molecule and the fragments as well as of the origin of the IETS signals, Lauhon and Ho concluded

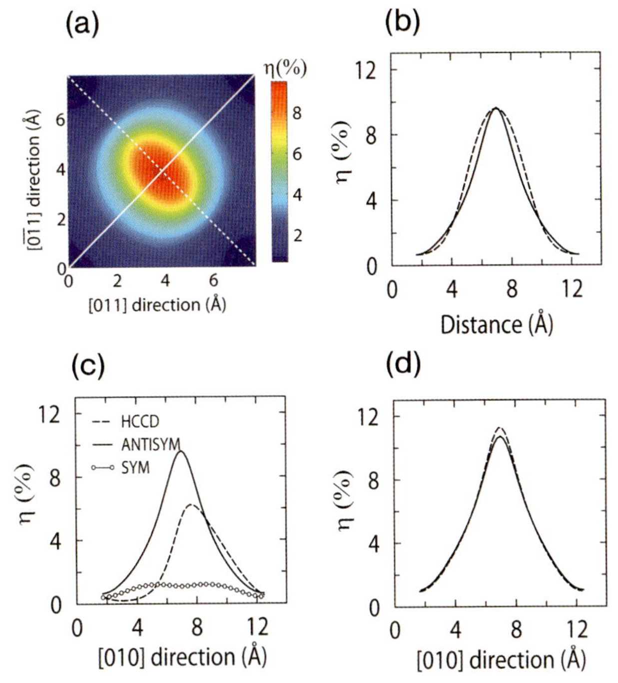


**Figure 11.3** Spatial distribution of the change in conductance when the C–H stretch of the acetylene molecule on Cu(100) is excited. (a) Inelastic efficiencies for the antisymmetric C–D stretch stretch mode of $C_2D_2$ over the 3×3 supercell of Cu(100), 7.5×7.5 Å. The C–C axis is oriented along the [010] direction, represented in the figure as a solid line. (b) Inelastic profiles along the [010] direction (full line) and the [001] direction (dashed line) as depicted in (a). (c) Profiles of the inelastic efficiency along the C–C axis for the antisymmetric and symmetric modes of DCCD, and the C–D stretch of HCCD. (d) Inelastic profile (full line) along the C–C axis for the sum of the antisymmetric and symmetric modes compared with the sum of the inelastic efficiency of two independent C–D stretch modes corresponding to the two bonds of DCCD (dashed line). (Reprinted with permission from Ref. [21], copyright American Physical Society.)

that the disrupted molecules were dehydrogenation fragments. Due to the apparent symmetry of the constant current image and to the enhanced stability of benzyne ($C_6H_4$) fragments, the molecular fragment was identified as a benzyne fragment. Kim and co-workers [20] worked on benzene molecules on Cu(110). The different substrate might lead to different results, and they concluded that due to the dehydrogenation process itself, the final fragment might rather be phenyl ($C_6H_5$).

Counting on the availability of experimental IETS of the fragments, a simulation could be the definitive way of finding the actual dehydrogenation fragment. In order to achieve it, an initial total energy calculation to determine geometries, vibrational energies, and constant current images was mandatory. The results of such a calculation are presented in Figure 11.4.

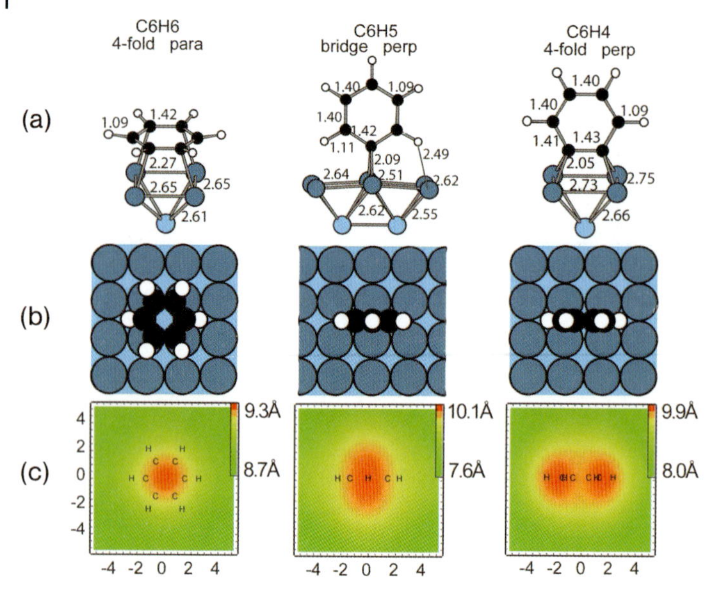


**Figure 11.4** (a) Sketch of the optimized adsorption geometry for benzene and derivatives on Cu(100) including key bonding distances in Å. Dark blue (blue, black, white) spheres correspond to surface Cu (sublayer Cu, C, H) atoms. (b) Large-scale top view of the adsorption configurations on a p(4 × 4) unit cell. (c) Associated STM simulations (10.32 Å × 10.32 Å) at constant current. Positions of C, H atoms are indicated with letters in the plots. (Reprinted with permission from Ref. [21], copyright American Physical Society.)

Figure 11.4 shows the constant current image simulations using the Tersoff and Hamman approach. There we see that the fragments are very different from the original benzene molecule image. However, the images of the fragments are not very different. Corrugations are in the same range and the small symmetry differences will be erased by the broadening induced by the tip as seen above for Bardeen's simulations. Hence, constant current images are not sufficient to discriminate between both fragments.

Figure 11.5 shows the voltage-resolved spectra of different fragments, including the deuterated species. The phenyl spectra are in excellent agreement with the experiment. However, the benzyne spectra are composed of a single peak, very different from the experimental ones. Phenyl fragments show that a splitting of 20 meV between the highest C–H stretch mode, $\nu_1$, and the lowest stretch $\nu_5$ is in very good agreement with the 20-meV difference between the two peaks found in the experimental IETS spectra [19]. The same computed difference is only of 4 meV for the benzyne fragment. When the deuterated species are used, the phenyl fragments yield a 16-meV splitting while the deuterated benzyne fragments maintain the 4-meV difference. The experimental peak distance for the deuterated case is 12 meV. The experimental changes in conductance are 3% and 2%, while the theoretical contribution at 387 meV is 2.5% and the sum of the two contributions

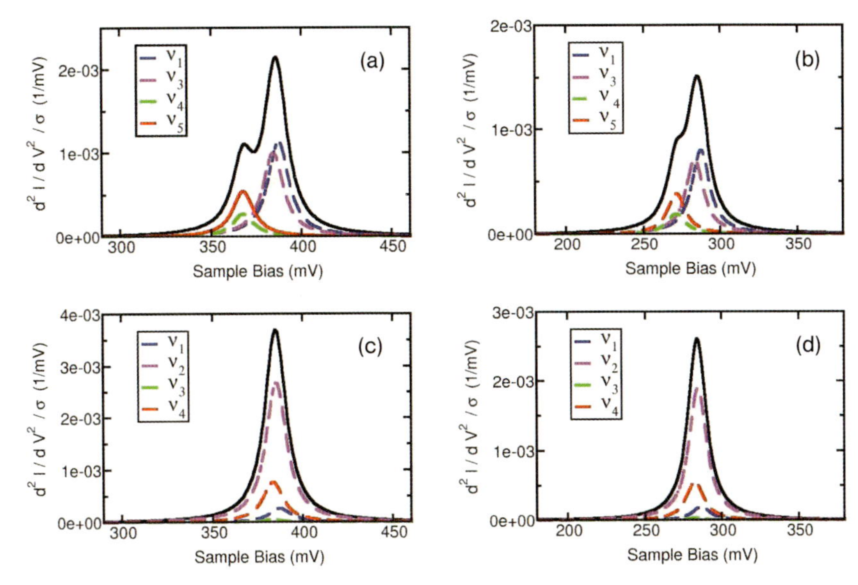


**Figure 11.5** Simulated IETS spectra of $C_6H_5$ (a),$C_6D_5$ (b), $C_6H_4$ (c), and $C_6D_4$ (d): change of conductance as a function of the tip–surface bias voltage above the center of each fragment. The black line is the total spectra, and the dashed lines are the contributions of each mode to the total spectra. For (a) and (b) phenyl derivatives, the blue (violet, green, red) line corresponds to the $\nu_1$ ($\nu_3$,$\nu_4$,$\nu_5$) mode, respectively. The individual contributions display a 14-meV width corresponding to the experimental one [19]. For (c) and (d) benzyne derivatives, the blue (violet, red) line corresponds to the $\nu_1$ ($\nu_2$,$\nu_4$) mode, respectively, but strictly localized in energy. (Reprinted with permission from Ref. [12], copyright American Physical Society.)

at 368 and 367 meV is 1.8%. The same values for the deuterated species are 2% and 1% in the experiment and 1.75% and 1.3% in the simulation.

From the above calculations on $C_2H_2$ and the benzene molecules and dehydrogenated products, as well as more recent calculations on the pyridine C–H stretch spectra [22], we have evidence that detectable IETS signals originate in a molecular electronic structure with a given mode character at the Fermi level. In the case of acetylene, we have seen that the $\pi^*$ LUMO straddles the Fermi energy. This molecular orbital shows some hyperconjugation due to the rehybridization with the surface electronic structure during the chemisorption process (see above); hence, there is an importance of the hydrogen electronic structure in the formation of the orbital. This leads to an important electron–vibration coupling with the C–H stretch mode. Similarly, pyridine and benzyne have a large $\sigma_{CH}$ electron structure at the Fermi energy due to the dehydrogenation process [22], and a sizeable IETS signal is obtained. However, benzene has no C–H electronic structure at the Fermi energy during its chemisorption [22]. As a consequence, the C–H stretch modes show no change in conductance and hence no IETS signal is detectable. Pyridine [22] similarly presents a small charge depletion from its HOMO ($\sigma_{CH}$) to the metal, yielding a weaker electron–vibration coupling. From our IETS simulation, we infer that the upright position of the molecule, and in general, its conformation play a minor role except in the spatial distribution of the IETS signal.

The IETS signal distribution is a direct consequence of the electronic symmetry at the Fermi level in combination with the mode symmetry. Hence, IETS yields information not only on the vibrational properties of the adsorbed molecule but also on its electronic properties.

## 11.6 Conclusions

In this chapter, we have shown that current quantum chemistry implementations, mainly based on density functional theory, permit us to reach quantitative agreement with experiment providing us with a privileged standpoint for the understanding of the atomic constituents of matter, the identification of substances on the atomic scale, and the actual reaction pathways in surface-assisted reactions.

Plane-wave methods for the calculation of both electronic structure and geometrical conformations of adsorbates on surfaces are particularly accurate due to the completeness of the basis set even for the discretion of extended vacuum regions. These vacuum regions are the basic characteristic that differentiates surface-based problems and simulations from other quantum chemistry problems.

In order to simulate STM experiments, plane-wave-based codes are then very indicated. Here we have shown how to accurately simulate constant current images by making use of the Bardeen approximation. We have shown a particular simulation of STM images of acetylene molecules on Cu(100) where the correctness of the Bardeen approach is emphasized. The main caveat of the Bardeen approximation is neglecting tip–substrate mutual interactions. This has to be considered before any simulation is undertaken by this method. For most STM imaging applications, the Bardeen approximation is largely justified, leading to accurate simulations.

When the tunneling current excites vibrations, the tunneling properties change. A simple theory can take into account the relative changes of conductance. Here, we have presented an extension of the Tersoff and Hamman approach where the tip's detailed structure is neglected, which is very accurate in the description of inelastic effects taking place on the substrate. By evaluating relative changes in conductance, the poor description of tip-based effects is basically canceled out, emphasizing the local aspects of the vibrating adsorbate. This approach has permitted us to develop an understanding of the experimental signals, ranging from the amount of change in conductance to the actual symmetry of the excited modes. Moreover, the simulations give important insight of the underlying electronic structure that leads to the possibility of exciting certain vibrational modes. These analyses are fundamental for understanding of the experimental data as well as for future applications of inelastic effects in electronic currents.

## References


1 G. Binnig, H. Rohrer, C. Gerber, and E. Weibel, *Phys. Rev. Lett.* **49**, 57 (**1982**).
2 B. C. Stipe, M. A. Rezaei, and W. Ho, *Science* **280**, 1732 (**1998**).
3 R. C. Jacklevic and J. Lambe, *Phys. Rev. Lett.* **17**, 1139 (**1966**).
4 J. Bardeen, *Phys. Rev. Lett.* **6**, 57 (**1961**).
5 W. A. Hofer, A. Foster, and A. Shluger, *Rev. Mod. Phys.* **75**, 1287 (**2003**).
6 O. Paz, I. Brihuega, J. M. Gómez-Rodríguez, and J. M. Soler, *Phys. Rev. Lett.* **94**, 056103 (**2005**).
7 S. Monturet, N. Lorente, and A. Arnau, unpublished.
8 J. Tersoff and D. R. Hamann, *Phys. Rev. Lett.* **50**, 1998 (**1983**); *Phys. Rev. B* **31**, 805 (**1985**).
9 N. Lorente and M. Persson, *Phys. Rev. Lett.* **85**, 2997 (**2000**).
10 N. Lorente, *App. Phys. A – Mater. Sci. & Processing* **78**, 799 (**2004**).
11 N. Lorente and M. Persson, *Faraday Discuss.* **117**, 277 (**2000**).
12 N. Lorente, M. Persson, L. J. Lauhon, and W. Ho, *Phys. Rev. Lett.* **86** 2593 (**2001**).
13 C. J. Chen, *Introduction to Scanning Tunneling Microscopy* (Oxford: Oxford University Press).
14 N. Mingo and K. Makoshi, *Phys. Rev. Lett.* **84**, 3694 (**2000**).
15 F. E. Olsson, M. Persson, N. Lorente, L. J. Lauhon, and W. Ho, *J. Phys. Chem. B* **106**, 8161 (**2002**).
16 K. Stokbro, U. Quaade, and F. Grey, *Appl. Phys. A – Mater. Sci. Process* **66**, S907 (**1998**).
17 L. D. Landau and E. M. Lifshitz, *Quantum Mechanics*, 3rd edn. (Oxford: Pergamon), **1991**.
18 J. I. Pascual and N. Lorente, Single-molecule vibrational spectroscopy and chemistry, in *Properties of Single-Molecules on Crystal Surfaces*, editors: P. Gruetter, W. Hofer, and F. Rosei, Imperial College Press, London (**2006**).
19 L. J. Lauhon and W. Ho, *J. Phys. Chem. A* **104**, 2463 (**2000**).
20 T. Komeda, Y. Kim, Y. Sainoo, and M. Kawai, *J. Chem. Phys.* **120**, 5347 (**2004**).
21 M.-L. Bocquet, H. Lesnard, and N. Lorente, *Phys. Rev. Lett.* **96**, 096101 (**2006**).
22 H. Lesnard, M.-L. Bocquet, and N. Lorente, *J. Am. Chem. Soc.* **129**, 4298 (**2007**).

# 12
# X-Ray Spectroscopy Calculations Within Kohn–Sham DFT: Theory and Applications

*Mikael Leetmaa, Mathias Ljungberg, Anders Nilsson, and Lars Gunnar Moody Pettersson*

## 12.1
## Introduction

In heterogeneous catalysis, molecules impinge on the catalyst, adsorb, diffuse, and undergo reactions transforming to new species. As the molecule binds to the surface, its electronic structure rearranges to form the required new bonds. Since the internal electronic structure of the molecule in gas phase has arranged itself to maximize the intramolecular bonding, the interaction with the surface by necessity involves weakening of internal bonds in order to allow new bonds to be formed; in heterogeneous catalysis, this is exploited to prepare the molecule for the desired reaction. Knowledge of these electronic structure rearrangements is thus key to a fundamental understanding of these processes, and a simple molecular orbital picture may here be extremely helpful. In this chapter, we shall describe a very powerful combination of experimental and theoretical tools that, when used in conjunction, allows the development of such a picture.

Theoretical studies of surface reactions and processes traditionally focus on structure and energetics, i.e., finding active sites where reactions occur and their associated barriers, which determine the overall rate. Much less emphasis has been put within the DFT community on analyzing the electronic structure in terms of a molecular orbital picture connected to the origin of the surface chemical bond. This may partly be due to the lack of a solid, theoretical connection between the Kohn–Sham (KS) orbitals and "real" physical variables in that in the KS theory, the orbitals are formally introduced as describing fictitious noninteracting electrons while still generating the same electron density as the fully interacting many-body system. In the present chapter, we will review recent work to extend the formal KS theory to excited states and endeavor to demonstrate, at least empirically, how the KS orbitals connect with spectroscopic observables. We will in particular focus on various surface-sensitive inner-shell spectroscopies, relevant for heterogeneous catalysis, and on how we can compute the corresponding spectra and binding energies based on the KS orbitals.

*Computational Methods in Catalysis and Materials Science*. Edited by Rutger A. van Santen and Philippe Sautet

ISBN: 978-3-527-32032-5

There exist a large number of spectroscopic techniques that are, or can be made, surface sensitive and which can be used to study the electronic and geometric structure of adsorbed molecules on surfaces [1]; here we will focus on spectroscopies involving the inner-core electrons such as X-ray photoelectron spectroscopy (XPS), X-ray emission spectroscopy (XES), and X-ray absorption spectroscopy (XAS), also denoted near-edge X-ray absorption fine structure (NEXAFS), while in chemistry the acronym XANES (X-ray absorption near-edge structure) is typically used.

Figure 12.1 illustrates the different core-level spectroscopies in a phenomenological way. The different techniques can be classified in terms of either creation or decay of core–holes. The core electron can be excited to a bound state or to the continuum where it becomes a free particle. Measurement of the kinetic energy of the outgoing photoelectron in the latter case gives the binding energy of the core level; this forms the basis of photoemission or PES (or XPS) [2–6]. If the excitation energy is not high enough to reach the ionization continuum, we can instead populate bound states and resonances above the Fermi level ($E_F$). The method is generally denoted as XAS and can be divided into two regimes: NEXAFS or XANES, for bound states and low-energy resonances in the continuum, and

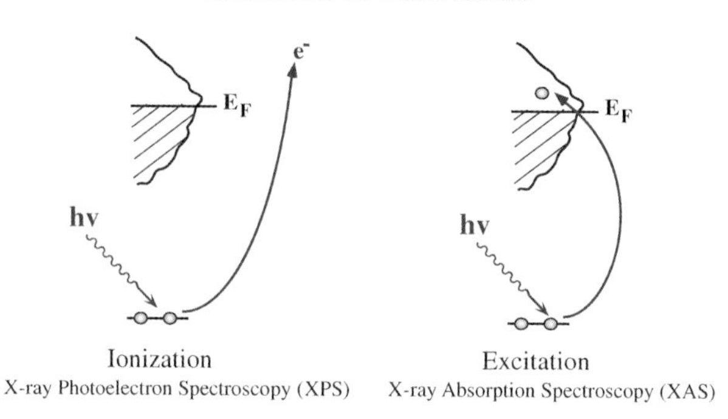


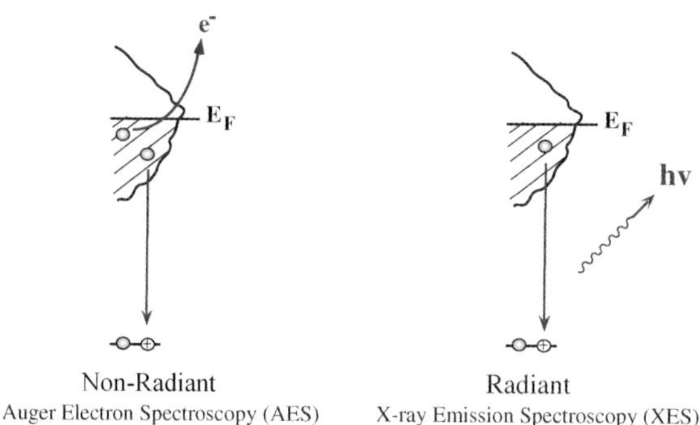


**Figure 12.1** A schematic illustration of core-level ionization, excitation, and decay processes.

extended X-ray absorption fine structure (EXAFS), when the outgoing electron has energy well into the ionization continuum. In the NEXAFS regime, XAS provides information on the symmetry and character of the empty electronic states above the Fermi level [7], while the EXAFS oscillations provide information on interatomic distances [8].

The core–hole can decay through either nonradiative or radiative processes. In nonradiative (Auger) decay, an electron from one of the outer shells fills the core–hole and a second electron takes up the excess energy and is emitted. In the radiative XES process, the excess energy is instead emitted in the form of a photon. For all core levels, both types of process contribute with relative occurrence dependent on the nuclear charge $Z$. In Auger electron spectroscopy (AES) [9–11], the emitted electrons are analyzed while the photons are registered in XES [12–14]. Note that for low-$Z$ elements, the Auger process dominates by several orders of magnitude, which makes XES a very demanding technique when applied to surface adsorbates for which the number of molecules/atoms of interest constitutes a very minute fraction of the sample.

The final states of the decay processes in AES and XES are rather different. The two-valence-hole state in AES often leads to strong interaction of the two holes making it difficult to deduce a simple interpretation in a one-electron molecular orbital picture. XES, on the other hand, provides a direct tool for studying the local valence electronic structure in a one-electron picture.

There is a close coupling between XES and XAS. The former, as an emission spectroscopy, gives information on the occupied orbitals, while XAS relates to the character and symmetry of the unoccupied levels, albeit in the presence of the core–hole that can lead to major deviations from a simple ground-state picture. For XES, on the other hand, the final state is similar to the valence hole state in ultraviolet photoelectron spectroscopy (UPS) [15]. Both XAS and XES are governed by the dipole selection rule, and through the localized character of the core orbital, they both provide a simple atom-selective projection of the electronic structure; this is due to the very large energy difference between inner shells of different elements as well as from smaller chemical shifts between atoms of the same element in different chemical surroundings. From the thus measured local electronic structure, strong conclusions can be drawn on the chemical environment, oxidation state, etc. of selected elements in the sample under study [16, 17]. We finally emphasize the close connection between XAS and XES in that, for a system where the constituent components contribute differently to the XAS energy-dependent cross section, XAS can be used to select different species for study using XES; this was exploited in the case of $N_2$/Ni(100) to be discussed next.

In Figure 12.2, we compare the occupied electronic structure of molecular $N_2$ adsorbed on Ni(100) measured using UPS [18] and XES [19, 20]. UPS, or valence band photoemission, probes the overall electronic structure through ionization of the valence electrons [21]. The initial (ground state) and final valence hole state are the same in the two spectroscopies, but the involvement of the intermediate core–hole in the case of XES leads to a spatially localized picture of the electronic

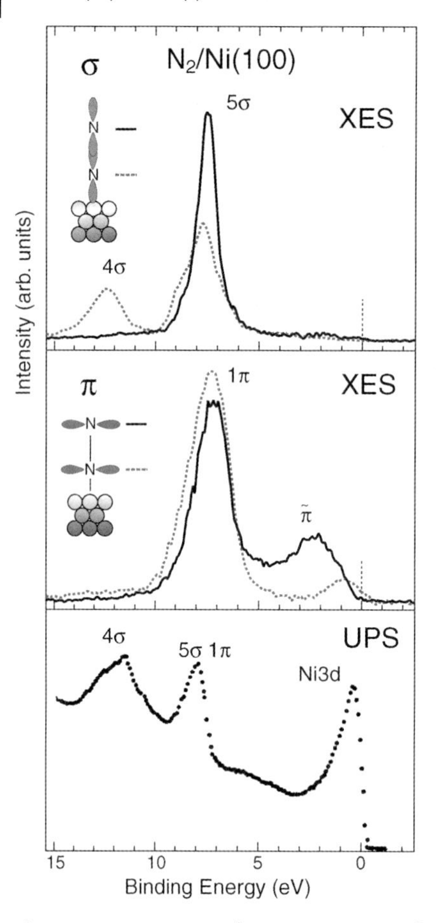


**Figure 12.2** Comparison between atom-specific and symmetry-resolved XES spectra [20] with UPS spectrum measured at a photon energy of 35 eV [18] of $N_2$ adsorbed on Ni(100).

structure, projected onto the core–hole site, in contrast to the spatially delocalized $k$-dependent periodic band structure measured in UPS.

$N_2$ adsorbs in a vertical end-on geometry making the two nitrogen atoms inequivalent. There are adsorbate-induced features in the 7–12 eV binding energy region in both spectroscopies. However, in the 3d band region around 0–5 eV the UPS spectrum only shows strong substrate emission, whereas the XES spectra clearly reveal adsorbate-derived states. The XES additionally provides separate projections on the two inequivalent nitrogen atoms by exploiting the chemical shift of the 1s level to select on which of the two nitrogen atoms the core–hole should be created; $\sigma$ and $\pi$ states can furthermore be separated by appropriate positioning of the detector. XES thus measures the spatially localized, atom-projected aspect of the electronic structure, while UPS determines the more collective, nonlocal, band-structure aspects. In this regard, the two techniques provide complementary information.

We expect these core-level-based experimental techniques to become ever more important where, e.g., recent advances in extending measurements to higher pressures allow "bridging the pressure-gap" [22], which makes much more realistic applications possible. The imminent introduction of free-electron lasers (FELs) in the X-ray regime will furthermore open a completely new area of physics and chemistry, where electronic structure changes can be followed through inner-shell spectroscopy on a femtosecond time scale. This is also expected to provide a very strong incentive for further theoretical developments to match the new experimental techniques and approaches. In the following, we will discuss how these processes and spectroscopies can be modeled theoretically and how spectrum simulations can be used to derive additional information both on geometrical and electronic structure through relating closely to experiment.

Due to the importance of XAS in probing the electronic structure of materials, surface adsorbates, etc., several computational approaches to reproduce the spectra have been developed and applied to assist with the analysis [23–35]. Specific techniques for molecules and clusters [25, 26, 28–30, 33–35] as well as for periodic systems [23, 24, 27] are available. In this chapter, we will focus on the KS theory in a molecular orbital picture as exemplified by the implementation of the transition potential (TP), half-core–hole (HCH) approach [36] for XAS in the StoBe-deMon code [37], and will furthermore discuss calculation of XPS-binding energies and XES spectra.

For more in-depth studies of XAS or NEXAFS we refer to the excellent book by Stöhr [8], for XES to the review by Nilsson and Pettersson [14] and for a more extensive, technical review of the general case of resonant inelastic X-ray scattering (RIXS) to that of Gel'mukhanov and Ågren [13]. Here we will begin by discussing the validity of excited states within the KS formalism as a prerequisite for the treatment of spectra.

## 12.2 Excited States in Kohn–Sham DFT

In physics, chemistry, engineering, and science in general, it often happens that intuitively reasonable methods are used on a pragmatic basis before they are theoretically completely justified. This is true for KS DFT which, since early days, has been used in a similar way to wavefunction methods although the theoretical basis has been lacking. The KS wavefunction is by definition the wavefunction of a fictitious noninteracting system introduced with the purpose to reproduce the electron density of the real, interacting system. Consequently KS orbitals have often been regarded as nothing but formal, auxiliary quantities without specific physical meaning. Nevertheless, the KS wavefunction has often been used as if it were the real interacting wavefunction; the KS orbital eigenvalues have been taken as ionization energies (band structure), their differences as excitation energies, and the orbitals themselves have been used to compute transition moments. To compute excitation energies in a less approximate way, the Delta Kohn–Sham ($\Delta$KS) method has been devised where an excited state of the interacting system is represented by the KS system with one or more electrons excited and where the orbitals have been allowed to relax; the excitation energy is obtained by subtracting the total energy of the so-computed excited state from that of the ground state.

Today we have more theoretical knowledge of DFT, and we shall see that the KS orbitals really have much more meaning than just auxiliary quantities that only serve to reproduce the density: that the eigenvalues can be considered as an approximation to ionization energies (and their differences to excitation energies), that they can be used to compute transition moments, and that the $\Delta$KS method can be viewed as an approximation to more advanced time-independent theories for excited states.

Contrary to popular belief, time-dependent DFT (TDDFT) is not the only rigorous way to treat excited states within DFT. As early as 1979, Theophilou [38] made the first attempt to treat excited states by a subspace formalism. In this formalism, equal mixtures of the $N$ first states with energy $E_{1,N} = 1/N(E_0 + E_1 + \cdots + E_N)$ could be calculated and the $N$th excitation energy obtained by subtraction as $E_N = NE_{1,N} - (N-1)E_{1,(N-1)}$. In 1988, Gross, Oliveira, and Kohn in a series of papers [39–41] extended the subspace formalism to general ensembles and further developments have been given by Levy and Nagy [42], Görling [43], and Sahni *et al.* [44].

The theory of Levy and Nagy [42] is based on the constrained search method. An energy functional $F[\rho_0, \rho]$ is defined that depends on the ground-state density $\rho_0$, the density of the excited state $\rho$, and the sequence number of the excited state. The numbering is important because the $N$th state of the interacting system should be orthogonal to the $N-1$ lower states. Now the KS Hamiltonian is defined as $\hat{H}_w = \hat{T} + \sum_i w(\rho_k, \rho_0; \mathbf{r}_i)$, which has the excited-state density $\rho_k$, and whose ground state has a density the closest to $\rho_0$ in a least-squares sense. The excited state in question is built up from a single KS determinant, which is constrained to be orthogonal to all lower KS states. The energy functional $F[\rho_0, \rho]$ is partitioned as

$$F[\rho_0, \rho] = T[\rho_0, \rho] + Q[\rho_0, \rho] + E_C[\rho_0, \rho], \tag{12.1}$$

where $T[\rho_0, \rho]$ is the noninteraction kinetic energy, $Q[\rho_0, \rho]$ is the Coulomb plus exchange, and $E_C[\rho_0, \rho]$ is the remaining correlation energy. Provided that one knows these state-dependent functionals, one could calculate an excited state by first doing a ground-state calculation, then finding the first excited state by solving for the lowest-energy KS wavefunction orthogonal to the ground state, and then finding the next that is orthogonal to both lower states, etc. The value of $Q[\rho_0, \rho]$ can be easily calculated, since it has the same form as the Coulomb and exchange in the Hartree–Fock (HF) theory. The problematic thing is to construct the potential $w$ in the KS Hamiltonian, $w = v_{\text{ext}} + \delta Q/\delta\rho + \delta E_C/\delta\rho$. The exchange part of $\delta Q/\delta\rho$ can be computed with the optimized effective potential (OEP) method [45,46], which is a method to compute a *local*, multiplicative potential that most closely reproduces exact non-local exchange.

In the beginning, OEP could only be used for atoms. It has been extended to plane wave and grid basis sets, but only recently to Gaussian basis sets because of numerical problems [47]. A widely used approximation to OEP is the Krieger, Li, and Iafrate (KLI) method [48]. However in practical applications, since corresponding correlation functionals have been lacking, most of the calculations published to date have been performed using exact exchange only, neglecting or approximating correlation. An exception is a paper by Glushkov and Gidopoulos [49] where an MP2-like scheme was used for the correlation energy with improved results. Now, if we do not have good functionals, or do not want to bother with cumbersome OEP and MP2 procedures, we could just try to use the exchange-correlation functionals that have been developed for the ground state. In the light of the above, we see that we can view a regular ΔKS calculation as an excited-state calculation in the theory of Levy and Nagy, only that an approximate functional is used.

In the generalized adiabatic connection Kohn–Sham (GAC-KS) approach of Görling [43], the Hohenberg and Kohn theorem [50] is bypassed by using the adiabatic connection in combination with the constrained search method. A mapping between the interacting system and the noninteracting one is made by continuously switching off the interaction between the electrons while keeping the density fixed, thus constructing a path between the two states. An interacting state (ground state or excited) may be mapped to either a ground state or an excited state of the noninteracting system. Because of this, the functionals only depend on the excited-state density, the state ordering parameter, and the coupling strength. Thus in this theory, the ground state and excited states are treated on an equal footing. The working KS equations turn out to be very similar to the ones in the Levy–Nagy theory [42]. However in the GAC-KS theory, the energies for excited states in the KS system (with the same external potential) do not have to be ordered. Also in the GAC-KS theory, for each excited state the effective potential is thus different and the OEP method must be used, as well as some state-dependent correlation functional. Görling notes, however, that the regular ΔKS procedure is justified as an approximation to GAC-KS [43].

Also the KS orbitals and eigenvalues have been found to have more physical meaning than was thought before as discussed by, e.g., Hamel *et al.* [51]. It is known that the highest occupied KS orbital energy corresponds to the first ionization energy in the case of a correct asymptotic behavior of the exchange potential [52]. Chong *et al.* [53] have shown that also the other eigenvalues correspond to approximate relaxed ionization energies (HF gives approximate *unrelaxed* ionization energies). By constructing KS potentials from accurate *ab initio* wavefunctions, they showed that the eigenvalues are in excellent agreement with experiment for outer-valence states, while the approximation becomes worse for more strongly bound orbitals; for core eigenvalues, the discrepancy was similar to HF.

Casida and Salahub have shown [54] from TDDFT that if one neglects relaxation effects and configurational mixing, the KS eigenvalue difference will lie between the triplet and the singlet excitation energy. This is the same result as is obtained in the static exchange method in HF, where the virtual orbitals are constructed with one electron removed from a specific occupied orbital [55]. In the HF theory, occupied and virtual orbitals do not feel the same potential because the exchange contribution cancels out the self-Coulomb term for an occupied orbital so that occupied orbitals feel an $N-1$ electron potential, while a virtual orbital feels the full $N$ electron potential. This gives the occupied and virtual orbital energies different meanings: the occupied are approximate ionization energies and the virtual are approximate electron affinities (Koopmans' theorem). This makes the eigenvalue difference for a ground-state HF calculation a bad approximation to the excitation energy, hence the need to calculate the virtual orbitals with a hole in the static exchange (STEX) approximation. In the KS case, however, the occupied and virtual orbitals feel the same potential, and the eigenvalue difference works well as an approximate excitation energy, provided that configuration mixing can be neglected and that the triplet and singlet excitation energies are close enough [54]. The last aspect is here a simple recognition of the fact that a description using a single Slater determinant gives an equal mixture of singlet and triplet couplings.

There is thus theoretical justification for using the KS orbital energies to estimate binding energies and excitation energies, but we also need to consider whether the KS orbitals can reliably describe transition moments and cross sections.

## 12.3 X-Ray Absorption Spectroscopy (XAS)

In an independent-particle picture, the XAS process can be thought of as an incoming X-ray photon being absorbed by a core-level electron, which gets excited. The process then has a certain probability depending on the properties of the initial and final states. Within the important dipole approximation, certain selection rules apply: total spin is conserved and the total angular momentum $l$ is changed by $\pm 1$. Thus, due to the very local atomic character of the process, for excitations from a 1s level only unoccupied states with local $p$-character are probed, but still with spatial sensitivity due to the extended character of the final states. The spectroscopic process is ultrafast (on an attosecond time scale) so that nuclear movement during the excitation can be completely neglected.

Assuming that at time $t=0$, the system is in an initial state $|i^{(0)}\rangle$, a standard time-dependent perturbation treatment gives the expansion coefficient $c_{\mathrm{fi}}(t)$ of the final state $|f\rangle$ under the perturbation $\hat{H}_1$ at time $t$ to first order as

$$c_{\mathrm{fi}}(t) = -\frac{i}{\hbar}\int_0^t dt'\, e^{i\omega_{fi}t'}\, \langle f^{(0)}|\hat{H}_1|i^{(0)}\rangle, \qquad (12.2)$$

where

$$\omega_{\mathrm{fi}} = \frac{\left(E_{\mathrm{f}}^{(0)} - E_{\mathrm{i}}^{(0)}\right)}{\hbar} \qquad (12.3)$$

and $E_{\mathrm{i}}^{0}$ and $E_{\mathrm{f}}^{0}$ are the energies of the initial and final states, respectively.

We assume that the full Hamiltonian of the system interacting with an electromagnetic field can be partitioned as $\hat{H} = \hat{H}_0 + \hat{H}_1$, where $\hat{H}_0$ is the *sum* of the Hamiltonian of the electromagnetic field *and* that of the unperturbed system, while $\hat{H}_1$, governing the interaction of the system with the electromagnetic field, is given by

$$\hat{H}_1 = \frac{e}{m_e c}\hat{\mathbf{A}} \cdot \hat{\mathbf{p}} + \frac{e^2}{2m_e c^2}\hat{\mathbf{A}}^2, \tag{12.4}$$

where $\hat{\mathbf{p}}$ is the linear momentum operator, and $e$ and $m_e$ are the electron charge and mass, respectively. The vector potential operator $\hat{\mathbf{A}}$ characterizing the electromagnetic field is in the case of a plane electromagnetic wave with wave vector $\boldsymbol{k}$, unit X-ray polarization vector $\boldsymbol{e}_p$, and frequency $\omega_\mathbf{k}$ given by

$$\hat{\mathbf{A}} = A_0[\hat{a}_{\mathbf{k},p}\mathbf{e}_p e^{i(\mathbf{k}\cdot\mathbf{r})} + \hat{a}^+_{\mathbf{k},p}\mathbf{e}_p e^{-i(\mathbf{k}\cdot\mathbf{r})}], \tag{12.5}$$

where $A_0 = \sqrt{\hbar/2\omega_\mathbf{k} V\varepsilon_0}$ and $V$ is an arbitrarily large volume connected to the normalization of the plane wave and $\varepsilon_0$ is the vacuum permittivity. $\hat{a}^+_{\mathbf{k},p}$ and $\hat{a}_{\mathbf{k},p}$ are the creation and annihilation operators for photons with wave vector $\mathbf{k}$ and polarization direction $p$. $\hat{\mathbf{A}}$ changes the number of photons by $\pm 1$. The second term of Eq. (12.4), which is quadratic in $\hat{\mathbf{A}}$, changes the number of photons by 0 or $\pm 2$ and will to first order thus not contribute to absorption or emission of photons. We furthermore realize that only the first of the two conjugate terms in Eq. (12.5) gives absorption of photons (the second term gives emission). We thus obtain the effective interaction Hamiltonian

$$\hat{H}_1^{\text{eff}} = \frac{e}{m_e} A_0 \left[\hat{a}_{\mathbf{k},p}\hat{\mathbf{p}} \cdot \mathbf{e}_p e^{i(\mathbf{k}\cdot\mathbf{r})}\right]. \tag{12.6}$$

Since we have chosen to work entirely in the Schrödinger picture, the interaction Hamiltonian is time independent. It can still, however, cause the state of the system to change with time. Performing the integration in Eq. (12.2) gives the probability of finding the system in state $|f\rangle$ in the limit of large $t$ as

$$|c_{fi}(t)|^2 = \frac{2\pi}{\hbar}\frac{e^2}{m_e^2}\frac{t}{2\omega_\mathbf{k} V\varepsilon_0}\delta(\omega_{fi})|\langle f^{(0)}|\hat{H}_1^{\text{eff}}|i^{(0)}\rangle|^2. \tag{12.7}$$

Both the initial $|i^{(0)}\rangle$ and the final state $|f^{(0)}\rangle$ in Eq. (12.7) are eigenstates of the unperturbed Hamiltonian, i.e., of the Hamiltonian of the unperturbed electronic system *and* the electromagnetic field. The initial state $|i^{(0)}\rangle$ can then be represented as a product state of an electronic, $|a\rangle$, and a photon state, $|n_{\mathbf{k},p}\rangle$, as $|i^{(0)}\rangle = |a\rangle|n_{\mathbf{k},p}\rangle$ and similarly for the final state $|f^{(0)}\rangle = |b\rangle|n_{\mathbf{k},p} - 1\rangle$ where one photon has been absorbed. The energies of the initial and final states are then given, respectively, by $E_i^{(0)} = E_a + \hbar\omega_\mathbf{k} n_{\mathbf{k},p}$ and $E_f^{(0)} = E_b + \hbar\omega_\mathbf{k}(n_{\mathbf{k},p} - 1)$, where $E_a$ and $E_b$ are the energies of the *electronic* initial and final states, respectively. Note here that we have taken the energies of the zeroth order Hamiltonian $\hat{H}_0$ and that $\hat{H}_0$ is the *sum* of the Hamiltonian of the electromagnetic field *and* that of the unperturbed system. Equation (12.3) for the transition frequency now becomes

$$\omega_{\rm fi} = \frac{(E_{\rm f}^{(0)} - E_{\rm i}^{(0)})}{\hbar} = \frac{E_b - E_a - \hbar\omega_{\mathbf{k}}}{\hbar}, \tag{12.8}$$

which, together with $\hat{a}_{\mathbf{k},\rm p}|n_{\mathbf{k},\rm p}\rangle = \sqrt{n_{\mathbf{k},\rm p}}|n_{\mathbf{k},\rm p} - 1\rangle$ and $\delta(\alpha x) = 1/\alpha\delta(x)$, gives for the transition rate $w_{\rm fi}$ between the initial and final state

$$w_{\rm fi} = \frac{d|c_{\rm fi}(t)|^2}{dt} = 2\pi\frac{e^2}{m_{\rm e}^2}\frac{n_{\mathbf{k},\rm p}}{2\omega_{\mathbf{k}}V\varepsilon_0}\delta(E_b - E_a - \hbar\omega_{\mathbf{k}})|M_{ba}|^2, \tag{12.9}$$

where

$$M_{ba} = \langle b|\hat{\mathbf{p}}\cdot\mathbf{e}_{\rm p}e^{i(\mathbf{k}\cdot\mathbf{r})}|a\rangle. \tag{12.10}$$

Let us now turn our attention to the matrix element $M_{ba}$. The exponential in Eq. (12.10) can be expanded in a Taylor series as $e^{i(\mathbf{k}\cdot\mathbf{r})} \approx 1 + i\mathbf{k}\cdot\mathbf{r} + 1/2$ $(i\mathbf{k}\cdot\mathbf{r})^2 + \cdots$. If $\mathbf{k}\cdot\mathbf{r} \ll 1$ or equivalently, if $\mathbf{r} \ll \lambda/2\pi$ where $\lambda$ is the X-ray wavelength, we can to a good approximation retain only the first nonzero term in the expansion and obtain for the matrix element the much simplified expression $M_{ba} = \langle b|\hat{\mathbf{p}}\cdot\mathbf{e}_{\rm p}|a\rangle = \mathbf{e}_{\rm p}\cdot\langle b|\hat{\mathbf{p}}|a\rangle$ known as the dipole approximation [8].

To make an estimate of the validity of the dipole approximation when calculating XAS, e.g., at the oxygen $K$-edge we can approximate $|\mathbf{r}|$ with the $K$-shell diameter from the Bohr radius $a_0 = 0.53$ Å and the atomic number $Z$ as $|\mathbf{r}| \approx 2a_0/Z$. With a photon energy $\hbar\omega \approx 545$ eV, representative of the oxygen 1s energy, we have the wavelength $\lambda = 22.75$ Å and the condition $\mathbf{r} \ll \lambda/2\pi$ for the dipole approximation to be valid is well satisfied with $0.13 \ll 3.62$.

If the initial and final states $|a\rangle$ and $|b\rangle$ are exact eigenstates of the full Hamiltonian $\hat{H} = \hat{H}_0 + \hat{H}_1^{\rm eff}$, the operator equality of the (total linear) momentum operator

$$\hat{\mathbf{p}} = \frac{im_{\rm e}}{\hbar}[\hat{H}, \mathbf{r}] \tag{12.11}$$

can be used to rewrite $M_{ba}$ from the "velocity form" above, to the "length" form

$$\mathbf{e}_{\rm p}\cdot\langle b|\hat{\mathbf{p}}|a\rangle = \frac{\omega_{\mathbf{k}}\,m_{\rm e}}{e}\mathbf{e}_{\rm p}\cdot\langle b|e\hat{\mathbf{r}}|a\rangle. \tag{12.12}$$

The difference between using the "velocity" (Eq. (12.10)) or the "length" form of the matrix element may sometimes be large in HF. In DFT, however, typically only minor differences are observed.

In KS DFT, we map the real many-body wavefunction onto that of a noninteracting system and thus need to connect the resulting KS orbitals to the fully interacting system. In Chapter 3, this was done through the derivation of Eq. (12.12) for optical absorption in DFT based on response theory. Since experimental X-ray absorption spectra are typically not normalized on an absolute scale, the relevant part to compute is the transition moments in Eq. (12.12). We then have the Fermi's Golden Rule expression for the energy-dependent absorption cross section $\sigma(\omega)$ as

$$\sigma(\omega) \propto \omega\sum_j|\mathbf{e}_{\rm p}\cdot\langle\Psi_0|\hat{\mathbf{r}}|\Psi_j\rangle|^2\delta(\hbar\omega - (E_j - E_0)). \tag{12.13}$$

For ordered systems, such as the surface adsorbates of interest in this chapter, using linearly polarized synchrotron light allows obtaining specific projections by varying the polarization vector $\mathbf{e}_p$ relative to the oriented molecule. For randomly oriented molecules and unpolarized light, e.g., gas-phase species and disordered systems, one needs to average over all orientations to obtain (as also given in Chapter 3)

$$\sigma(\omega) \propto \frac{2}{3}\omega \sum_j |\langle \Psi_0 | \hat{\mathbf{r}} | \Psi_j \rangle|^2 \delta(\hbar\omega - (E_j - E_0)). \tag{12.14}$$

These results are valid for arbitrary wavefunctions. We now want to investigate how to evaluate this expression in the case of single-determinant wavefunctions to connect with KS states.

We write a Hartree product of molecular orbitals as $|\phi_1(\mathbf{r}_1)\phi_2(\mathbf{r}_2)\cdots\phi_n(\mathbf{r}_n)\rangle$. In terms of Hartree products, the wavefunction can be written as $\sum_P (-1)^p \hat{P}|\phi_1(\mathbf{r}_1)\phi_2(\mathbf{r}_2)\cdots\phi_n(\mathbf{r}_n)\rangle$, where $\hat{P}$ is a permutation operator that interchanges the coordinates of all electrons and $p$ is the parity of $\hat{P}$. There are $n!$ $\hat{P}$:s for $n$ particles. The alternating signs make the wavefunction antisymmetric with respect to interchange of particles as required by the Pauli principle. $\hat{P}$ has the property that when applied on the coordinates the effect is the same as $\hat{P}^{-1}$ applied on the orbitals, since both of these, put together, would bring back the starting Hartree product.

For a general one-electron operator $\sum_k^n \hat{h}_k$, with $n$ being the number of electrons, we have

$$\begin{aligned}
\langle \Phi | \sum_k^n \hat{h}_k | \Psi \rangle &= \frac{1}{n!} \sum_P^{n!} \sum_{P'}^{n!} (-1)^{P+P'} \langle \hat{P}\phi_1(\mathbf{r}_1)\phi_2(\mathbf{r}_2)\cdots\phi_n(\mathbf{r}_n)| \\
&\quad \times \sum_k^n \hat{h}_k \hat{P}' |\psi_1(\mathbf{r}_1)\psi_2(\mathbf{r}_2)\cdots\psi_n(\mathbf{r}_n)\rangle \\
&= \frac{1}{n!} \sum_P^{n!} \sum_{P'}^{n!} (-1)^{P+P'} \langle \phi_1(\mathbf{r}_1)\phi_2(\mathbf{r}_2)\cdots\phi_n(\mathbf{r}_n)| \\
&\quad \times \left[ \hat{P}^{-1} \sum_k^n \hat{h}_k \right] \hat{P}^{-1}\hat{P}' |\psi_1(\mathbf{r}_1)\psi_2(\mathbf{r}_2)\cdots\psi_n(\mathbf{r}_n)\rangle.
\end{aligned} \tag{12.15}$$

Products of permutations are also permutations and we call the product $\hat{P}^{-1}\hat{P}' = \hat{Q}$. Note that although $\hat{P}^{-1}$should also be applied on $\sum_k^n \hat{h}_k$, as indicated in (12.15), it will not have any effect on the total sum that can be moved outside the integrals

$$= \frac{1}{n!} \sum_k^n \sum_Q^{n!} (-1)^Q \langle \phi_1(\mathbf{r}_1)\phi_2(\mathbf{r}_2)\cdots\phi_n(\mathbf{r}_n)| \hat{h}_k \hat{Q} |\psi_1(\mathbf{r}_1)\psi_2(\mathbf{r}_2)\cdots\psi_n(\mathbf{r}_n)\rangle \tag{12.16}$$

and applying the permutation

$$= \sum_k^n \sum_Q^{n!} (-1)^Q \langle \phi_1 | \psi_{a_1} \rangle \langle \phi_2 | \psi_{a_2} \rangle \cdots \langle \phi_k | \hat{h}_k | \psi_{a_k} \rangle \cdots \langle \phi_n | \psi_{a_n} \rangle. \tag{12.17}$$

This is a sum over determinants containing orbital overlap integrals and one row of matrix elements over the one-electron operator such that (with compound indices simplified)

$$\langle \Phi | \sum_k^n \hat{h}_k | \Psi \rangle = \sum_k \begin{vmatrix} \langle \phi_1 | \psi_1 \rangle & \langle \phi_1 | \psi_2 \rangle & \cdots & \langle \phi_1 | \psi_n \rangle \\ \langle \phi_2 | \psi_1 \rangle & \langle \phi_2 | \psi_2 \rangle & \cdots & \langle \phi_2 | \psi_n \rangle \\ \vdots & \vdots & \vdots & \vdots \\ \langle \phi_k | \hat{h} | \psi_1 \rangle & \langle \phi_k | \hat{h} | \psi_2 \rangle & \cdots & \langle \phi_k | \hat{h} | \psi_n \rangle \\ \vdots & \vdots & \vdots & \vdots \\ \langle \phi_n | \psi_1 \rangle & \langle \phi_n | \psi_2 \rangle & \cdots & \langle \phi_n | \psi_n \rangle \end{vmatrix}. \tag{12.18}$$

We can expand these determinants along the row with the matrix element in it (row $k$) as

$$\langle \Phi | \sum_k^n \hat{h}_k | \Psi \rangle = \sum_k \sum_l \langle \phi_l | \hat{h} | \psi_k \rangle C_{kl}, \tag{12.19}$$

where $C_{kl}$ is the cofactor defined as $(-1)^{k+l} M_{kl}$, where $M_{kl}$ is the determinant with row $k$ and column $l$ deleted. This is the general case when each state is optimized separately resulting in nonorthogonal single-particle orbitals, but with orthogonality maintained for the resulting wavefunctions, i.e., $\langle \Psi_i | \Psi_j \rangle = \delta_{ij}$. In terms of calculating inner-shell spectra within the transition-potential formalism where the same orbitals are used to build both states, the expression simplifies considerably and becomes simply $\langle \Psi_i | \sum \hat{h}_k | \Psi_j \rangle = \langle \phi_i | \hat{h} | \phi_j \rangle$, where the excitation is from original orbital $\phi_i$ to excited orbital $\phi_j$; all cofactors $C_{kl}$ are zero except for $C_{ij}$ which is one.

In DFT, we map the real many-body wavefunction to the wavefunction of a noninteracting system. We shall see that we can use the KS wavefunctions in the framework of the TDDFT response theory (and approximations to it) to calculate absorption cross sections.

Following Chapter 3, we write for the KS system as the response to the external perturbation

$$\delta\rho(\mathbf{r}, \omega) = \int \chi_{KS}(\mathbf{r}, \mathbf{r}', \omega) v_1^{\text{eff}}(\mathbf{r}', \omega) d^3\mathbf{r}' \tag{12.20}$$

in terms of the density–density response function $\chi_{KS}(\mathbf{r}, \mathbf{r}', \omega)$ and an effective perturbation, $v_1^{\text{eff}}(\mathbf{r}, \omega)$, giving the same change in the density in the noninteracting KS system as in the interacting many-body case. If the same orbitals are used for all states, but with differing occupation numbers $f_k$, the density–density response function can be written in terms of these orbitals and associated orbital energies $\varepsilon_k$ as (see above for the case of different orbitals where appropriate cofactors must be included)

$$\chi_{\mathrm{KS}}(\mathbf{r},\mathbf{r}',\omega)=\sum_{i,j}(f_i-f_j)\frac{\phi_i^*(\mathbf{r})\phi_j(\mathbf{r})\phi_j^*(\mathbf{r}')\phi_i(\mathbf{r}')}{\omega-(\varepsilon_j-\varepsilon_i)+i\gamma}. \tag{12.21}$$

The difficulty of obtaining the response in the density is now moved to the effective perturbation $v_1^{\mathrm{eff}}(\mathbf{r},\omega)$, which is defined as

$$v_1^{\mathrm{eff}}(\mathbf{r},\omega)=v_1(\mathbf{r},\omega)+\int\delta\rho(\mathbf{r}',\omega)\left[\frac{1}{|\mathbf{r}-\mathbf{r}'|}+f_{xc}(\mathbf{r},\mathbf{r}',\omega)\right]d^3\mathbf{r}', \tag{12.22}$$

where $f_{xc}(\mathbf{r},\mathbf{r}',\omega)$ is the exchange-correlation kernel that depends only on the unperturbed density $\rho_0$. If $f_{xc}(\mathbf{r},\mathbf{r}',\omega)$ is set to zero, this is called the time-dependent Hartree approximation or the random phase approximation (to confuse matters, the latter is also used as a synonym for the time-dependent HF approximation). Neglecting the Coulomb term as well in Eq. (12.22) returns the KS orbitals in which case the Fermi's Golden Rule becomes (same orbitals for both states)

$$\sigma\propto\hbar\omega\sum_i^{\mathrm{occ}}\sum_j^{\mathrm{unocc}}\left|\langle\phi_i|\mathbf{r}|\phi_j\rangle\right|^2\delta(\hbar\omega-E_i+E_j), \tag{12.23}$$

where $|\phi_i\rangle$ and $|\phi_j\rangle$ are KS orbitals thus providing the desired connection between KS orbitals and cross sections, but so far with unknown quality of the calculated cross sections. We can, however, estimate the importance of the neglected integral in Eq. (12.22) in the case of a core-excitation where we assume a closed-shell initial state. For the change in density, we find (suppressing arguments)

$$\begin{aligned}\delta\rho_{i0}=\rho_i-\rho_0&=\left(2\sum_{k=1}^{n-1}|\phi_k'|^2+|\phi_0'|^2+|\phi_n'|^2\right)-\left(2\sum_{k=0}^{n-1}|\phi_k|^2\right)\\&=|\phi_0'|^2+|\phi_n'|^2-2|\phi_0|^2+2\sum_{k=1}^{n-1}(|\phi_k'|^2-|\phi_k|^2).\end{aligned} \tag{12.24}$$

The same occupation is retained for all other orbitals, but the excited-state orbitals $\{\phi_k'\}$ will be different from those of the ground-state $\{\phi_k\}$ due to relaxation effects. We can write the relaxed final-state orbitals as $\phi_k'=\alpha_k\phi_k+\beta_k\phi_k^u$, where $\phi_k^u$ is a linear combination of virtual orbitals from the ground-state KS solution and we assume that the pairwise overlaps between the occupied orbitals in the two sets have been maximized. This gives for the density difference between corresponding pairs

$$|\phi_k'|^2-|\phi_k|^2=(\alpha_k^2-1)|\phi_k|^2+2\alpha_k\beta_k\phi_k\phi_k^u+\beta_k^2|\phi_k^u|^2, \tag{12.25}$$

where the admixture given by $\beta_k$ may be significant. Particularly considering the full occupation change in Eq. (12.24), i.e., $|\phi_n'|^2-|\phi_0|^2$, the change in density relative to the ground state can be substantial making procedures such as TDDFT necessary to describe the excited state with the ground state as reference. However, if instead of the ground state one takes as reference orbitals the self-consistent solution for a half-occupied core–hole state, the main effect of the core-occupation change is already included in the orbitals giving the response in Eq. (12.21). The excited orbital still contributes to the integral in Eq. (12.22), but for highly excited and diffuse states, as in XAS, particularly the exchange-correlation part can typically be neglected.

In this case, we can express the ground- and excited-state orbitals in terms of orbitals $\{\chi_k\}$ determined in the presence of the half-occupied core–hole as $\phi_k = \alpha_k \chi_k + \beta_k \chi_k^u$ with $\alpha_k \approx 1$ giving contribution to the density difference

$$|\phi_k'|^2 - |\phi_k|^2 = (\alpha_k'^2 - \alpha_k^2)|\chi_k|^2 + 2\chi_k(\alpha_k'\beta_k'\chi_k'^u - \alpha_k\beta_k\chi_k^u) + (\beta_k'^2|\chi_k'^u|^2 - \beta_k^2|\chi_k^u|^2), \quad (12.26)$$

which typically will be rather small, and it will be a good approximation to neglect the integral in Eq. (12.22) noting, however, that the occupation change is still effectively included through the orbital relaxation in Eq. (12.21).

## 12.4
## Practical Excited State Calculations

For a nondegenerate 1s level, we may in practice construct the first KS core-excited state as a single-determinant wavefunction built from the KS orbitals by simply requiring that the 1s level be unoccupied and occupying instead the lowest unoccupied molecular orbital (LUMO) [56]. The resulting state differs in two orbitals from the ground-state determinant, but the variational relaxation of the orbitals can still introduce a small overlap with the ground state; the orthogonality to the ground state is thus not strict but typically the overlap is negligible. Removing the variationally determined LUMO from the variational space and instead occupying LUMO+1 then gives the second core-excited state as the lowest neutral state obeying the constraint that the 1s be unoccupied; the state is strictly orthogonal to the first since they now differ in two orthogonal orbitals. The LUMO is removed by simply transforming the KS matrix to MO basis and eliminating all off-diagonal matrix elements involving the LUMO before diagonalizing. Repeating this gives a sequence of variationally determined strictly orthogonal core-excited states. As an example of the accuracy that can be achieved, we show in Table 12.1 a comparison between computed and measured core ionization potential (IP) and excitation energies at the nitrogen K-edge in gas-phase pyridine [56].

The entry TP in the table gives the values obtained from the difference in orbital energies when using the TP or HCH method of Triguero *et al.* [36], where the unoccupied states are determined in the field of a half-occupied core–hole; this will be discussed in connection with calculations of XAS spectra in the next section. Here we simply note that the $\Delta$KS correction to the TP excitation energy ranges between 1.5 and 2.4 eV and results in final excitation energies that differ from experiment by 0.3–0.5 eV. An overall offset of 0.3 eV from experiment can be

**Table 12.1** Theoretical (transition potential (TP) and Delta Kohn–Sham ($\Delta$KS)) and experimental N 1s ionization potential (IP) and excitation energies for pyridine (all values in eV) [56].[a]

| Peak | Resonance | DFT | | | Experiment |
|---|---|---|---|---|---|
| | | TP | $\Delta$KS | Diff $\Delta$KS-TP/ Experiment | Gas phase |
| | IP | 406.1 | 404.5 | −1.6/−0.3 | 404.8[a] |
| A | N $1s^{-1}$ | 400.5 | 398.4 | −2.1/−0.4 | 398.8 |
| | $\pi^*(b_1)$ | 5.6 | 6.1 | | 6.0 |
| B | N $1s^{-1}$ | 401.2 | 399.7 | −1.5/−0.5 | 400.2 |
| | $\pi^*(a_2)$ | 4.9 | 4.8 | | 4.6 |
| C | N $1s^{-1}$ | 403.7 | 401.5 | −2.2/−0.5 | 402.0 |
| | $\sigma^*(a_1)$ | 2.4 | 3.0 | | 2.8 |
| D | N $1s^{-1}$ | 404.7 | 402.3 | −2.4/−0.3 | 402.6 |
| | $3\pi^*(b_1)$ | 1.4 | 2.2 | | 2.2 |

[a]The first line of the excitation energies shows the absolute position, and the second the energy relative to the corresponding ionization potential (IP). Computed energies include a relativistic correction (+0.3 eV).

expected from the difference between computed and experimental core IP (XPS energy) since the core–hole is included in all core-excited states; when this constant offset is accounted for, we find computed energies in very close agreement with experiment [57].

Calculating core IP and first few low-lying core-excited states for a set of 18 molecules and comparing the results for all 27 combinations of 9 exchange functionals with 3 correlation functionals, we find significant dependence of the calculated values on which functional is used [57]. In Figure 12.3, we show the results for the average absolute deviation between experimental and computed XPS energies demonstrating a strong dependence on which functional is used. However, as we see from the corresponding results for the computed term values, i.e. energy differences between the core-excited states, the problems with the functionals are mainly associated with the electron-dense inner core level; for the relative energies, the 1s contribution cancels out and the results become very stable and rather insensitive to which functional is selected [57].

In terms of XPS, which we calculate as the energy difference between ground and variationally relaxed core–hole state, this indicates that *relative shifts* in XPS energy can be expected to be very reliably computed, while a single value may have a significant error dependent on functional and molecule. This realization opens the possibility, however, to calibrate the calculations if a suitable comparison can be found where the error in the functional can be evaluated against some other calculations or experiments. In the case of, e.g., water we can calculate the gas-phase XPS value and compare with experiment to determine an empirical computational correction for functional deficiency and relativistic effects (as well

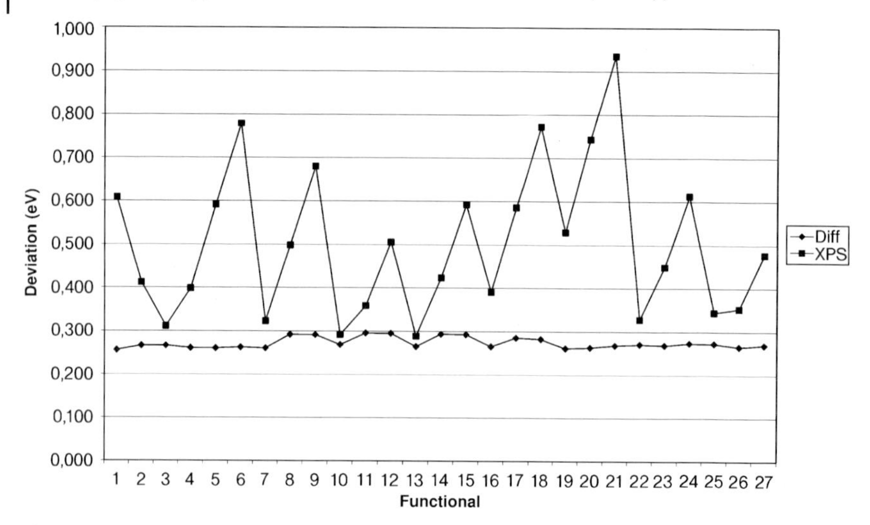


**Figure 12.3** Average absolute deviation of the energy difference (in eV) between experimental and theoretical XPS energies (squares) and term-values (Diff, diamonds) for the target molecules using 27 different functional combinations ($x$-axis).

as basis set deficiencies); this will be applicable independent of hydrogen-bonding situation [58]. The correction depends on the chemical environment in a molecule such that it is not possible to define a unique and generally applicable correction for each atom and functional [57]; a sufficiently similar system must be found. Still, chemical shifts between, e.g., different oxygens at an oxide surface [59], water and OH on a metal, or inequivalent atoms in a molecule can be very reliably obtained since the functional dependency largely cancels out when taking differences.

As discussed above, the localized character of the core orbitals makes it possible to define a variationally lowest first core-excited KS state by simply constraining the occupation number of the 1s KS orbital. Higher excited states are possible to get by starting from the lowest excited state, removing the orbital containing the excited electron from the variational space and adding an electron in the next lowest unoccupied orbital. This state-by-state excitation procedure yields accurate relative energies between the excited states [56, 57]. Oscillator strengths can be obtained by taking the (somewhat involved) nonorthogonal matrix elements between the KS ground-state determinant and each excited-state determinant. This procedure gives an accurate description of the XAS spectrum, but only a few lowest states are accessible due to convergence and basis set limitations.

Note also that due to the single-determinant form of the KS state, the resulting states are not pure spin states; to describe a singlet-coupled excitation from a closed-shell ground state two determinants are needed. A single determinant with two open shells with total spin projection $M_S = 0$ corresponds to an equal mixture of the triplet and singlet coupling of the excited orbital with the remaining

1s electron. The triplet and singlet energies differ by an exchange integral with the core level, which is normally negligible. In some cases, e.g., C1s $\rightarrow 2\pi^*$ in CO where the $2\pi^*$ is a valence level strongly polarized toward the carbon, the singlet–triplet splitting induced by the interaction with the 1s can, however, be significant. The correct singlet energy must then be determined by computing both the high-spin pure triplet state (which is describable as a single determinant) and the mixed state for which the obtained energy can be written as $E_{\mathrm{Mix}} = (E_{\mathrm{T}} + E_{\mathrm{S}})/2$. From these two calculations, the correct singlet energy $E_{\mathrm{S}}$ can be obtained as $E_{\mathrm{S}} = 2E_{\mathrm{Mix}} - E_{\mathrm{T}}$. This is reminiscent of the approach of Theophilou to obtain excited states as specific mixtures discussed in the preceding section [38].

## 12.5 Slater Transition-State Method

Slater devised a method for calculating excitation energies, the transition-state method [60, 61], where the excitation energy is estimated as the orbital energy difference between two levels of a variationally determined transition state, with one-half electron excited. To see the importance and limitations of this, we first note (Janak's theorem [62]) that the orbital energy, $\varepsilon_i$, is given by the derivative of the KS energy with respect to the occupation number $n_i$ in the expression for the density, $\rho(\mathbf{r}) = \sum_i n_i \phi_i^2(\mathbf{r})$, i.e.,

$$\frac{\partial E}{\partial n_{\mathrm{i}}} = \varepsilon_i^{\Delta\vec{n}}, \tag{12.27}$$

where a self-consistent solution of the KS equation is assumed for the variation of the occupation numbers so that the effect on the KS orbitals is directly included.

Following Slater [60], we can now show that the orbital energy difference $\varepsilon_{\mathrm{f}}^{(1/2,-1/2)} - \varepsilon_{\mathrm{i}}^{(1/2,-1/2)}$ gives an estimate of the excitation energy correct to second order in the change in occupation. The notation stands for a state where half an electron has been removed from the orbital i and placed in the orbital f; the number of electrons is thus conserved, i.e. $\sum_i n_i = N$. Considering only the orbitals that change occupation, we write the energy as $E(n_{\mathrm{i}}, n_{\mathrm{f}})$, where we take for the reference state $n_{\mathrm{i}} = n_{\mathrm{f}} = 1/2$. With respect to this reference, our desired ground state then has energy $E(n_{\mathrm{i}} + 1/2, n_{\mathrm{f}} - 1/2)$ and the excited state $E(n_{\mathrm{i}} - 1/2, n_{\mathrm{f}} + 1/2)$. Taylor expanding to second order in the occupation numbers and using Eq. (12.27), we obtain for the ground state (all derivatives taken at the variationally determined reference state with the two orbitals half-occupied)

$$E\left(n_{\mathrm{i}} + \frac{1}{2}, n_{\mathrm{f}} - \frac{1}{2}\right) = E\left(n_{\mathrm{i}}, n_{\mathrm{f}}\right) + \frac{1}{2}\left(\varepsilon_{\mathrm{f}}^{(1/2,-1/2)} - \varepsilon_{\mathrm{i}}^{(1/2,-1/2)}\right) + \frac{1}{2!}\left(\frac{1}{2}\right)^2 \times \left(\frac{\partial^2 E}{\partial^2 n_{\mathrm{i}}} - 2\frac{\partial^2 E}{\partial n_{\mathrm{i}} \partial n_{\mathrm{f}}} + \frac{\partial^2 E}{\partial^2 n_{\mathrm{f}}}\right) + \cdots \tag{12.28}$$

For the desired excited state, we instead obtain

$$E\left(n_{\rm i}-\frac{1}{2},n_{\rm f}+\frac{1}{2}\right)=E\left(n_{\rm i},n_{\rm f}\right)-\frac{1}{2}\left(\varepsilon_{\rm f}^{(1/2,-1/2)}-\varepsilon_{\rm i}^{(1/2,-1/2)}\right)$$
$$+\frac{1}{2!}\left(\frac{1}{2}\right)^2\left(\frac{\partial^2 E}{\partial^2 n_{\rm i}}-2\frac{\partial^2 E}{\partial n_{\rm i}\partial n_{\rm f}}+\frac{\partial^2 E}{\partial^2 n_{\rm f}}\right)+\cdots. \qquad (12.29)$$

For the transition energy $E_{\rm fi}=E_{\rm f}-E_{\rm i}=E(n_{\rm i}-1/2,n_{\rm f}+1/2)-E(n_{\rm i}+1/2,n_{\rm f}-1/2)$, we then find (to second order)

$$E_{\rm fi}=\varepsilon_{\rm f}^{(1/2,-1/2)}-\varepsilon_{\rm i}^{(1/2,-1/2)}+O(E_{\bf n}^{(3)}). \qquad (12.30)$$

The derivation assumes that the second-order response of the orbitals to the change in the occupation can be neglected, i.e., the half-occupied orbitals represent respectively the initial and final state characters; the first-order contribution is eliminated through self-consistency.

## 12.6
## Transition Potential Approach

The transition-state approach of Slater [60,61] still requires a state-by-state calculation of the spectrum but leads to a much simplified evaluation of the oscillator strengths since the final and initial state use the same orbitals but differ in one orbital occupation. The dipole transition moment reduces to involve only the initial and the excited KS orbitals, but the specific involvement of the half-occupied excited level still makes the approach impractical for generating a complete X-ray absorption spectrum including valence, Rydberg, and continuum states.

The TP approach introduced by Triguero *et al.* [36] avoids the state-by-state calculations by simply removing half an electron from the core level of interest. The variationally relaxed density of the resulting molecular ion core is then considered to provide a potential from which all excited states are obtained in *one* global diagonalization. Since the initial and final states use the same orthogonal orbital set, it is necessary only to take the matrix elements between the KS orbitals corresponding to the 1s level and the unoccupied states to obtain the whole spectrum. By furthermore using a double basis set technique [36,55,63], where the occupied density of the HCH state is determined with a normal molecular basis set and used to build the KS matrix in a much larger, very extended and diffuse basis set, the description of Rydberg and continuum states can be much improved.

Although the choice of 0.5 electrons as fractional occupation number of the 1s level minimizes the error made in terms of the change in occupation, this is only strictly valid when the upper level is half-occupied. The TP approach neglects the excited half electron in the final level. This is a good approximation when the orbital character of the level as empty and half-occupied relaxed are similar and when the interaction with the remaining orbitals is weak, i.e., when $\partial\phi_k/\partial n_j\approx 0$

and $\phi_k$ is an orbital in the TP solution and $n_j$ refers to the excited level. The half core–hole approximation is thus not strict [64, 65], but has been shown to give overall reliable and accurate results and a fair balance of spectral intensity over the whole energy range of interest for a large range of systems. Next we shall consider some examples and begin with gas-phase water molecule.

## 12.7
## Applications of XAS Calculations

### 12.7.1
### Water in Gas Phase and as a Liquid

For well-ordered systems such as surface adsorbates and smaller gas-phase molecules [56], it is often sufficient to calculate XAS for one single geometry and shift the spectrum to match experiment as an alternative to the $\Delta$KS energy calibration. When there are several inequivalent sites, such as is the case for oxygen species at various metal oxide surfaces [59], the relative energy scale needs to be improved through the full $\Delta$KS calculation at least for the lowest core-excited state; this determines the onset of the spectrum (after correcting also for relativity and optionally for the functional) and accounts for differential relaxation effects that may occur between species in different chemical environments. This is particularly critical when studying, e.g., water where the large variation in local structures requires calculating and summing together several hundred individual spectra to yield a total XAS spectrum.

Figure 12.4 illustrates the accuracy in energy position and spectral shape obtained when the full energy calibration procedure is applied. One thousand water gas-phase monomer spectra were calculated and summed (middle), with geometries sampled from the zero-point vibrational distribution obtained from a path-integral

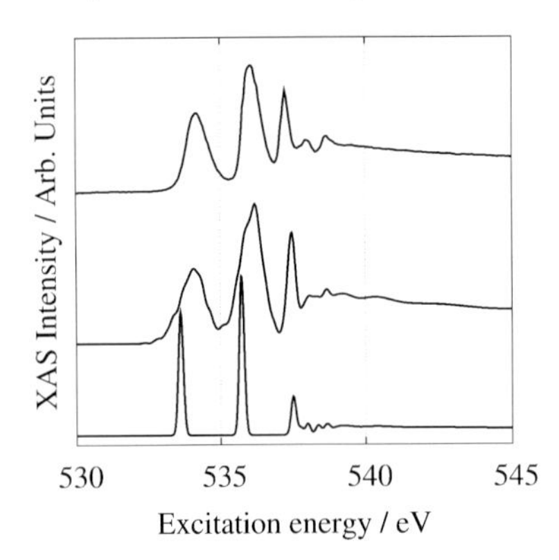


**Figure 12.4** Comparison of computed and experimental XAS spectra of gas phase water. (Top) Experiment, (middle) sum of 1000 computed individual spectra where a typical spectrum is shown in the bottom graph.

MD simulation [66]. All spectra were individually ΔKS shifted, and the O 1s XPS for the equilibrium geometry was compared to the experimental value [67] to obtain the empirical energy shift used for all 1000 spectra. Figure 12.4 (bottom) shows XAS for one single geometrical structure at a snapshot geometry from the full vibrational geometry distribution; for all spectra in the figure, a Gaussian broadening with a full width at half maximum (FWHM) of 0.2 eV was used up to 538.7 eV, after which the broadening was linearly increased to 4.25 at 544.7 eV. It is interesting to note the great sensitivity of the computed XAS spectrum to changes in the geometrical structure. Note especially the variations in the antibonding $4a_1$ and $2b_2$ states giving the observed broadening, while the first Rydberg state (the third peak in energy) in the summed spectrum (middle) is seemingly unaffected by changes in geometry, in accordance with the experiment (top).

Experimentally the XAS spectrum is broadened due to instrumental, core–hole life-time and vibrational Franck–Condon effects [8]. For gas-phase molecules, resolved vibrational fine structure can be seen for bound states such as the $\pi^*$ resonance of $N_2$ [68]. Dissociative or strongly antibonding final states, such as the $4a_1$ and $2b_2$ states of gas-phase water (the first two peaks in Figure 12.4), show instead an inhomogeneous vibrational broadening.

A theoretical TP XAS spectrum consists in reality of a collection of energy positions with associated oscillator strengths; these are the computed "hard data." To be able to communicate with experimentalists and analyze not only a few low-lying excitations but also the important overall spectral shape necessitates broadening of the theoretical spectra. Although it is possible to account for most experimental broadening effects by simply applying a broadening scheme to the discrete theoretical spectrum, this has to be done with great care. Broadening should be designed in such a way as to mimic known experimental broadening effects, as well as compensating for the discrete sampling of the continuum, without affecting the overall shape of the spectrum. Using a finite Gaussian basis set, as in TP DFT calculations with the StoBe-deMon code [37], spurious basis set dependent states may appear at higher energies due to the limited size of the basis set. They can be eliminated either by extending the basis set or by applying a larger broadening.

Figure 12.5 compares (a) experimental [69] and (b)–(c) computed XAS spectra summing computed spectrum contributions from all 128 molecules in a snapshot from a periodic Car–Parrinello molecular dynamics (CPMD) simulation [70] of room-temperature liquid water using different broadening schemes combined with the ΔKS plus core–electron binding energy calibration scheme described above; the mean ΔKS energy calibration shift for these 128 structures is −1.72 eV with a standard deviation of 0.33 eV. The smallest and largest shifts are −0.93 and −2.51 eV, respectively. Since the statistics is rather good with 128 summed structures, and the large 46 molecule clusters used in the calculations provide a good density of states, a constant Gaussian broadening of 0.5 eV FWHM is enough to give a reliable theoretical spectrum (b). This can be regarded as a computational reference point when comparing broadening and energy calibration schemes. (c) Uses a Gaussian broadening of 0.5 eV up to 536 eV, and then linearly increased

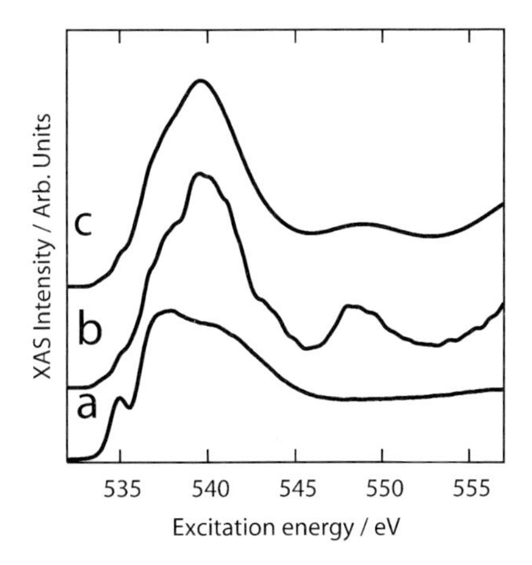


**Figure 12.5** Comparison of (a) experimental room temperature liquid water XAS spectrum from Ref. [69] with computed spectra for all 128 molecules in a CPMD snapshot from Ref. [70]. (b) Constant broadening of 0.5 eV applied for all transitions and (c) linearly increased broadening as described in the text.

to 4.25 at 548 eV, i.e., over an interval of 12 eV. This broadening scheme smoothes out some of the bumpiness in the spectrum, while leaving the overall shape of the spectrum intact compared with the fixed 0.5 eV broadening.

We note in Figure 12.5 a rather poor comparison between experiment and the computed spectrum based on the CPMD dump. Neither the pre-edge peak at 535 eV nor the main edge at 537–538 eV are reproduced and instead the spectrum is dominated by intensity around 539–540 eV (postedge), similar to ice. This relates to the application of XAS to liquid water by Wernet *et al.* [69] and the analysis of the resulting spectrum as due to a predominant population of molecules in very asymmetric H-bonding situations with, on average, only one strong donating H-bond. In contrast, all present molecular dynamics simulations result in very highly H-bonded structure models with predominantly fully H-bonded molecules. This discrepancy is attracting strong attention [58, 64, 69, 71–88] and further underlines the need to be able to reliably predict a theoretical XAS spectrum given the structure of the system.

Since it is precisely the structure of liquid water that has been questioned, it does not provide a good calibration of the calculations. Even for normal hexagonal ice, Ih there are, however, difficulties in preparing a good experimental sample, which causes uncertainties in terms of reliable reference spectra. The difficulty here arises from the fact that layer-by-layer growth of ice on, e.g., Pt(111) [89, 90], Ru(0001) [91, 92], and Pd(111) [90] does not occur and instead three-dimensional ice crystallites are formed on top of the first monolayer as one anneals the deposited initially amorphous multilayers. This is due to the fact that the first monolayer has an unexpected structure, where molecules bonding through the oxygen lone-pair are alternated with waters with one donating and two accepting H-bonds and with the non-H-bonded O–H group instead bonding to the metal [93]; this results in a rather flat overlayer, which turns out to be hydrophobic to additional layers [90].

#### 12.7.1.1 Glycine on Cu(110)

As an alternative, we turn to the ordered (3 × 2) overlayer of glycine on Cu(110), which forms as a result of H-bonding between the amino and deprotonated carboxylic groups on neighboring molecules [94]. In particular, the N 1s XAS is of interest since the amino group is the H-bond donor and as such holds empty states of 2p character; the electron-rich accepting oxygens of the carboxylic group show much less sensitivity due to the nearly completely filled H-accepting lone-pairs. Cu(110) is a corrugated surface with atomic rows along the [110] direction. By using linearly polarized X-rays with E-vector parallel to the surface along or orthogonal to the rows, respectively with the E-vector in the vertical direction, the local, atom-projected electronic structure can be completely decomposed in the three directions. In Figure 12.6, we show the structure and resulting computed N 1s XAS spectra for models with either a single glycine adsorbed or with two H-bonded glycine molecules in comparison with experiment [94].

Glycine adsorbs in a lying-down geometry with the carboxylic oxygens binding on-top to two coppers in one atomic row and the amino end near-on-top on the next row [94–97]. We take a coordinate system with the $x$-axis along the row and $y$-axis orthogonal to the rows in the surface plane. We have the amino-carboxyl H-bond mainly along the $x$-direction and the free N–H pointing mainly in the $y$-direction. With the single adsorbate, we find that the computed $p_x$ and $p_y$ spectra bear little resemblance to the corresponding experimental spectrum; in the $x$-direction, we see a split in the main peak and a too low energy position of the broad maximum, while in the $y$-direction we have too much intensity accumulated toward the onset

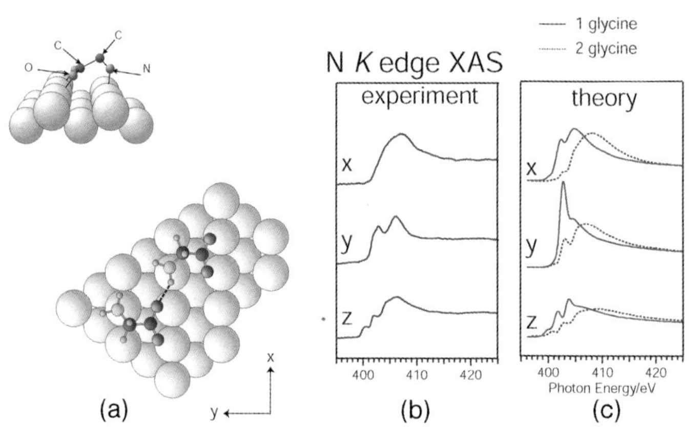


**Figure 12.6** Adsorption structure (a) and experimental (b) and computed (c) symmetry decomposed N 1s XAS spectra for glycine chemisorbed on Cu(110). Two computational models were used for the theoretical spectra with either one (full line) or two H-bonded adsorbates (dotted line).

of the spectrum. Introducing the H-bond by adding the second glycine has a very large effect on the spectral features in spite of the rather low H-bond strength. In the H-bond direction, along the row, the intensity now shifts up in energy and we obtain a single broad peak, which is characteristic of H-bonding [69, 98]; a very similar upward shift of intensity is also seen in XAS spectra of the surface of ice where the prominent pre-edge peak associated with non-H-bonded OH groups at the surface is quenched and intensity shifted up by 5–6 eV when $NH_3$ is used to complete the coordination at the ice surface [69].

In the $y$-direction, across the rows, we now get a clear split in the N 1s intensity where the low-energy pre-edge feature is associated with the non-H-bonded N–H group on which the excitation localizes due to the asymmetric H-bonding; this is again similar to what is found at the ice surface and suggested to be the origin of the pre-edge peak also in liquid water [69, 98, 99]. However, for the purpose of the present chapter, we conclude that H-bonding, in spite of a low resulting binding energy, produces a strong and measurable effect on the spectra. When the appropriate H-bonds are introduced in the computational model, we can furthermore, within the TP KS formalism, generate reliable representations of the corresponding spectra.

#### 12.7.1.2 Contact Layer of Ice on Pt(111)

We will now illustrate one use of spectrum calculations in structure determination of the wetting layer of water on Pt(111) using a small single-layer $Pt_7$ cluster model to describe the surface [93]; this model is actually too small to give a reliable geometry in an energy optimization, but the spectrum is generated through orbital overlaps, which are very sensitive to the distance and less dependent on the cluster size. Through the essentially exponential dependence on distance for the orbital overlaps, computed spectra provide a very sensitive measure of the structure, under the assumption that they can be reliably computed.

From the lack of a pre-edge feature at 535 eV in the measured out-of-plane XAS spectrum (Figure 12.7(b)), it could immediately be concluded that the structure of the water overlayer was very different from the traditional picture assumed to correspond to the termination of an ice Ih layer [93]; as mentioned above, a non-H-bonded OH pointing toward vacuum, as in the traditional model of the contact layer, would give rise to this characteristic pre-edge peak. In the previously accepted picture, the structure was alternating between molecules binding through the oxygen lone-pair and molecules only H-bonding to other waters through two accepting and one donating bond with the remaining non-H-bonded O–H pointing toward vacuum. The absence of the pre-edge signal directly indicates that the free O–H actually is not free but points down toward the surface and contacts the metal. The strong peak at 533 eV on the other hand is due to charge depletion in the oxygen lone-pair through interaction with the Pt 5d states with the formation of bonding and antibonding states; since the antibonding state is seen in absorption, we can conclude that it is only partially occupied and thus expect a net bonding contribution from the lone-pair interaction with the 5d states.

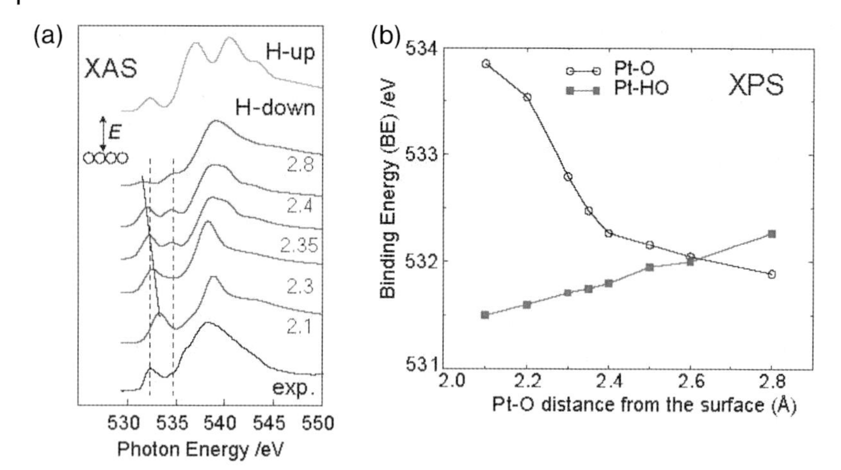


**Figure 12.7** (a) X-ray absorption spectra of water on Pt(111) corresponding to the $p_z$ component. Computed X-ray absorption spectra of water on Pt(111) with free OH toward the vacuum (H-up) and toward the substrate (H-down). The H-down spectrum is shown for different Pt–O distances (bottom). (b) O1s binding energies (relative to the Fermi level) for Pt–O and Pt–OH species for the H-down model as a function of Pt–O distance. The binding energies were obtained by subtracting the work function of Pt metal (6.4 eV) from the computed O1s IPs.

A straightforward geometry optimization of the water overlayer on the small $Pt_7$ cluster with four waters leads to a rather long O–Pt distance for the central O-bonded water, 2.83 Å [93]. Computing the XAS for this structure leads to a much too low intensity in the important signature peak at 533 eV. Since the appearance of this peak is due to molecular orbital interaction, it can be tuned by varying the distance to the substrate; this is allowed since the small cluster is a rather limited-size model of the extended surface, since van der Waals interaction is not included and the energy surface is very shallow leading to uncertainties in the geometry determination based on energy gradients even if the computational model was significantly more extended. We find indeed that the overlayer needs to be brought closer to the surface for the specific feature at 533 eV to obtain significant intensity and at the correct energy position (Figure 12.7(a)). Simultaneously monitoring also the XPS shift (Figure 12.7(b)) between O-bonded and H-down molecules, which, based on the FWHM of the measured peak could not exceed 0.8 eV, we could conclude an overlayer to surface distance of ~2.35 Å from the spectroscopic calculations and furthermore provide a decomposition of the spectrum into contributions from the two bonding situations [93].

## 12.8 X-Ray Emission Spectroscopy

As additional examples, we would like to return to Figure 12.2 and discuss the bond formation in the important textbook examples of chemisorption of $N_2$ and

CO on metal surfaces and also consider the adsorption of saturated hydrocarbons, exemplified by methane on Pt(111) [100,101] and *n*-octane on Cu(110) [102–104], where we will demonstrate how the electronic structure becomes strongly perturbed in spite of these molecules, in terms of resulting bond strength, being only physisorbed on the respective surfaces. This will show that a weak bond to the surface does not necessarily imply a weak interaction ($CH_4$, *n*-octane) and the same resulting bond strength does not at all imply similar interaction strengths (CO in different sites). Chemistry can be viewed as a "capitalistic venture": there is always a rehybridization cost (*investment*) which is overcome by the bond formation (*return*), but it is only the difference between the two that shows up as the resulting binding energy (*net profit*). However, since we here would like to use X-ray emission spectra in the characterization of the bonding and electronic structure changes, we will need to introduce also this spectroscopy before continuing.

We will use XES mainly to decompose experimentally and theoretically the occupied molecular orbital structure of surface adsorbates, as in the case of $N_2$/Ni(100) in Figure 12.2, which formed the starting point for the present chapter. The aim is to derive a simple one-electron picture of the ground-state molecular orbitals and to this end we need to distinguish between nonresonant and resonant XES, as well as the more general RIXS process as illustrated in Figure 12.8. XES is a photon-in/photon-out spectroscopy corresponding to inelastic scattering of photons by the molecule of interest. From a perturbation treatment of RIXS, we find the modified Kramers–Heisenberg scattering formula for the spectral distribution [13,105]:

$$I_{\mathrm{RIXS}}(\omega', \omega) \propto \sum_{F} \left| \sum_{M} \frac{\langle F|\mathbf{D}\cdot\mathbf{E}'|M\rangle\langle M|\mathbf{D}\cdot\mathbf{E}|G\rangle}{\hbar\omega_{-}(E_{\mathrm{M}} - E_{\mathrm{F}}) + i\Gamma_{M/2}} \right|^{2} \delta(\hbar\omega - \hbar\omega' + E_{\mathrm{G}} - E_{\mathrm{F}}). \qquad (12.31)$$

Here $|F\rangle$, $|M\rangle$, and $|G\rangle$ are the total wavefunctions with energies $E_{\mathrm{F}}$, $E_{\mathrm{M}}$, and $E_{\mathrm{G}}$ of the final, intermediate, and ground states, respectively. In the resonant case, the intermediate states are core excitations with lifetime broadening $\Gamma_M$. The energies of the incoming and outgoing photons are given by $h\omega$ and $h\omega'$ with the electric field vectors $\boldsymbol{E}$ and $\boldsymbol{E}'$, respectively, and $\mathbf{D}$ represents the dipole operator. The summation over different possible intermediate states can lead to channel interference effects if more than one intermediate state can be reached, which in such cases makes the cross section strongly energy dependent. This may occur due to closely spaced virtual orbitals, closely spaced core-orbitals, i.e., symmetry-adapted core–hole wavefunctions [106] overlapping due to lifetime broadening, or to vibrationally excited states. RIXS thus provides a wealth of information not only on the occupied valence levels but also on the symmetry and character of the unoccupied states [13].

For the purpose of connecting the observed spectral features back to the ground-state electronic structure, however, the key points are whether the excited electron is still present and affects the decay process or whether it can be disregarded in the analysis of the spectrum and furthermore if the intermediate core–hole affects the

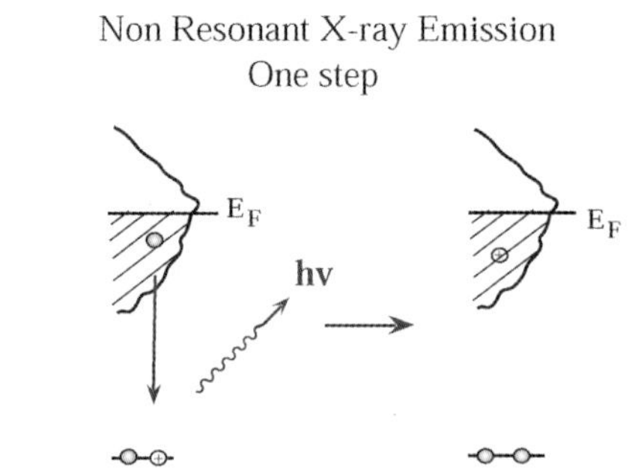


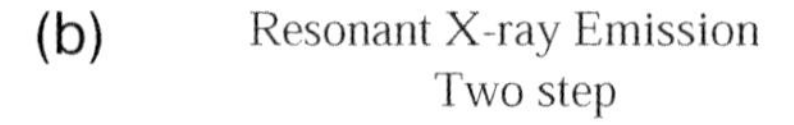


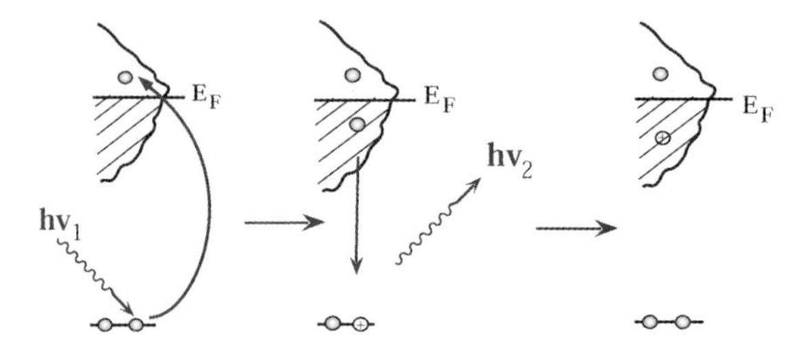


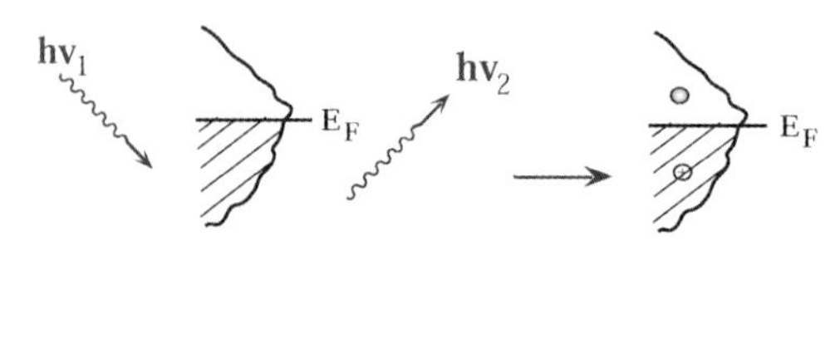


**Figure 12.8** Schematic illustration of (a) nonresonant XES, (b) resonant XES in a two-step description, and (c) one-step resonant inelastic X-ray scattering (RIXS).

measured electronic structure. Using nonresonant excitation, the excited electron has energy sufficiently high above the core ionization threshold to make the XES spectral profile practically independent of the excitation energy. As illustrated in Figure 12.8a, we can consider this case as a one-step process where we only need to describe the decay of the core–hole. In the case of resonant XES (Figure 12.8b), a specific core-excited state is prepared, which then, in a second and independent step, decays with the emission of a photon; the specifics of the excited intermediate state could, in this case, be expected to affect the resulting spectral shape. In this two-step picture, the intermediate state furthermore contains a core–hole, which could affect the transition intensities.

For all chemisorbed systems, the strong coupling to the substrate allows for an extremely fast delocalization of the excited electron prior to the decay of the core–hole state [107–110]. Even for an insulating system, such as ice and liquid water, excitation into the postedge conduction band region results in delocalization of the excited electron on a time scale faster than 500 attoseconds [111]. On the other hand, excitation into the pre-edge of liquid water and ice reaches a localized state that remains localized on a time scale (>20 fs) substantially longer than the O 1s core–hole lifetime (3.6 fs [112]) in which case a downward screening shift of the resulting emission energy of the order of 1 eV is observed [88]. On metals we can, however, in most cases disregard the presence of the excited electron due to strong coupling with the substrate corresponding to excitation into the conduction band.

If we want to derive a one-electron picture of the occupied electronic states, we should consider the X-ray emission process where one valence electron simply fills the inner-hole vacancy and an X-ray photon is emitted. This can be achieved when only a single core–hole intermediate state $|M\rangle$ is reached. The spectral distribution is then given by the spontaneous radiative decay of valence electrons into the core–hole state, described by Fermi's Golden Rule,

$$I_{\mathrm{XE}}(\omega') \propto \sum_{F} (E_{\mathrm{M}} - E_{\mathrm{F}})^3 |\langle F|\mathbf{D} \cdot \mathbf{E}'|M\rangle|^2 \delta(E_{\mathrm{M}} - E_{\mathrm{F}} - \hbar\omega'). \tag{12.32}$$

We refer to this situation as resonantly excited X-ray emission (XE) (see Figure 12.8) as an approximation to RIXS when channel interference can be effectively neglected. This is the case for most adsorbate systems where the excited electron couples to the continuum of the delocalized metal states during the core–hole lifetime, and the same fully screened core-excited intermediate state is reached locally irrespective of excitation energy [107–110]. In a simple picture, we can think in terms of an ultrafast charge transfer of the excited electron from the adsorbate into the substrate prior to the decay. This may not necessarily be the case for weakly adsorbed species where the charge-transfer process can be much slower [107, 110, 113].

In the following, we will only consider the simple XES process according to Eq. (12.32) and also neglect the influence of the intermediate core–hole in the calculations. Ideally, one should take as the initial state in the XES process the fully relaxed and screened core–hole state and as final state the equally relaxed and screened valence hole state. The core–hole state can be easily treated in a ΔKS approach, particularly when the remaining atoms of the same element are described by an effective core potential (ECP) eliminating their core levels [114]. The sequence of final valence-hole states is, however, not as easily treated while maintaining the requirement of orthogonal final states. As a balanced approximation, we neglect relaxation effects both in the initial core–hole and the final valence-hole states. This assumes similar effects of charge transfer and response of the metal to screen a hole in a core level as in the valence. Empirically, we find this to be a valid approximation for strongly coupled adsorbates and thus use the ground-state orbitals to describe both states [115].

### 12.8.1 $N_2$ and CO on Ni(100)

Let us begin by returning to Figure 12.2 and discuss the bonding mechanism of CO and $N_2$ to metal surfaces as revealed experimentally through XES combined with DFT spectrum calculations [14,19,20,115–120]. We find neither experimental nor computational support for the traditional textbook picture of $\sigma$-donation and $\pi^*$ back donation in the sense of a frontier orbital picture; all molecular orbitals change as a result of the interaction with the metal substrate. We find an overall $\sigma$-repulsion, which is similar to the case of water on metals where significant removal of charge in the substrate ("digging the s-band hole") [121] is required to minimize the repulsion and generate a dative bond involving the closed-shell lone-pair of water. The significant bond formation for CO and $N_2$ instead occurs in the $\pi$-system where the involvement of the $\pi^*$ is to allow a polarization of the $1\pi$ toward the metal to form the new bond; as a consequence of the bond formation and weakening of the internal N–N bond, a new lone-pair state appears with intensity only on the outer nitrogen, while the $\pi^*$ as a specific now occupied molecular level is hardly observed.

The homonuclear $N_2$ molecule is ideal to start the investigation of the bond formation using XES since any asymmetry between the two atoms by necessity must be induced by the interaction with the metal [19,20]. Returning to Figure 12.2, we first focus on the XES spectra of the $\pi$ states and read off the relative amounts of atomic 2p character in the molecular levels projected onto each inequivalent nitrogen atom and construct an interaction-level diagram directly from the spectrum.

Starting at the highest binding energy, we see that the $1\pi$ has become polarized toward the surface with somewhat higher intensity on the innermost nitrogen. Going toward lower binding energy, we find in the energy region from 5 up to 2 eV intensity only on the outer atom corresponding to a partial lone-pair state being formed through the interaction with the substrate; this state has no correspondence in the gas-phase molecule and is not at all considered in the frontier orbital picture with $\sigma$-donation and $\pi^*$ back donation. Close to the Fermi level, we begin to regain intensity also on the inner nitrogen indicative of some slight $\pi^*$ occupation. However, a fully occupied $\pi^*$ should have intensity similar to that of the $1\pi$, so we conclude that involvement of the $\pi^*$ in terms of the gas-phase orbital becoming occupied is indeed slight. Instead, we have a three-orbital interaction involving the $3d_\pi$ of the metal and $\pi$ and $\pi^*$ of the molecule forming bonding, nonbonding, and totally antibonding states. This is the same picture as was deduced by Blyholder [122] from his extended Hückel calculations, but not quite $\pi^*$ back donation. In Figure 12.9, we illustrate the formation of this so-called allylic configuration in a three-level interaction diagram for the $\pi$-system. In the

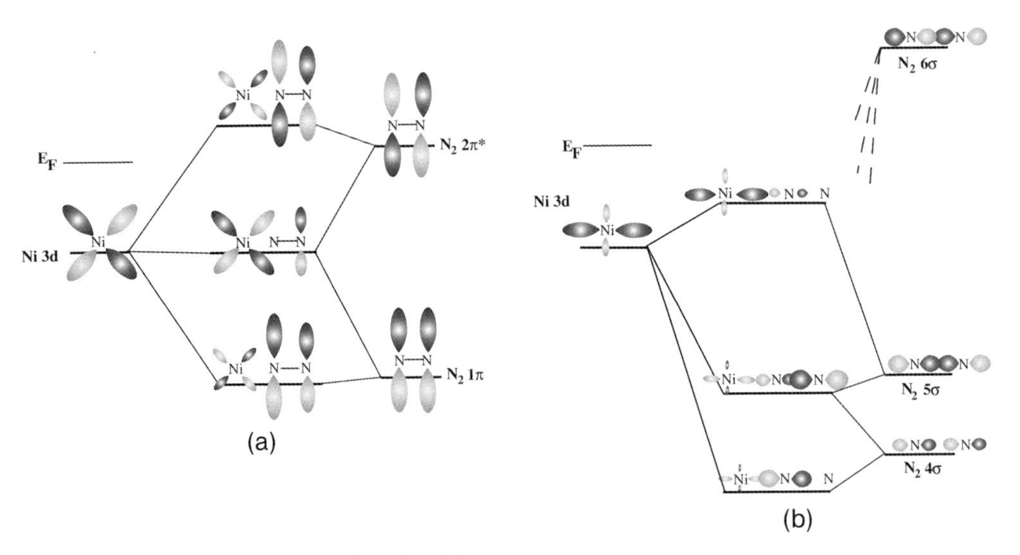


**Figure 12.9** Schematic illustrations of the bond formation in the $\pi$ (a) and $\sigma$ (b) systems of $N_2$ chemisorbing atop on Ni(100). In the $\pi$ system, the allylic bond formation involves the Ni $3d\pi$ and $N_2 1\pi$ and $2\pi^*$ levels, while in the $\sigma$ system the interaction involves Ni $3d\sigma$ and $N_2 4\sigma - 6\sigma$, i.e., all relevant orbitals are involved and affected by the bond formation.

$\sigma$-system, the internally antibonding unoccupied $6\sigma$-orbital of the molecule, which could move charge between the atoms, lies significantly higher in energy than the $\pi^*$ making the $\sigma$-system of the molecule much "stiffer" and less responsive to the interaction with the metal.

Where the $1\pi$ mixes with $\pi^*$ to polarize toward the surface to improve bonding to the metal d-states, we find the opposite polarization in the $\sigma$-system. Due to closed-shell repulsion against the electronic states of the substrate, the spatially extended $5\sigma$ polarizes away from the surface and gets the highest intensity on the outer nitrogen; as a consequence, the deeper lying $4\sigma$ localizes toward the inner nitrogen. This is indicative of a predominant $\sigma$-repulsion, rather than donation. This point will become even clearer when we compare CO chemisorption into different sites on Ni(100) next [118].

As a further indication of the bonding mode we can, from the XES spectra, follow how the electronic structure changes as the interaction strength is increased through increasing coordination. In Figure 12.10, we show experimental and computed XES spectra of CO adsorbed in on-top, bridge, and hollow sites on the Ni(100) surface where coadsorption with hydrogen has been used to force population of the energetically somewhat less favorable bridge and hollow sites [118]. Compared to the homonuclear $N_2$ molecule, CO is already polarized in the

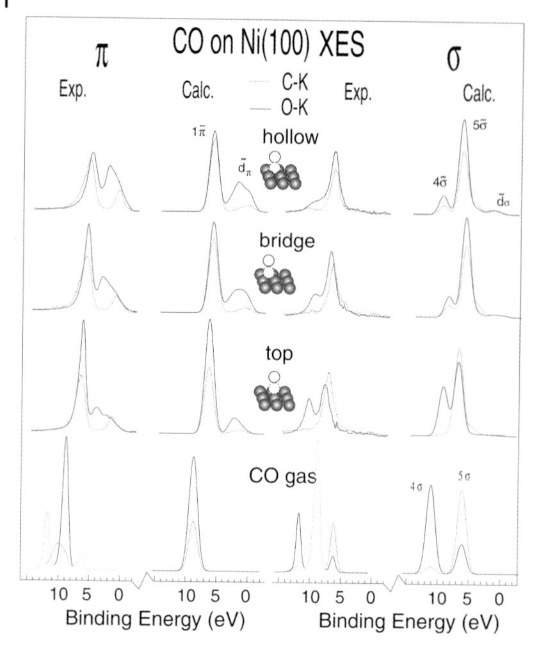


**Figure 12.10** CO Chemisorption into different sites on Ni(100). Left half shows experimental and computed XES spectra of $\pi$-states, while the right half shows the corresponding $\sigma$-states. The coordination increases from bottom to top in order: gas phase reference spectra, and then on-top, bridge and hollow adsorption.

gas phase with the $1\pi$ polarized toward the more electronegative oxygen and the $2\pi^*$ toward the carbon; $5\sigma$ is mainly localized on the carbon while $4\sigma$ is more associated with the oxygen.

Already in the on-top site where the CO interacts (through the carbon) mainly with a single Ni atom, we see the appearance in the $\pi$-system of the characteristic nonbonding state, similar to $N_2$, which is localized only on the oxygen. The low intensity shows that much of the electron density here lies in the Ni 3d. The resulting bonding $\pi$-state is still mainly polarized toward the oxygen, but this changes as the coordination is increased and in the fourfold hollow site we find equal intensity on oxygen and carbon; the nonbonding, lone-pair state has increased in intensity to become similar to the bonding $\pi$-state, which indicates a strong orbital interaction. Simultaneously, the $\sigma$-system polarizes away from the surface where the $5\sigma$, which is initially found mainly on the carbon, gains more intensity on the oxygen for the highest coordination. The p-character in the $4\sigma$ is greatly quenched with increasing coordination, corresponding to an increased 2s involvement as revealed by the calculations. The experiment only measures the local 2p character, so remaining contributions must be taken from the calculations where we use the comparison of computed and experimental spectra to ensure that a relevant structure and bonding situation is actually modeled.

From the charge movements in the $\pi$ and $\sigma$ systems, respectively, it seems quite clear that the bonding of these unsaturated molecules to the surface occurs via the formation of a $\pi$-bond with partial breaking of the internal $\pi$-bonding (formation of the lone-pair state of $\pi$ symmetry) coupled with a polarization of the $\sigma$-system away from the surface to reduce the closed-shell Pauli repulsion. The resulting binding energies in the three sites are rather similar, while the electronic structure as measured by XES shows significant effects. This is corroborated by the vibrational frequencies of CO adsorbed at different sites, which also show a significantly increased red shift with increasing coordination. This is typical of a balance between attraction ($\pi$-bonding) and repulsion ($\sigma$-system) where both increase strongly with increasing coordination, but where the net chemisorption energy, i.e., the "profit" in terms of "capitalistic chemistry," remains rather constant.

### 12.8.2 Saturated Hydrocarbons: Determine Structures and Rehybridization

Our final examples of the use of the combination of experimental X-ray spectroscopy measurements and DFT-based spectrum calculations will be the physisorption of alkanes [102], $CH_4$/Pt(111) [100, 101], and *n*-octane on Cu(110) [103, 104]. We will show that in spite of both molecules being only physisorbed at the respective metal surface, i.e., only very weakly bonded in terms of resulting binding energy, the interaction with the metal still leads to significant changes in the electronic structure and a rehybridization at the carbon atom(s). In this respect, the molecules are prepared for reaction already in the "physisorbed" state thus providing further examples that a low resulting binding energy does not necessarily imply a weak interaction.

The *n*-octane molecule adsorbs aligned along [1–10] on the Cu(110) surface, with its CCC plane parallel to the surface and with a slight preference for adsorption on-top the protruding ridges [101]. The alkanes are closed-shell molecules, and van der Waals forces will be important in determining the interaction and association with the surface. Dispersion forces are, however, not included in the DFT approach [123, 124] so that geometries of systems characterized by these weak interactions will not be described correctly. The total energy will thus not be a reliable measure for the geometric structure, but the electronic structure should be. This follows from the fact that the dispersion interaction does not in a direct way alter the orbital structure except for the indirect change caused by distortions of the geometry. We can exploit this fact to perform a systematic investigation of how the XAS and XES spectra are influenced by changes of different structural parameters and then, in comparison with the experimental spectra, draw conclusions on the structure of the adsorbed molecule.

We simulate the adsorption of *n*-octane atop the protruding ridges of the Cu(110) surface using a 23-atom single-layer copper cluster, where the 7 copper atoms closest to the molecule are described at the all-electron level and the rest by one-electron ECPs [125]. By performing a full geometry optimization of this structure, where the octane molecule was allowed to fully relax while the Cu atoms were kept fixed, a potential energy minimum was found at a carbon–metal distance of

3.0 Å. For this geometry, theoretical XAS and XES spectra were calculated finding, however, only qualitative agreement with experiment.

The energy position of the main $CC_{\sigma*}$ resonance seen in the [1–10] projection of the calculated XAS spectra in Figure 12.11 is too low by 0.9 eV compared to experiment and the intensity of the low-energy $CC_{\sigma*}$ resonance in the [001] projection is too high. According to the empirical "bond length with a ruler" concept of Stöhr and coworkers [8, 126], a shift of the $CC_{\sigma*}$ resonance to higher energy corresponds to a shortening of the CC bond length. This suggests that this bond length might be shorter in the experiment than in the calculations. We can understand this in a rehybridization picture. If a bond between the hydrogen atoms and the metal surface is to be formed, the internal CH bond must be weakened

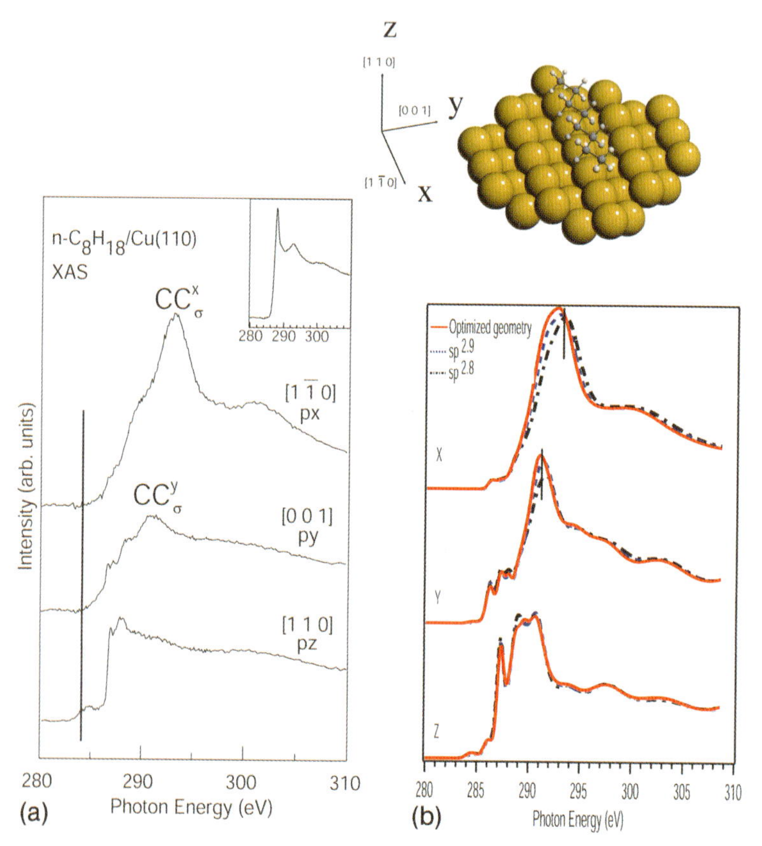


**Figure 12.11** Experimental (a) and computed (b) symmetry decomposed XAS spectra of *n*-octane physisorbed on Cu(110) with adsorption structure as illustrated in the top figure. The spectra in the right panels were calculated for a sequence of geometries corresponding to varying hybridization at the carbons ranging from $sp^3$ to $sp^{2.8}$.

and, as a result, the CC bond becomes stronger corresponding to the molecule becoming slightly unsaturated.

We can determine to what extent the molecule becomes rehybridized through the interaction with the surface by interpolating the structure between the initial C $sp^3$ hybridization and a C $sp^2$ hybridization. Artificially shortening the CC bonds, corresponding to lower $sp^n$ hybridization, and calculating the XAS spectra for the adsorbed molecule, we find that the $CC_{\sigma*}$ resonance shifts to higher energy in the [1−10] spectra and the intensity decreases in the [001] projection. By calibrating the position and intensity of the $CC_{\sigma*}$ resonances in both in-plane orbital projections against the experimental spectra, the degree of rehybridization was determined [103] to $sp^{2.8}$. This corresponds to a CC bond length of 1.49 Å, with a CCC bond angle of 112.6°, as compared to 1.53 Å and a CCC bond angle of 111.9° for the optimized structure. The decreased intensity in the [001] direction is consistent with the increased CCC bond angle. XES spectra were calculated for the same structures, but were found to be insensitive to these changes in the internal structure. On the other hand, the calculated XES spectra showed significant sensitivity to the adsorbate–substrate distance and in particular to CH agostic interactions and bond weakening through the interaction with the metal; the specific indicator of this interaction is the appearance of new states below the Fermi level where the computed intensity at the optimized distance, however, was too low. Since the missing van der Waals interaction is always attractive, it was clear that a shorter distance to the surface was required.

Pushing the molecule closer to the surface, there is a significant increase in the intensity of the new occupied states in the simulated XES spectra. However, even at a carbon–metal distance of 2.4 Å, the intensity is still much smaller than in the experiment suggesting that additional distortion of the molecular geometry, opening up the CH bonds and allowing for bonding to the surface, should be considered. This would be consistent with the rehybridization deduced from the XAS spectra and suggesting that the CC bonds and the CH bonds pointing from the surface contract slightly relative to the gas phase, whereas the CH bonds pointing toward the surface get slightly longer. Since the $sp^2$ structure is flat, the CH bonds can also be assumed to rotate down toward the surface.

By simply taking the carbon skeleton of the XAS-determined $sp^{2.8}$ geometry at a carbon–metal distance of 2.7 Å, and rotating the CH bonds that point down toward the surface, and then optimizing the remaining hydrogen atoms in the gas phase, these assumptions can be tested. By rotating the CH bonds by 10°, we find that the intensity at 5 eV increases while the intensity at higher binding energy decreases in the out-of-plane direction. The in-plane directions are essentially unchanged. The next step is to stretch the CH bonds. In organometallic chemistry, CH bonds taking part in agostic interactions show lengthening of up to as much as 10% [127]. By taking the previous geometry and stretching the CH bonds pointing down toward the surface by 5%, and again optimizing the rest of the hydrogens in the gas phase, a better agreement with the experimental spectra is achieved, as can be seen in the bottom right panel of Figure 12.12.

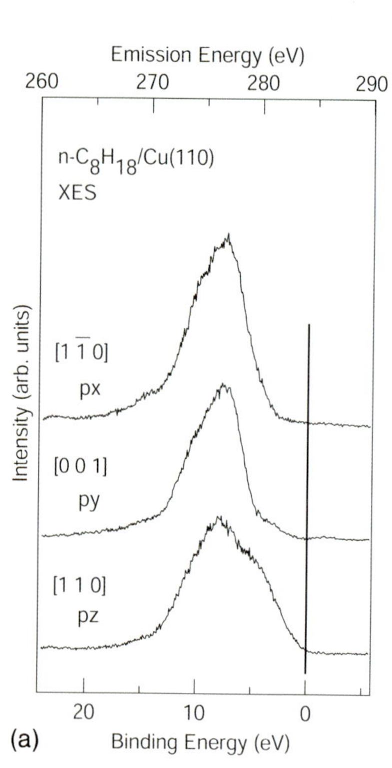


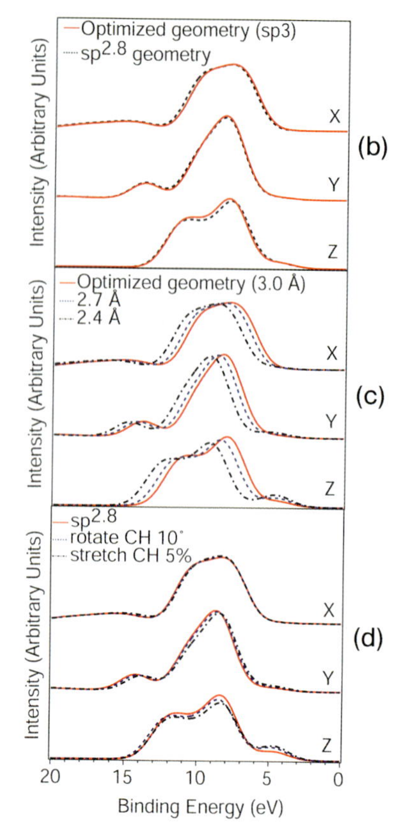


**Figure 12.12** Symmetry resolved XES spectra of *n*-octane physisorbed on Cu(110) comparing experiment (a) and spectra computed for systematic variations of the geometry and structure. In the right panel, effects of different hybridization at the carbons (b), varying distance to the surface (c) and CH rotation and bond elongations on the computed XES spectra are shown in (d).

Thus, there must be large changes in the molecular geometry upon adsorption, both in the CC and CH bonds, and the molecule has to be closer to the surface than the optimized structure. To determine the feasibility of the deduced distortions, we need to compare the energy cost with what could be expected from the missing dispersion forces. If we assume that this is the same as the total adsorption energy for alkanes, the energy is approximately 2–4 kcal/mole per carbon atom [128]. This means that the dispersion energy will be between 0.7 and 1.4 eV for *n*-octane, which is less than the computed energy cost of 1.92 or 0.24 eV per carbon atom. Considering, however, that the structural changes were performed arbitrarily without any further geometry optimizations on the surface, the energy cost for the proposed distortions is still reasonable and within the theoretical error bars considering the size of the system. A definite result of the combination of XAS and XES measurements and DFT simulations is that the molecule rehybridizes through the interaction with the surface with a trend toward CH activation, in spite of the fact that the molecule is only physisorbed.

As our second example of saturated hydrocarbons interacting with metals, we will consider the interaction of methane with a Pt substrate [100,101]; this is of particular importance since the activated dissociation of methane forms the rate-limiting step in industrial hydrogen production through steam reforming [129]. Molecular beam experiments [130–137] show that methane dissociation at surfaces can be activated equally well by incident kinetic energy as by vibrational energy added to the incident methane. From our present viewpoint of considering changes in the electronic structure, however, it is of particular interest that dissociation of adsorbed methane can also be activated by irradiation using 193 nm (6.4 eV) photons [138]; in the gas phase, methane is transparent to light at this wavelength [139, 140], which indicates that new electronic states on the molecule have been created through the interaction with the metal in spite of the molecule being only physisorbed to the Pt surface.

Present DFT calculations lack a description of van der Waals interactions and, as a consequence, do not find any total energy minimum [141]. Again we underline that this does not indicate that the molecule is interacting only weakly with the substrate. On the contrary, we can again use the combination of experimental X-ray spectroscopy measurements and DFT spectrum calculations to show that the molecule already in the physisorbed state is prepared for dissociation through CH bond elongation by ~0.09 Å due to mixing between bonding and antibonding CH orbitals. The underlying mechanism behind this is minimization of the Pauli repulsion, and the small resulting binding energy is again the result of only a small net balance between repulsive and attractive interactions.

The LUMO in gas-phase methane is the $3a_1$, which is purely of atomic s-character on the carbon and, therefore, symmetry-forbidden in C 1s XAS. The gas-phase methane XAS spectrum [142] (solid line) is shown at the bottom of the left panel in Figure 12.13. The triply degenerate $2t_2$ LUMO+1 orbital shows up as a main peak at 288.0 eV followed by vibrational progression and Rydberg orbitals at higher energy. The small feature at 286.9 eV corresponds to transitions into the symmetry-forbidden $3a_1$ LUMO orbital, which become weakly allowed due to vibronic coupling. The middle and top solid-line XAS spectra in the left panel were recorded for methane physisorbed at a Pt(977) surface [100, 101]. With the E-vector of the incoming light in the surface plane (middle spectrum), orbitals with p-character pointing parallel to the surface are probed. The interaction among the physisorbed methane molecules is weak, and the resulting spectrum shows the same main features as the gas-phase spectrum, albeit broadened through the interaction with the metal surface. The interaction with the substrate is instead probed by aligning the E-vector orthogonal to the surface plane as in the topmost solid-line spectrum in Figure 12.13 (left panel). This spectrum has the same features as the in-plane spectrum, but in addition there is a prominent signal from the LUMO orbital ("$A_1$"), which now becomes visible due to the symmetry being broken at the metal surface [143]. In addition, similar to octane on Cu(110), a

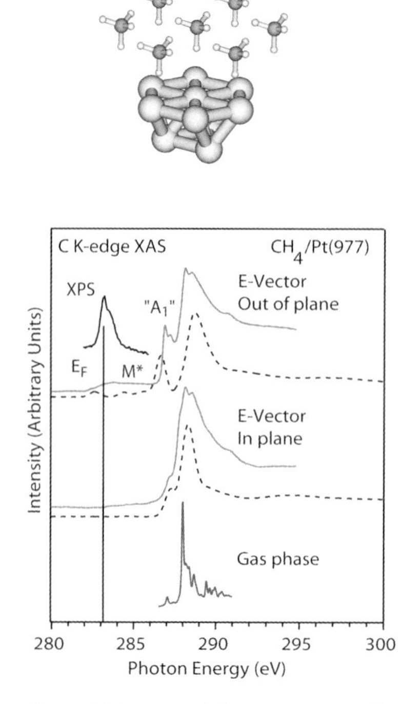


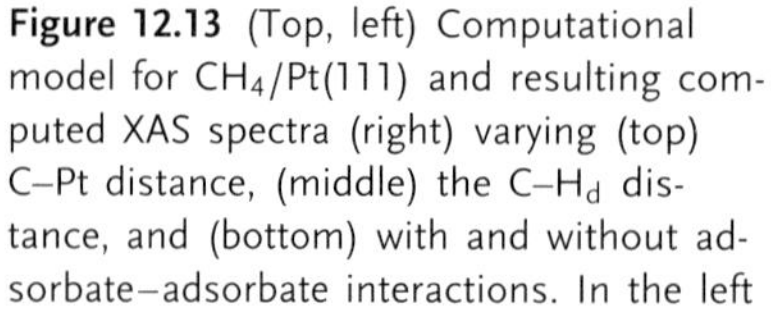

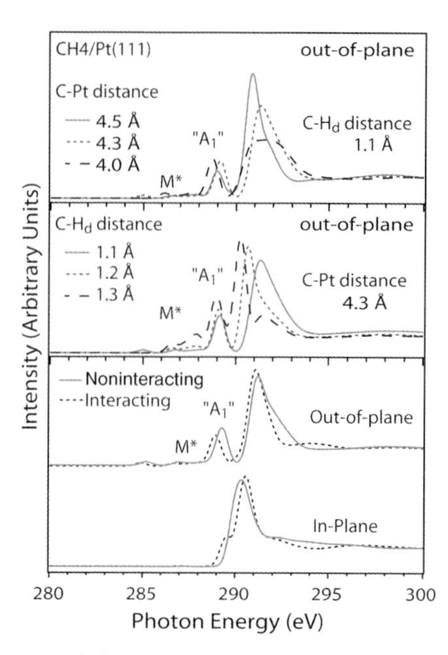


**Figure 12.13** (Top, left) Computational model for $CH_4$/Pt(111) and resulting computed XAS spectra (right) varying (top) C–Pt distance, (middle) the C–$H_d$ distance, and (bottom) with and without adsorbate–adsorbate interactions. In the left panel, the experimental gas phase (bottom), in-plane (middle), and out-of-plane (top) spectra are shown (full lines). Dashed lines show best computed spectra. The C1s XPS spectrum is included to position the Fermi-level of the system.

broad band, denoted by M*, ranging from the gas-phase LUMO all the way down to the Fermi level appears as a consequence of the interaction with the metal.

For the calculations of XAS spectra, the system was modeled using the simpler (111) surface since the (977) surface is a stepped (111) surface and no effects from the steps were observed in the experiment [100, 101]. A $Pt_{10}$ cluster was used and several different molecular geometries tested for the adsorbate. The calculations reproduce both the appearance of the gas-phase symmetry-forbidden $3a_1$ state as well as the metal-induced states above the Fermi level. In the calculations, methane was adsorbed with one hydrogen atom pointing toward the surface and with adsorbate–adsorbate interaction taken into account, as indicated at the top of Figure 12.13; alternative structures resulted in less satisfactory agreement. To get a good quantitative agreement, the molecule–metal (C–Pt) distance as well as the distance between the carbon and hydrogen atom pointing toward the surface (C–$H_d$) were varied. The main sensitivity was found for the relative intensity of the main peak to the region between the Fermi level and symmetry-forbidden peak but also in the splitting between the peak corresponding to the symmetry-forbidden state and the main peak as shown in the top and middle sections of the right panel in Figure 12.13.

The splitting is sensitive to full variational relaxation of the corresponding core-excited states as found for the gas-phase methane molecule where the TP approach gave a 50% larger splitting than experiment, but the state-by-state [56] calculations resolve this discrepancy. These more accurate calculations are, however, not possible for the adsorbate system where the band structure of the metal contributes too many low-lying states for this procedure to be practical. Instead, it was assumed that the TP approach would produce a similarly enhanced splitting also for the physisorbed species. In the left panel of Figure 12.13, the experimental spectra are plotted with solid and the final theoretical spectra with dashed lines where the adsorption geometry has been adjusted to give the derived from the gas phase 50% larger splitting compared with the experiment. The absolute energy scale in relation to experiment is a difficulty here since the work function of the cluster may be far from experiment. Based on the overall good shape of the theoretical spectra, a 2.3 eV shift toward lower energy was instead applied to match the experiment. Based on these spectrum calculations, we deduce a geometry where one C–H bond is stretched 8% of the original bond length from 1.09 Å to 1.18 Å with the molecule at a distance of 4.15 Å between the C and closest Pt atom. The computed energy cost ($\sim$0.1 eV) for the distortions is small enough to be compensated by dispersion interactions that are not included in the theory.

Important consequences of the orbital mixing are both the elongation of the C–$H_d$ bond and the broken symmetry, which lead to the appearance of the gas-phase symmetry-forbidden state in the adsorbate XAS spectrum. The driving force is minimization of the Pauli repulsion through charge polarization away from the Pt–H interaction region. This rehybridization involves the CH $\sigma^*$ antibonding orbital and is completely consistent with the strong energy dependence of methane dissociation in molecular beam experiments of methane [133, 136]. From our results, we can understand this since the molecules with larger incident kinetic energy are pushed further into the repulsive barrier and thus respond with a larger polarization of the electron density away from the metal in order to minimize the Pauli repulsion, leading to an increased elongation of the CH bond and subsequent dissociation. The adsorbed methane, for which the CH bond is elongated already in the adsorbed state, can thus be viewed as a precursor step to the dissociation [100, 101].

## 12.9
## Summary and Outlook

In heterogeneous catalysis, the system is often rather complex with nanosized particles on a supporting material. There is clear evidence that many catalytic surfaces are not uniformly active, but the activity occurs only at specific sites where a special arrangement of surface atoms or special chemical composition exists. Variation of particle size gives rise to changes in the surface structure where the relative concentrations of atoms in step, kink, and terrace sites are altered. Nearly all catalytic reactions that involve breaking of molecular bonds as a rate-limiting step take place at step or kink sites. In a similar manner, many important reactions on naturally occurring mineral surfaces can be related to defect sites. It will be important to investigate in the future the bonding interaction of adsorbates at these active sites using XES and DFT where it is expected that new spectrometer developments will make this at-present very difficult experimental technique more accessible and of greater applicability. As further examples of what can be achieved in terms of detailed experimental resolution of the electronic structure with these techniques, we consider a stepped single-crystal surface that represents a system with lowered symmetry. The E-vector of the incoming light can be aligned in three directions, along the step, perpendicular to the step, and perpendicular to the surface allowing for a precise determination of the orientation and geometry of an adsorbate in conjunction with DFT calculations. Both stepped and in particular kinked surfaces have unidirectional asymmetry perpendicular to the step or along the kink direction. Furthermore, for small particles, the electronic structure could deviate substantially from the bulk and new unique electronic states related to quantum confinement could appear. The theoretical description of the spectroscopic information upon adsorption and reaction of molecules on these particles including interaction with the support could become a major challenge.

There is an enormous and fast-growing activity using theoretical studies to model catalytic reactions using total-energy-based DFT calculations. Several theoretical estimations of activation barriers for the rate-limiting steps seem to be in good agreement with kinetic measurements of the total reaction rate. However, there are experimental indications of strong nonadiabatic coupling in adsorption and dissociation processes pointing at normally not considered important energy dissipation channels. The nonadiabatic damping of vibrational motion close to the activation barrier is due to creation of electron–hole pairs in the substrate. It will be extremely important to experimentally resolve the elementary steps during a surface reaction and thereby reach a mechanistic understanding on a microscopic level that can be compared with theory. The theoretical challenges will be to develop techniques which in a realistic manner take into account the coupling also with the phonon modes in the substrate and not only in the sense of the lowest energy response of the substrate atoms to the interaction with the impinging adsorbate, but also taking into account the full dynamics of the surface where instantaneous properties of specific sites may deviate significantly from the average. An example of this is given by the discussion of the influence of phonon modes on the reduction of $SO_2$ over Ca-doped $CeO_2$, where large fluctuations of ions out from their most stabilized lattice positions have been shown to affect charge-transfer energies significantly [144].

The development of ultrafast FEL X-ray sources will offer new experimental capabilities in the near future. We can envisage that an ultrafast laser can be used to initiate a chemical reaction, which can subsequently be probed with an extremely short X-ray pulse to generate core–holes for XES measurements. By changing the delay between the pump and probe pulses, a new dynamic range will become feasible. This is a standard procedure for femtosecond laser spectroscopy where the excitation and probe steps involve valence electrons that are delocalized over many atomic centers making it difficult to analyze complex systems in detail. It would be an important development if we can probe how molecular orbitals are transformed during a chemical reaction in an atom-specific way around the center where the interesting chemistry takes place. This will open new prospects to locally study changes in molecular orbitals on surfaces during surface reactions. The ultimate chemist dream experiment to watch a chemical reaction in real time could become a reality.

Excitation of phonons, frustrated rotational and translational motions of molecular adsorbates play important roles in processes at surfaces that are driven by kT, i.e., temperature. This accounts for nearly all the essential processes of economic interest. Femtosecond visible laser pulses have been used to trigger the reaction. Laser pulses heat the electrons, leading to a very high transient electronic temperature followed by the subsequent energy transfer to phonons and frustrated vibrational modes. As vibrational temperature rises, the reaction is initiated and the time evolution is followed through products released into the gas phase. At the moment, there exists no direct way to pump a surface reaction by exciting the motion of the nuclei of an adsorbed molecule on an ultrafast time scale. The ultrashort electron pulse obtained in the linear accelerator to feed the X-ray FEL can, however, also be used for generation of coherent synchrotron radiation in the low-energy THz regime to be used as a pump. The coherent THz radiation will be an electric field pulse with a certain direction that can collectively manipulate atoms or molecules on surfaces. In this respect, a chemical reaction can be initiated by collective atomic motion along a specific reaction coordinate. If the coherent THz radiation is generated from the same source as the X-ray FEL radiation, full-time synchronization for pump-probe experiments will be possible. The combination of THz and X-ray spectroscopy could be a unique opportunity for FEL facilities to conduct ultrafast chemistry studies at surfaces. It will be essential to have a theoretical description of the dynamics that is induced by the ultrafast THz source. For this to become a reality, we will need improved methods that describe the dynamics together with the electronic structure and are capable of handling large unit cells or locally disordered coadsorbate systems, and for these reliably predict the effect of the varying electric field associated with the THz pulse.

Synchrotrons today have a multitude of beam-lines and end stations where many different types of experiments are conducted simultaneously during 24-hour around-the-clock working shifts. The new FEL installations in the X-ray regime will have one or at most two experiments running at the same time, and competition for beam time will be fierce. It is essential that the experiments when actually performed are likely to succeed in discovering something exciting and of interest; to this end, theoretical simulations will become even more essential both as analytical and in the future even more importantly as exploratory predictive tools in the planning of experiments.

## Acknowledgments

We gratefully acknowledge all the people involved in the various projects on which this chapter is based and particularly acknowledge fruitful discussions with Tomasz Wesolowski and Mark Casida on the DFT aspects. This work was supported by the Swedish Foundation for Strategic Research and the Swedish Natural Science Research Council, the National Science Foundation under Contract No. CHE-0431425 (Stanford Environmental Molecular Science Institute), the U.S. Department of Energy, Office of Basic Energy Sciences, through the Stanford Synchrotron Radiation Laboratory, Contract No. DE-AC02-05CH11231, and under the auspices of the President's Hydrogen Fuel Initiative. We are grateful for generous allotments of CPU time at the National Swedish Computer Centers at NSC and PDC.

## References

1 Woodruff, D.P., Delchar, T.A. *Modern Techniques of Surface Science*; Cambridge University Press: New York, **1986**.
2 Siegbahn, K., Nordling, C., Johansson, G., Hedman, J., Heden, P.F., Hamrin, K., Gelius, U., Bergmark, T., Werme, L.O., Manne, R., Baer, Y. *ESCA Applied to Free Molecules*; North-Holland: Amsterdam, **1969**.
3 Fadley, C.S. In *Electron Spectroscopy: Theory, Techniques and Applications*; Brundle, C.R., Baker, A.D., Eds.; Academic Press: New York, **1978**; Vol. 2.
4 Hufner, S. *Photoelectron Spectroscopy*; Springer: Berlin Heidelberg, **1995**; Vol. 82.
5 Egelhoff, W.F. *Surf. Sci. Reps*. **1986**, *6*, 253.
6 Mårtensson, N., Nilsson, A. In *Applications of Synchrotron Radiation; High Resolution Studies of Molecules and Molecular Adsorbates*; Eberhardt, W., Ed.; Springer: Berlin, Heidelberg, **1995**; Vol. 35.
7 Nilsson, A., Zdansky, E., Tillborg, H., Björneholm, O., Mårtensson, N., Andersen, J.N., Nyholm, R. *Chem. Phys. Lett*. **1992**, *197*, 12.
8 Stöhr, J. *NEXAFS Spectroscopy*; Springer: Berlin, Heidelberg, **1992**; Vol. 25.
9 Fuggle, J.C. In *Electron Spectroscopy: Theory, Techniques and Applications*; Brundle, C.R., Baker, A.D., Eds.; Academic Press: London, **1981**; Vol. 4.
10 Mårtensson, N., Nilsson, A. *J. El. Spec. Rel. Phenom*. **1995**, *72*, 1.
11 Eberhardt, W. In *Applications of Synchrotron Radiation; High Resolution Studies of Molecules and Molecular Adsorbates*; Eberhardt, W., Ed.; Springer: Berlin, Heidelberg, **1995**; Vol. 35.
12 Nordgren, J. *J. El. Spec. Rel. Phenom*. **2000**, *110 and 111*, ix.
13 Gel'mukhanov, F., Ågren, H. *Phys. Reps*. **1999**, *312*, 87–330.
14 Nilsson, A., Pettersson, L.G.M. *Surf. Sci. Rep*. **2004**, *55*, 49–167
15 Nilsson, A., Mårtensson, N. *Physica B* **1995**, *208 and 209*, 19.
16 Stöhr, J. *NEXAFS Spectroscopy*; Springer: Berlin, Heidelberg, **1992**.
17 Brown Jr., G.E., Sturchio, N.C. *Rev. Mineral. Geochem*. **2002**, *49*, 1.
18 Horn, K., Dinardo, J., Eberhardt, W., Freund, H.J. *Surf. Sci*. **1982**, *118*, 465.
19 Bennich, P., Wiell, T., Karis, O., Weinelt, M., Wassdahl, N., Nilsson, A., Nyberg, M., Pettersson, L.G.M., Stöhr, J., Samant, M. *Phys. Rev. B* **1998**, *57*, 9274.
20 Nilsson, A., Weinelt, M., Wiell, T., Bennich, P., Karis, O., Wassdahl, N.,

Stöhr, J., Samant, M. *Phys. Rev. Lett.* **1997**, *87*, 2847.

21 Kevan, S.D. *Angle-Resolved Photoemission*; Elsevier: Amsterdam, **1992**; Vol. 74.

22 Bluhm, H., Andersson, K., Araki, T., Benzerara, K., Brown, G.E., Dynes, J.J., Ghosal, S., Gilles, M.K., Hansen, H.-C., Hemminger, J.C., Hitchcock, A.P., Ketteler, G., Kilcoyne, A.L.D., Kneedler, E., Lawrence, J.R., Leppard, G.G., Majzlam, J., Mun, B.S., Myneni, S.C.B., Nilsson, A., Ogasawara, H., Ogletree, D.F., Pecher, K., Salmeron, M., Shuh, D.K., Tonner, B., Tyliszczak, T., Warwick, T., Yoon, T.H. *J. Electron Spectr.* **2006**, *150*, 86.

23 Rehr, J.J., Mustre de Leon, J., Zabinsky, S.I., Albers, R.C. *J. Am. Chem. Soc.* **1991**, *113*, 5135.

24 Zabinsky, S.I., Rehr, J.J., Ankudinov, A., Albers, R.C., Eller, M.J. *Phys. Rev. B* **1995**, *52*, 2995.

25 Kosugi, N., Kuroda, H. *Chem. Phys. Lett.* **1980**, *74*, 490.

26 Kosugi, N. *Theo. Chim. Acta* **1987**, *72*, 149.

27 Shirley, E.L. *Phys. Rev. Lett.* **1980**, *80*, 794.

28 Casida, M.E. In *Recent Advances in Density-Functional Methods*; Chong, D.P., Ed.; World Scientific: Singapore, **1995**; Vol. 1; p. 155.

29 Casida, M.E. In *Recent Developments and Applications of Modern Density Functional Theory, Theoretical and Computational Chemistry*; Seminario, J.M., Ed.; Elsevier: Amsterdam, **1996**; Vol. 4.

30 Stratmann, R.E., Scuseria, G.E., Frisch, M.J. *J. Chem. Phys.* **1998**, *109*, 8218.

31 Stener, M., Fronzoni, G., Toffoli, D., Decleva, P. *Chem. Phys.* **2002**, *282*, 337.

32 Stener, M., Fronzoni, G., de Simone, M. *Chem. Phys. Lett.* **2003**, *373*, 115.

33 Triguero, L., Pettersson, L.G.M., Ågren, H. *Phys. Rev. B* **1998**, *58*, 8097.

34 Ågren, H., Carravetta, V., Vahtras, O., Pettersson, L.G.M. *Chem. Phys. Lett.* **1994**, *222*, 75.

35 Ågren, H., Carravetta, V., Vahtras, O., Pettersson, L.G.M. *Theo. Chem. Acc.* **1997**, *97*, 14.

36 Triguero, L., Pettersson, L.G.M., Ågren, H. *Phys. Rev. B* **1998**, *58*, 8097.

37 Hermann, K., Pettersson, L.G.M., Casida, M.E., Daul, C., Goursot, A., Koester, A., Proynov, E., St-Amant, A., Salahub, D.R., Carravetta, V., Duarte, A., Godbout, N., Guan, J., Jamorski, C., Leboeuf, M., Leetmaa, M., Nyberg, M., Pedocchi, L., Sim, F., Triguero, L., Vela, A.; 5.3 ed.; deMon Software: Stockholm, Berlin, **2005**.

38 Theophilou, A.K. *J. Phys. C: Solid State Phys.* **1979**, *12*, 5419–5430.

39 Gross, E.K.U., Oliveira, L.N., Kohn, W. *Phys. Rev. A* **1988**, *37*, 2805–2808.

40 Gross, E.K.U., Oliveira, L.N., Kohn, W. *Phys. Rev. A* **1988**, *37*, 2809–2820.

41 Oliveira, L.N., Gross, E.K.U., Kohn, W. *Phys. Rev. A* **1988**, *37*, 2821–2833.

42 Levy, M., Nagy, A. *Phys. Rev. Lett.* **1999**, *83*, 4361–4364.

43 Görling, A. *Phys. Rev. A* **1999**, *59*, 3359–3374.

44 Sahni, V., Massa, L., Singh, R., Slamet, M. *Phys. Rev. Lett.* **2001**, *87*, 113002.

45 Sharp, R.T., Horton, G.K. *Phys. Rev.* **1953**, *90*, 317.

46 Talman, J.D., Shadwick, W.F. *Phys. Rev. A* **1976**, *14*, 36.

47 Heßelmann, A., Götz, A.W., Della Sala, F., Görling, A. *J. Chem. Phys.* **2007**, *127*, 054102.

48 Krieger, J.B., Li, Y., Iafrate, G.J. *Phys. Rev. A* **1992**, *45*, 101.

49 Glushkov, V.N., Gidopoulos, V. *Int. J. Quant. Chem.* **2007**, *107*, 2604–2615.

50 Hohenberg, P., Kohn, W. *Phys. Rev.* **1964**, *136*, B864–B871.

51 Hamel, S., Duffy, P., Casida, M.E., Salahub, D.R. *J. El. Spec. Rel. Phen.* **2002**, *123*, 345–363.

52 Jones, R.O., Gunnarsson, O. *Rev. Mod. Phys.* **1989**, *61*, 689.

53 Chong, D.P., Gritsenko, O.V., Baerends, E.J. *J. Chem. Phys.* **2002**, *116*, 1760.

54 Casida, M.E., Salahub, D.R. *J. Chem. Phys.* **2000**, *113*, 8918–8935.

55 Ågren, H., Carravetta, V., Vahtras, O., Pettersson, L.G.M. *Theoret. Chem. Acc.* **1997**, *97*, 14.

56 Kolczewski, C., Püttner, R., Plashkevych, O., Ågren, H., Staemmler, V., Martins, M., Snell, G., Sant'anna, M., Kaindl, G., Pettersson, L.G.M. *J. Chem. Phys.* **2001**, *115*, 6426–6437.

57 Takahashi, O., Pettersson, L.G.M. *J. Chem. Phys.* **2004**, *121*, 10339–10345.

58 Leetmaa, M., Ljungberg, M., Ogasawara, H., Odelius, M., Näslund, L.-Å., Nilsson, A., Pettersson, L.G.M. *J. Chem. Phys.* **2006**, *125*, 244510.

59 Kolczewski, C., Hermann, K. *J. Chem. Phys.* **2003**, *118*, 7599.

60 Slater, J.C. *Adv. Quant. Chem.* **1972**, 6, 1.

61 Slater, J.C., Johnsson, K.H. *Phys. Rev. B* **1972**, *5*, 844.

62 Janak, J.F. *Phys. Rev. B* **1978**, *18*, 7165.

63 Ågren, H., Carravetta, V., Vahtras, O., Pettersson, L.G.M. *Chem. Phys. Lett.* **1994**, *222*, 75.

64 Cavalleri, M., Nordlund, D., Odelius, M., Nilsson, A., Pettersson, L.G.M. *Phys. Chem. Chem. Phys.* **2005**, *7*, 2854–2858.

65 Nyberg, M., Luo, Y., Triguero, L., Pettersson, L.G.M., Ågren, H. *Phys. Rev. B* **1999**, *60*, 7956

66 Stern, H.A., Berne, B.J. *J. Chem. Phys.* **2001**, *115*, 7622.

67 Sankari, R., Ehara, M., Nakatsuji, H., Senba, Y., Hosokawa, K., Tyoshida, H., De Fanis, A., Tamenori, Y., Aksela, S., Ueda, K. *Chem. Phys. Lett.* **2003**, *380*, 647.

68 Chen, C.T., Ma, Y., Sette, F. *Phys. Rev. A* **1989**, *40*, 6737–6740.

69 Wernet, P., Nordlund, D., Bergmann, U., Cavalleri, M., Odelius, M., Ogasawara, H., Näslund, L.Å., Hirsch, T.K., Ojamäe, L., Glatzel, P., Pettersson, L.G.M., Nilsson, A. *Science* **2004**, *304*, 999.

70 Kuo, I.F.W., Mundy, C.J., McGrath, M.J., Siepmann, J.I., VandeVondele, J., Sprik, M., Hutter, J., Chen, B., Klein, M.L., Mohamed, F., Krack, M., Parrinello, M. *J. Phys. Chem. B* **2004**, *108*, 12990.

71 Fernandez-Serra, M.V., Artacho, E. *Phys. Rev. Lett.* **2006**, *96*, 016404.

72 Head-Gordon, T., Johnson, M.E. *Proc. National Acad. Sci. (USA)* **2006**, *103*, 7973–7977.

73 Head-Gordon, T., Johnson, M.E. *Proc. National Acad. Sci. (USA)* **2007**, *103*, 16614.

74 Head-Gordon, T., Rick, S.W. *Phys. Chem. Chem. Phys.* **2007**, *9*, 83–91.

75 Hetényi, B., De Angelis, F., Giannozzi, P., Car, R. *J. Chem. Phys.* **2004**, *120*, 8632–8637.

76 Ludwig, R. *Chem. Phys. Chem.* **2007**, *8*, 938–943.

77 Nilsson, A., Wernet, P., Nordlund, D., Bergmann, U., Cavalleri, M., Odelius, M., Ogasawara, H., Näslund, L.-Å., Hirsch, T.K., Ojamäe, L., Glatzel, P., Pettersson, L.G.M. *Science* **2005**, *308*, 793a.

78 Näslund, L.Å., Lüning, J., Ufuktepe, Y., Ogasawara, H., Wernet, P., Bergmann, U., Pettersson, L.G.M., Nilsson, A. *J. Phys. Chem. B* **2005**, *109*, 13835–13839.

79 Odelius, M., Cavalleri, M., Nilsson, A., Pettersson, L.G.M. *Phys. Rev. B* **2006**, *73*, 024205.

80 Paesani, F., Iuchi, S., Voth, G.A. *J. Chem. Phys.* **2007**, *127*, 074506.

81 Prendergast, D., Galli, G. *Phys. Rev. Lett.* **2006**, *96*, 215502.

82 Smith, J.D., Cappa, C.D., Messer, B.M., Cohen, R.C., Saykally, R.J. *Science* **2005**, *308*, 793b.

83 Smith, J.D., Cappa, C.D., Messer, B.M., Drisdell, W.S., Cohen, R.C., Saykally, R.J. *J. Phys. Chem. B* **2006**, *110*, 20038–20045.

84 Smith, J.D., Cappa, C.D., Wilson, K.R., Cohen, R.C., Geissler, P.L., Saykally, R.J. *Proc. National Acad. Sci. (USA)* **2005**, *102*, 14171.

85 Smith, J.D., Cappa, C.D., Wilson, K.R., Messer, B.M., Cohen, R.C.,

Saykally, R.J. *Science* **2004**, *306*, 851–853.

86 Soper, A.K. *J. Phys.: Condens. Matter* **2005**, *17*, S3273–S3282.

87 Soper, A.K. *J. Phys.: Condens. Matter* **2007**, *19*, 335206.

88 Tokushima, T., Harada, Y., Takahashi, O., Senba, Y., Ohashi, H., Pettersson, L.G.M., Nilsson, A., Shin, S. *Chem. Phys. Lett.* **2008**, *460*, 387.

89 Kimmel, G.A., Petrik, N.G., Dohnalek, Z., Kay, B.D. *Phys. Rev. Lett.* **2005**, 95, 166102.

90 Kimmel, G.A., Petrik, N.G., Dohnálek, Z., Kay, B.D. *J. Chem Phys.* **2007**, *126*, 114702.

91 Kondo, T., Kato, H.S., Bonn, M., Kawai, M. *J. Chem. Phys.* **2007**, *126*, 181103.

92 Haq, S., Hodgson, A. *J. Phys. Chem. C* **2007**, *111*, 5946.

93 Ogasawara, H., Brena, B., Nordlund, D., Nyberg, M., Pelmenschikov, A., Pettersson, L.G.M., Nilsson, A. *Phys. Rev. Lett.* **2002**, *89*, 276102.

94 Nyberg, M., Odelius, M., Pettersson, L.G.M., Nilsson, A. *J. Chem. Phys.* **2003**, *119*, 12577–12585.

95 Hasselström, J., Karis, O., Nyberg, M., Pettersson, L.G.M., Weinelt, M., Wassdahl, N., Nilsson, A. *J. Phys. Chem. B* **2000**, *104*, 11480.

96 Hasselström, J., Karis, O., Weinelt, M., Wassdahl, N., Nilsson, A., Nyberg, M., Pettersson, L.G.M., Samant, M., Stöhr, J. *Surf. Sci.* **1998**, *407*, 221.

97 Booth, N.A., Woodruff, D.P., Schaff, O., Giessl, T., Lindsay, R., Baumgärtel, P., Bradshaw, A.M. *Surf. Sci.* **1998**, *397*, 258.

98 Cavalleri, M., Ogasawara, H., Pettersson, L.G.M., Nilsson, A. *Chem. Phys. Lett.* **2002**, *364*, 363–370.

99 Myneni, S., Luo, Y., Näslund, L.Å., Cavalleri, M., Ojamäe, L., Ogasawara, H., Pelmenschikov, A., Wernet, P., Väterlein, P., Heske, C., Hussain, Z., Pettersson, L.G.M., Nilsson, A. *J. Phys. Condens. Matter* **2002**, *14*, L213.

100 Öström, H., Ogasawara, H., Näslund, L.-Å., Pettersson, L.G.M., Nilsson, A. *Phys. Rev. Lett.* **2006**, *96*, 146104.

101 Öström, H., Ogasawara, H., Näslund, L.-Å., Andersson, K., Pettersson, L.G.M., Nilsson, A. *J. Chem. Phys.* **2007**, *127*, 144702.

102 Öström, H., Triguero, L., Nyberg, M., Ogasawara, H., Pettersson, L.G.M., Nilsson, A. *Phys. Rev. Lett.* **2003**, *91*, 046102.

103 Öström, H., Triguero, L., Weiss, K., Ogasawara, H., Garnier, M.G., Nordlund, D., Pettersson, L.G.M., Nilsson, A. *J. Chem. Phys.* **2003**, *118*, 3782.

104 Weiss, K., Öström, H., Triguero, L., Ogasawara, H., Garnier, M.G., Pettersson, L.G.M., Nilsson, A. *J. El. Spec. Rel. Phenom.* **2003**, *128*, 179.

105 Sakurai, J. *Advanced Quantum Mechanics*; Addison-Wesley: Menlo Park, CA, **1967**.

106 Schöffler, M.S., Titze, J., Petridis, N., Jahnke, T., Cole, K., Schmidt, L.P.H., Czasch, A., Akoury, D., Jagutzki, O., Williams, J.B., Cherepkov, N.A., Semenov, S.K., McCurdy, C.W., Rescigno, T.N., Cocke, C.L., Osipov, T., Lee, S., Prior, M.H., Belkacem, A., Landers, A.L., Schmidt-Böcking, H., Weber, T., Dörner, R. *Science* **2008**, *320*, 920.

107 Karis, O., Nilsson, A., Weinelt, M., Wiell, T., Puglia, C., Wassdahl, N., Mårtensson, N., Samant, M., Stöhr, J. *Phys. Rev. Lett.* **1996**, *76*, 1380.

108 Keller, C., Stichler, M., Comelli, G., Esch, F., Lizzit, S., Wurth, W., Menzel, D. *Phys. Rev. Lett.* **1998**, *80*, 1774.

109 Föhlisch, A., Hasselström, J., Karis, O., Menzel, D., Mårtensson, N., Nilsson, A. *J. El. Spec. Rel. Phenom.* **1999**, *101–103*, 303.

110 Nilsson, A. *J. El. Spec. Rel. Phenom.* **2002**, *126*, 3.

111 Nordlund, D., Ogasawara, H., Bluhm, H., Takahashi, O., Odelius, M., Nagasono, M., Pettersson, L.G.M., Nilsson, A. *Phys. Rev. Lett.* **2007**, 99, 217406.

112 Neeb, M., Rubensson, J.E., Biermann, M., Eberhardt, W. *J. El. Spec. Rel. Phenom.* **1994**, *67*, 261–274.

113 Keller, C., Stichler, M., Comelli, G., Esch, F., Lizzit, S., Menzel, D., Wurth, W. *Phys. Rev. B* **1998**, *57*, 11951.

114 Pettersson, L.G.M., Wahlgren, U., Gropen, O. *J. Chem. Phys.* **1987**, *86*, 2176.

115 Föhlisch, A., Bennich, P., Hasselström, J., Karis, O., Nilsson, A., Triguero, L., Nyberg, M., Pettersson, L.G.M. *Phys. Rev. B* **2000**, *16*, 16229.

116 Bennich, P. Ph.D., Uppsala University, Sweden, **1996**.

117 Föhlisch, A., Nyberg, M., Bennich, P., Triguero, L., Hasselström, J., Karis, O., Pettersson, L.G.M., Nilsson, A. *J. Chem Phys.* **2000**, *112*, 1946.

118 Föhlisch, A., Nyberg, M., Hasselström, J., Karis, O., Pettersson, L.G.M., Nilsson, A. *Phys. Rev. Lett.* **2000**, *85*, 3309.

119 Föhlisch, A., Wurth, W., Stichler, M., Keller, C., Nilsson, A. *J. Chem. Phys.* **2004**, *121*, 4848.

120 Nyberg, M., Föhlisch, A., Triguero, L., Bassan, A., Nilsson, A., Pettersson, L.G.M. *J. Mol. Struct.* (THEOCHEM) **2006**, *762*, 123.

121 Schiros, T., Takahashi, O., Andersson, K., Öström, H., Pettersson, L.G.M., Nilsson, A., Ogasawara, H., **2008**, to be published.

122 Blyholder, G. *J. Phys. Chem.* **1964**, *68*, 2772.

123 Müller-Dethlefs, K., Hobza, P. *Chem. Rev.* **2000**, *100*, 143.

124 Rappé, A.K., Bernstein, E.R. *J. Phys. Chem. A* **2000**, *104*, 6117.

125 Mattsson, A., Panas, I., Siegbahn, P., Wahlgren, U., Åkeby, H. *Phys. Rev. B* **1987**, *36*, 7389.

126 Stöhr, J., Sette, F., Johnson, A.L. *Phys. Rev. Lett.* **1984**, *53*, 1684.

127 Brookhart, M., Green, M.L.H. *J. Organometallic Chem.* **1983**, *250*, 395.

128 Madey, T.E., Yates, J.T. *Surf. Sci.* **1978**, *76*, 397.

129 Bengaard, H.S., Nørskov, J.K., Sehested, J., Clausen, B.S., Nielsen, L.P., Molenbroek, A.M., Rostrup-Nielsen, J.R. *J. Catal.* **2002**, *209*, 365.

130 Beck, R.D., Maroni, P., Papageorgopoulos, D.C., Dang, T.T., Schmid, M.P., Rizzo, T.R. *Science* **2003**, *302*, 98.

131 Beebe, T.P.J., Goodman, D.W., Kay, B.D., Yates, J.T.J. *J. Chem. Phys.* **1987**, *87*, 2305.

132 Holmblad, P.M., Wambach, J., Chorkendorff, I. *J. Chem. Phys.* **1995**, *102*, 8255.

133 Lee, M.B., Yang, Q.Y., Ceyer, S.T. *J. Chem. Phys.* **1987**, *87*, 2724.

134 Luntz, A.C., Bethune, D.S. *J. Chem. Phys.* **1989**, *90*, 1274.

135 Rettner, C.T., Pfnür, H.E., Auerbach, D.J. *Phys. Rev. Lett.* **1985**, *54*, 2716.

136 Lee, M.B., Yang, Q.Y., Tang, S.L., Ceyer, S.T. *J. Chem. Phys.* **1986**, *85*, 1693.

137 Fuhrmann, T., Kinne, M., Tränkenschuh, B., Papp, C., Zhu, J.F., Denecke, R., Steinrück, H.-P. *New J. Phys.* **2005**, *7*, 107.

138 Watanabe, K., Sawabe, K., Matsumoto, Y. *Phys. Rev. Lett.* **1996**, *76*, 1751.

139 Lee, L.C., Chiang, C.C. *J. Chem. Phys.* **1983**, *78*, 688.

140 Wainfan, N., Walker, W.C., Weissler, G.L. *Phys. Rev.* **1955**, *99*, 542.

141 Kratzer, P., Hammer, B., Nørskov, J.K. *J. Chem. Phys.* **1996**, *105*, 5595.

142 Schirmer, J., Trofimov, A.B., Randall, K.J., Feldhaus, J., Bradshaw, A.M. *Phys. Rev. A* **1993**, *47*, 1136.

143 Yoshinobu, J., Ogasawara, H., Kawai, M. *Phys. Rev. Lett.* **1995**, *75*, 2176.

144 Triguero, L., de Carolis, S., Baudin, M., Wójcik, M., Hermansson, K., Nygren, M.A., Pettersson, L.G.M. *Faraday Discussions* **1999**, *114*, 351.

# 13
# Basics of Crystallography

*Klaus Hermann*

Research in heterogeneous catalysis and materials science focuses in many cases on detailed studies of crystalline solid-state systems on an atomic scale where the systems may act as meaningful models, for example, simulating reactive sites of catalysts, or may represent real materials such as complex semiconductors. Here physical and chemical insight requires always a detailed knowledge about the geometry of local environments near atom centers as well as about possible periodic atom arrangements inside the crystal under study. As examples, we mention that

- chemical binding depends on local geometry (coordination) inside the crystal,
- magnetic bulk and surface properties depend on the crystal structure,
- isotropic electrical conductivity is connected with dense atom packing inside the crystal,
- the changed atomic environment near surfaces compared with the bulk substrate affects electronic and magnetic behavior, and
- the atomic surface geometry characterizes adsorption and reaction sites that are relevant for catalytic reactions.

In many experimental and theoretical studies, the real crystalline systems are, for the sake of simplicity, approximately described by ideal single crystals with a well-defined atomic composition and an unperturbed three-dimensional periodicity where, in addition, surfaces of the single crystals are assumed to be bulk-terminated and of unperturbed two-dimensional periodicity. With this approximation in mind, a rigorous mathematical description of all geometric parameters becomes possible and is one of the basic subjects of classical crystallography.

This chapter discusses basic elements and mathematical methods used in crystallography to evaluate geometric parameters concerning atom environments inside single crystals as well as at their surfaces. Theoretical concepts will be illustrated by example applications for further understanding. As a result of space limitations, the discussion cannot cover all aspects of the field and may, in some cases, be quite brief. Further, the selection of topics as well as their presentation is, to some degree, determined by the author's personal preferences. However,

*Computational Methods in Catalysis and Materials Science*. Edited by Rutger A. van Santen and Philippe Sautet

ISBN: 978-3-527-32032-5

the interested reader is referred to the extensive crystallographic literature, see e.g. Refs. [?, ?, ?, ?], to explore additional details.

## 13.1
## Single Crystals and Bulk Lattices

The ideal single crystal is described geometrically by a periodic arrangement (lattice) of identical regions (elementary cells) each containing a set of $p$ atoms at positions $\underline{r}_i, i = 1, \ldots, p$, where the periodicity points in three independent directions along the vectors $\underline{R}_1$, $\underline{R}_2$, $\underline{R}_3$. Thus, the positions of all atoms in the crystal are given by

$$\underline{R} = \underline{r}_i + n_1 \underline{R}_1 + n_2 \underline{R}_2 + n_3 \underline{R}_3, \quad i = 1, \ldots, p, \tag{13.1}$$

where $n_1, n_2, n_3$ are coefficients of any integer value. This will be expressed in the following by a shorthand notation of the **crystal lattice**

$$L = \{\underline{R}_1, \underline{R}_2, \underline{R}_3; \underline{r}_1, \ldots, \underline{r}_p\}, \tag{13.2}$$

where the three lattice vectors $\underline{R}_1$, $\underline{R}_2$, $\underline{R}_3$ describe the lattice periodicity and the lattice basis vectors $\underline{r}_i, i = 1, \ldots, p$, denote atom positions inside the elementary cell of volume

$$V_{\mathrm{el}} = |(\underline{R}_1 \times \underline{R}_2)\underline{R}_3|. \tag{13.3}$$

For $p = 1$, the lattice contains only one atom in the elementary cell and will be termed **primitive**, while for $p > 1$ the lattice is called **non-primitive**. Single crystals of many metals, such as Cu, Fe, Au, form primitive lattices, while oxides or other compound materials crystallize in nonprimitive lattices. As an example, Figure 13.1 shows a section of the $YBa_2Cu_3O_7$ lattice where $\underline{R}_1$, $\underline{R}_2$, $\underline{R}_3$ refer to mutually perpendicular lattice vectors (forming a tetragonal lattice) and the building

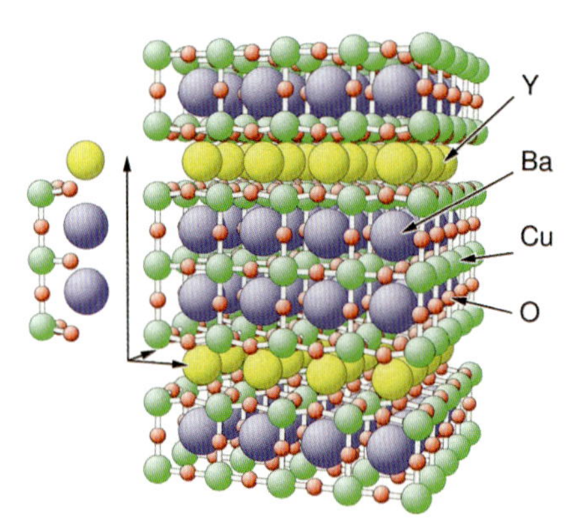


**Figure 13.1** Crystal lattice of tetragonal $YBa_2Cu_3O_7$. The atoms are shown as colored balls and labeled accordingly. In addition, the building unit (elementary cell) of 13 atoms and lattice vectors are included to the left.

block contains $p = 13$ atoms (one Y, two Ba, three Cu, and seven O atoms). This material has become famous because of its high-temperature superconducting property.

Lattice vectors $\underline{R}_1, \underline{R}_2, \underline{R}_3$ can be given in different ways where the choice depends on the type of application. While for numerical calculations it may be preferable to define $\underline{R}_1, \underline{R}_2, \underline{R}_3$ with respect to an absolute Cartesian coordinate system as

$$\underline{R}_i = (x_i, y_i, z_i), \quad i = 1,\ 2,\ 3, \tag{13.4}$$

it is common in the crystallographic literature to define these vectors by their lengths (lattice constants) $a$, $b$, $c$ and by their mutual angles $\alpha, \beta, \gamma$ where

$$a = |\underline{R}_1|,\ b = |\underline{R}_2|,\ c = |\underline{R}_3| \tag{13.5a}$$

$$\underline{R}_1\underline{R}_2 = ab\cos(\gamma),\ \underline{R}_1\underline{R}_3 = ac\cos(\beta),\ \underline{R}_2\underline{R}_3 = bc\cos(\alpha). \tag{13.5b}$$

As examples, Table 13.1 lists parameters of the seven crystal systems from which the three-dimensional Bravais lattices (primitive lattices by definition) are derived as will be discussed later.

The six parameters, lattice constants $a, b, c$, and angles $\alpha, \beta, \gamma$, defining the lattice periodicity can be used to build lattice vectors in an absolute Cartesian coordinate system. One example representation is

$$\begin{aligned} \underline{R}_1 &= a(1,0,0), \underline{R}_2 = b(\cos(\gamma), \sin(\gamma), 0), \\ \underline{R}_3 &= c(\cos(\beta), (\cos(\alpha) - \cos(\beta)\cos(\gamma))/\sin(\gamma), v_3/\sin(\gamma)) \\ \text{with } v_3 &= \{(\cos(\beta-\gamma) - \cos(\alpha))(\cos(\alpha) - \cos(\beta+\gamma))\}^{1/2}. \end{aligned} \tag{13.6}$$

**Table 13.1** List of the seven crystal systems with their Bravais lattice members and relationships of the six lattice parameters, lattice constants $a$, $b$, $c$, and angles $\alpha$, $\beta$, $\gamma$[a].

| Crystal system | Bravais lattices | Relationships |
|---|---|---|
| Triclinic | 1 (–P) | $a \neq b \neq c,\ \alpha \neq 90^\circ,\ \beta \neq 90^\circ,\ \gamma \neq 90^\circ$ |
| Monoclinic | 2 (–P, –C) | $a \neq b \neq c,\ \alpha = 90^\circ,\ \beta \neq 90^\circ,\ \gamma = 90^\circ$ |
| Orthorhombic | 4 (–P, –C, –I, –F) | $a \neq b \neq c,\ \alpha = \beta = \gamma = 90^\circ$ |
| Tetragonal | 2 (–P, –I) | $a = b \neq c,\ \alpha = \beta = \gamma = 90^\circ$ |
| Hexagonal | 1 (–P) | $a = b \neq c,\ \alpha = \beta = 90^\circ,\ \gamma = 60^\circ, 120^\circ$ |
| Trigonal, rhombohedral | 1 (–R) | $a = b = c,\ \alpha = \beta = \gamma \neq 90^\circ$ |
| Cubic | 3 (–P, –I, –F) | $a = b = c,\ \alpha = \beta = \gamma = 90^\circ$ |

[a] For each crystal system, the number of distinct Bravais lattices with corresponding crystallographic labels (in parentheses) is given.

Since the lattice vectors $\underline{R}_1, \underline{R}_2, \underline{R}_3$ are linearly independent, all lattice basis vectors $\underline{r}_i$ inside the elementary cell can be written as linear combinations

$$\underline{r}_i = x_i\underline{R}_1 + y_i\underline{R}_2 + z_i\underline{R}_3, \quad i = 1, \ldots, p, \tag{13.7}$$

where $x_i, y_i, z_i$ are real-valued coefficients with $x_i < 1, y_i < 1, z_i < 1$. The notation of "relative coordinates" $x_i, y_i, z_i$ used to describe non-equivalent atoms inside the elementary cell is common practice in the crystallographic literature. It must be emphasized that these coefficients are in general not connected with the Cartesian coordinate system but with that of the lattice vectors.

In the following, we mention two important theorems concerning relationships between primitive and non-primitive lattices, which are quite useful for practical purposes. First, non-primitive lattices can always be decomposed into sets of primitive sublattices of the same periodicity. This may be written as

$$\boldsymbol{L} = \{\underline{R}_1, \underline{R}_2, \underline{R}_3; \underline{r}_1, \ldots, \underline{r}_p\} = \Sigma \boldsymbol{L}_i, \quad \boldsymbol{L}_i = \{\underline{R}_1, \underline{R}_2, \underline{R}_3; \underline{r}_i\}, \tag{13.8}$$

where $\boldsymbol{L}_i$ denotes a primitive lattice of periodicity $\underline{R}_1, \underline{R}_2, \underline{R}_3$ originating at $\underline{r}_i$. As an example, we mention the $YBa_2Cu_3O_7$ crystal lattice shown in Figure 13.1, which can be decomposed into 13 primitive tetragonal sublattices, one Y, two Ba, three Cu, and seven O sublattices.

Second, primitive (and non-primitive) lattices can be represented by alternative non-primitive lattices of different periodicity and consequently different numbers of atoms inside the alternative elementary cell. This feature can simplify both conceptual thinking and computations in practical applications. The initial lattice description $\boldsymbol{L} = \{\underline{R}_1, \underline{R}_2, \underline{R}_3; \underline{r}_1, \ldots, \underline{r}_p\}$ is equivalent to an alternative description ("superlattice description") $\boldsymbol{L}' = \{\underline{R}'_1, \underline{R}'_2, \underline{R}'_3; \underline{r}'_1, \ldots, \underline{r}'_{p'}\}$ if the alternative lattice vectors $\underline{R}'_1, \underline{R}'_2, \underline{R}'_3$ are integer-valued linear combinations of the initial lattice vectors $\underline{R}_1, \underline{R}_2, \underline{R}_3$ and if the set of lattice basis vectors $\underline{r}'_i$ is completed according to the change of the elementary cell. Examples are the three cubic Bravais lattices sketched by finite sections in Figure 13.2. The primitive simple cubic (sc, cubic-P) lattice is defined by orthogonal lattice vectors $\underline{R}_1, \underline{R}_2, \underline{R}_3$ of identical lengths

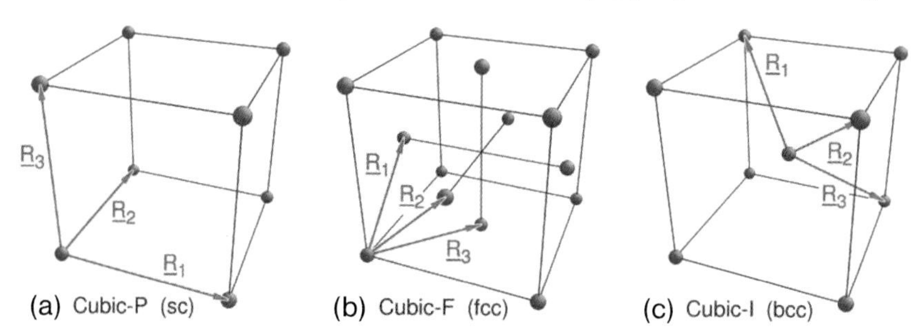


**Figure 13.2** Sections of cubic lattices with corresponding lattice vectors $\underline{R}_1, \underline{R}_2, \underline{R}_3$, cf. Table 13.2. (a) Simple cubic (cubic-P, sc) lattice, (b) face-centered cubic (cubic-F, fcc) lattice, (c) body-centered cubic (cubic-I, bcc) lattice. Lattice atoms are shown as shaded balls and the simple cubic framework is sketched by connecting sticks.

**Table 13.2** Cubic Bravais lattices with generic Bravais and simple cubic description.

| Crystal lattice | Generic Bravais $L = \{\underline{R}_1, \underline{R}_2, \underline{R}_3; \underline{r}_1\}$ | Simple cubic $L = \{\underline{R}_1, \underline{R}_2, \underline{R}_3; \underline{r}_1, \ldots, \underline{r}_p\}$ |
|---|---|---|
| Simple cubic (cubic-P) | $\underline{R}_1 = a\,(1,0,0)$<br>$\underline{R}_2 = a\,(0,1,0)$<br>$\underline{R}_3 = a\,(0,0,1);\; p = 1$<br>$\underline{r}_1 = a\,(0,0,0)$ | See generic Bravais |
| Face-centered cubic (cubic-F) | $\underline{R}_1 = a/2\,(0,1,1)$<br>$\underline{R}_2 = a/2\,(1,0,1)$<br>$\underline{R}_3 = a/2\,(1,1,0);\; p = 1$<br>$\underline{r}_1 = a\,(0,0,0)$ | $\underline{R}_1 = a\,(1,0,0)$<br>$\underline{R}_2 = a\,(0,1,0)$<br>$\underline{R}_3 = a\,(0,0,1);\; p = 4$<br>$\underline{r}_1 = a\,(0,0,0),\, \underline{r}_2 = a/2\,(0,1,1)$<br>$\underline{r}_3 = a/2\,(1,0,1),\, \underline{r}_4 = a/2\,(1,1,0)$ |
| Body-centered cubic (cubic-I) | $\underline{R}_1 = a/2\,(-1,1,1)$<br>$\underline{R}_2 = a/2\,(1,-1,1)$<br>$\underline{R}_3 = a/2\,(1,1,-1);\; p = 1$<br>$\underline{r}_1 = a\,(0,0,0)$ | $\underline{R}_1 = a\,(1,0,0)$<br>$\underline{R}_2 = a\,(0,1,0)$<br>$\underline{R}_3 = a\,(0,0,1);\; p = 2$<br>$\underline{r}_1 = a\,(0,0,0),\, \underline{r}_2 = a/2\,(1,1,1)$ |

$|\underline{R}_i| = a$ and as such of particular simplicity. Therefore, scientists prefer to use these lattice vectors also to describe the other primitive cubic lattices, face-centered cubic (fcc, cubic-F) and body-centered cubic (bcc, cubic-I) at the expense of having to deal with non-primitive lattices. Table 13.2 shows lattice vectors and lattice basis vectors of the three cubic lattices given as generic Bravais and simple cubic descriptions.

Based on similar arguments as for cubic lattices, scientists prefer to think of trigonal (rhombohedral) lattices as being hexagonal, which results in lower lattice symmetry with a unit cell of triple size, however, with a seemingly more obvious periodicity. The correspondence between the triclinic lattice and the hexagonal lattice

$$L^{\text{tri}} = \{\underline{R}_1, \underline{R}_2, \underline{R}_3; \underline{r}_1, \ldots, \underline{r}_p\} \Longrightarrow L^{\text{hex}} = \{\underline{R}'_1, \underline{R}'_2, \underline{R}'_3; \underline{r}'_1, \ldots, \underline{r}'_{p'}\} \tag{13.9}$$

can be obtained if one sets

$$\begin{aligned} &\underline{R}'_1 = \underline{R}_2 - \underline{R}_1, && p' = 3p: \quad \underline{r}'_1 = \underline{r}_1, \ldots, \underline{r}'_p = \underline{r}_p, \\ &\underline{R}'_2 = \underline{R}_3 - \underline{R}_1, && \underline{r}'_{p+1} = \underline{r}_1 + \underline{R}_1, \ldots, \underline{r}'_{2p} = \underline{r}_p + \underline{R}_1, \\ &\underline{R}'_3 = \underline{R}_1 + \underline{R}_2 + \underline{R}_3, && \underline{r}'_{2p+1} = \underline{r}_1 + 2\underline{R}_1, \ldots, \underline{r}'_{3p} = \underline{r}_p + 2\underline{R}_1. \end{aligned} \tag{13.9a}$$

The concept of describing a given lattice by using an artificially increased elementary cell is applied quite frequently in theoretical supercell studies dealing, for example, with imperfections (approximated by periodic imperfection arrangements with large separations) inside crystals or with local lattice distortions (frozen phonon calculations). These subjects will not be discussed further.

A wide area of crystallography concerns the classification of all possible types of crystal lattices based on their symmetry behavior. This subject will be treated only very briefly and the reader is referred to the literature [?, ?, ?, ?, ?, ?] for further details. Based on its original definition, every lattice has translational symmetry along the directions of its vectors $\underline{R} = n_1\underline{R}_1 + n_2\underline{R}_2 + n_3\underline{R}_3$ ($n_i$ integer-valued) where a translation

$$\underline{r} \rightarrow \underline{r}' = \underline{r} + \underline{R} \tag{13.10}$$

reproduces the lattice. In addition, lattices may exhibit point symmetry with respect to given lattice points $\underline{r}_o$, i.e., symmetry operations

$$\underline{r} \rightarrow \underline{r}' = \underline{r}_o + \underline{\underline{P}}\,(\underline{r} - \underline{r}_o) \tag{13.11}$$

reproduce the lattice where $\underline{\underline{P}}$ refers to any of four types of operations,

- inversion with respect to $\underline{r}_o$,
- rotation by an angle $\varphi$ about an axis along $\underline{e}$ through $\underline{r}_o$,
- mirror operation with respect to a plane of normal vector $\underline{e}$ through $\underline{r}_o$,
- rotoreflection (equal to a rotation followed by mirror operation) by an angle $\varphi$ about an axis along $\underline{e}$ through $\underline{r}_o$ (where $\underline{e}$ coincides with the normal vector of the mirror plane through $\underline{r}_o$).

Translational symmetry and point symmetry elements of a lattice are subject to compatibility constraints, which limit the number of possible point symmetry operations and determine the different types of lattices. Using group theoretical methods, it can be shown that each primitive three-dimensional lattice can be characterized by one of 14 different types of lattices, the three-dimensional **Bravais lattices** listed in Table 13.1 and sketched in Figure 13.3. Further, adding lattice basis vectors to all Bravais lattices to yield nonprimitive lattices and applying all possible point symmetry operations results in altogether 230 different three-dimensional **space groups** describing all lattice types [?, ?]. The mathematical proof of this result is one of the longest ever published in the mathematical literature, see [?] and references therein.

One very important finding of lattice theory is that the mathematical representation of a lattice is not unique. A two-dimensional example for this is given by the square lattice in Figure 13.4, where both $\boldsymbol{L} = \{\underline{R}_1, \underline{R}_2\}$ and $\boldsymbol{L}' = \{\underline{R}_1', \underline{R}_2'\}$ give a mathematically correct description of the lattice with, however, quite different lattice vectors. The non-uniqueness adds some ambiguity to the formal description of lattices. However, in practical cases it can be used to adapt the lattice description to other geometric constraints based on chemical or physical considerations. As an example, we mention lattice descriptions adapted to ideal single crystal surfaces, which will be discussed in greater detail below.

There are different possibilities of alternative mathematical descriptions of a given lattice. Here we focus on transformations of the lattice vectors $\underline{R}_1$, $\underline{R}_2$, $\underline{R}_3$, which may, however, also result in changed lattice basis vectors $\underline{r}_1, \ldots, \underline{r}_p$. A lattice vector transformation

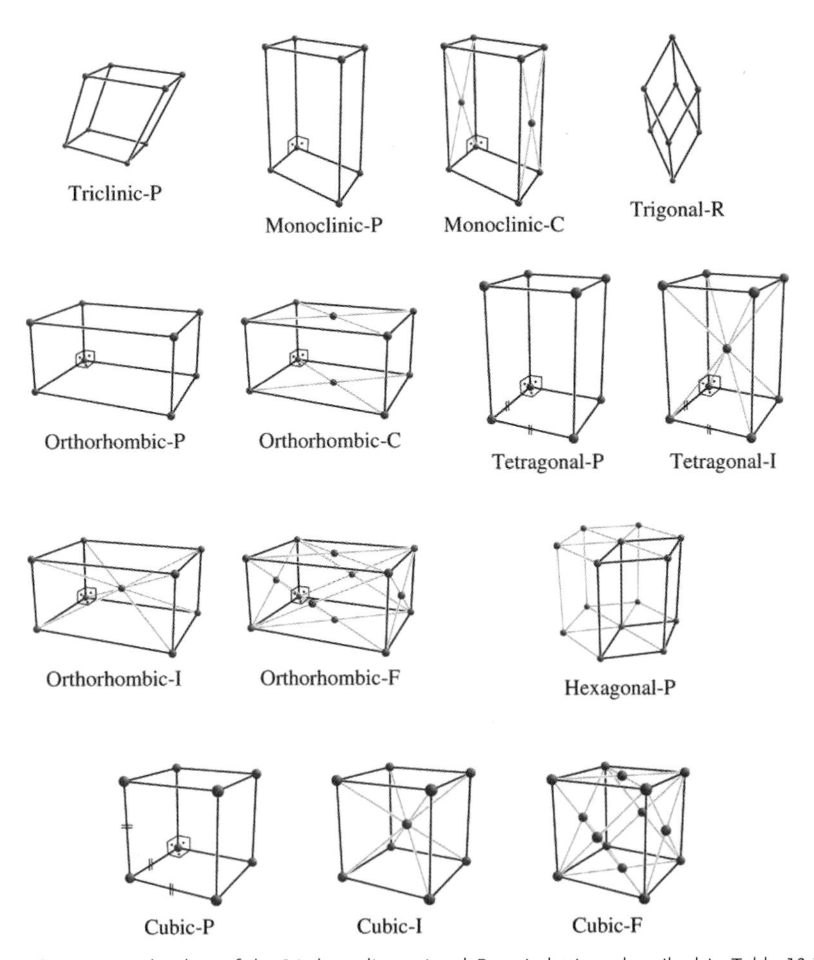


**Figure 13.3** Sketches of the 14 three-dimensional Bravais lattices described in Table 13.1.

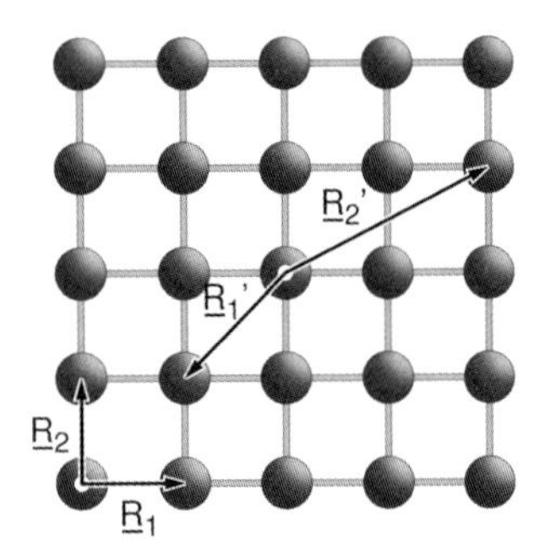


**Figure 13.4** Equivalent mathematical descriptions of the two-dimensional square lattice by lattice vectors $\underline{R}_1$, $\underline{R}_2$ and $\underline{R}'_1$, $\underline{R}'_2$, respectively.

$$\{\underline{R}_{o1}, \underline{R}_{o2}, \underline{R}_{o3}\} \xrightarrow{\text{T}} \{\underline{R}_1, \underline{R}_2, \underline{R}_3\} \tag{13.12}$$

results in an equivalent lattice description if each lattice position is given uniquely by both the descriptions, i.e., if

$$\underline{R} = n_{o1}\underline{R}_{o1} + n_{o2}\underline{R}_{o2} + n_{o3}\underline{R}_{o3} = n_1\underline{R}_1 + n_2\underline{R}_2 + n_3\underline{R}_3 \tag{13.13}$$

for all integer-valued $n_{o1}, n_{o2}, n_{o3}$ and corresponding integer-valued $n_1, n_2, n_3$. This means in particular that there is a linear transformation between the two sets of lattice vectors with $(3 \times 3)$ transformation matrices $\underline{\underline{T}} = (t_{ij})$ and $\underline{\underline{T}}' = (t_{ij}')$ that are both integer-valued, i.e.,

$$\begin{pmatrix} \underline{R}_1 \\ \underline{R}_2 \\ \underline{R}_3 \end{pmatrix} = \begin{pmatrix} t_{11} & t_{12} & t_{13} \\ t_{21} & t_{22} & t_{23} \\ t_{31} & t_{32} & t_{33} \end{pmatrix} \cdot \begin{pmatrix} \underline{R}_{o1} \\ \underline{R}_{o2} \\ \underline{R}_{o3} \end{pmatrix} = \underline{\underline{T}} \cdot \begin{pmatrix} \underline{R}_{o1} \\ \underline{R}_{o2} \\ \underline{R}_{o3} \end{pmatrix} \tag{13.14a}$$

and

$$\begin{pmatrix} \underline{R}_{o1} \\ \underline{R}_{o2} \\ \underline{R}_{o3} \end{pmatrix} = \begin{pmatrix} t'_{11} & t'_{12} & t'_{13} \\ t'_{21} & t'_{22} & t'_{23} \\ t'_{31} & t'_{32} & t'_{33} \end{pmatrix} \cdot \begin{pmatrix} \underline{R}_1 \\ \underline{R}_2 \\ \underline{R}_3 \end{pmatrix} = \underline{\underline{T}}' \cdot \begin{pmatrix} \underline{R}_1 \\ \underline{R}_2 \\ \underline{R}_3 \end{pmatrix}. \tag{13.14b}$$

Combining Eqs. (13.14a) and (13.14b) we obtain

$$\underline{\underline{T}}' \cdot \underline{\underline{T}} = \underline{\underline{T}} \cdot \underline{\underline{T}}' = \underline{\underline{1}} = \begin{pmatrix} 1 & 0 & 0 \\ 0 & 1 & 0 \\ 0 & 0 & 1 \end{pmatrix}, \quad \underline{\underline{T}}' = \underline{\underline{T}}^{-1}. \tag{13.15}$$

Thus, the two transformation matrices $\underline{\underline{T}}$ and $\underline{\underline{T}}'$ are inverse with respect to each other. This means for the determinants of these matrices

$$\det(\underline{\underline{T}} \cdot \underline{\underline{T}}') = \det(\underline{\underline{T}})\det(\underline{\underline{T}}') = \det(\underline{\underline{1}}) = 1 \rightarrow \det(\underline{\underline{T}}') = 1/\det(\underline{\underline{T}}) \tag{13.15a}$$

and since the determinant values must both be nonzero and integer, i.e., $|\det(\underline{\underline{T}}')| \geq 1$, one obtains

$$\det(\underline{\underline{T}}) = \det(\underline{\underline{T}}') = \pm 1. \tag{13.16}$$

Here the value $-1$ can be ignored since it affects only the handedness between the two lattice vector sets. Since according to Eq. (13.3) and (13.16)

$$V_{el} = |(\underline{R}_{o1} \times \underline{R}_{o2})\underline{R}_{o3}| = |\det(\underline{\underline{T}}')(\underline{R}_1 \times \underline{R}_2)\,\underline{R}_3| = |(\underline{R}_1 \times \underline{R}_2)\,\underline{R}_3| = V'_{el}, \tag{13.17}$$

the elementary cell volume $V_{el}$ will not be changed by the lattice vector transformation.

## 13.2 Netplanes, Miller Indices

The concept of netplanes in crystals is a central issue in many theoretical studies of physical/chemical effects. As examples, we mention the detailed atomic structure at single crystal surfaces or electron or photon diffraction from crystals. Starting from an initial three-dimensional lattice $\boldsymbol{L}_o = \{\underline{R}_{o1}, \underline{R}_{o2}, \underline{R}_{o3}; \underline{r}_{o1}, \ldots, \underline{r}_{op}\}$ and any equivalent description $\boldsymbol{L} = \{\underline{R}_1, \underline{R}_2, \underline{R}_3; \underline{r}_1, \ldots, \underline{r}_p\}$ according to Eq. (13.12), one can define an infinitely large set of two-dimensional sublattices given each by lattice points

$$\underline{R} = (\underline{r}_i + n_3\underline{R}_3) + n_1\underline{R}_1 + n_2\underline{R}_2, \tag{13.18}$$

where $\underline{r}_i$ and $n_3$ are kept fixed and can be used as labels of the sublattices, while $n_1$ and $n_2$ can assume any integer value. These two-dimensional sublattices are usually called **netplanes**. Obviously, the netplanes $\boldsymbol{N} = \{\underline{R}_1, \underline{R}_2 \mid \underline{R}_3; \underline{r}_i\}$ exhibit the same two-dimensional periodicity described by lattice vectors $\underline{R}_1$, $\underline{R}_2$ and different netplanes are always parallel to each other. Further, the complete set of all netplanes $\boldsymbol{N}$ spans the three-dimensional lattice $\boldsymbol{L}$. While netplanes $\boldsymbol{N} = \{\underline{R}_1, \underline{R}_2 \mid \underline{R}_3; \underline{r}_i\}$ originating from the same atom type at position $\underline{r}_i$ inside the elementary cell are shifted with respect to each other by multiples of lattice vector $\underline{R}_3$, netplanes of different $\underline{r}_i$ may lie on the same plane giving rise to primitive and non-primitive two-dimensional sublattices depending on the choice of the lattice vectors. This can have important physical and chemical consequences as will be discussed below.

The netplane definition $\boldsymbol{N} = \{\underline{R}_1, \underline{R}_2 \mid \underline{R}_3; \underline{r}_i\}$ given in this chapter is based on transformed lattice vectors $\underline{R}_1$, $\underline{R}_2$, $\underline{R}_3$ where the two vectors $\underline{R}_1$, $\underline{R}_2$ determine the periodicity of each netplane. Thus, the corresponding lattice description $\boldsymbol{L} = \{\underline{R}_1, \underline{R}_2, \underline{R}_3; \underline{r}_1, \ldots, \underline{r}_q\}$, connected with a transformation $\underline{\underline{T}}$ given by Eq. (13.14a), can be considered as adapted for a given set of netplanes. As a result, transformation $\underline{\underline{T}}$ may be used to characterize the netplanes and vectors $\underline{R}_1$, $\underline{R}_2$, $\underline{R}_3$ will be called netplane-adapted lattice vectors. The normal direction of the netplanes $\boldsymbol{N} = \{\underline{R}_1, \underline{R}_2 \mid \underline{R}_3; \underline{r}_i\}$, which can be viewed as the stacking direction inside the crystal lattice, is given by a vector $\underline{n}$ where, using transformation $\underline{\underline{T}}$ from Eq. (13.14a)

$$\begin{aligned}\underline{n} &= \alpha\,(\underline{R}_1 \times \underline{R}_2) = \alpha \sum_{i=1}^{3}\sum_{j=1}^{3} t_{1i}\,t_{2j}\,(\underline{R}_{oi} \times \underline{R}_{oj}) \\ &= \alpha\,\{h\,(\underline{R}_{o2} \times \underline{R}_{o3}) + k\,(\underline{R}_{o3} \times \underline{R}_{o1}) + l\,(\underline{R}_{o1} \times \underline{R}_{o2})\}\end{aligned} \tag{13.19}$$

with coefficients

$$h = t_{12}\,t_{23} - t_{13}\,t_{22},\; k = t_{13}\,t_{21} - t_{11}\,t_{23},\; l = t_{11}\,t_{22} - t_{12}\,t_{21}, \tag{13.20}$$

where $\alpha$ is a normalization constant to guarantee that $|\underline{n}| = 1$. Since all elements of the transformation matrix $\underline{\underline{T}}$ are integers, the coefficients $h, k, l$, which are commonly named **Miller indices**, must also be integer-valued. This means in particular that according to Eq. (13.19), normal directions of netplanes in a lattice are always discrete and Miller indices $(h\ k\ l)$ can be used to characterize sets of netplanes for a given direction. In this spirit, transformation matrices $\underline{\underline{T}}$, which are connected with a netplane stacking direction, will be labeled as $\underline{\underline{T}}^{(hkl)}$ in the following. The close relationship between transformation matrices $\underline{\underline{T}}^{(hkl)}$ and Miller indices $(h\ k\ l)$ is also obvious in the $(3 \times 3)$ set of linear equations

$$\underline{\underline{T}}^{(hkl)} \cdot \begin{pmatrix} h \\ k \\ l \end{pmatrix} = \begin{pmatrix} t_{11} & t_{12} & t_{13} \\ t_{21} & t_{22} & t_{23} \\ t_{31} & t_{32} & t_{33} \end{pmatrix} \cdot \begin{pmatrix} h \\ k \\ l \end{pmatrix} = \begin{pmatrix} 0 \\ 0 \\ 1 \end{pmatrix}, \tag{13.21}$$

which is obtained by using Eq. (13.20) together with Eq. (13.16). This shows that for any transformation matrix $\underline{\underline{T}}^{(hkl)}$, corresponding Miller indices $(h\ k\ l)$ result from solving linear equations (13.21). In fact, the solutions are already given in Eq. (13.20). On the other hand, for given $(h\ k\ l)$ values example transformations $\underline{\underline{T}}^{(hkl)}$ can be evaluated from Eq. (13.21) using number theoretical methods as has been shown elsewhere [?].

Netplane normal vectors as given by Eq. (13.19) can be thought of as belonging to a primitive lattice with lattice vectors arising – up to a global scaling factor – from vector products of the initial lattice vectors. This defines the so-called reciprocal lattice $\mathbf{G} = \{\underline{G}_{o1}, \underline{G}_{o2}, \underline{G}_{o3}\}$ where

$$\underline{G}_{o1} = \beta(\underline{R}_{o2} \times \underline{R}_{o3}),\ \underline{G}_{o2} = \beta(\underline{R}_{o3} \times \underline{R}_{o1}),\ \underline{G}_{o3} = \beta(\underline{R}_{o1} \times \underline{R}_{o2}) \qquad \beta = 2\pi / V_{el}, \tag{13.22}$$

with scaling factor $\beta$ containing the volume $V_{el}$ of the elementary cell of the initial lattice as defined in Eq. (13.3). Reciprocal lattice vectors $\underline{G}_{oi}$ are of dimension [inverse length] and fulfill orthogonality relations

$$\underline{G}_{oi}\,\underline{R}_{oi} = 2\pi \quad \text{for} \quad i = 1,\ 2,\ 3;\ \underline{G}_{oi}\,\underline{R}_{oj} = 0 \quad \text{for} \quad i \neq j, \tag{13.23}$$

which is obvious from Eq. (13.22). By definition the reciprocal lattice $\mathbf{G}$ is a primitive lattice and concerns only the initial lattice vectors $\underline{R}_{o1}, \underline{R}_{o2}, \underline{R}_{o3}$ of the real lattice $\mathbf{L}$. Therefore, different non-primitive lattices of the same periodicity have the same reciprocal lattice. According to Eqs. (13.19) and (13.22), the normal direction $\underline{n}$ of a $(h\ k\ l)$ indexed netplane points along vector

$$\underline{G}_{(hkl)} = 2\pi / V_{el} \cdot (\underline{R}_1 \times \underline{R}_2) = h\,\underline{G}_{o1} + k\,\underline{G}_{o2} + l\,\underline{G}_{o3} \tag{13.24}$$

of the reciprocal lattice. This vector is quite useful in describing numerous properties of netplanes. As an example, we mention that the distance $d_{(hkl)}$ between two adjacent translationally equivalent $(h\ k\ l)$ netplanes, which are connected by a lattice vector $\underline{R}_3$. It can, according to Eqs. (13.17) and (13.24), be written as

$$d_{(hkl)} = \underline{R}_3\,\underline{n} = \underline{R}_3\,\underline{G}_{(hkl)}/|\underline{G}_{(hkl)}| = \underline{R}_3\,2\pi/V_{el} \cdot (\underline{R}_1 \times \underline{R}_2)/G_{(hkl)} = 2\pi/G_{(hkl)}. \tag{13.25}$$

Thus, if the length of $\underline{G}_{(hkl)}$, determined by the size of the Miller indices, gets large, the distance between translationally equivalent netplanes gets small. Netplanes belonging to large Miller indices lie close together. Further, the average atom density $\rho_{(hkl)}$ of a $(h\ k\ l)$ netplane with one atom type, according to Eqs. (13.24) and (13.25), is given by

$$\rho_{(hkl)} = 1/|\underline{R}_1 \times \underline{R}_2| = (2\pi/V_{el})/G_{(hkl)} = d_{(hkl)}/V_{el}. \tag{13.26}$$

Thus, for large vectors $\underline{G}_{hkl}$ the atom density of corresponding netplanes gets small. Netplanes belonging to small Miller indices are the densest planes.

Miller indices $(h\ k\ l)$ are by definition based on reciprocal lattice vectors $\underline{G}_{o1}$, $\underline{G}_{o2}$, and $\underline{G}_{o3}$ as given by Eq. (13.22) for the corresponding Bravais lattice (Bravais–Miller indices). In the case of cubic lattices, scientists often prefer to use real space and reciprocal lattice vectors of the simple cubic (sc, cubic-P) lattice due to their geometric simplicity even in studies on face-centered cubic (fcc, cubic-F) and body-centered cubic (bcc, cubic-I) lattices. This choice also affects Miller index values. According to Table 13.2 and Eq. (13.22), the simple cubic lattice is described by real space and reciprocal lattice vectors (in Cartesian coordinates)

$$\begin{aligned}
\underline{R}^{sc}_{o1} &= a\,(1,\ 0,\ 0) & \underline{G}^{sc}_{o1} &= 2\pi/a\,(1,\ 0,\ 0)\\
\underline{R}^{sc}_{o2} &= a\,(0,\ 1,\ 0) & \underline{G}^{sc}_{o2} &= 2\pi/a\,(0,\ 1,\ 0)\\
\underline{R}^{sc}_{o3} &= a\,(0,\ 0,\ 1) & \underline{G}^{sc}_{o3} &= 2\pi/a\,(0,\ 0,\ 1),
\end{aligned} \tag{13.27}$$

while the face-centered cubic lattice in Bravais notation, see Table 13.2 and Eq. (13.22), is given by

$$\begin{aligned}
\underline{R}^{fcc}_{o1} &= a/2\,(0,1,1) & \underline{G}^{fcc}_{o1} &= 2\pi/a\,(-1,1,1) = -\underline{G}^{sc}_{o1} + \underline{G}^{sc}_{o2} + \underline{G}^{sc}_{o3}\\
\underline{R}^{fcc}_{o2} &= a/2\,(1,0,1) & \underline{G}^{fcc}_{o2} &= 2\pi/a\,(1,-1,1) = \underline{G}^{sc}_{o1} - \underline{G}^{sc}_{o2} + \underline{G}^{sc}_{o3}\\
\underline{R}^{fcc}_{o3} &= a/2\,(1,1,0) & \underline{G}^{fcc}_{o3} &= 2\pi/a\,(1,1,-1) = \underline{G}^{sc}_{o1} + \underline{G}^{sc}_{o2} - \underline{G}^{sc}_{o3}.
\end{aligned} \tag{13.28}$$

As a consequence, netplane normal directions point along vectors

$$\begin{aligned}
\underline{G}_{hkl} &= h^{fcc}\,\underline{G}^{fcc}_{o1} + k^{fcc}\,\underline{G}^{fcc}_{o2} + l^{fcc}\,\underline{G}^{fcc}_{o3}\\
&= (-h^{fcc}+k^{fcc}+l^{fcc})\,\underline{G}^{sc}_{o1} + (h^{fcc} - k^{fcc} + l^{fcc})\,\underline{G}^{sc}_{o2} + (h^{fcc}+k^{fcc} - l^{fcc})\,\underline{G}^{sc}_{o3}\\
&= h^{sc}\,\underline{G}^{sc}_{o1} + k^{sc}\,\underline{G}^{sc}_{o2} + l^{sc}\,\underline{G}^{sc}_{o3},
\end{aligned}$$

suggesting, in addition to the Bravais notation $(h^{fcc}\ k^{fcc}\ l^{fcc})$, a simple cubic notation $(h^{sc}\ k^{sc}\ l^{sc})$ for Miller indices of the face-centered cubic lattice where there is a linear transformation between the indices

$$\begin{pmatrix} h^{sc}\\ k^{sc}\\ l^{sc}\end{pmatrix} = \begin{pmatrix} -1 & 1 & 1\\ 1 & -1 & 1\\ 1 & 1 & -1\end{pmatrix} \cdot \begin{pmatrix} h^{fcc}\\ k^{fcc}\\ l^{fcc}\end{pmatrix}, \tag{13.29a}$$

$$\begin{pmatrix} h^{fcc}\\ k^{fcc}\\ l^{fcc}\end{pmatrix} = \frac{1}{2}\begin{pmatrix} 0 & 1 & 1\\ 1 & 0 & 1\\ 1 & 1 & 0\end{pmatrix} \cdot \begin{pmatrix} h^{sc}\\ k^{sc}\\ l^{sc}\end{pmatrix}. \tag{13.29b}$$

While the Bravais–Miller indices, $h^{fcc}, k^{fcc}, l^{fcc}$, can assume any integer values, the factor 1/2 in the transformation (13.29b) shows that corresponding Miller indices, $h^{sc}, k^{sc}$ and $l^{sc}$, of the simple cubic notation are constrained by the requirement that all three values must be either even or all odd.

According to Table 13.2 and Eq. (13.22), the body-centered cubic lattice is described in Bravais notation by real space and reciprocal lattice vectors (in Cartesian coordinates)

$$\begin{aligned} \underline{R}_{o1}^{bcc} &= a/2\,(-1,\ 1,\ 1) & \underline{G}_{o1}^{bcc} &= 2\pi/a\,(0,\ 1,\ 1) = \underline{G}_{o2}^{sc} + \underline{G}_{o3}^{sc} \\ \underline{R}_{o2}^{bcc} &= a/2\,(1,\ -1,\ 1) & \underline{G}_{o2}^{bcc} &= 2\pi/a\,(1,\ 0,\ 1) = \underline{G}_{o1}^{sc} + \underline{G}_{o3}^{sc} \\ \underline{R}_{o3}^{bcc} &= a/2\,(1,\ 1,\ -1) & \underline{G}_{o3}^{bcc} &= 2\pi/a\,(1,\ 1,\ 0) = \underline{G}_{o1}^{sc} + \underline{G}_{o2}^{sc}. \end{aligned} \tag{13.30}$$

Therefore, netplane normal directions point along vectors

$$\begin{aligned} \underline{G}_{hkl} &= h^{bcc}\,\underline{G}_{o1}^{bcc} + k^{bcc}\,\underline{G}_{o2}^{bcc} + l^{bcc}\,\underline{G}_{o3}^{bcc} \\ &= (k^{fcc} + l^{fcc})\,\underline{G}_{o1}^{sc} + (h^{fcc} + l^{fcc})\,\underline{G}_{o2}^{sc} + (h^{fcc} + k^{fcc})\,\underline{G}_{o3}^{sc} \\ &= h^{sc}\,\underline{G}_{o1}^{sc} + k^{sc}\,\underline{G}_{o2}^{sc} + l^{sc}\,\underline{G}_{o3}^{sc}, \end{aligned}$$

which suggests, in addition to the Bravais notation $(h^{bcc}\,k^{bcc}\,l^{bcc})$, a simple cubic notation $(h^{sc}\,k^{sc}\,l^{sc})$ where there is a linear transformation between the corresponding Miller indices

$$\begin{pmatrix} h^{sc} \\ k^{sc} \\ l^{sc} \end{pmatrix} = \begin{pmatrix} 0 & 1 & 1 \\ 1 & 0 & 1 \\ 1 & 1 & 0 \end{pmatrix} \cdot \begin{pmatrix} h^{bcc} \\ k^{bcc} \\ l^{bcc} \end{pmatrix}, \tag{13.31a}$$

$$\begin{pmatrix} h^{bcc} \\ k^{bcc} \\ l^{bcc} \end{pmatrix} = \frac{1}{2} \begin{pmatrix} -1 & 1 & 1 \\ 1 & -1 & 1 \\ 1 & 1 & -1 \end{pmatrix} \cdot \begin{pmatrix} h^{sc} \\ k^{sc} \\ l^{sc} \end{pmatrix}. \tag{13.31b}$$

While the Bravais–Miller indices, $h^{bcc}, k^{bcc}, l^{bcc}$, can assume any integer values, the factor 1/2 in transformation in Eq. (13.31b) shows that corresponding Miller indices, $h^{sc}, k^{sc}, l^{sc}$ of the simple cubic notation are constrained by the requirement that the sum of all three values, i. e., $g = h^{sc} + k^{sc} + l^{sc}$ must always be even.

The numerical constraints on Miller indices, $h^{sc}, k^{sc}, l^{sc}$, of simple cubic notation for face- or body-centered cubic lattices become important when Miller indices (and corresponding reciprocal lattice vectors $\underline{G}_{hkl}$) are used in numerical evaluations such as determining netplane distances $d_{hkl}$. or in decomposing Miller indices discussed later on. As an example, we mention the face-centered cubic lattice where according to Eqs. (13.25) and (13.27) netplane distances $d_{hkl}$.are given by

$$d_{hkl} = 2\pi/|\underline{G}_{hk}| = a/[(h^{sc})^2 + (k^{sc})^2 + (l^{sc})^2]^{1/2}, \tag{13.32}$$

with all three Miller indices being either even or all odd. Thus, netplanes with (1 1 2) orientation in simple cubic notation must use $h^{sc} = 2, k^{sc} = 2, l^{sc} = 4$ in the evaluation of Eq. (13.32). However, if Miller indices $(h\ k\ l)$ are used to denote only netplane orientations in the crystal, common integer factors in the indices are usually left out. In practical applications, $(h\ k\ l)$ Miller indices given in simple cubic notation for face-centered cubic lattices and being mixtures of even and odd numbers must all be multiplied by a factor of 2 ((1 2 3) → (2 4 6)) to be used in quantitative numerical evaluations. In analogy, $(h\ k\ l)$ Miller indices given in simple cubic notation for body-centered cubic lattices where $(h + k + l)$ is an odd number must all be multiplied by a factor of 2 ((1 1 1) → (2 2 2)) for numerical evaluations.

In an alternative concept of characterizing netplanes, one assumes two adjacent translationally equivalent $(h\ k\ l)$ netplanes in a primitive lattice $\mathbf{L} = \{\underline{R}_{o1}, \underline{R}_{o2}, \underline{R}_{o3}\}$, where one netplane originates at $\underline{R} = (0, 0, 0)$ of the lattice. Then the adjacent translationally equivalent netplane will in general cut the three lattice vectors at

$$\underline{A} = \alpha_1\, \underline{R}_{o1}, \quad \underline{B} = \alpha_2\, \underline{R}_{o2}, \quad \text{and} \quad \underline{C} = \alpha_3\, \underline{R}_{o3} \tag{13.33}$$

as shown in Figure 13.5.

Thus, the intercept factors $\alpha_1, \alpha_2$ and $\alpha_3$ can be used to characterize the netplane uniquely. The distance between the two adjacent netplanes with normal vector $\underline{n}$, together with Eqs. (13.19) and (13.24), is given by

$$d_{(hkl)} = \alpha_1\, \underline{R}_{o1}\, \underline{n} = \alpha_1\, \underline{R}_{o1}\, \underline{G}_{(hkl)}/|\underline{G}_{(hkl)}| = \alpha_1\, \underline{R}_{o1}\, (h\, \underline{G}_{o1} + k\, \underline{G}_{o2} + l\, \underline{G}_{o3})/G_{(hkl)}$$

and using orthogonality relations (13.23) together with (13.25)

$$d_{(hkl)} = \alpha_1\, h\, 2\pi/G_{(hkl)} = 2\pi/G_{(hkl)}$$

from which the intercept factor $\alpha_1$ follows as

$$\alpha_1\, h = 1 \quad \text{or} \quad \alpha_1 = 1/h. \tag{13.34a}$$

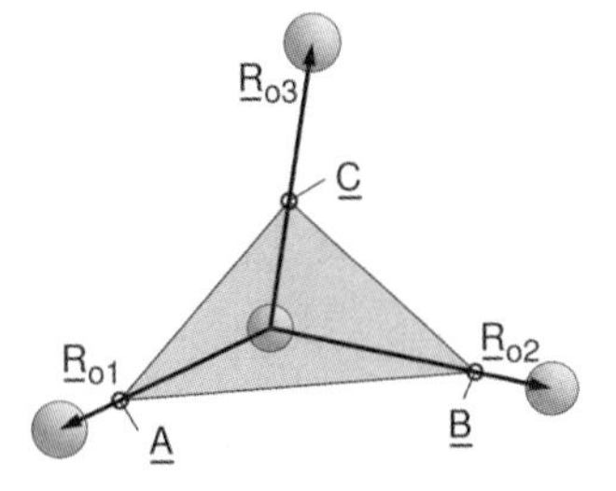


**Figure 13.5** Netplane definition by its cuts through the three lattice vectors at $\underline{A}$, $\underline{B}$, $\underline{C}$. The lattice vectors, $\underline{R}_{o1}, \underline{R}_{o2}, \underline{R}_{o3}$, are sketched accordingly. The netplane is indicated by light gray.

In analogy, relations

$$d_{(hkl)} = \alpha_2\, \underline{R}_{o2}\, \underline{n} = \alpha_2\, k\, \underline{R}_{o2}\underline{G}_{o2}/G_{(hkl)} = \alpha_3\, \underline{R}_{o3}\, \underline{n} = \alpha_3\, l\, \underline{R}_{o3}\underline{G}_{o3}/G_{(hkl)} \tag{13.34b}$$

result in

$$\alpha_2 = 1/k, \quad \alpha_3 = 1/l \tag{13.34c}$$

connecting between inverse Miller indices $h, k, l$ and the intercept factors $\alpha_i$ of the three lattice vectors cutting the netplane. Since Miller indices are all integer-valued relations, Eqs. (13.34a)–(13.34c) show that nonzero values of $h, k, l$ yield intercept factors $\alpha_i$ that are bound to $0 \leq |\alpha_i| \leq 1$. In addition, according to Eq. (13.34a) $h = 0$ can be considered a result of the limiting case $\alpha_1 \to \infty$ such that the corresponding netplanes lie parallel to the lattice vector $\underline{R}_{o1}$. Analogously, $k = 0$ and $l = 0$ refer to netplanes that lie parallel to the vectors $\underline{R}_{o2}$ and $\underline{R}_{o3}$, respectively.

In the hexagonal Bravais lattice $\boldsymbol{L}^{hex} = \{\underline{R}_{o1}, \underline{R}_{o2}, \underline{R}_{o3}\}$, lattice vectors $\underline{R}_{o1}$ and $\underline{R}_{o2}$ form a two-dimensional hexagonal lattice with angles $\angle(\underline{R}_{o1}, \underline{R}_{o2}) = 120^\circ$ (obtuse representation) or $= 60^\circ$ (acute representation), while $\underline{R}_{o3}$ is perpendicular to both $\underline{R}_{o1}$ and $\underline{R}_{o2}$. Assuming an obtuse representation, the threefold symmetry of the sublattice $\{\underline{R}_{o1}, \underline{R}_{o2}\}$ induces a third vector $\underline{R}'_{o2} = -\underline{R}_{o1} - \underline{R}_{o2}$, see Figure 13.6, which can be used as an alternative basis vector replacing $\underline{R}_{o1}$ or $\underline{R}_{o2}$. Therefore, crystallographers treat the three basis vectors $\underline{R}_{o1}$, $\underline{R}_{o2}$ and $\underline{R}'_{o2}$ on an equal footing (ignoring overcompleteness) and characterize netplanes by intercepts of the three lattice vectors $\underline{R}_{o1}, \underline{R}_{o2}, \underline{R}_{o3}$ and of vector $\underline{R}'_{o2}$, i.e., by

$$\underline{A} = \alpha_1\, \underline{R}_{o1}, \quad \underline{B} = \alpha_2\, \underline{R}_{o2}, \quad \underline{C} = \alpha_3\, \underline{R}_{o3}, \quad \text{and} \quad \underline{D} = \alpha'_2\, \underline{R}'_{o2} \tag{13.35}$$

as shown in Figure 13.6 where simple algebra yields

$$1/\alpha'_2 = -1/\alpha_1 - 1/\alpha_2. \tag{13.36}$$

This is the basis of the so-called four-index notation of the Miller indices where the initial definition

$$(h\; k\; l) = (1/\alpha_1\; 1/\alpha_2\; 1/\alpha_3)$$

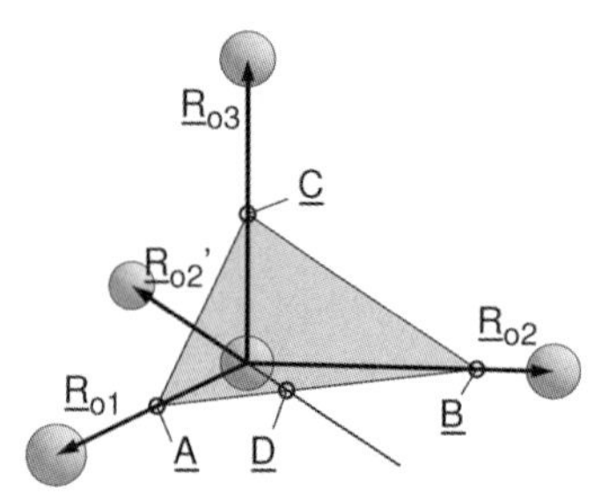


**Figure 13.6** Netplane definition by its cuts through the three lattice vectors at $\underline{A}$, $\underline{B}$, $\underline{C}$, and through $\underline{D}$ for the hexagonal lattice. The lattice vectors, $\underline{R}_{o1}$, $\underline{R}_{o2}$, $\underline{R}'_{o2}$, $\underline{R}_{o3}$, are sketched accordingly, see text. The netplane is indicated by light gray.

is, with the help of Eq. (13.36), replaced by

$$(l\ m\ n\ q) = (1/\alpha_1 \quad 1/\alpha_2 \quad 1/\alpha_2' \quad 1/\alpha_3) = (h \quad k \ -h-k \quad l). \qquad (13.37)$$

Examples for the hexagonal graphite lattice are listed in the following table.

**Table 13.3** Examples of Miller indices in 3- and 4-index notation, (h k l) and (l m n q), for hexagonal lattices.

| *(h k l)* | *(l m n q)* |
|---|---|
| (1 0 0) | (1 0 −1 0) |
| (0 1 0) | (0 1 −1 0) |
| (0 0 1) | (0 0 0 1) |
| (1 1 0) | (1 1 −2 0) |
| (1 0 1) | (1 0 −1 1) |
| (0 1 1) | (0 1 −1 1) |
| (1 1 1) | (1 1 −2 1) |

Analogous to the procedure for three-dimensional lattices, netplanes as two-dimensional lattices can be classified according to their symmetry behavior. In addition to translational symmetry, netplanes may exhibit point symmetry with respect to given points $\underline{r}_o$ on the planes allowing operations

- inversion with respect to $\underline{r}_o$,
- rotation by an angle $\varphi$ about an axis along $\underline{e}$ through $\underline{r}_o$, and
- mirror operation with respect to a plane of normal vector $\underline{e}$ through $\underline{r}_o$.

Compatibility constraints limit the number of possible point symmetry operations and confine the different types of netplanes. Using group theoretical methods, it can be shown that each primitive two-dimensional lattice can be characterized by one of five different types of lattices, the two-dimensional Bravais lattices [?, ?]

- *oblique* lattices that allow only inversion and a twofold rotation axis,
- *primitive rectangular* lattices with $|\underline{R}_1| \neq |\underline{R}_2|$ and $\underline{R}_1\underline{R}_2 = 0$ allowing inversion, a twofold rotation axis, and two mirror planes,
- *centered rectangular* lattices with $|\underline{R}_1| \neq |\underline{R}_2|$ and $\underline{R}_1\underline{R}_2 = 1/2\ R_1^2$ or $\underline{R}_1\underline{R}_2 = 1/2\ R_2^2$ allowing inversion, a twofold rotation axis, and two mirror planes,
- *square* lattices with $|\underline{R}_1| = |\underline{R}_2|$ and $\underline{R}_1\underline{R}_2 = 0$ allowing inversion, two- and fourfold rotation axes, and four mirror planes, and
- *hexagonal* lattices with $|\underline{R}_1| = |\underline{R}_2|$ and $\underline{R}_1\underline{R}_2 = \pm\ 1/2\ R_1^2$ allowing inversion, two-, three-, and sixfold rotation axes, and six mirror planes.

These netplanes are sketched in Figure 13.7.

Adding lattice basis vectors to yield non-primitive netplanes and applying all possible point symmetry operations results in altogether 17 different two-dimensional space groups, which classify all netplane types [?, ?].

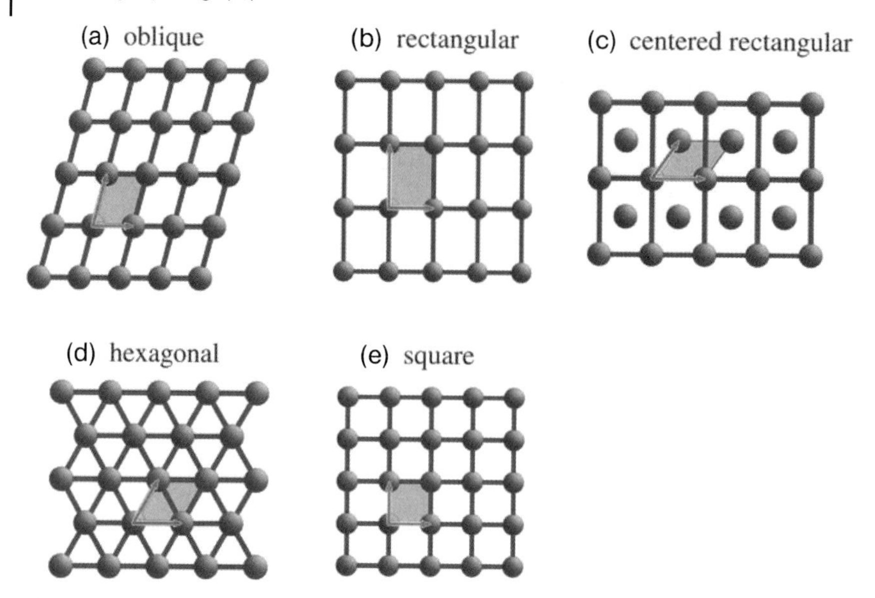


**Figure 13.7** The five different Bravais lattices: (a) oblique, (b) rectangular, (c) centered rectangular, (d) hexagonal, (e) square. Corresponding lattice vectors together with unit cells (shaded in gray) are included.

## 13.3 Ideal Single Crystal Surfaces

Ideal single crystal surfaces are defined by two-dimensionally periodic bulk truncations of three-dimensional lattices parallel to $(h\ k\ l)$ netplanes yielding the bulk lattice below and vacuum above as illustrated in Figure 13.8 for different surfaces of MgO. Therefore, one can use Miller indices $(h\ k\ l)$ to characterize the surface orientation and apply mathematical descriptions of netplanes also to ideal single crystal surfaces. Nonprimitive lattices $\boldsymbol{L} = \{\underline{R}_1, \underline{R}_2, \underline{R}_3; \underline{r}_1, \ldots, \underline{r}_p\}$ contain $p$ non-equivalent $(h\ k\ l)$ netplanes of which, depending on the actual $(h\ k\ l)$ values, some may fall on the same spatial plane yielding non-primitive sublattices. As a result, there are $q \leq p$ different terminations of corresponding ideal single crystal surfaces described by $(h\ k\ l)$. As examples, we mention

- the ***MgO* lattice** (NaCl type) shown above, which is described as a superposition of two face-centered cubic lattices (see Table 13.2), one for each element, shifted by $a/2$ (1, 1, 1) with respect to each other, yielding $p = 2$. For (1 0 0) netplanes (in simple cubic notation), one obtains $q = 1$ since $Mg^{2+}$ and $O^{2-}$ ions fall on the same plane. This results in only one possible (1 0 0) surface termination where $Mg^{2+}$ and $O^{2-}$ ions exist in equal amounts giving rise to nonpolar surfaces, see Figure 13.8. On the other hand, (1 1 1) netplanes of $Mg^{2+}$ and $O^{2-}$ ions are distinct, i.e., $q = 2$, which results in two possible (1 1 1) surface terminations,

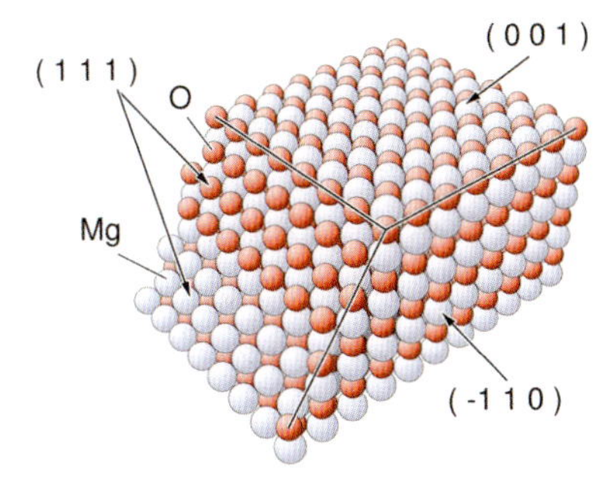


**Figure 13.8** Crystal lattice section of MgO (NaCl lattice) confined by different single crystal surfaces. The atoms are shown as colored balls and labeled accordingly. The section is enclosed by nonpolar (0 0 1), (−1 1 0) and by the polar (1 1 1) oriented surfaces.

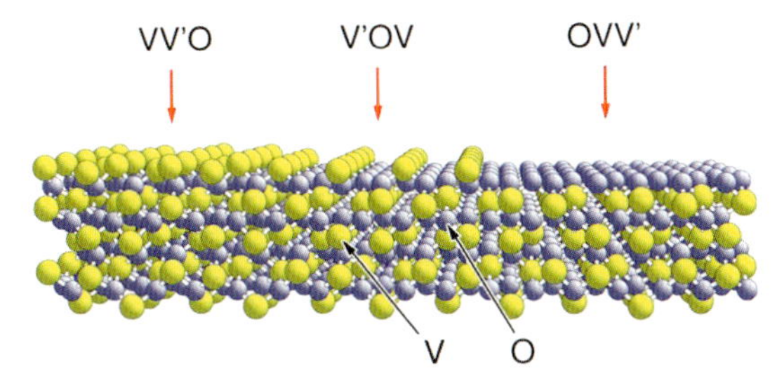


**Figure 13.9** Crystal lattice section of $V_2O_3$ (corundum lattice, trigonal-R type) lattice with differently terminated (0001) surfaces, VV′O, V′OV, OVV′. The atoms are shown as colored balls and labeled accordingly.

one with $Mg^{2+}$ and one with $O^{2-}$ ions at the top, see Figure 13.8, giving rise to highly polar surfaces that are quite difficult to prepare experimentally.

- the $\mathbf{V_2O_3}$ **lattice** (corundum, trigonal-R type [?]) shown in Figure 13.9, which contains $p = 10$ atoms (four vanadium, six oxygen) in the elementary cell. This lattice can be described along the (1 1 1) direction (corresponding to (0001) in the hexagonal four-index notation) by two sets of three different hexagonal netplanes ($q = 6$) where the two sets are equivalent by inversion symmetry. Each set contains two netplanes with $V^{3+}$ ions (one each of the elementary cell) and one with $O^{2-}$ ions (three of the elementary cell). This results in three different (1 1 1) surface terminations: the full metal termination VV′O, the half metal termination V′OV, and the oxygen termination OVV′, as shown in Figure 13.9. Here V′OV seems to be energetically favorable since it is the least polar of the three terminations [?].

The overall shape (morphology) of $(h\ k\ l)$ oriented single crystal surfaces is only partly determined by the geometric structure of corresponding $(h\ k\ l)$ netplanes. Local binding between atoms, which may involve several netplanes, will also become important. Thus, the shape of $(h\ k\ l)$ oriented surfaces is characterized in many cases by sections of densest netplanes of the crystal, forming terraces (microfacets) and being separated by steps that may straight steps or broken ("stepped steps," also called kinks). Since the $(h\ k\ l)$ orientation of these surfaces is often quite close to those of the densest netplanes, they are usually called vicinal surfaces. As an example, we mention the (10 12 16) surface of a face-centered cubic copper lattice, see Figure 13.10, which is described by (1 1 1) terraces (the (1 1 1) netplanes of the crystal are the densest) separated by periodically broken steps (kinked steps).

Vicinal surfaces with large Miller index values $(h\ k\ l)$ correspond, according to Eq. (13.26), to rather open netplanes and can be described in some cases by combinations of terraces with $(h_t\ k_t\ l_t)$ orientation separated by steps with $(h_s\ k_s\ l_s)$ orientation where the Miller index triples $(h\ k\ l)$, $(h_t\ k_t\ l_t)$, and $(h_s\ k_s\ l_s)$ are connected by an additivity theorem. Starting from an initial primitive lattice described by $\{\underline{R}_{o1}, \underline{R}_{o2}, \underline{R}_{o3}\}$ with a stepped surface as shown, for example, in Figure 13.11 (sketching the stepped face-centered cubic (3 3 5) surface), let us construct step-adapted lattice vectors $\underline{R}_1, \underline{R}_2, \underline{R}_3$ where $\underline{R}_1$ and $\underline{R}_2$ describe the periodicity of the terrace netplanes with $\underline{R}_1$ pointing along the step edges and $\underline{R}_3$ along the connection between the lower and upper edge of each step, see Figure 13.11. Let us further assume that terraces are $n_2$ vector lengths $\underline{R}_2$ "wide" ($n_2 = 3$ in Figure 13.11) and the steps $n_s$ vector lengths $\underline{R}_3$ "high" ($n_s = 1$ in Figure 13.11). Then the connection of lattice points $A, B, C$ determines the $(h\ k\ l)$ direction of the surface ((3 3 5) in Figure 13.11), while $\underline{R}_1$ and $\underline{R}_2$ refer to $(h\ k\ l)_t$ of the terrace ((1 1 1) in Figure 13.11) and $\underline{R}_1$ and $\underline{R}_3$ to $(h\ k\ l)_s$ of the step side ((0 0 2) in Figure 13.11). Figure 13.11 shows that

$$\begin{aligned}\underline{G}_{(hkl)} &= 2\pi/V_{\text{el}}\,(\underline{AB} \times \underline{AC}) = \underline{R}_1 \times (n_2\,\underline{R}_2 - n_s\,\underline{R}_3)\\ &= 2\pi/V_{\text{el}}\,\{n_2\,(\underline{R}_1 \times \underline{R}_2) + n_s\,(\underline{R}_3 \times \underline{R}_1)\}\\ &= n_2\underline{G}_{(hkl)_t} + n_s\underline{G}_{(hkl)_s}\end{aligned} \tag{13.38}$$

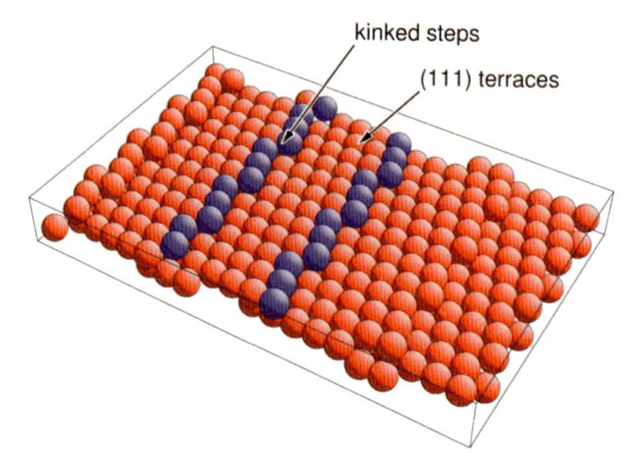


**Figure 13.10** Shape of the kinked (10 12 16) surface of a face-centered cubic copper lattice. The atoms are shown as red balls where the kinked step lines are emphasized by blue.

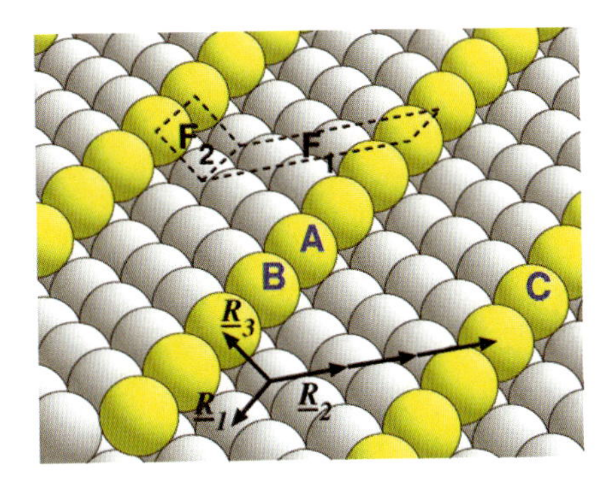


**Figure 13.11** Scheme of a Miller index decomposition for the stepped (3 3 5) surface of a face-centered cubic lattice. The atoms are shown as colored balls where the step lines, forming a (3 3 5) netplane, are emphasized by different colors. The step-adapted lattice vectors $\underline{R}_1$, $\underline{R}_2$, $\underline{R}_3$ are sketched accordingly. The elementary terrace, step sections of area $F_1$, $F_2$ are indicated by dashed lines.

or after the three reciprocal lattice vectors have been decomposed into their Miller index combinations, one obtains the **additivity theorem**

$$(h\ k\ l) = n_2\,(h\ k\ l)_t + n_s\,(h\ k\ l)_s. \tag{13.39}$$

This is the basis of the so-called **step notation** [?, ?] of vicinal surfaces according to which the $(h\ k\ l)$ surface is denoted as

$$(h\ k\ l) \equiv [p_1\,(h\ k\ l)_t \times p_2\,(h\ k\ l)_s], \quad p_1 = n_2 + 1, \quad p_2 = n_s. \tag{13.40}$$

Here the terrace width of $n_2\ \underline{R}_2$ used above corresponds to $(n_2 + 1)$ rows of terrace atoms used in the definition of the step notation. This definition was initially proposed for cubic lattices (face- and body-centered) with Miller indices of simple cubic notation and single steps ($n_s = 1$) only [?] where each of the Miller index triples $(h\ k\ l)$, $(h\ k\ l)_t$, and $(h_s,\ k_s,\ l_s)$ is scaled such that the indices do not have a common multiple, i.e., for example (2 2 0) is written as (1 1 0). Examples of the additivity theorem (13.39) for simple (sc), face-centered (fcc), and body-centered (bcc) cubic lattices together with the corresponding step notations are given in Table 13.4. Here [sc] and [gen] refer to the Miller index notations, simple cubic, and generic Bravais, see above, and constant $z$ can assume any positive integer value.

Surfaces, where the decomposition (13.39) yields multiple-height steps, $n_s > 1$, can give rise to more complex structural behavior depending on local binding. For strong nearest neighbor binding, like in metals, these surfaces still form single-height steps even if $n_s > 1$. Here the multiple-height step structure of $n_2\ \underline{R}_2$ "wide" terraces with $n_s\ \underline{R}_3$ "high" steps is filled with $n_s$ additional subterraces,

$$\begin{aligned} ((p+1)n_s - n_2) &\ \text{terraces of width } p\ \underline{R}_3 \text{ and} \\ (n_2 - p\,n_s) &\ \text{terraces of width } (p+1)\ \underline{R}_3,\ p = [n_2/n_s], \end{aligned} \tag{13.41}$$

**Table 13.4** Examples of Miller index decompositions and step notation, see text.

| $(h,\ k,\ l) = n_2\ (h\ k\ l)_t + n_s\ (h\ k\ l)_s$ | Step notation |
|---|---|
| fcc [sc]: | |
| $(7\ 7\ 5) = 6(1\ 1\ 1) + (1\ 1\ -1)$ | $7(1\ 1\ 1) \times (1\ 1\ -1)$ |
| $(3\ 3\ 5) = 3(1\ 1\ 1) + (0\ 0\ 2)$ | $4(1\ 1\ 1) \times (0\ 0\ 1)$ |
| $(9\ 1\ 1) = 4(2\ 0\ 0) + (1\ 1\ 1)$ | $5(1\ 0\ 0) \times (1\ 1\ 1)$ |
| $(z{+}2\ z{+}2\ z) = (z{+}1)(1\ 1\ 1) + (1\ 1\ -1)$ | $(z{+}2)(1\ 1\ 1) \times (1\ 1\ -1)$ |
| $(z{+}2\ z\ z) = z(1\ 1\ 1) + (2\ 0\ 0)$ | $(z{+}1)(1\ 1\ 1) \times (1\ 0\ 0)$ |
| $(2z{+}1\ 1\ 1) = z(2\ 0\ 0) + (1\ 1\ 1)$ | $(z{+}1)(1\ 0\ 0) \times (1\ 1\ 1)$ |
| fcc [gen]: | |
| $(5\ 5\ 1) = 4(1\ 1\ 0) + (1\ 1\ 1)$ | $5\ (1\ 1\ 0) \times (1\ 1\ 1)$ |
| $(4\ 3\ 2) = 3(1\ 1\ 1) + (1\ 0{-}1)$ | $4\ (1\ 1\ 1) \times (1\ 0\ -1)$ |
| bcc [sc]: | |
| $(5\ 5\ 2) = 5(1\ 1\ 0) + (0\ 0\ 2)$ | $6\ (1\ 1\ 0) \times (0\ 0\ 1)$ |
| $(6\ 6\ 10) = 5(1\ 1\ 2) + (1\ 1\ 0)$ | $6\ (1\ 1\ 2) \times (1\ 1\ 0)$ |
| $(8\ 1\ 1) = 4(2\ 0\ 0) + (0\ 1\ 1)$ | $5\ (2\ 0\ 0) \times (0\ 1\ 1)$ |
| $(z\ z\ 2) = z(1\ 1\ 0) + (0\ 0\ 2)$ | $(z{+}1)\ (1\ 1\ 0) \times (0\ 0\ 1)$ |
| $(z{+}1\ z{+}1\ 2z) = z(1\ 1\ 2) + (1\ 1\ 0)$ | $(z{+}1)\ (1\ 1\ 2) \times (1\ 1\ 0)$ |
| $(2z\ 1\ 1) = z(2\ 0\ 0) + (0\ 1\ 1)$ | $(z{+}1)\ (1\ 0\ 0) \times (0\ 1\ 1)$ |
| bcc [gen]: | |
| $(4\ 1\ 1) = 4(1\ 0\ 0) + (0\ 1\ 1)$ | $5\ (1\ 0\ 0) \times (0\ 1\ 1)$ |
| $(1\ 1\ 2) = 2(0\ 0\ 1) + (1\ 1\ 0)$ | $3\ (0\ 0\ 1) \times (1\ 1\ 0)$ |
| sc [gen]: | |
| $(9\ 1\ 1) = 9(1\ 0\ 0) + (0\ 1\ 1)$ | $10\ (1\ 0\ 0) \times (0\ 1\ 1)$ |

where $[x]$ denotes the well-known *integer function*. As an example, we mention the (15 15 23) surface of the face-centered cubic lattice (in simple cubic notation), see Figure 13.12, which is decomposed in (1 1 1) terraces and (0 0 2) steps according to

$$(15\ 15\ 23) = 15\ (1\ 1\ 1) + 4\ (0\ 0\ 2)$$

with four subterraces, one of width $3\underline{R}_2$ and three of widths $4\underline{R}_2$ filling the initial multiple-step structure with $4\underline{R}_3$ "high" steps and $15\underline{R}_2$ "wide" terraces.

Vicinal surfaces with large Miller index values $(h\,k\,l)$ can also be described in some cases by combinations of terraces with $(h_t\ k_t\ l_t)$ orientation separated by kinked steps with $(h_{s1}\ k_{s1}\ l_{s1})$ and $(h_{s2}\ k_{s2}\ l_{s2})$ orientation where the Miller index triples $(h\ k\ l)$, $(h_t\ k_t\ l_t)$, $(h_{s1}\ k_{s1}\ l_{s1})$, and $(h_{s2}\ k_{s2}\ l_{s2})$ are connected by an additivity theorem. Starting from an initial primitive lattice $\{\underline{R}_{o1}, \underline{R}_{o2}, \underline{R}_{o3}\}$ with a kinked surface as shown, for example in Figure 13.13 (sketching the kinked face-centered cubic (11 13 19) surface), let us construct kink-adapted lattice vectors $\underline{R}_1$, $\underline{R}_2$, $\underline{R}_3$ where $\underline{R}_1$ and $\underline{R}_2$ describe the periodicity of the terrace netplanes with these two vectors pointing along the two kink directions and $\underline{R}_3$ along the connection between the lower and upper edge of the kink, see Figure 13.13. Let us further assume that the

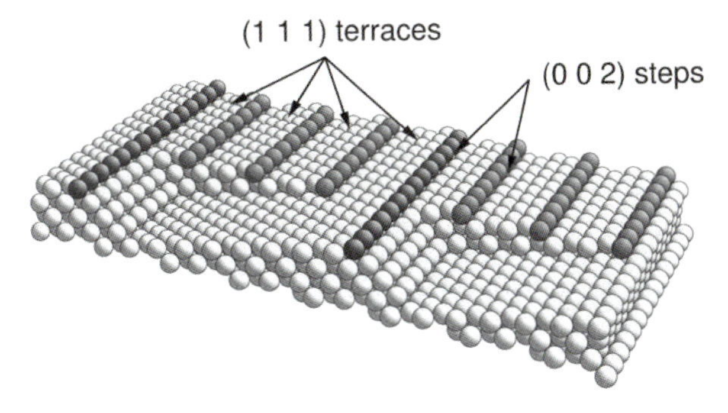


**Figure 13.12** Sketch of the stepped (15 15 23) surface of the face-centered cubic lattice, with multiple-height steps (front), with single-height steps, and subterraces (back). The atoms in dark color indicate the step lines.

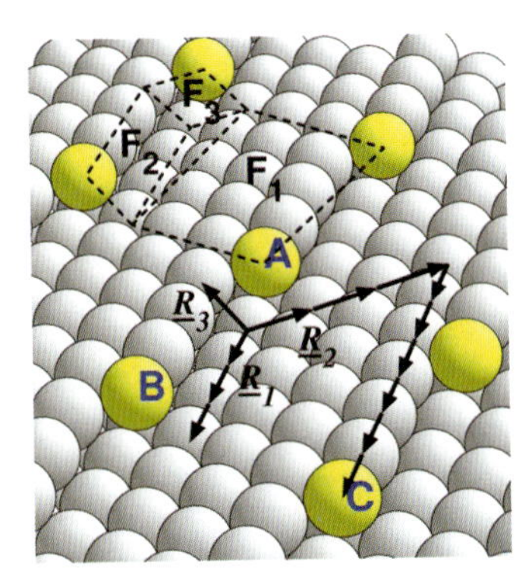


**Figure 13.13** Scheme of a Miller index decomposition for the kinked (11 13 19) surface of a face-centered cubic lattice. The atoms are shown as colored balls where the kink centers, forming a (11 13 19) netplane, are emphasized by yellow. The kink adapted lattice vectors $\underline{R}_1$, $\underline{R}_2$, $\underline{R}_3$ are sketched accordingly. The elementary terrace and two kink sections of area $F_1, F_2, F_3$ are indicated by dashed lines.

two kink edges are $m_1$ and $m_2$ vector lengths $\underline{R}_1$ and $\underline{R}_2$ "long" ($m_1 = 3,\ m_2 = 1$ in Figure 13.13), and the terrace width between kinked steps is described by a vector $\underline{R}_t = n_1\ \underline{R}_1 + n_2\ \underline{R}_2$ connecting the lower kink edge of one kink line with an upper edge of the adjacent kink line ($n_1 = 3,\ n_2 = 6$ in Figure 13.13). In addition, the kinks are assumed to be $n_s$ vector lengths $\underline{R}_3$ "high" ($n_s = 1$ in Figure 13.13). Then the connection of $A$, $B$, and $C$ determines the $(h\ k\ l)$ direction of the surface ((11 13 19) in Figure 13.13), while $\underline{R}_1$ and $\underline{R}_2$ refer to $(h\ k\ l)_t$ of the terrace ((1 1 1) in Figure 13.13), $\underline{R}_3$ and $\underline{R}_1$ to $(h\ k\ l)_{s1}$ of one kink side ((−1 1 1) in Figure 13.13), and $\underline{R}_2$ and $\underline{R}_3$ to $(h\ k\ l)_{s2}$ of the other kink side((0 0 2) in Figure 13.13). Then Figure 13.13 shows that

$$\begin{aligned}\underline{G}_{(h\,k\,l)} &= 2\pi/V_{\text{el}}\ (\underline{AB} \times \underline{AC})\\ &= (m_1\ \underline{R}_1 - m_2\ \underline{R}_2)\cdot(n_1\ \underline{R}_1 + n_2\ \underline{R}_2 - m_2\ \underline{R}_2 - n_s\ \underline{R}_3)\end{aligned}$$

$$= 2\pi/V_{el}\ \{(m_1\ n_2 + m_2\ n_1 - m_1\ m_2)(\underline{R}_1 \times \underline{R}_2) + m_1\ n_s(\underline{R}_3 \times \underline{R}_1) + m_2\ n_s(\underline{R}_2 \times \underline{R}_3)\}$$
$$= (m_1\ n_2 + m_2\ n_1 - m_1\ m_2)\ \underline{G}_{(hkl)_t} + m_1\ n_s\ \underline{G}_{(hkl)_{s1}} + m_2\ n_s\ \underline{G}_{(hkl)_{s2}} \quad (13.42)$$

or after the three reciprocal lattice vectors have been decomposed into their Miller index combinations, one obtains the additivity theorem

$$(h\ k\ l) = (m_1\ n_2 + m_2\ n_1 - m_1\ m_2)\ (h\ k\ l)_t + m_1\ n_s\ (h\ k\ l)_{s1} + m_2\ n_s\ (h\ k\ l)_{s2}. \quad (13.43)$$

The scalar factors in this equation have a simple geometric meaning, which is obvious from Figure 13.13. The elementary terrace section defined as the periodic repeat cell along the terrace, sketched by dashed lines in Figure 13.13, has an area $F_1$ where

$$F_1 = p_1|\underline{R}_1 \times \underline{R}_2|, \quad p_1 = (m_1 n_2 + m_2 n_1 - m_1 m_2), \quad (13.44a)$$

while the areas $F_2$ and $F_3$ of the two kink steps, also sketched by dashed lines in Figure 13.13, are given by

$$F_2 = p_2|\underline{R}_3 \times \underline{R}_1|, \quad p_2 = m_1 n_s, \quad (13.44b)$$
$$F_3 = p_3|\underline{R}_2 \times \underline{R}_3|, \quad p_3 = m_2 n_s, \quad (13.44c)$$

where $|\underline{R}_i \times \underline{R}_j|$ are the elementary cell areas of the corresponding netplanes.

The additivity theorem (13.43) is the basis of the so-called microfacet notation [?,?] of vicinal surfaces according to which the $(h\ k\ l)$ surface is, in its general form, denoted as

$$(h\ k\ l) = a_\lambda (h\ k\ l)_t + b_\mu (h\ k\ l)_{s1} + c_\nu (h\ k\ l)_{s2}, \quad (13.45)$$

where for general Bravais lattices, using Eqs. (13.43) and (13.44a)–(13.44c), one obtains

$$a_\lambda = p_1, \quad b_\mu = p_2, \quad c_\nu = p_3. \quad (13.46)$$

Notation (13.45) was initially proposed for cubic lattices (face- and body-centered) with Miller indices of simple cubic notation and single steps only where each of the Miller index triples $(h\ k\ l)$, $(h_t\ k_t\ l_t)$, $(h_{s1}, k_{s1}, l_{s1})$, and $(h_{s1}, k_{s1}, l_{s1})$ is scaled such that the indices do not have a common multiple, i.e., for example (12 8 4) is written as (3 2 1). In this case, the factors $a_\lambda, b_\mu$ and $c_\nu$ are written as indexed numbers where

$$\lambda = p_1, \quad \mu = p_2, \quad \nu = p_3 \quad (13.47)$$

and $a, b, c$ are coefficients that guarantee the additivity (13.45) of the Miller indices. Thus the additivity theorem for the (20 16 14) indexed surface of the face-centered cubic lattice in simple cubic notation reads as

$$(20\ 16\ 14) = 15\,(1\ 1\ 1) + 2\,(2\ 0\ 0) + (1\ 1\ -1),$$

while the corresponding microfacet notation reads as

$$(10\ 8\ 7) = (15/2)_{15}(1\ 1\ 1) + 2_2(1\ 0\ 0) + (1/2)_1(1\ 1\ -1).$$

Obviously, using Bravais Miller indices or simple cubic Miller indices with the correct numerical constraints, see Eqs. (13.29b) and (13.31b), yields

$$a = \lambda = p_1, \quad b = \mu = p_2, \quad c = \nu = p_3, \tag{13.47a}$$

making the indexed numbers unnecessary and resulting in the initial additivity theorem (13.43).

Other examples of the additivity theorem for face- (fcc) and body-centered (bcc) cubic lattices together with corresponding microfacet notations (in curly brackets) are given in Table 13.5. Here [sc] and [gen] refer to the Miller index notations, simple cubic and generic Bravais, see above, and constant $z$ can assume any positive integer value.

**Table 13.5** Examples of Miller index decompositions and Microfacet notation, see text.

| $(h, k, l) = a\ (h\ k\ l)_t + b\ (h\ k\ l)_{s1} + c\ (h\ k\ l)_{s2}$ | Microfacet notation |
|---|---|
| fcc [sc]: | |
| $(17\ 11\ 9) = 10(1\ 1\ 1) + 3(2\ 0\ 0) + (1\ 1\ -1)$ | $10_{10}(1\ 1\ 1) + 6_3(1\ 0\ 0) + 1_1(1\ 1\ -1)$ |
| $(11\ 3\ 1) = 4(2\ 0\ 0) + 2(1\ 1\ 1) + (1\ 1\ -1)$ | $8_4(1\ 0\ 0) + 2_2(1\ 1\ 1) + 1_1(1\ 1\ -1)$ |
| $(17\ 15\ 1) = 7(2\ 2\ 0) + (1\ 1\ 1) + (2\ 0\ 0)$ | $14_7(1\ 1\ 0) + 1_1(1\ 1\ 1) + 2_1(1\ 0\ 0)$ |
| $(2z{+}7\ 2z{+}1\ 2z{-}1) = 2z(1\ 1\ 1) + 3(2\ 0\ 0) + (1\ 1\ -1)$ | $(2z)_{2z}(1\ 1\ 1) + 6_3(1\ 0\ 0) + 1_1(1\ 1\ -1)$ |
| $(2z{+}1\ 3\ 1) = (z{-}1)(2\ 0\ 0) + 2(1\ 1\ 1) + (1\ 1\ -1)$ | $(2z{-}2)_{(z-1)}(1\ 0\ 0) + 2_2(1\ 1\ 1) + 1_1(1\ 1\ -1)$ |
| $(2z{+}1\ 2\ z{-}1\ 1) = (z{-}1)\ (2\ 2\ 0) + (1\ 1\ 1) + (2\ 0\ 0)$ | $(2z{-}2)_{(z-1)}(1\ 1\ 0) + 2_1(1\ 0\ 0) + 1_1(1\ 1\ 1)$ |
| fcc [gen]: | |
| $(10\ 13\ 14) = 10(1\ 1\ 1) + 3(0\ 1\ 1) + (0\ 0\ 1)$ | $10_{10}(1\ 1\ 1) + 3_3(0\ 1\ 1) + 1_1(0\ 0\ 1)$ |
| $(2\ 6\ 7) = 4(0\ 1\ 1) + 2(1\ 1\ 1) + (0\ 0\ 1)$ | $4_4(0\ 1\ 1) + 2_2(1\ 1\ 1) + 1_1(0\ 0\ 1)$ |
| bcc [sc]: | |
| $(8\ 7\ 3) = 6(1\ 1\ 0) + 2(1\ 0\ 1) + (0\ 1\ 1)$ | $6_6(1\ 1\ 0) + 2_2(1\ 0\ 1) + 1_1(0\ 1\ 1)$ |
| $(15\ 10\ 3) = 10(1\ 1\ 0) + 3(1\ 0\ 1) + (2\ 0\ 0)$ | $10_{10}(1\ 1\ 0) + 3_3(1\ 0\ 1) + 2_1(1\ 0\ 0)$ |
| $(18\ 16\ 4) = 15(1\ 1\ 0) + 3(1\ 0\ 1) + (0\ 1\ 1)$ | $(15/2)_{15}(1\ 1\ 0) + (3/2)_3(1\ 0\ 1) + (1/2)_1(0\ 1\ 1)$ |
| $(2z{+}2\ 2z{+}1\ 3) = 2z(1\ 1\ 0) + 2(1\ 0\ 1) + (0\ 1\ 1)$ | $(2z)_{2z}(1\ 1\ 0) + 2_2(1\ 0\ 1) + 1_1(0\ 1\ 1)$ |
| $(2z{+}5\ 2z\ 3) = 2z(1\ 1\ 0) + 3(1\ 0\ 1) + (2\ 0\ 0)$ | $(2z)_{2z}(1\ 1\ 0) + 3_3(1\ 0\ 1) + 2_1(1\ 0\ 0)$ |
| bcc [gen]: | |
| $(1\ 2\ 6) = 6(0\ 0\ 1) + 2(0\ 1\ 0) + (1\ 0\ 0)$ | $6_6(0\ 0\ 1) + 2_2(0\ 1\ 0) + 1_1(1\ 0\ 0)$ |
| $(1\ 3\ 15) = 15(0\ 0\ 1) + 3(0\ 1\ 0) + (1\ 0\ 0)$ | $15_{15}(0\ 0\ 1) + 3_3(0\ 1\ 0) + 1_1(1\ 0\ 0)$ |

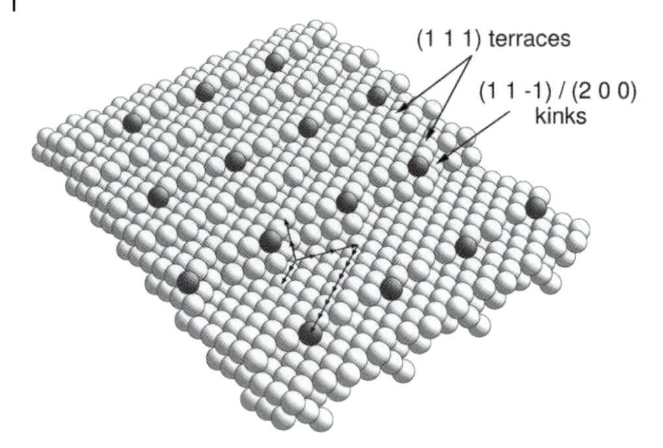


**Figure 13.14** Sketch of the kinked fcc (37 25 17) surface, with alternating single-height kinks (back) and one line of multiple-height kinks (front). The atoms in dark color illustrate the periodicity of the topmost netplane.

Surfaces, where the decomposition (13.42) yields multiple-height kinks, $n_s > 1$, can give rise to much more complex structural behavior depending on local binding. This is analogous to stepped surfaces described above. For strong nearest neighbor binding, like in metals, these surfaces form kinks with single atom steps even if $n_s > 1$. As an example, we mention the (37 25 17) surface of the face-centered cubic lattice (in simple cubic notation), see Figure 13.14, which is decomposed in (1 1 1) terraces and (1 1 −1) / (2 0 0) kinks according to

$$(37\ 25\ 17) = 21(1\ 1\ 1) + 2 \cdot 2(1\ 1\ -1) + 2 \cdot 3(2\ 0\ 0)$$

with $m_1 = 2, m_2 = 3, n_s = 2, n_1 = 7, n_2 = 3$ according to Eq. (13.42).

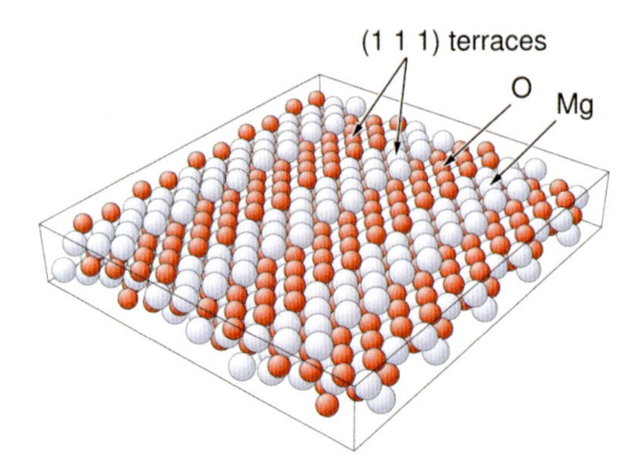


**Figure 13.15** Sketch of the kinked (15 11 9) surface of cubic MgO with alternating (1 1 1) terraces of the two elements, Mg and O. The atoms are painted in different colors and labeled accordingly.

So far, the discussion of the shape of vicinal surfaces was restricted to primitive lattices of Bravais type. A generalization to non-primitive lattices is formally straightforward since according to Eq. (13.8), any non-primitive lattice can be decomposed into a set of primitive sublattices of the same periodicity. Thus, vicinal surfaces of non-primitive lattices can be considered as superpositions of those of their primitive sublattices. However, the detailed local geometric structure at the surfaces may be rather complicated depending on the lattice type. As an illustration, Figure 13.15 shows an example of moderate complexity, the kinked (15 11 9) surface of cubic MgO discussed earlier. This surface can be decomposed in (1 1 1) terraces and (1 1 −1) / (2 0 0) kinks where the terrace sections alternate between the two atom types.

## 13.4 Real Crystal Surfaces, Relaxation, Reconstruction, Adsorbates

Atoms at real crystal surfaces appearing in nature experience a different local binding environment (connected with different atom coordination) as compared to atom sites inside the bulk crystal. This leads to geometric structures of real surfaces, which differ from those of simple bulk truncations discussed for ideal single crystal surfaces. The differences may be quite small (examples are many elemental metal surfaces) but can also be substantial (examples are semiconductor surfaces). The real surface can be restructured locally by bond changes including making and breaking of bonds, which may result in an overall disordered geometry. In many other cases, the surface will still exhibit a two-dimensionally periodic atom arrangement where, however, the periodicity, specific atom positions, and the placement of atom planes may be different from those of bulk netplanes. These effects are usually described by surface relaxation and reconstruction where details and nomenclature have been interpreted differently in the literature [?]. However, the basic concepts discussed below are universal.

The effect of **surface relaxation** assumes that $(h\ k\ l)$ indexed surfaces of a lattice with netplane-adapted definition $\{\underline{R}_1, \underline{R}_2, \underline{R}_3; \underline{r}_1, \ldots, \underline{r}_p\}$ are terminated by atom layers described as $(h\ k\ l)$ netplanes with their periodicity given by lattice vectors $\underline{R}_1$, $\underline{R}_2$ of the bulk. However, the relative positioning of the netplanes near the surface, expressed by interlayer distances and lateral shifts, deviates from that in the bulk. Thus, atom positions of layers near the surface of a real crystal may be described mathematically by

$$\underline{R}^{(m)} = \underline{r}_i + n_1\, \underline{R}_1 + n_2\, \underline{R}_2 + \underline{s}^{(m)} \quad \text{for layer } m \text{ near the surface,} \tag{13.48}$$

where $\underline{r}_i$ refer to positions of atoms inside the elementary cell of the bulk lattice, $n_1$ and $n_2$ are integer-valued coefficients accounting for the layer (netplane) periodicity, and $\underline{s}^{(m)}$ is a shift vector corresponding to the absolute positioning of layer $m$. Obviously, according to Eq. (13.18), vectors $\underline{s}^{(m)}$ can be set equal to integer multiples of lattice vector $\underline{R}_3$ for ideal single crystal surfaces. Further, $\underline{s}^{(m)}$ is expected to approach the bulk value $n_3\, \underline{R}_3$ as the layers go deeper into the bulk. As an illustration, Figure 13.16 shows a section near the (0 0 1) surface of a fictitious

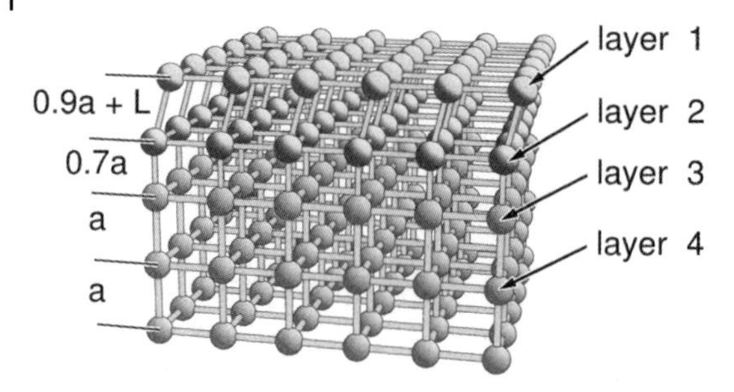


**Figure 13.16** Sketch of the simple cubic (001) surface with the two topmost netplane layers relaxed, see text.

simple cubic lattice (lattice constant $a$) where the topmost layer 1 is relaxed inward by 10% and shifted sideways (L denoting a lateral shift vector) and layer 2 is relaxed inward by 30%.

Surface relaxation occurs for most metal surfaces where, so far, mainly netplane shifts perpendicular to the surface have been considered with only few examples of lateral shifts in cases of stepped surfaces [?, ?, ?].

Real surfaces, which differ structurally from simple bulk truncations other than described by relaxation, are considered **reconstructed surfaces** where the effect may result in surface disorder or yield a periodic surface geometry with sizeable displacements of the atoms and/or with additional or fewer atoms in the layer-building cells compared with bulk truncation. In the periodic case, the $(h\ k\ l)$ indexed surface of a lattice with netplane-adapted definition $\{\underline{R}_1, \underline{R}_2, \underline{R}_3; \underline{r}_1, \ldots, \underline{r}_p\}$ is terminated by atom layers that exhibit a two-dimensional periodicity given by vectors $\underline{R}'_1, \underline{R}'_2$. These vectors can be identical to those of the initial $(h\ k\ l)$ netplanes, $\underline{R}_1, \underline{R}_2$, or can be different forming so-called **superlattices**. Surface reconstruction is usually combined with relaxation such that atom positions of layers near the surface are described mathematically by

$$\underline{R}^{(m)} = \underline{r}'_i + n_1\,\underline{R}'_1 + n_2\,\underline{R}'_2 + \underline{s}^{(m)} \quad \text{for layer } m \text{ near the surface,} \tag{13.49}$$

where $\underline{r}'_i$ refer to positions of atoms inside the reconstructed layer (which may or may not include positions of the elementary cell of the bulk lattice), $n_1$ and $n_2$ are integer-valued coefficients accounting for the layer periodicity, and $\underline{s}^{(m)}$ is a shift vector that includes relaxation. The periodicity vectors $\underline{R}'_1$ and $\underline{R}'_2$ can be connected with those of the initial $(h\ k\ l)$ netplanes, $\underline{R}_1$ and $\underline{R}_2$, by linear $(2 \times 2)$ transformations, written in matrix form as

$$\begin{pmatrix} \underline{R}'_1 \\ \underline{R}'_2 \end{pmatrix} = \begin{pmatrix} m_{11} & m_{12} \\ m_{21} & m_{22} \end{pmatrix} \cdot \begin{pmatrix} \underline{R}_1 \\ \underline{R}_2 \end{pmatrix} = \underline{\underline{M}} \cdot \begin{pmatrix} \underline{R}_1 \\ \underline{R}_2 \end{pmatrix}. \tag{13.50}$$

As a consequence, the elementary cell area $F'$ of the reconstructed layer is given by

$$\begin{aligned} F' &= |\underline{R}'_1 \times \underline{R}'_2| = |(m_{11}\,\underline{R}_1 + m_{12}\,\underline{R}_2) \times (m_{21}\,\underline{R}_1 + m_{22}\,\underline{R}_2)| \\ &= |(m_{11}\,m_{22} - m_{12}\,m_{21})\,(\underline{R}_1 \times \underline{R}_2)| = |det(\underline{\underline{M}})|\ F, \end{aligned} \tag{13.51}$$

where $F$ is the elementary cell area of the initial $(h\ k\ l)$ netplane. Thus, $|\det(\underline{\underline{M}})|$ gives the ratio of the elementary cell area $F'$ of the new periodicity and that, $F$, of the initial periodicity. The mathematical type of the $(2 \times 2)$ transformation matrix $\underline{\underline{M}}$ in Eq. (13.50) allows a classification of reconstructed surfaces into two categories:

- Reconstruction with **commensurate superlattices** is described by matrix $\underline{\underline{M}}$ containing only integer-valued elements $m_{ij}$. In this case, the periodicity vectors $\underline{R}'_1$ and $\underline{R}'_2$ are also vectors of the corresponding $(h\ k\ l)$ netplane lattice, and the elementary cell area of the reconstructed layer is an integer multiple of that of the netplane. (This includes systems where matrix $\underline{\underline{M}}$ equals the unit matrix and the reconstructed layer is of the same periodicity as the corresponding $(h\ k\ l)$ netplane.) As a simple example, Figure 13.17 compares the ideal (1 1 0) surface of face-centered cubic palladium with the so-called "(2 × 1)-missing-row" reconstructed surface where every second row of atoms of the topmost (1 1 0) layer is missing resulting in a (2 × 2) transformation matrix

$$\underline{\underline{M}} = \begin{pmatrix} 2 & 0 \\ 0 & 1 \end{pmatrix}. \tag{13.52}$$

- Reconstruction with **incommensurate superlattices** is described by matrix $\underline{\underline{M}}$ containing elements $m_{ij}$, which are real valued in the mathematical sense. In this case, the periodicity vectors $\underline{R}'_1$ and $\underline{R}'_2$ are not vectors of the corresponding $(h\ k\ l)$ netplane lattice. As an example, Figure 13.18 shows a (1 0 0) surface of face-centered cubic gold where the topmost layer is reconstructed with hexagonal geometry, while the (1 0 0) netplanes are of square geometry. (This reconstruction is taken from the NIST Surface Structure Database (SSD) collecting all experimentally identified surface structures [?, ?, ?].) This results in a (2 × 2) transformation matrix

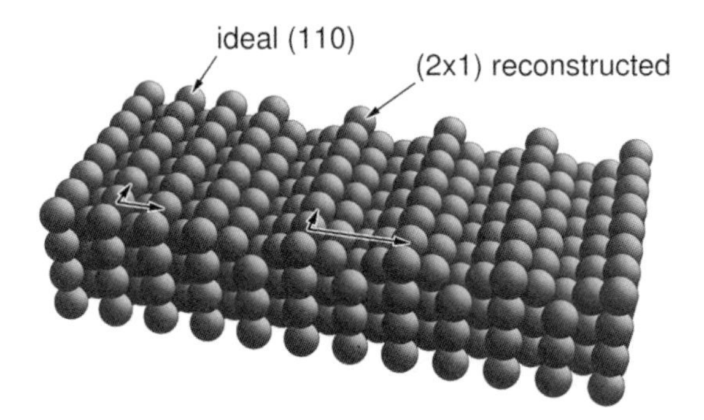


**Figure 13.17** Sketch of the ideal (a) and the (2 × 1) reconstructed Pd(110) surface (b). The layer periodicity vectors are indicated for both geometries.

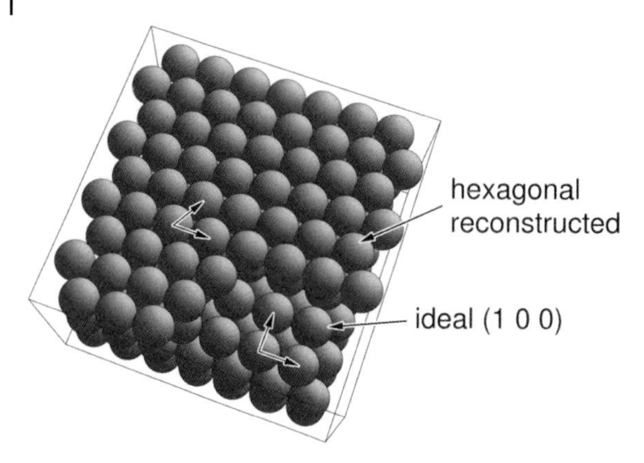


**Figure 13.18** Sketch of the hexagonal reconstructed Au(100) surface taken from the Surface Structure Database [?, ?, ?]. The periodicity vectors of the top reconstructed and the underlying substrate layers are sketched accordingly.

$$\underline{\underline{M}} = \eta \begin{pmatrix} 1 & 0 \\ 1/2 & \sqrt{3}/2 \end{pmatrix}, \tag{13.53}$$

where $\eta$ is the ratio of the lattice constants of the reconstructed layer and the initial (1 0 0) netplane.

In addition to periodicity transformations discussed above, reconstruction includes major displacements of individual atoms near the surface. As a rather complex example, Figure 13.19 shows the dimer-adatom-stacking-fault (DAS) model of the $(7 \times 7)$ reconstructed (1 1 1) surface of silicon [?, ?, ?], where the Si adatoms stick out of the surface and $Si_2$ dimers stabilize in long trenches, which cross to form open holes.

The discussion of geometry at real crystals where the topmost layers are relaxed or reconstructed can also be applied to adsorbate systems where foreign atoms and/or molecules stabilize at a $(h\ k\ l)$ oriented surface of a single crystal. In general, adsorbate particles may bind at substrate surfaces in a completely random fashion (*lattice gas systems*) or at specific surface sites in disordered patches (*disordered adsorption*). These two scenarios depend very much on details of local adsorbate–substrate binding and are not relevant for general crystallographic considerations. In addition, there are many adsorbate systems [?] where the adparticles form two-dimensionally periodic overlayers. Thus, they exhibit, from a crystallographic point of view, the same structural elements that have been discussed before in connection with surface reconstruction. The only difference compared with surface reconstruction is due to the fact that adsorbate systems contain overlayers at the surface where the atom types (elements) differ from those of the underlying substrate. Therefore, one can distinguish between commensurate and incommensurate periodic overlayers based on the $(2 \times 2)$ transformation matrix $\underline{\underline{M}}$ connecting between the periodicity vectors $\underline{R}_1'$ and $\underline{R}_2'$ of the adsorbate layer

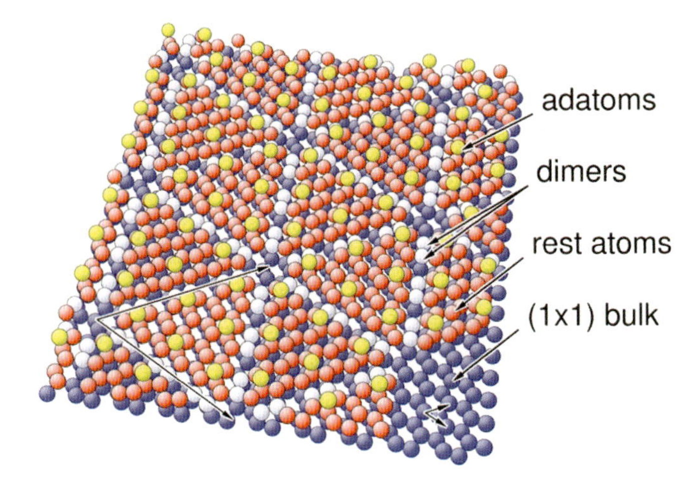


**Figure 13.19** Sketch of the reconstructed Si(111)–(7 × 7) surface according to the dimer-adatom-stacking fault (DAS) model. The overlayer structure is removed at the bottom right to reveal the ideal bulk termination of Si(111). The different types Si atoms are painted differently, and the periodicity vectors of the overlayers and the ideal bulk termination are sketched accordingly.

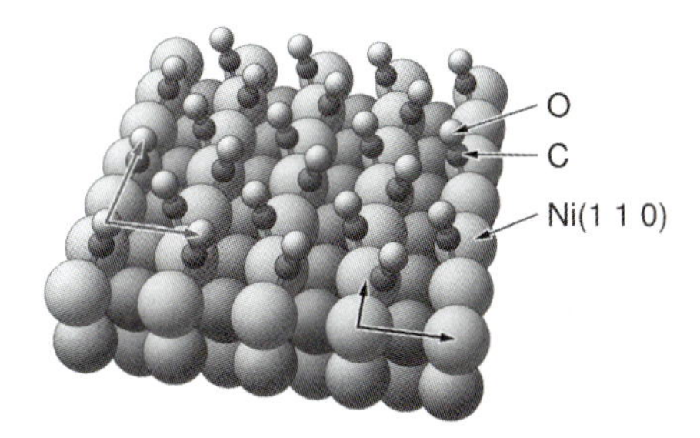


**Figure 13.20** Sketch of the Ni(110)–p(2×1)–2CO adsorbate system. The different atoms are painted differently, and the periodicity vectors of the CO adsorbate layer and of the Ni substrate are sketched accordingly.

and those, $\underline{R}_1$ and $\underline{R}_2$, of the underlying substrate surface as given in Eq. (13.50). As an illustration, Figure 13.20 shows the adsorption geometry of CO molecules at the Ni(1 1 0) surface, formally described as Ni(1 1 0)–$p(2\times1)$–2CO (referring to the so-called Wood notation [?] that will not be discussed in this chapter). The molecules stabilize between nickel substrate atoms in a tilted geometry forming a two-dimensional periodic array at the surface.

Further crystallographic information on adsorbate systems can be found in the SSD database published by the National Institute of Standards and Technology [?]. In addition, graphical information on many surface structures is available from the web surface structure gallery published by the author [?].

## References


1. C. Giacovazzo, H.L Monaco, D. Viterbo, F. Scordari, G. Gilli, G. Zanotti and M. Catti *"Fundamentals of Crystallography"*, IUCr Texts on Crystallography 2, Oxford Science Publishing, Oxford (**1998**).
2. H.D. Megaw, *"Crystal Structures: A Working Approach"*, W. B. Saunders, Philadelphia (**1973**).
3. R.J.D. Tilley, *"Crystals and Crystal Structures"*, John Wiley Sons, Chichester, (**2006**).
4. R.W.G. Wyckoff, *"Crystal Structures"*, Vols. I–VI, Interscience, New York, (**1963**).
5. *"International Tables for Crystallography"*, Vol. A, T. Hahn (Ed.), Reidel Publishing, Boston, (**1965**, **1983**, **1987**).
6. G. Burns and A.M. Glazer, *"Space Groups for Solid State Scientists"*, 2nd Ed., Academic Press, New York, (**1990**).
7. K. Hermann, *Surf. Rev. Lett.* 4 (**1997**) 1063.
8. I. Czekaj, K. Hermann, and M. Witko, *Surf. Sci.* 525 (**2003**) 33.
9. B. Lang, R.W. Joyner, and G.A. Somorjai, *Surf. Sci.* 30 (**1972**) 454.
10. M.A. Van Hove, W.H. Weinberg, and C.M. Chan, *"Low Energy Electron Diffraction"*, Springer Series in Surface Science, Vol. 6, Heidelberg, (**1986**).
11. M.A. Van Hove and G.A. Somorjai, *Surf. Sci.* 92 (**1980**) 489.
12. P.R. Watson, M.A. Van Hove, and K. Hermann, *"Atlas of Surface Crystallography based on the NIST Surface Structure Database (SSD)"*, Vols.1a, b, Monograph No. 5, *J. Phys. Chem. Ref. Data*, American Chemical Society, (**1994**).
13. M.A. Van Hove, K. Hermann, and P.R. Watson, *Acta Cryst. B* 58 (**2002**) 338.
14. P.R. Watson, M.A. Van Hove, and K. Hermann, "NIST Surface Structure Database (SSD), Standard Reference Database 42", Software Version 5, (**2003**), available at http://www.nist.gov/srd/nist42.htm.
15. E.A. Wood, *J. Appl. Phys.* 35 (**1964**) 1306.
16. K. Hermann, http://www.fhi-berlin.mpg.de/~hermann/Balsac/pictures.html

# 14
# Adsorption and Diffusion in Porous Systems

*Kourosh Malek, Thijs J.H. Vlugt, and Berend Smit*

## 14.1
## Introduction

In this chapter, we review adsorption and diffusion processes in some selected porous systems. In particular, we will discuss the transport properties of water in protein crystals (Section 14.2), as well as the adsorption (Section 14.3) and diffusion (Section 14.4) of hydrocarbons in zeolites. We end with a discussion on the simulation of diffusion and reaction in functionalized, amorphous nanoporous catalysts, and membranes in Section 14.5.

## 14.2
## Transport in Protein Crystals: Insights from Molecular Simulations

### 14.2.1
### Introduction

This section provides simulation of transport processes in protein crystals and highlights the importance of protein–solvent and protein–solute interactions. It is of great importance to know whether the catalytic properties of proteins are the same in a crystal and in solution. In their crystalline form, proteins have a highly ordered 3D structure in which the molecules are strongly bound to each other via strong intermolecular interactions. Crosslinked protein crystals (CLPCs) consist of an extensive regular matrix of chiral nanopores, through which ions and solutes travel in and out. CLPCs are especially interesting materials for biotechnological applications to device highly selective biocatalysts, biosensors, and bioseparations. A fundamental understanding of the factors that control the diffusion rate of solutes and solvents in protein crystals is vital for improving and extending the biotechnological applications of such materials. This section focuses on the dynamic properties of water in protein crystals using a lysozyme lattice as a simple model. We also look at distinct features of water and counter-ion diffusion in lysozyme crystals. For a more detailed review of this topic, we refer the reader to Refs. [4, 7–10, 14].

*Computational Methods in Catalysis and Materials Science.* Edited by Rutger A. van Santen and Philippe Sautet

ISBN: 978-3-527-32032-5

### 14.2.2
### Crosslinked Protein Crystal Technology

Protein crystals are important elements in purification and structure determination of enzymes [1–4]. Such crystals contain pores that range in width from approximately 0.3 nm up to 10 nm, and their porosity is comparable to that of inorganic catalysts and sorbents such as zeolites and silica gel [5]. The complex crystal structure of a protein also contains many functional active sites, where substrates are bound to the protein surface [6]. Recently, CLPCs have been successfully applied as extremely stable biocatalysts and as selective (chiral) separation media. The analysis of water motion in protein crystals is very useful because proteins and other essential biological molecules are in contact through an aqueous medium. In addition to advanced experimental techniques, versatile computational tools are generally needed to correlate the protein's reactivity and the solute's transport with the enzyme crystals' nanoporosity at an atomistic level. MD simulations with explicit representations of molecules and ions should, in principle, provide realistic information about the diffusive motion of water, individual solute molecules, and ions at atomic resolution. Yet, these simulations are practical only at longer time and length scales.

### 14.2.3
### Computational Methodology

A number of computational approaches have been utilized to understand the nanostructure and transport properties of protein crystals. In the following, we dwell a computational approach based on MD simulations to examine molecular motions in orthorhombic and tetragonal lysozyme lattices (LYZO and LYZT, respectively).

Lysozyme consists of 129 amino acids with 1001 nonhydrogen atoms. In a simulation of a single-unit cell at pH = 7, one assumes that the amino acids Glu and Asp are deprotonated, while the Lys, Arg, His residues are protonated. The latter leads to +8 electron charges per protein molecule. The lysozyme crystal structure comes from the Brookhaven Protein Database and serves as a starting point for the simulations. In the case of the orthorhombic crystal (LYZO), there are four protein molecules per unit cell related via crystallographic orthorhombic symmetry ($P2_12_12_1$) with $a = 5.9062$ nm, $b = 6.8451$ nm, and $c = 3.0517$ nm. For the tetragonal lattice of symmetry $P4_32_12$ (LYZT), there are eight protein molecules in a unit cell of size $a = 7.91$ nm, $b = 7.91$ nm, and $c = 3.79$ nm. Thus, for the single-unit cell simulations described here, the LYZO system consists of 5372 protein atoms, 13,576 water molecules, and 32 chloride ions, totaling 46,132 atoms. The LYZT system comprised 10,744 protein atoms, 11,005 water molecules, and 64 chloride ions, totaling 43,823 atoms. Figure 14.1 shows the instantaneous configuration of an atomic model for the fully hydrated LYZO and its crystal structure. We can determine channels and cavities in LYZO by using a procedure explained more fully elsewhere [7–9]. Essentially, we start by

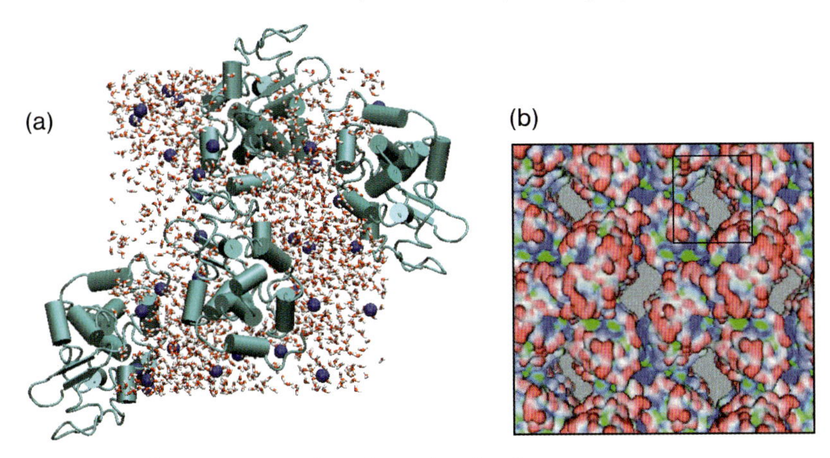


**Figure 14.1** All-atom representation of a single-unit cell of orthorhombic lysozyme lattice (LYZO). (a) Blue balls represent chloride ions, and thick ropes represent water molecules. (b) On the lattice surface, hydrophilic and hydrophobic regions are shown in blue and red, respectively. The pore region is in the square [10].

constructing a simulation box, and then we determine a pore radius by calculating the maximum size for diffusing spherical probe to still fit in that pore without overlapping the van der Waals radii of the atoms in the pore wall.

We divide interactions among atoms into nonbonded (between any pair of atoms in a given cut-off radius) and bonded (between atoms connected by chemical bonds). For nonbonded interactions (electrostatic and van der Waals), we assign a partial charge and parameters for repulsion and attraction to each atom. The bonded interaction consists of bond, angle, and dihedral terms, with bond and angle bending given by simple harmonic potentials. The torsional–rotational potential for the dihedral angle is a periodic function with a threefold barrier; a typical effective potential looks like this,

$$
\begin{aligned}
U = {} & \sum_{\text{bonds}} \frac{k_{b,ij}}{2}\left(r_{ij} - b_{0,ij}\right)^2 + \sum_{\text{angles}} \frac{k_{\theta,ijk}}{2}\left(\theta_{ijk} - \theta_{0,ijk}\right)^2 \\
& + \sum_{\text{dihedrals}} k_{\phi}\left[1 + \cos\left(n\left(\phi - \phi_0\right)\right)\right] + \\
& + \sum_{i<j}\left[4\varepsilon\left[\left(\frac{\sigma}{r_{ij}}\right)^{12} - \left(\frac{\sigma}{r_{ij}}\right)^{6}\right] + \frac{q_i q_j \operatorname{erfc}(\alpha r_{ij})}{4\pi\varepsilon_0 r_{ij}}\right] \\
& + \frac{1}{2V\varepsilon_0}\sum_{\mathbf{k}\neq\mathbf{0}} \frac{\exp[-k^2/4\alpha^2]}{k^2} \sum_{j=1}^{N}\left|q_j \exp[-i\mathbf{k}\cdot\mathbf{r}_j]\right|,
\end{aligned}
\qquad (14.1)
$$

where $r_{ij}$ is the distance between atoms $i$ and $j$ (or united atoms when $CH_n$ groups are treated as a single interaction site); $q_i$ is the partial charge on atom $i$; $\alpha$ is a damping parameter; erfc is the complementary error function; $\sigma$ and $\varepsilon$ are Lennard–Jones parameters; $k_b$, $k_\theta$, and $k_\phi$ are force constants for bonds, angles, and dihedrals; $n$ is the dihedral multiplicity, and $b_0$, $\theta_0$, and $\phi_0$ are equilibrium values for the bond lengths, angles, and dihedrals. Here, we model bonds and angles as harmonic oscillators and represent the dihedral term with a cosine expansion. Most importantly, we take into account only the pair interactions; we neglect nonbonded interactions between three or more atoms, and because atoms are represented as point charges, we also neglect electronic polarizability. The last term in Eq. (14.1) corresponds to the Fourier part of the Ewald summation [11]. This term was calculated using a Particle Mesh Ewald (PME) with a grid spacing of 0.12 nm and fourth-order interpolation. All MD simulations described here are performed in a canonical ensemble (*NVT*; given number of particles $N$, volume $V$, and temperature $T$). The Berendsen algorithm controls the temperature by mimicking a weak coupling to an external heat bath at a given temperature $T_0$ [12]. In the simulations, the weak-coupling algorithm is separately applied for protein, solute, and solvent plus ions with a time constant $\tau = 0.1$ ps and a temperature $T_0 = 300$ K. We use the profile of the root-mean-square deviation (RMSD) from the initial configuration to determine the structure and stability of the protein. Hydrogen atoms are treated as dummy atoms with an increased mass of 4 Da (Dalton number), which can push the integration time step to 5 fs.

The HOLE and CHANNEL algorithms [123, 124] can help us determine cavities and channels in the LYZO and LYZT lattices. Figure 14.2(a) visualizes part of the typical pores along the $z$-axis in LYZO; the average pore radius in LYZO ($0.68 + 0.02$ nm) is bigger than that in LYZT ($0.52 + 0.02$ nm). Figure 14.2(b) shows the pore radius as a function of pore axis and that each pore has constricted zones inside it. The pore radius in LYZO, for example, slowly decreases from over 0.75 nm to slightly less than 0.61 nm at its narrowest point, whereas in LYZT, it changes from 0.6 nm to 0.45 nm at its narrowest part. An alternative, more indirect, way to look at pore size is to calculate water density as a function of pore

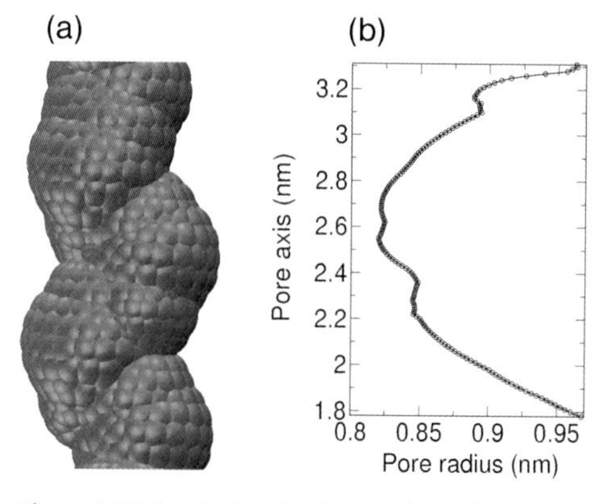


**Figure 14.2** (a) A visualized pore along the $z$-axis in LYZO system. (b) The pore radius profile [10].

axis $z$. Such calculations yield similar results to those obtained with the HOLE algorithm.

### 14.2.4
### Dynamic Properties of Water Motion

A recent quasielastic neutron-scattering (QENS) study suggested that the water molecules inside the pore region of a triclinic lysozyme crystal could be divided into two populations [13]. The first mainly corresponds to the first hydration shell (surface zone), in which water molecules reorient themselves 5–10 times slower than in the bulk solvent and diffuse by jumps from hydration site to hydration site. The second group (core zone) corresponds to water molecules further away from the protein surface, in an incomplete hydration layer. In protein crystals, this second layer is actually confined *between* hydrated proteins. The self-diffusion coefficient is the most common measure of water mobility because one can use it more readily to compare simulation accuracy to experimental values. For the protein–water interface in a protein nanopore, the dynamic properties of surface water molecules show deviations from those of core zones. QENS studies show that diffusion behavior in the core zone corresponds to a self-diffusion coefficient, reduced approximately 50-fold compared to the diffusivity of free water and almost 10-fold compared with the self-diffusion coefficient of water molecules in the surface zone. Experimental studies also suggest that water mobility is highest close to the protein surface, and, under some crowding conditions, a 2D motion dominates.

Hydration sites are high-density regions in the 3D time-averaged solvent structure in diffraction experiments and in MD simulations of hydrated proteins. The hydration sites are characterized by a combination of an average occupancy and residence time of water molecules. Residence time determines the relaxation time of water molecules in the hydration layer around a given hydration site. The average site occupancy is defined as the average fraction of time that any water molecule occupies the site. We can characterize temporal ordering of the water molecules in a hydration shell of a protein via a population analysis of the hydration sites from which we derive average water residence times. Characterization of hydration sites at the protein surface can also explain the site-to-site jumping mechanism of water motion in a protein nanopore's surface zone. In MD simulations, however, protein hydration sites are defined as local maxima in the water number density map that satisfy certain conditions. They should be no further than 5 Å away from any protein atom and no closer than at least 1 Å. We determine the hydration structure of a lysozyme molecule by computing the water density on the entire surface of the enzyme in solution. By comparing the latter to the experimental hydration structure, we identified a total of 245 and 185 hydration sites around lysozyme molecules in orthorhombic and tetragonal unit cells, respectively. The number of hydration sites can be smaller than that of free proteins in solution because many of the sites in a crystal are buried. In general, the first hydration shell follows the shape of the protein, but in some regions, the hydration sites are

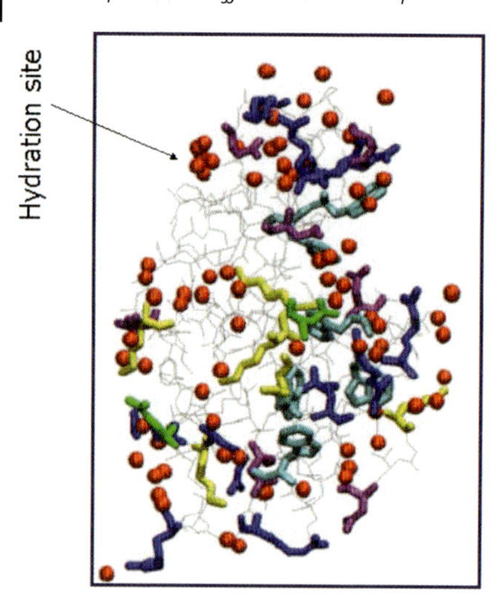


**Figure 14.3** Lysozyme hydration sites (red balls) calculated from the solvent-density map data (Lys is yellow, Arg is blue, Asp is purple, Glu is grebe, and Trp is light blue) [10].

clustered. Figure 14.3 shows that charged (Lys, Asp, Glu, and Arg) as well as polar (Trp) residues are located near high-density sites.

The residence times of water molecules on each hydration site are calculated from trajectories of 5 ns. The mean residence time varies for different proteins, but we use a survival probability function $P_{ij}(t, t+\tau)$ to determine the relaxation of water molecules in hydration layers around a protein atom. This probability adopts a value of one if the water molecule labeled $j$ is in the first hydration layer around site $i$, from time $t$ to time $t+\tau$. The site is considered occupied if a water molecule is found in a spherical volume with radius 1.5 Å (the van der Waals radius of water) centered at this site. The average of the survival probability over all the time configurations and the sum over all water molecules gives the so-called survival function, $P_i(\tau)$. This function is, in fact, the average number of $N$ water molecules belonging to hydration site $i$ that remain on the site after time $t$. Using this procedure, we calculate the residence time and site occupancy of water molecules in the first hydration layer of lysozyme molecules in LYZO and LYZT systems. Figure 14.4 shows that typical residence time varies between 20 and 700 ps for LYZO and between 10 and 1000 ps for LYZT. For both LYZO and LYZT systems, a Poisson-like residence time distribution is evident. The broad peak on the right-hand side of LYZT's residence-time distributions indicates that there are many hydration sites with long residence time that prolong the life of bound water molecules by up to two orders of magnitude, but such sites are few for LYZO. Residence times of the order of hundreds of picoseconds (strong sites) in a 5 ns simulation are likely due to the visit of a water molecule to a particular site, which is rare over the whole simulation. On average, more than 90% of hydration water molecules have short residence times

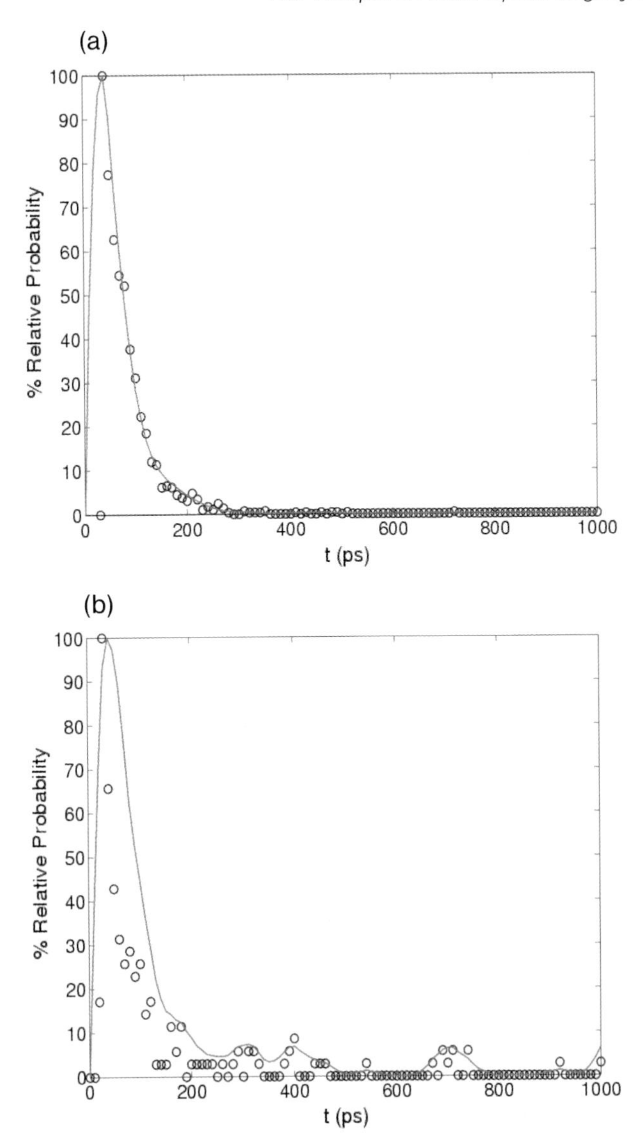


**Figure 14.4** Water residence time distributions for (a) LYZO and (b) LYZT systems [10].

(<100 ps), roughly 5% have intermediate residence times (100–500 ps), and the rest have long residence times (>500 ps). The water molecule that enters a site could stay there for a long time period or enter and immediately exit within a short time. Ultimately, the chemical nature of the protein residue close to the hydration shell, the number of hydrogen-binding opportunities, and the protein surface's local geometry (whether a site is buried or exposed) control the residence times.

### 14.2.5
### Water and Ion Diffusion

MD simulations provide lots of information about the static and dynamical pictures found in protein hydration structures. The self-diffusion coefficient $D$, which is widely used in both spectroscopic experiments and MD simulations, is a suitable parameter for characterizing water's dynamical behavior at the protein–solvent interface. The self-diffusivity $D$ is computed from the slope of the mean-square displacement (MSD) of water molecules by the Einstein relation. The use of the Einstein equation to determine $D$ requires the MSD to be linearly dependent with time. In practice, to determine the water diffusivity, a linear relation between MSD and time must be fulfilled on time scales longer than a few hundred picoseconds. In contrast to the bulk water molecules far from the protein surface (the core zone), water mobility in the protein's vicinity (the surface zone) is restricted. Researchers have also found this in the radial profiles of local diffusion coefficients for a free protein in solution. (In some simulations, translational diffusion in the surface zone is considerably reduced compared to that of bulk water). To analyze water diffusion at both the surface and core zones inside a pore in LYZO and LYZT systems, we consider a selected box of water in the crystal's pore region at a given time and then average the results from five selected boxes in different simulation frames every nanosecond [14]. For each frame of the trajectory, the selected water molecules reorder according to their distance from the water's center of mass and all the other atoms in the protein. An algorithm writes the output trajectories to the surface zone's water trajectory (within 0.3 nm from the protein surface) and that of the core zone (beyond 0.3 nm).

We analyze the MSD versus time for water molecules in the surface and core zones for the two lattices. In all the cases, the log–log behavior is linear with a slope close to one. This slope reaches to one for the bulk water molecules indicating that in fully hydrated pores, the Einstein equation describes the diffusion of water in the core zone. The diffusion coefficient of water molecules in the core zone of LYZT is approximately one half of that in LYZO. For LYZO, the diffusion coefficients in the core ($0.21 \pm 0.03$ nm$^2$ ns$^{-1}$) zones are roughly 4% of the diffusion coefficient $D_0$ of free water ($D_0 = 5.2$ nm$^2$ ns$^{-1}$ at 310 K). The previous studies show that the rate of solvent diffusion differs in the directions parallel and perpendicular to the surface. In particular, the change in the parallel diffusion rate could be a manifestation of pore-surface irregularity.

To show the differences in the protein–water attachment mechanism for three different ranges of residence times ($\tau < 100$ ps; $100$ ps $< \tau < 500$ ps; and $\tau >$ 500 ps), we collect the number of residues $n_r$ that each diffusing water molecule visited and then computed the probability distribution of $n_r$ for the three time ranges. We collect the number of times water molecules visited a hydration site. For molecules with relatively short residence times (<100 ps), the corresponding probability is higher, meaning that water molecules with longer residence times visit a given site less frequently. For hydration sites near the protein surface, it seems that a water molecule reorients itself much slower than in the core zone.

The molecules in the first hydration shell not only reorient themselves, but they also jump between hydration sites with a broad range of residence time on each site, from 10 ps up to 500 ps; few water molecules with residence times higher than 0.5 ns bound to a protein surface very strongly. We calculate the average jumping length (the average displacement between two successive jumps) for these water molecules in LYZO and LYZT, and found that to be five- and eightfold less, respectively, than that for free water. Water molecules further from the protein surface undergo a long-range translation and are often confined within the pore wall's hydrated proteins. The average MSD of these water molecules has a long-range diffusion coefficient 25- and 40-fold less than $D_0$ for LYZO and LYZT, respectively. Therefore, such molecules move five times slower than water molecules in the surface zone. Note that water molecules in the second layer act as a water reservoir for the first hydration layer. A mobile and rapidly exchanging hydrogen-bonding network that interacts with protein surface atoms allows for this dynamic behavior. Also note that no part of the solvent in a pore behaves like free water does. Water molecules close to the protein surface show a higher mobility, most likely because of the reduced space available for the solvent at the water–protein interface. For a highly confined system (such as ion channels), the diffusion rate of water is fastest on the surface; conversely, in free protein in water, diffusion is slower close to the protein surface and faster in the bulk. Although a complete understanding is still forthcoming [15], molecular simulations highlight the importance of the water dynamics on transport processes in CLPCs. Despite simplifications, the computer simulations address key issues such as residence time distribution, occupancy, and self-diffusion of water molecules in CLPCs. The essential parameters such as electrostatic interactions should be adjusted for better understanding the principle mechanisms of water and solvent motion in CLPCs. Moreover, a theoretical framework should be developed that can complement and refine simulation approaches.

## 14.3 Adsorption of Hydrocarbons in Zeolites

### 14.3.1 Introduction

Zeolites are microporous crystalline materials with pores of about the same size of a small molecule like water or *n*-hexane. The structure of a zeolite consists of covalently bonded $TO_4$ units, in which the T atom is usually a silicon (Si) or aluminum (Al) atom. To obey charge neutrality, the substitution of a silicon atom by an aluminum atom requires the presence of a nonframework cation (usually $Na^+$ or $K^+$) or a proton ($H^+$). There are approximately 170 different zeolite framework types that have been synthesized [16]. Figure 14.5 shows the pore structure of MFI-type zeolite. Zeolitic materials are widely used as water softener, selective adsorber, and catalyst for hydrocarbon conversions (catalytic cracking and isomerization) [17].

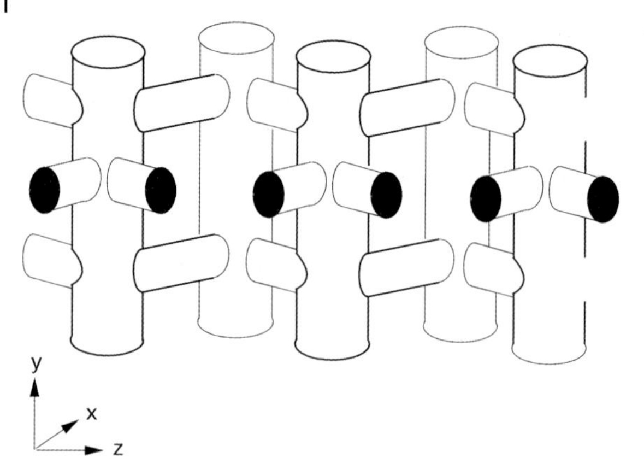


**Figure 14.5** Schematic representation of the pore structure of MFI-type zeolite. Straight channels (along the *y*-axis) are connected by zigzag channels (*xz*-plane) that cross at the intersections. The all-silica version of MFI-type zeolite is often called silicalite.

As molecular simulations can provide a fundamental understanding of processes and properties at the molecular scale, in the past few years these type of simulations have become an important tool for investigating the adsorption properties of small guest molecules in zeolite hosts. In particular, simulations have been used to study the adsorption of alkanes [18–27], alkenes [28–31], water [32–35], and aromatics [36–42]. For the comprehensive review on this topic, we refer the reader to Ref. [125].

As guest–zeolite interactions are often dominated by the dispersive interactions of oxygen atoms with the guest [43], classical force fields based on Lennard–Jones (LJ) interactions have become very popular in this field of research, especially for alkanes. Intramolecular interactions usually consist of bond stretching, bond bending, and torsion potentials, as well as nonbonded interactions for atoms separated more than three bonds. For adsorption of polar molecules such as benzene and $CO_2$, as well as for systems with nonframework cations, it is important to take the partial charges on the guest and the zeolite atoms into account; these are usually handled by the Ewald summation [11]. Nonframework cations cannot be neglected as they have a strong effect on adsorption properties, see for example Refs. [44–52]. To enable the use of efficient grid interpolation techniques to compute guest–zeolite interactions [53], the zeolite framework is often kept rigid. Framework flexibility usually does not have a large influence on adsorption properties [54]. However, this can be different when studying transport properties like the diffusion [55–57].

As explained in Chapter 7, the configurational-bias Monte Carlo (CBMC) technique can be used to efficiently compute the thermodynamic properties of guest molecules inside the pores of zeolites [11, 22, 58]. In a typical MC simulation, it is chosen at random with a fixed probability where trial move is performed: translation or rotation of a randomly selected guest molecule, (partial) regrow of a randomly selected guest molecule using CBMC, insertion or deletion of a guest molecule using CBMC (only for simulations in the grand-canonical ensemble, it is decided at random to insert (50%) of delete (50%) a guest molecule), identity change of a guest molecule (only for mixtures in the grand-canonical ensemble). For more simulation details, we refer the reader to Refs [22, 25].

### 14.3.2
### Obtaining Force-Field Parameters by Fitting Experimental Adsorption Isotherms

Suitable force-field parameters are a key ingredient for computing thermodynamic properties of hydrocarbons in zeolites that agree well with experiments. Dubbeldam *et al.* [25, 26] have shown that reliable interaction parameters can be obtained by fitting simulations to a set of experimental adsorption isotherms (average loading $\langle N \rangle$ as a function of the pressure $P$ or fugacity $f$). The force-field parameters are especially sensitive to isotherms with *inflection behavior*, e.g., a kink in the adsorption isotherm. While the isotherms of most alkanes for silicalite (all-silica MFI-type framework) show a conventional Langmuir isotherm, i.e.,

$$\frac{\langle N \rangle}{\langle N \rangle_{\max}} = \frac{P}{a + P}, \tag{14.2}$$

isobutane and heptane show an inflection behavior at a loading of approximately four molecules per unit cell of the zeolite (corresponding to approximately 0.7 mol per kg zeolite). For heptane, this is due to the so-called *commensurate freezing* effect [20, 59]. In the case of isobutane, the inflection behavior occurs because of the size and shape of an isobutane molecule relative to the size and shape of the channels and intersection of silicalite [21] (see Figure 14.6). To investigate this inflection, we have plotted the siting of isobutane at a pressure of 0.1 kPa (before the inflection) and 200 kPa (after the inflection) at 308 K and compared this with the siting of butane (see Figure 14.7). The differences are striking. While *n*-butane has approximately an equal probability to be in the straight channel, zigzag channel or intersection, isobutane has a strong preference for the intersection as most space is available there. Let us now compare the siting of isobutane before (low loading, Figure 14.7(b)) and after (high loading, Figure 14.7(c)) the inflection point in the isotherm. Below a loading of four molecules per unit cell, isobutane occupies only the intersections. At a loading of four molecules per unit cell, the intersections are fully occupied and to achieve higher loadings, isobutane must also seek residence in the other channels. This, however, is energetically very demanding and requires a significantly higher driving force (pressure) resulting in the inflection behavior. Only a *single* pair of LJ parameters ($\varepsilon$, $\sigma$) is able to correctly describe the experimental adsorption isotherms (see Figure 14.6). Similar inflection behavior is found for other branched alkanes. The inflection behavior of branched alkanes has severe consequences for adsorption isotherms of mixtures of linear and branched alkanes [60–65].

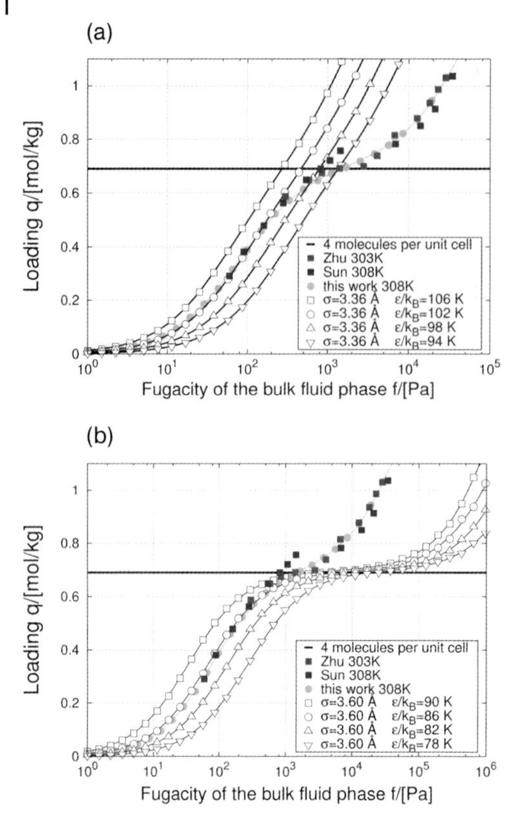


**Figure 14.6** Isotherms of isobutane at 308 K in silicalite. The O–CH parameters remain fixed at $\sigma = 3.92$ Å and $\varepsilon/k_B = 40$ K, while $\varepsilon_{O-CH_3}$ is examined over a range of reasonable values for two fixed values of $\sigma_{O-CH_3}$ (a) a rather too small of $\sigma_{O-CH_3} = 3.36$ Å and (b) a too high value of $\sigma_{O-CH_3} = 3.60$ Å. Only a single parameter pair, $\varepsilon_{O-CH_3}/k_B = 93$ K and $\sigma_{O-CH_3} = 3.48$ Å is able to describe the experimental data of Sun *et al.* [66] and Zhu *et al.* [67]. Figure reproduced with kind permission of Ref. [25].

### 14.3.3
### Adsorption of Alkanes at Low Loading

In Table 14.1, we have plotted the Henry coefficient and the heat of adsorption of linear alkanes (with CN carbon atoms) in silicalite. The heat of adsorption $\Delta H$ is a linear function of the carbon number, CN. The simulations using the force field of Dubbeldam *et al.* [25] are in excellent agreement with the experiments of Denayer *et al.* [68]. It is important to note that the experiments of Denayer *et al.* have *not* been used to calibrate the force field. In fact, the force field of Dubbeldam *et al.* is transferable to many other all-silica zeolite structures that have not been used to calibrate the force field [25, 27].

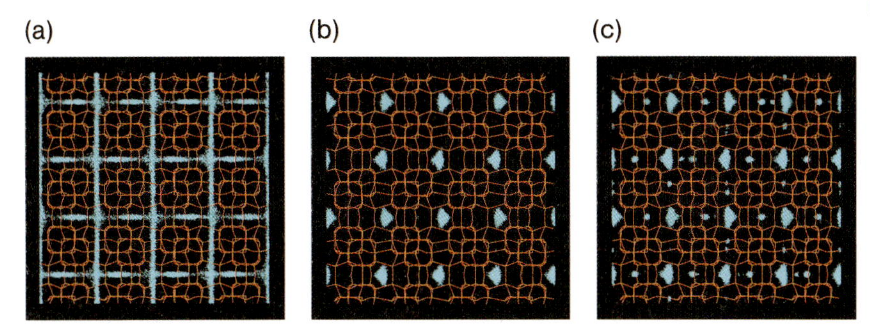


**Figure 14.7** Probability distributions of *n*-butane at 0.1 kPa (a), isobutane at 0.1 kPa (b) and isobutane at 200 kPa (c) on silicalite at 300 K. The zigzag channels are from the left to the right and the straight channels are perpendicular to the zigzag channels (projection on the *xy* plane), see also Figure 14.5. These figures were obtained by plotting the centers of mass of the molecules (blue dots) every 200 MC cycles until 10,000 points were collected.

**Table 14.1** Comparison of computed low-coverage properties in MFI with the experimental results of Denayer *et al.* [68].[a]

| CN | $K_H$ 573 K (mol kg$^{-1}$ Pa$^{-1}$) | | $K_\infty$ (mol kg$^{-1}$ Pa$^{-1}$) | | $-\Delta H$ (kJ mol$^{-1}$) | |
|---|---|---|---|---|---|---|
| | Simulation | Experiment | Simulation | Experiment | Simulation | Experiment |
| 5 | $3.04 \times 10^{-6}$ | $2.99 \times 10^{-6}$ | $2.33 \times 10^{-11}$ | $2.64 \times 10^{-11}$ | 56.13 | 55.7 |
| 6 | $6.10 \times 10^{-6}$ | $5.93 \times 10^{-6}$ | $6.07 \times 10^{-11}$ | $6.07 \times 10^{-11}$ | 65.87 | 66.0 |
| 7 | $1.23 \times 10^{-5}$ | $1.22 \times 10^{-5}$ | $1.53 \times 10^{-12}$ | $1.29 \times 10^{-12}$ | 75.77 | 76.7 |
| 8 | $2.43 \times 10^{-5}$ | $2.49 \times 10^{-5}$ | $3.67 \times 10^{-13}$ | $3.25 \times 10^{-13}$ | 85.82 | 86.6 |
| 9 | $4.61 \times 10^{-5}$ | $4.73 \times 10^{-5}$ | $8.59 \times 10^{-14}$ | $8.41 \times 10^{-14}$ | 95.81 | 96.1 |

| Relation | Simulation | Experiment |
|---|---|---|
| $-\Delta H = \alpha \text{CN} + \beta$ | $\alpha = 9.93$ | $\alpha = 10.1$ |
| $-\ln(K_\infty) = -A\Delta H + B$ | $A = 0.141, B = 16.54$ | $A = 0.143, B = 16.4$ |

[a] Both the Denayer and the simulation Henry coefficients $K_H$ of the linear alkanes have been fitted to $K_H = K_\infty e^{\frac{-\Delta H}{RT}}$ in the temperature range $T = 473$–$673$ K. Here, $K_\infty$ denotes the pre-exponential Henry coefficient, $\Delta H$ the heat of adsorption, and $R = 8.31451$ J mol$^{-1}$K$^{-1}$ the gas constant (table reproduced with kind permission of Ref. [25]).

## 14.4 Simulating Loading Dependence of the Diffusion in Zeolites Using Rare-Events Simulations

Most of the rare-event simulations are performed in the limit of infinite dilution. At higher concentration, guest–guest interactions can become important. For example, a molecule can occupy a lattice site and hence prevent another molecule to jump to this site. This effect can be incorporated in a kinetic Monte Carlo simulation. More complicated is the case that the presence of other guest molecules changes the hopping rate. Techniques have been developed by Tunca and Ford [69, 70] or using an approximate theory by Auerbach and coworkers [71–74] to take into account these effects using a simulation approach.

An alternative approach is to assume a distribution of lattice sites in a zeolite and to obtain the hopping rates from fitting to experimental data or to assume certain values and investigate the effect of changes in the hopping rate. These simulations are particularly useful to obtain insight in the mechanism of diffusion. Applications of these simulations have recently been reviewed by Keil *et al.* [75].

A rigorous extension of transition theory to high loading (dcTST) has been proposed by Dubbeldam and coworkers [76–78]. In this technique one computes the free-energy barrier of a tagged particle. The contribution of the other particles in the systems is included in this free-energy barrier and in the recrossing rate. From this hopping rate one can compute the self-diffusion coefficient directly. This diffusion coefficient corresponds exactly with the one that would be obtained from MD simulations if the assumptions underlying the rare-events simulations hold, that is, once a particle has hopped over a free-energy barrier it remains sufficiently long in the free energy minimum such that it can fully equilibrate before jumping over the next barrier. This technique, however, only provides the self-diffusion coefficient and not the collective or Maxwell–Stefan diffusion coefficient [79].

### 14.4.1 Diffusion of Hydrocarbons in MFI

At present, many experimental and simulation data have been published on the diffusion of hydrocarbons in zeolites; yet it is difficult to open a text on diffusion in zeolites that does not start with a figure similar to Figure 14.8. This figure demonstrates that depending on the experimental technique diffusion coefficients of linear alkanes in MFI are found that can vary many orders of magnitude. Microscopic techniques like pulse field gradient NMR and quasielastic neutron experiments (QENS) are in reasonably good agreement [80]. The more macroscopic techniques that are based on measuring changes in the weight of the zeolite sample can deviate significantly from the microscopic techniques [81]. Many different possible explanations have been put forward, but we are still lacking a detailed understanding. However, the consensus is that molecular simulations are in reasonable agreement with microscopic techniques [82, 83].

It is interesting to discuss the most simple hydrocarbon, methane, in MFI in more detail. This system has been simulated by many groups [53, 56, 84–99, 99–106]. Figure 14.9 shows the diffusion of methane as a function of loading. The difference between Figures 14.8 and 14.9 is that in the former it is implicitly assumed that diffusion coefficients are independent of the loading. Clearly, this figure indicates a much better agreement of various experimental and simulation results. Essential to obtain this agreement is that in Figure 14.9, for each data point the loading has been carefully estimated for each experimental data point. The figure also shows some experimental data points that show large differences. This maybe caused

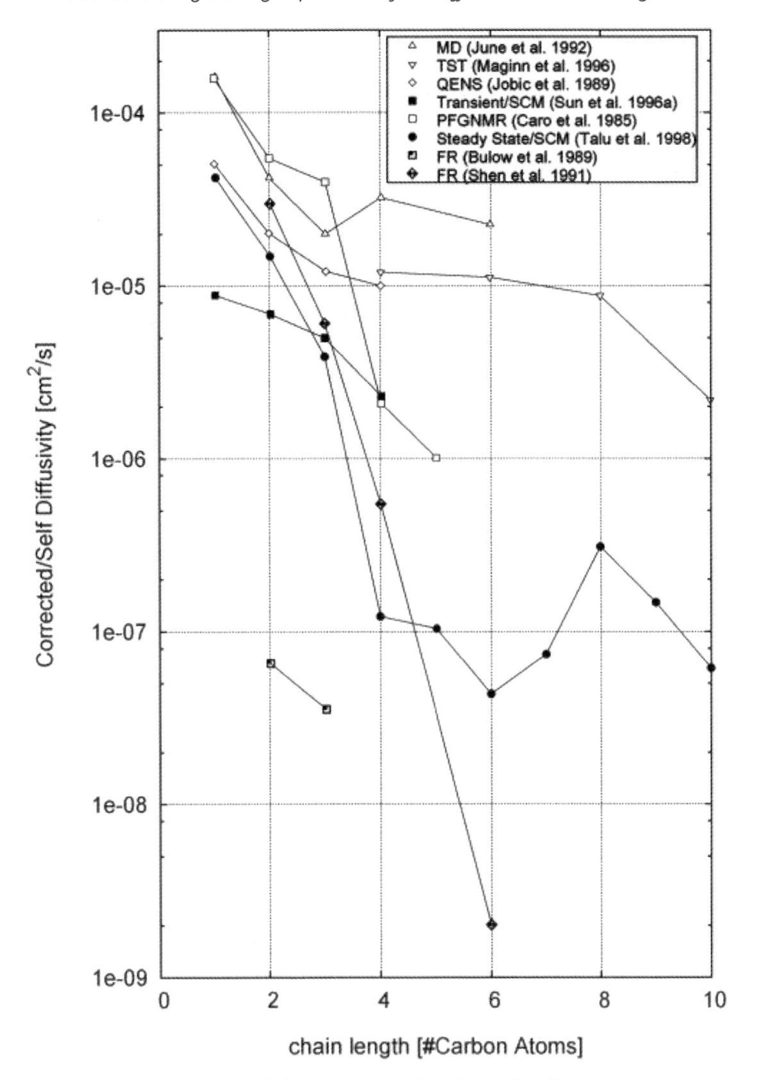


**Figure 14.8** Comparison of the experimental and simulated diffusion coefficients of linear alkanes in MFI as a function of chain length.

by a poor estimate of the loading, or that indeed macroscopic techniques also for methane show large deviations.

The interesting phenomena in Figure 14.9 is the peak in the collective diffusion coefficient at 16 molecules per unit cells; careful inspection of the collective diffusion coefficient shows more local maxima (e.g., loadings of 2, 8, or 24 molecules per unit cell). As the accuracy of the simulation results in Figure 14.9 is smaller than the symbol size, these "humps" do not disappear if simulations are extended for a very long time, but this irregular behavior is intrinsic to these systems [107].

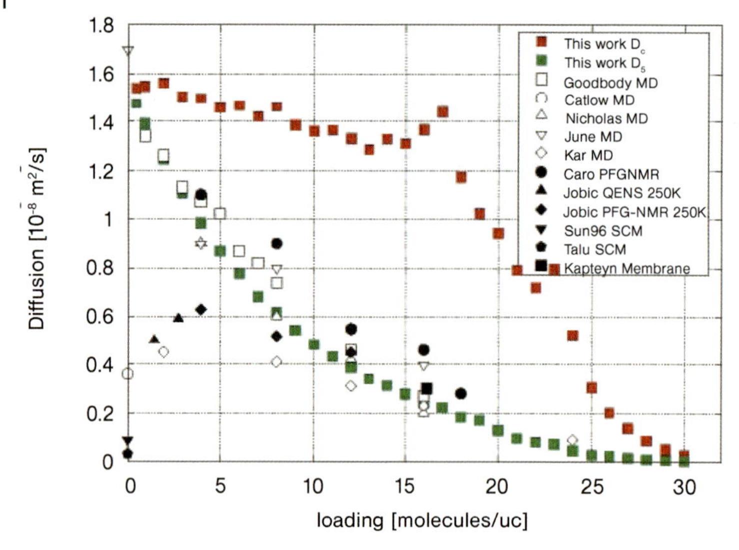


**Figure 14.9** Comparison of experimental and simulated diffusion coefficients of methane MFI as a function of loading.

The dcTST transition theory has been used to explain this irregular behavior. This method, however, is used in Ref. [107] in a "wrong way." Instead of using the technique to compute the diffusion coefficient, the technique is used to relate a diffusion coefficient that has been computed with a simple MD simulation to a corresponding free-energy profile. The dcTST method is used to justify that the changes in the free-energy profiles have a one-to-one correspondence to changes in the diffusion coefficient.

Figure 14.10(c) shows that each hump can be associated with a corresponding hump in only one of the three components of the diffusion coefficients. Figure 14.10(d) shows the loading dependence of the free-energy profile of a methane molecule moving along the straight channel ($y$-direction). At low loading there are three adsorption sites: two in each of the intersections and one in the middle of the channel (the figure shows two minima but only one can be occupied). These adsorption sites are visualized in Figure 14.10(a). If one further increases the loading to the point that all the low-loading adsorption sites are occupied, the system needs to create "space" for the additional molecules. The free-energy profiles show additional adsorption sites at high loading (see Figure 14.10(b)). One can visualize this at low loading the molecules hop on a lattice that suddenly "changes" as the loading is increased. As the number of lattice sites has changed the loading dependence will have a different slope. In addition, at the point the system "switches" from one lattice to the other, the free-energy profile becomes relatively flat and hence causing an increase in the diffusion coefficient. As such changes depend on the details of the channel one can understand that a similar effect occurs in the zigzag channel at a different loading.

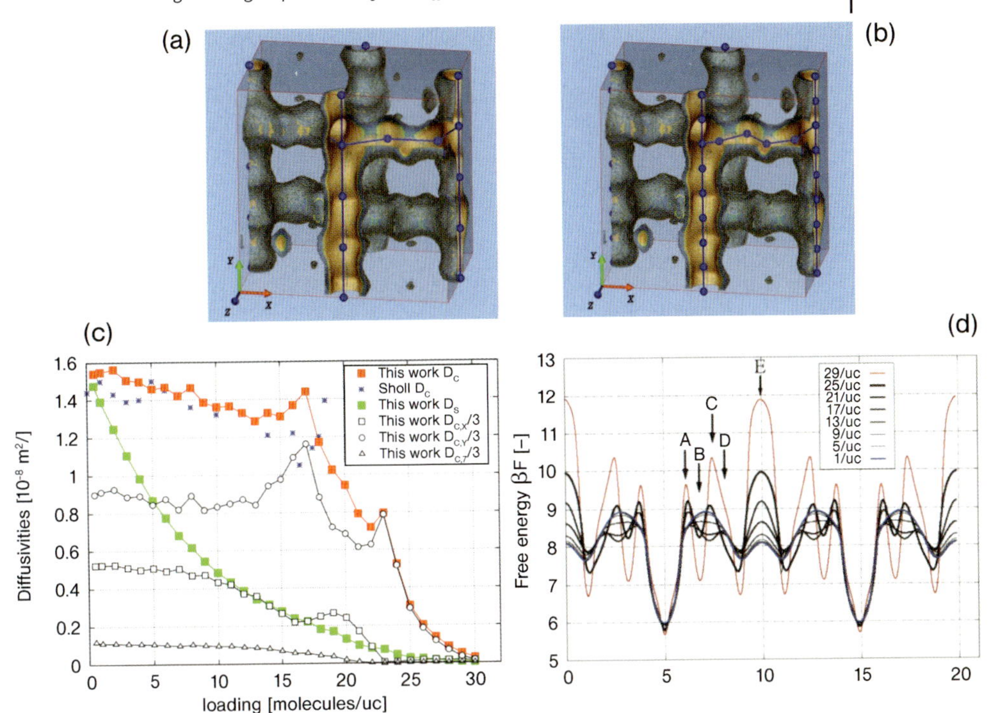


**Figure 14.10** Diffusion of methane in MFI: (a) and (b) show the adsorption sites of methane in MFI at low and high loading, respectively, (c) shows the diffusion coefficients in various directions, and (d) the free-energy profile along the straight channel (y-direction) at various loadings. Figure reproduced with kind permission of Refs [107, 108].

The lack of accurate and consistent experimental data, as is suggested in Figure 14.8, makes it very difficult to validate the force field and the assumptions underlying a given model. The fact that in Figure 14.8 all results are plotted in the same figure, irrespective of the loading, illustrates that for a very long time it was assumed that diffusion results in MFI, and many other structures, are independent of the loading. For methane, however, we have shown that a careful analysis of the experimental data illustrates that a part of inconsistencies can be attributed to this assumption. It would be interesting to perform a similar analysis for the other systems.

Beerdsen *et al.* [108, 109] assumed that diffusion can be described by a hopping process from one site to another on a lattice model of the zeolite. For a given molecule, Beerdsen *et al.* computed the free-energy profile of a molecule hopping from one site to another. These free-energy profiles allow to make a classification that helps us understanding the loading dependence. This classification is shown in Figure 14.11. The assumption is that one can model zeolite as ellipsoids. One can make one-dimensional zeolites by placing the cylinders on a line and connecting them. The circle connecting two cylinders defines the window diameter and depending on the orientation of the cylinders one gets one-dimensional tubes or a zeolite with a more cage-like character. By connecting the cylinders in alternating orientations one can mimic a two- or three-dimensional structure. The corresponding free-energy profiles show that the barrier for diffusion in the case of a tube is very small if the ratio of the window diameter and diameter of the middle part of the ellipsoid is close to one. A very different situation is the cage-like structure in which the windows form a barrier for diffusion, while for the two-dimensional structures the vertically oriented cylinders form entropic traps. The right part of the figure shows some examples of real zeolite structures and the corresponding free-energy profile of methane.

We now consider the effect of increasing the loading; as all ellipsoids are identical for all structures the "second" molecule will be preferentially placed at the same location in the ellipsoid. In fact, to a first approximation, this probability is the highest where the free energy in Figure 14.11 is the lowest. At this point, it is important to mention that the hopping rate and hence the diffusion coefficient is not only determined by the free-energy barrier, but also is the product of this free-energy barrier and the recrossing coefficient. However, for all systems that have been studied this coefficient is a monotonically decreasing function of the loading. We can now envision the following scenarios:

- *Tube-like zeolites.* There is little preference for the additional molecules to be adsorbed. However, as the molecules prefer to be in contact with the walls rather than with another molecule, as a consequence the free-energy profile shifts to higher values but there will be little difference in the shift of the top of the barrier and the bottom. To a first approximation one would expect the free-energy barrier to remain constant and hence the recrossing coefficient causes a decrease in the diffusion coefficient as a function of loading.
- *Cage-like zeolites.* For these zeolites, the preferential adsorption is in the cages. Hence, additional molecules will increase the bottom of the free-energy profile, but not the top. Hence, additional molecules will lower the free-energy barrier and one would expect an increase in the diffusion coefficient.
- *Two- or three-dimensional structures.* Here the adsorption will be both in the horizontal and vertically oriented ellipsoids. In the horizontal ellipsoids they will form an additional barrier and these additional molecules will make the vertically oriented ellipsoids less attractive. As the latter effect will be smaller and the net result is an increase in the free-energy barrier, resulting in a decrease in the diffusion coefficient.

The above free-energy arguments only apply for the self-diffusion coefficient. To understand the collective diffusion coefficient, one has to take into account the lattice topology and other factors that influence the collective behavior. For a more complete review of the diffusion of guest molecules in zeolites, we refer the reader to Refs [110–112].

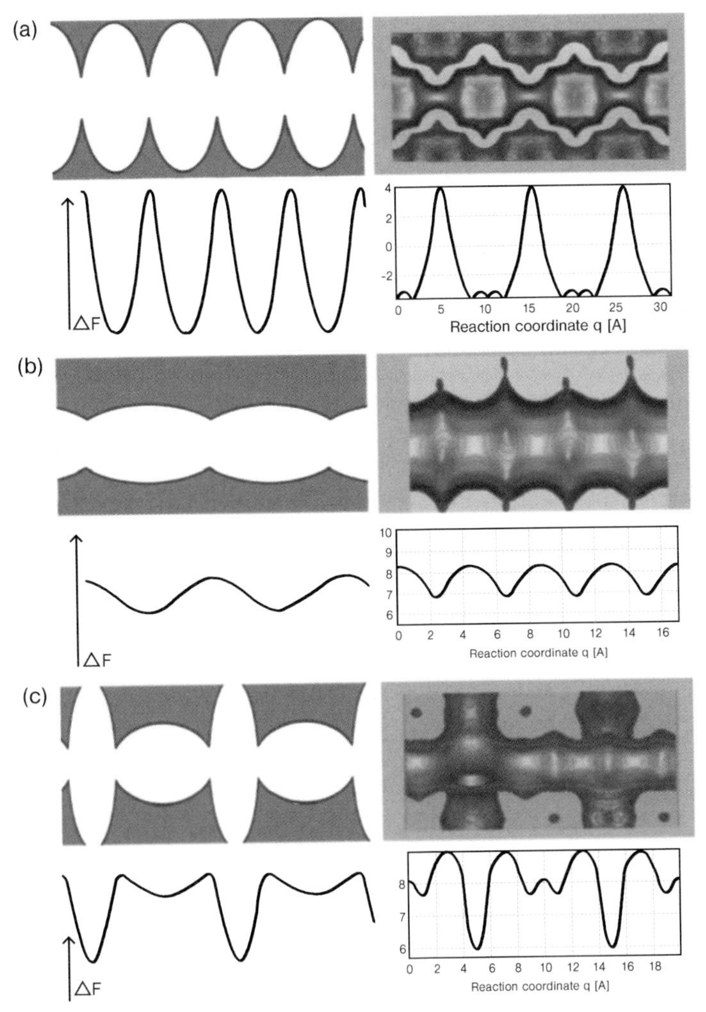


**Figure 14.11** Zeolites made out of ellipsoids; the left figures show three ways in which these cylinders can be connected and below the corresponding free-energy profile of a molecule diffusion through these structures. The right figures give an example of real zeolites ((a): SAS, (b): AFI, and (c): MFI) and methane diffusing through them at zero loading. Figure reprinted with kind permission from Ref. [109].

## 14.5
## Simulation of Diffusion and Reaction in Functionalized, Amorphous Nanoporous Catalysts, and Membranes

In a heterogeneous catalyst, reactant molecules diffuse through the pore network, collide with pore walls, and react on active sites on these walls. This implies that the topology of the pore network and the morphology of pores affect the molecular movement and the accessibility of the active sites. Hence the diffusivities of components and therefore the conversion and product distributions of the reaction depend on the catalyst geometry. On the other hand, porous amorphous catalysts and supports often have a random (fractal) internal surface down to molecular scales [113–116]. This has been confirmed by small-angle X-ray scattering and adsorption studies, and it is a consequence of the specific preparation conditions of these materials, which are often based on a sol–gel synthesis method. Numerous works on the effects of this fractal surface morphology on diffusion and reaction phenomena are performed, from the catalyst fractal pore scale [116], all the way up to the scale of industrial reactors [117, 118]. The attention was initially focused on Knudsen diffusion, since it is expected that it is most influenced by surface roughness, and it is a dominant diffusion mechanism in many gas reactions in mesoporous catalysts. It has been shown how the *gradientless* (self-)diffusion is strongly influenced by surface roughness [116, 119]. A model reaction has also been added to study the effective activity and selectivity as a function of surface roughness [120]. Especially interesting is to find out whether the diffusivities depend on the rates of reactant conversion. This is important for applications, because it shows when it is allowed to use the easier computations based on traditional continuity equations, using (nonfractal) reaction calculations and fractal diffusivities, which were calculated earlier in the absence of reactions. The results have been qualitatively compared to what is known for diffusion in microporous materials. Microporous materials like zeolites are frequently used for separations, adsorption, and catalyst supports. Since these processes are often diffusion limited, it is important to know the molecular diffusivities. The correlation between the chemical heterogeneity and diffusion still remains poorly understood. It is not feasible, if not impossible, to make exact predictions for the effective properties of composite materials with the simple models of morphology. Both continuum and discrete models have been applied to provide and describe the theoretical approaches for estimating the effective properties of random heterogeneous materials. As described above, continuum models represent the classical approach in describing and analyzing transport processes in materials of complex and irregular morphology. Thus, the effective properties of materials are defined as averages of the corresponding microscopic quantities. The shortcoming of discrete models such as random network models, Bethe lattice models, etc., compared to continuum models, is the demand for large computational effort to represent a realistic model to describe the material and simulate its effective properties. It is of course a challenging problem to generate a realization of a heterogeneous material, for which limited microstructure information is available. Based on density functional calculations (DFT and QM/MM), new insights were provided on the importance of electronic and steric effects for epoxidation reaction of *cis-* and *trans*-methylstyren catalyzed by anchored oxo-Mn–salen into MCM-41 channels (see Figure 14.12). All calculations were performed on a catalytic surface with triplet spin state to avoid the complications regard to the Mn-salen spin-crossing [126, 127]. Calculations showed that how immobilization relates to the linker and substrate choice as well as the interplay with the confined channel. Although a *trans*-substrate has a

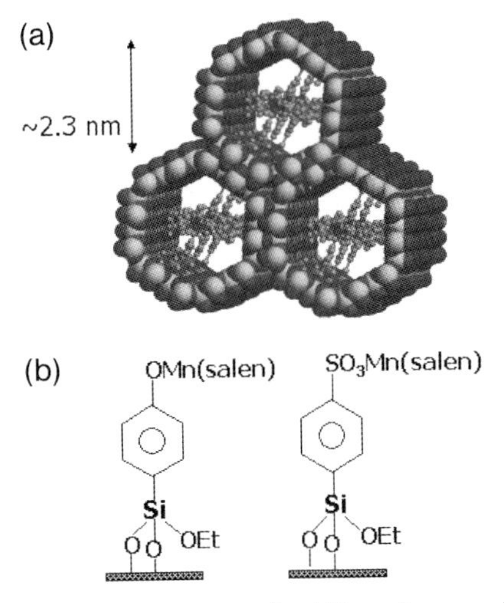


**Figure 14.12** (a) Visualized Mn–salen complexes anchored inside a MCM-41 channel using phenoxyl group as the immobilizing linker. (b) The model for the immobilized Mn–salen complexes using different axial linkages [126, 127].

higher level of asymmetric induction to the immobilized Mn–salen complex than that to a homogeneous catalyst, the reaction path is most likely more favorable for the *cis*-substrate. The MCM-41 channel reduces the energy barriers and enhances the enantioselectivity by influencing geometrical distortions of Mn–salen complex.

Novel functionalized nanotube membranes are recently developed and used to efficiently separate a chiral drug from its racemic mixture [121]. Enantiomeric separation in these materials strongly relates to modifier choices and the interplay with the nanopore confinement and substrate–modifier interactions. By means of molecular simulations we propose that the enantioselectivity of such membranes can be improved in a bioinspired way. MD simulations are used to evaluate the capability of a modified silica nanotube for enantiomeric separation of two amino acids, R- and S-2-phenylglycine. This smart nanotube is functionalized as an artificial protein channel in cell membranes. The biomimicry is performed through attaching functional residues (Arg, Glu, Asp) into the nanotube (see Figure 14.13). Simulations indicate that the selective transport of one of the enantiomers (S-) inside the modified channel is strongly affected by the presence of a special electrostatic field inside the channel. The mechanism of enantioselective passage depends on the internal degrees of freedom of the attached residues and interactions of phenylglycine molecules with these residues. The translational–rotational motion of chiral molecules as well as their average dipole orientation is responsible for selective chiral transport inside the nanotube. It is remarkable how configuration of the immobilized residues enhances the enantiomeric separation of the functionalized nanotube. The degree of stereoselectivity in these membranes is as

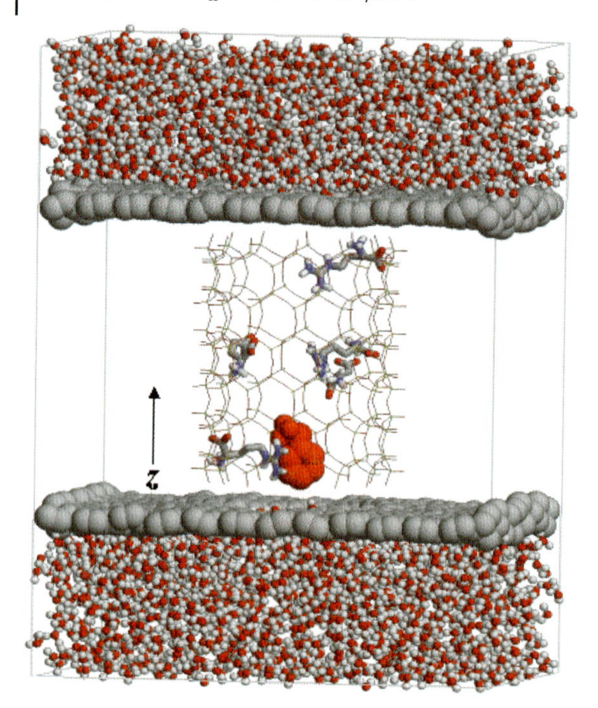


**Figure 14.13** The functionalized silica channel is fixed within a slab membrane opening to bulk of water molecules in the two reservoirs at both ends. The phenylglycin molecule is shown in red. Water molecules in the MCM-41 channel are removed for clarity [122].

important as enhancing fluxes across the membrane. We suggest that the enantioselectivity of these membranes can be improved in a nature-inspired way [122]. In order to demonstrate the importance of chiral interactions in atomic scale, within the framework of equilibrium and nonequilibrium MD simulations, we study the complete passage of enantiomers of R- and S-phenylglycine through a modified silica nanotube. The simulations reflect the channel's chiral selectivity for S-isomer. The shape of the electric field, the strength of the force, and translational–rotational motion of chiral molecules are the main reasons for selective chiral transport inside the nanotube. The present results guide future research on biomimic selective chiral membranes, which are able to separate chiral drugs from a mixture.

## Acknowledgments

TJHV acknowledges financial support from the Netherlands Organization for Scientific Research (NWO-CW) through a VIDI grant.

## References


1 Ohmine, I., Tanaka, H. *Chem. Rev.* **1993**, *95*, 2545.

2 Johnson, L.N., Phillips, D.C. *Nature* **1965**, *206*, 761–763.

3 Vilenchik, L.Z., Griffith, J.P., St. Clair, N., Navia, M.A., Margolin, A.L. *J. Am. Chem. Soc.* **1998**, *120*, 4290–4294.

4 Margolin, A.L., Navia, M.A. *Angew. Chem. (Int. Ed.)* **2001**, *40*, 2204–2222.

5 Morozov, V.N., Kachalova, G.S., Evtodienko, V.U., Lanina, N.F., Morozova, T.Y. *Eur. Biophys. J.* **1995**, *24*, 93–98.

6 Velev, O.D., Kaler, E.W., Lenhoff, A.M. *J. Phys. Chem. B* **2000**, *104*, 9267–9275.

7 Malek, K., Odijk, T., Coppens, M.O. *Nanotechnology* **2005**, *16*, S522.

8 Malek, K., Odijk, T., Coppens, M.O. *Chem. Phys. Chem.* **2004**, *5*, 1596.

9 Malek, K., Coppens, M.O. *J. Phys. Chem. B* **2008**, *112*, 1549.

10 Malek, K. *Comput. Sci. Eng.* **2007**, *9*, 90–95.

11 Frenkel, D., Smit, B., *Understanding Molecular Simulation: from Algorithms to Applications*, 2nd ed., Academic Press; San Diego, **2002**.

12 Berendsen, H.J.C., Postma, J.P.M., van Gunsteren, W.F., DiNola, A., Haak, J.R. *J. Chem. Phys.* **1984**, *81*, 3684–3690.

13 Bon, C., Dianoux, A.J., Ferrand, M., Lehmann, M.S. *Biophys. J.* **2002**, *83*, 1578.

14 Malek, K. *Comput. Sci. Eng.* **2007**, *9*, 70–75.

15 van Hijkoop, J., Dammers, A.J., Malek, K., Coppens, M.O. *J. Chem. Phys.* **2007**, *127*, 085101.

16 Baerlocher, Ch., McCusker, L.B., Olson, D.H., *Atlas of Zeolite Framework Types*, 6th ed., Elsevier; Amsterdam, **2007**.

17 Smit, B., Maesen, T.L.M. *Nature* **2008**, *451*, 671–678.

18 Smit, B., Siepmann, J.I. *Science* **1994**, *264*, 1118–1120.

19 Maginn, E.J., Bell, A.T., Theodorou, D.N. *J. Phys. Chem.* **1995**, *99*, 2057–2079.

20 Smit, B., Maesen, T.L.M. *Nature* **1995**, *374*, 42–44.

21 Vlugt, T.J.H., Zhu, W., Kapteijn, F., Moulijn, J.A., Smit, B., Krishna, R. *J. Am. Chem. Soc.* **1998**, *120*, 5599–5600.

22 Vlugt, T.J.H., Krishna, R., Smit, B. *J. Phys. Chem. B* **1999**, *103*, 1102–1118.

23 Pascual, P., Ungerer, P., Tavitian, B., Pernot, P., Boutin, A. *Phys. Chem. Chem. Phys.* **2003**, *5*, 3684–3693.

24 Chempath, S., Denayer, J. F. M., De Meyer, K. M. A., Baron, G. V., Snurr, R. Q. *Langmuir* **2004**, *20*, 150–156.

25 Dubbeldam, D., Calero, S., Vlugt, T.J.H., Krishna, R., Maesen, T.L.M., Smit, B. *J. Phys. Chem. B* **2004**, *108*, 12301–12313.

26 Dubbeldam, D., Calero, S., Vlugt, T.J.H., Krishna, R., Maesen, T.L.M., Beerdsen, E., Smit, B. *Phys. Rev. Lett.* **2004**, *93*, 088302.

27 Liu, B., Smit, B. *J. Phys. Chem. B* **2006**, *110*, 20166–20171.

28 Pascual, P., Ungerer, P., Tavitian, B., Boutin, A. *J. Phys. Chem. B* **2004**, *108*, 393–398.

29 Jakobtorweihen, Sven, Hansen, Niels, Keil, Frerich *J. Mol. Phys.* **2005**, *103*, 471–489.

30 Granato, M.A., Vlugt, T.J.H., Rodrigues, A.E. *Ind. Eng. Chem. Res.* **2007**, *46*, 7239–7245.

31 Liu, B., Smit, B., Rey, F., Valencia, S., Calero, S. *J. Phys. Chem. C* **2008**, *112*, 2492–2498.

32 Ramachandran, C. E., Chempath, S., Broadbelt, L. J., Snurr, R. Q. *Microporous Mesoporous Mat.* **2006**, *90*, 293–298.

33 Di Lella, A., Desbiens, N., Boutin, A., Demachy, I., Ungerer, P., Bellat, J. P., Fuchs, A. H. *Phys. Chem. Chem. Phys.* **2006**, *8*, 5396–5406.

34 Desbiens, N., Boutin, A., Demachy, I. *J. Phys. Chem. B* **2005**, *109*, 24071–24076.

35 Desbiens, N., Demachy, I., Fuchs, A.H., Kirsch-Rodeschini, H.,

Soulard, M., Patarin, J. *Angew. Chem. (Int. Ed.)* **2005**, *44*, 5310–5313.
36 Snurr, R.Q., Bell, A.T., Theodorou, D.R. *J. Phys. Chem.* **1993**, *97*, 13742–13752.
37 Lachet, V., Boutin, A., Tavitian, B., Fuchs, A. H. *Faraday Discuss.* **1997**, *106*, 307–323.
38 Lachet, V., Boutin, A., Tavitian, B., Fuchs, A. H. *Langmuir* **1999**, *15*, 8678–8685.
39 Chempath, S., Snurr, R.Q., Low, J.J. *AIChE J.* **2004**, *50*, 463–469.
40 Yue, X.P., Yang, X.N. *Langmuir* **2006**, *22*, 3138–3147.
41 Zeng, Y.P., Ju, S.G., Xing, W.H., Chen, C.L. *Ind. Eng. Chem. Res.* **2007**, *46*, 242–248.
42 Ban, S., van Laak, A., de Jongh, P.E., van der Eerden, J.P.J.M., Vlugt, T.J.H. *J. Phys. Chem. C* **2007**, *111*, 17241–17248.
43 Bezus, A.G., Kiselev, A.V., Lopatkin, A.A., Du, P.Q. *J. Chem. Soc., Faraday Trans. II* **1978**, *74*, 367–379.
44 Buttefey, S., Boutin, A., Fuchs, A. H. *Mol. Simul.* **2002**, *28*, 1049–1062.
45 Beerdsen, E., Smit, B., Calero, S. *J. Phys. Chem. B* **2002**, *106*, 10659–10667.
46 Beerdsen, E., Dubbeldam, D., Smit, B., Vlugt, T.J.H., Calero, S. *J. Phys. Chem. B* **2003**, *107*, 12088–12096.
47 Beauvais, C., Guerrault, X., Coudert, F. X., Boutin, A., Fuchs, A. H. *J. Phys. Chem. B.* **2004**, *108*, 399–404.
48 Calero, S., Dubbeldam, D., Krishna, R., Smit, B., Vlugt, T.J.H., Denayer, J.F., Martens, J. A., Maesen, T.L.M. *J. Am. Chem. Soc.* **2004**, *126*, 11377–11386.
49 García-Pérez, E., Dubbeldam, D., Maesen, T. L. M., Calero, S. *J. Phys. Chem. B* **2006**, *110*, 23968–23976.
50 Liu, B., García-Pérez, E., Dubbeldam, D., Smit, B., Calero, S. *J. Phys. Chem. C* **2007**, *111*, 10419–10426.
51 García-Pérez, E., Dubbeldam, D., Liu, B., Smit, B., Calero, S. *Angew. Chem.-Int. Edit.* **2007**, *46*, 276–278.
52 García-Sánchez, A., García-Pérez, E., Dubbeldam, D., Krishna, R., Calero, S. *Adsorpt. Sci. Technol.* **2007**, *25*, 417–427.
53 June, R.L., Bell, A.T., Theodorou, D.N. *J. Phys. Chem.* **1992**, *96*, 1051–1060.
54 Vlugt, T.J.H., Schenk, M. *J. Phys. Chem. B* **2002**, *106*, 12757–12763.
55 Kopelevich, D.I., Chang, H.C. *J. Chem. Phys.* **2001**, *114*, 3776–3789.
56 Leroy, F., Rousseau, B., Fuchs, A. H. *Phys. Chem. Chem. Phys.* **2004**, *6*, 775–783.
57 Zimmermann, N. E. R., Jakobtorweihen, S., Beerdsen, E., Smit, B., Keil, F. J. *J. Phys. Chem. C* **2007**, *111*, 17370–17381.
58 Smit, B., Siepmann, J.I. *J. Phys. Chem.* **1994**, *98*, 8442–8452.
59 van Well, W.J.M., Wolthuizen, J.P., Smit, B., van Hooff, J.H.C., van Santen, R.A. *Angew. Chem. (Int. Ed.)* **1995**, *34*, 2543–2544.
60 Krishna, R., Smit, B., Vlugt, T.J.H. *J. Phys. Chem. A* **1998**, *102*, 7727–7730.
61 Krishna, R., Vlugt, T.J.H., Smit, B. *Chem. Eng. Sci.* **1999**, *54*, 1751–1757.
62 Schenk, M., Vidal, S.L., Vlugt, T.J.H., Smit, B., Krishna, R. *Langmuir* **2001**, *17*, 1558–1570.
63 Krishna, R., Smit, B., Calero, S. *Chem. Soc. Rev.* **2002**, *31*, 185–194.
64 Krishna, R., Calero, S., Smit, B. *Chem. Eng. J.* **2002**, *88*, 81–94.
65 Smit, B., Krishna, R. *Chem. Eng. Sci.* **2003**, *58*, 557–568.
66 Sun, M.S., Shah, D.B., Xu, H.H., Talu, O. *J. Phys. Chem. B* **1998**, *102*, 1466–1473.
67 Zhu, W., Kapteijn, F., Moulijn, J.A. *Phys. Chem. Chem. Phys.* **2000**, *2*, 1989–1995.
68 Arik, I.C., Denayer, J.F.M., Baron, G.V. *Microporus Mesoporus Mat.* **2003**, *60*, 111–114.
69 Tunca, C., Ford, D. M. *Chem. Eng. Sci.* **2003**, *58*, 3373–3383.
70 Tunca, C., Ford, D.M. *J. Chem. Phys.* **1999**, *111*, 2751–2760.
71 Saravanan, C., Auerbach, S. M. *J. Chem. Phys.* **1997**, *107*, 8120–8131.
72 Saravanan, C., Auerbach, S. M. *J. Chem. Phys.* **1997**, *107*, 8132–8137.
73 Saravanan, C., Jousse, F., Auerbach, S.M. *Phys. Rev. Lett.* **1998**, *80*, 5754–5757.

74 Saravanan, C., Auerbach, S. M. *J. Chem. Phys.* **1999**, *110*, 11000–11011.
75 Keil, F. J., Krishna, R., Coppens, M. O. *Rev. Chem. Eng.* **2000**, *16*, 71–197.
76 Beerdsen, E., Dubbeldam, D., Smit, B. *Phys. Rev. Lett.* **2004**, *93*, 0248301.
77 Dubbeldam, D., Beerdsen, E., Vlugt, T.J.H., Smit, B. *J. Chem. Phys.* **2004**, *122*, 224712.
78 Dubbeldam, D., Beerdsen, E., Calero, S, Smit, B. *J. Phys. Chem. B.* **2006**, *110*, 3164–3172.
79 Krishna, R., Wesselingh, J.A. *Chem. Eng. Sci.* **1997**, *52*, 861–911.
80 Jobic, H., Schmidt, W., Krause, C. B., Karger, J. *Microporous Mesoporous Mat.* **2006**, *90*, 299–306.
81 Karger, J., Ruthven, D. M. *Zeolites* **1989**, *9*, 267–281.
82 Leroy, F., Jobic, H. *Chem. Phys. Lett.* **2005**, *406*, 375–380.
83 Jobic, H., Theodorou, D. N. *J. Phys. Chem. B* **2006**, *110*, 1964–1967.
84 June, R.L., Bell, A.T., Theodorou, D.N. *J. Phys. Chem.* **1990**, *94*, 8232–8240.
85 Demontis, P., Fois, E.S., Suffritti, G.B., Quartieri, S. *J. Phys. Chem.* **1990**, *94*, 4329–4334.
86 Goodbody, S.J., Watanabe, K., MacGowan, D., Walton, J.P.R.B., Quirke, N. *J. Chem. Soc., Faraday Trans.* **1991**, *87*, 1951–1958.
87 Snurr, R.Q., June, R.L., Bell, A.T., Theodorou, D.R. *Mol. Sim.* **1991**, *8*, 73–92.
88 Nowak, A.K., Ouden, C.J.J. den, Pickett, S.D., Smit, B., Cheetham, A.K., Post, M.F.M., Thomas, J.M. *J. Phys. Chem.* **1991**, *95*, 848–854.
89 Hufton, J.R. *J. Phys. Chem.* **1991**, *95*, 8836–8839.
90 Catlow, C. R. A., Freeman, C. M., Vessal, B., Tomlinson, S. M., Leslie, M. *J. Chem. Soc.-Faraday Trans.* **1991**, *87*, 1947.
91 Demontis, P., Suffritti, G.B., Fois, E.S., Quartieri, S. *J. Phys. Chem.* **1992**, *96*, 1482–1490.
92 Demontis, P., Suffritti, G.B., Mura, P. *Chem. Phys. Lett.* **1992**, *191*, 553–560.
93 Kawano, M., Vessal, B., Catlow, C.R.A. *J. Chem. Soc., Chem. Commun.* **1992**, *12*, 879–880.
94 Maginn, E.J., Bell, A.T., Theodorou, D.N. *J. Phys. Chem.* **1993**, *97*, 4173–4181.
95 Nicholas, J.B., Trouw, F.R., Mertz, J.E., Iton, L.E., Hopfinger, A.J. *J. Phys. Chem.* **1993**, *97*, 4149–4163.
96 Smirnov, K.S. *Chem. Phys. Lett.* **1994**, *229*, 250–256.
97 Skoulidas, A. I., Sholl, D. S. *J. Phys. Chem. B* **2001**, *105*, 3151–3154.
98 Skoulidas, A. I., Sholl, D. S. *J. Phys. Chem. B* **2002**, *106*, 5058–5067.
99 Ahunbay, M.G., Elliott, J.R., Talu, O. *J. Phys. Chem. B* **2002**, *106*, 5163–5168.
100 Fritzsche, S., Wolfsberg, M., Haberlandt, R. *Chem. Phys.* **2003**, *289*, 321–333.
101 Bussai, C., Fritzsche, S., Haberlandt, R., Hannongbua, S. *J. Phys. Chem. B* **2004**, *108*, 13347–13352.
102 Bussai, C., Fritzsche, S., Haberlandt, R., Hannongbua, S. *Langmuir* **2005**, *21*, 5847–5851.
103 Hussain, I., Titiloye, J. O. *Microporous Mesoporous Mat.* **2005**, *85*, 143–156.
104 Lopez, F., Perez, R., Ruette, F., Medina, E. *Phys. Rev. E* **2005**, *72*, 061111.
105 Beerdsen, E., Smit, B. *J. Phys. Chem. B* **2006**, *110*, 14529–14530.
106 Krishna, R., van Baten, J.M., García-Pérez, E., Calero, S. *Chem. Phys. Lett.* **2006**, *429*, 219–224.
107 Beerdsen, E., Dubbeldam, D., Smit, B. *Phys. Rev. Lett.* **2005**, *95*, 164505.
108 Beerdsen, E., Dubbeldam, D., Smit, B. *J. Phys. Chem. B* **2006**, *110*, 22754–22772.
109 Beerdsen, E., Dubbeldam, D., Smit, B. *Phys. Rev. Lett.* **2006**, *96*, 044501.
110 Demontis, P., Suffritti, G.B. *Chem. Rev.* **1997**, *97*, 2845–2878.
111 Fuchs, A.H., Cheetham, A.K. *J. Phys. Chem. B* **2001**, *105*, 7375–7383.
112 Dubbeldam, D., Snurr, R.Q. *Mol. Sim.* **2007**, *33*, 305–325.
113 Sahimi, M. *Rev. Mod. Phys.* **1993**, *65*, 1394.
114 Avnir, D., Farin, D., Pfeifer, P. *Nature* **1984**, *308*, 261.
115 Coppens, M.O., in *Fractals in Engineering*, edited by Lévy-Vehel, J., Lutton, E., Tricot, C. Springer; Berlin, **1997**.

116 Malek, K., Coppens, M.O. *Phys. Rev. Lett.* **2001**, *87*, 125505.
117 Sahimi, M., Gavalas, G.R., Tsotsis, T.T. *Chem. Eng. Sci.* **1990**, *45*, 1443.
118 Coppens, M.O. *Colloids Surf. A* **2001**, *187–188*, 257.
119 Coppens, M.O., Froment, G.F. *Fractals* **1995**, *3*, 807.
120 Malek, K., Coppens, M.O. *Chem. Eng. Sci.* **2003**, *58*, 4787.
121 Lee, S.B., Mitchell, D.T., Trofin, L., Nevanen, T.K., Soderlund, H., Martin, C.R. *Science* **2002**, *296*, 2198.
122 Malek, K., van Santen, R.A. *J. Membr. Sci.* **2008**, *311*, 192–199.
123 Smart, O.S. *et al. J. Molecular Graphics* **1996**, *14*, 354–376.
124 Kisljuk, O.S., Kachalova, G.S., Lanina, N.P. *J. Molecular Graphics* **1994**, *12*(4), 305–307.
125 Smit, B., Maesen, T.L.M. *Chem. Rev.* **2008**, *108*(10), 4125–4184.
126 Malek, K., Jansen, A.P.J., Li, C., van Santen, R.A. *J. Catal.* **2007**, *246*, 127–135.
127 Malek, K., Li, C. van Santen, R.A. *J. Mol. Catal. A* **2007**, *271*, 93–104.

# 15
# Transport Processes in Polymer Electrolyte Fuel Cells: Insights from Multiscale Molecular Simulations

*Kourosh Malek*

## 15.1
## Introduction

Nanoporous heterogeneous materials are nowadays of primary interest in chemistry and chemical engineering in view of the possible applications in catalysis and separation, electrochemistry, and biotechnology. Examples of pertinent materials range from nanoporous catalysts and amorphous, mesoporous organic or inorganic membranes (silicas and $\gamma$-aluminas) [1] to porous composite electrodes in electrochemical devices such as polymer electrolyte fuel cells (PEFC) [2–4]. When the functions of these materials depend on their porous composite morphology, the physics of diffusion and reaction processes in and/or across the material becomes an important issue.

During the last two decades, formidable progress has been made in understanding the fundamental physics of transport in porous media, particularly in structured microporous materials such as zeolites [5]. Amorphous nanoporous materials are technologically as important as crystalline microporous materials; however, their transport properties remain poorly understood, even *qualitatively*. A lack of structure in these materials, has led to some phenomenological treatments, lumping the geometrical and chemical heterogeneity into so-called geometrical or chemical correction terms.

The evolution of computational facilities and techniques has tremendously widened the feasible range of micro-scale models and calculations. The complexity of problems that can be solved has increased dramatically and is still increasing. According to Moore's law, the density of transistors on an integrated circuit doubles approximately every two years. The science of porous media is one of the areas that have benefited tremendously from the corresponding increase in computing power. Moreover, recent progress of a class of powerful theoretical and computational methods, mostly derived from a combination of electronic structure theory and statistical physics,has enabled us to interpret experimental observations and predict many properties of disordered materials. Included in this class are quantum/classical/statistical dynamics simulations (dynamic Monte Carlo, molecular dynamics (MD) and *ab initio* MD), renormalization group theory,

*Computational Methods in Catalysis and Materials Science.* Edited by Rutger A. van Santen and Philippe Sautet

ISBN: 978-3-527-32032-5

modern versions of the effective medium approximation, percolation theory, and more importantly, fractal geometry and multifractals.

Diffusion in disordered porous materials, possibly accompanied by adsorption or reaction (heterogeneous catalysis), is conventionally simulated using continuum approaches, where the inherently heterogeneous porous medium is treated as a pseudo-continuum. The pore interconnectivity, representing the random network topology, is usually accounted for as a key parameter in such continuum models. When the connectivity is low, viz. close to the percolation threshold, the simple continuum approach that treats the porous material as an effective medium becomes rather inaccurate. Continuum approaches rarely account for the effects of pore morphology and pore heterogeneity (chemical or geometrical). It is widely recognized that a more refined understanding of the transport processes in these materials involves collecting information over a broad range of length and timescale from micro- to meso- (single-pore approach) and up to macro-scales (pore network and continuum approaches).

Porous composite media constitute the major functional components in PEFC, which are promising electrochemical devices for the direct conversion of chemical energy of a fuel into useful electrical work [6–10]. Worldwide programs in research and technology work towards the objective of establishing PEFC as viable alternative power sources that could replace internal combustion engines in vehicles and provide power to portable and stationary applications. PEFC exhibit unrivaled thermodynamic efficiencies of chemical-to-electrical energy conversion, high energy densities, and ideal compatibility with hydrogen as a fuel supplied on the anode side. When operated on hydrogen the only chemical product of the fuel cell reaction is water. These assets distinguish PEFC as a primary solution to the global energy challenge. Figure 15.1 shows the principal layout and basic processes in PEFCs under standard operation with hydrogen as a fuel. Anodic oxidation of $H_2$ produces protons that move through the polymer electrolyte membrane (PEM) to the cathode, where reduction of $O_2$ produces water. Meanwhile, electrons, produced at the anode, perform work in the external electrical load. The spatial separation of anodic and cathodic reactions enables the direct conversion of the enthalpy released in the net reaction $H_2 + 1/2O_2 \rightarrow H_2O$ into electrical energy, with the only by-products of this process being waste heat and water.

Nowadays, it is widely recognized that progress towards the commercialization of PEFC technology hinges on breakthroughs in design, fabrication, and implementation of innovative materials. In operational PEFCs, all components have to cooperate well in order to optimize the complex interplay of transport and reaction. It can be estimated that this optimization involves more than 50 parameters [11]. The multilayered design of a single operational PEFC is depicted in Figure 15.1. The seven-layer structure consists of a proton-conducting PEM, sandwiched between anode and cathode. Each electrode compartment consists of an active catalyst layer (CL), a gas diffusion layer (GDL), and a flow field (FF) plate. Fuel cell operation entails proton migration and water fluxes in PEM; circulation and electrochemical conversion of electrons, protons, reactant gases, and water in PEM and CLs; and vaporization/condensation in pores and channels of CLs, GDLs, and FFs. All

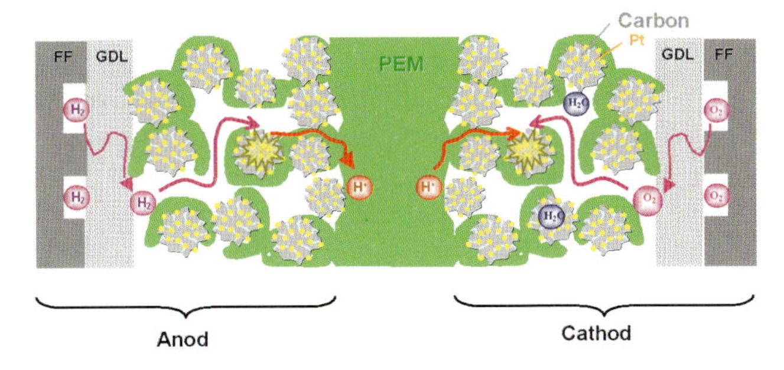


**Figure 15.1** Principal layout of polymer electrolyte fuel cells under standard operation with hydrogen-depicted seven-layer structure and basic processes.

components in Figure 15.1 have to cooperate well in order to optimize the highly nonlinear interplay of transport and reaction.

The toughest competitions between random nanocomposite morphologies and complex coupled processes unfold in catalyst layers (CL) and polymer electrolyte membranes of PEFC. These materials fulfill key functions in the cell and at the same time they offer to most compelling opportunities for innovative materials design. The main function of the membrane is to provide high proton conductivity ($>0.1\ \mathrm{S\,cm^{-1}}$). Moreover, it should be impermeable to gases and possess sufficient chemical and mechanical robustness. The currently utilized PEMs, i.e. Nafion and similar perfluorinated sulfonic acid ionomers, attain high rates of proton conduction only if they are well-hydrated. Migrating protons pull along water molecules from anode to cathode in a process known as electroosmotic drag [12]. Taken alone this process would rapidly dehydrate the membrane under operation at viable fuel current densities which are in the range of $0.5-1.0\ \mathrm{A\,cm^{-2}}$. A backflux of water via diffusion or hydraulic permeation counterbalances this electro-osmotic water flux. The balance of the two opposing fluxes establishes the non-linear profile of the water distribution in the PEM under operation in PEFC. The state of water and the resulting water flux depend strongly on the phase-segregated membrane morphology. Due to the humidification requirements and the coupled proton and water fluxes via electro-osmosis, PEMs play a key role for the water balance in the complete cell. The generation of electrical power in PEFC requires the conversion of fluxes of gaseous reactants, i.e. hydrogen on the anode side streams in anode and cathode.

The CLs, the cathode side in particular, are powerhouses of the cell. In conventional PEFCs, CLs are fabricated as random heterogeneous composites. Two significant steps in their development were the advent of highly dispersed Pt catalysts with particle sizes in the range of 1–10 nm, deposited on high surface area carbon, as well as the impregnation with Nafion ionomer. These steps enabled the reduction of catalyst loadings from about $4-10\ \mathrm{mg\,Pt\,cm^{-2}}$ (in the 1980s) to about $0.2\ \mathrm{mg\,Pt\,cm^{-2}}$. In the current design, CL consists of platinum (hereafter Pt) nanoparticles, dispersed on porous carbonaceous substrates, and Nafion ionomer as the proton-conductor. Mixing ratios and distribution of these ingredients control the interplay of the transport of gaseous reactants, protons, electrons, and product water in the relevant phases with interfacial electrochemical processes that occur at wetted Pt nanoparticles. The target of many structure vs. property studies is to identify compositions and porous morphologies that provide uniform distributions of reactants and reaction rates.

Our focus in this chapter is on providing an overview of the state of affairs in molecular modeling of transport processes in materials for PEFC. As discussed in Ref. [13], the theoretical framework fulfills an integrating function in this complex interrelated endeavor, linking the various disciplines in fuel cell research. At the fundamental level, theory helps to unravel complex relations between chemical and morphological structures and properties, bridging scales from molecular to macroscopic resolutions. Understanding these relations could facilitate the design of novel, fuel cell materials.

## 15.2 Relevant Approaches in Materials Modeling

Figure 15.2 illustrates the general strategy along with the relevant processes for multiscale simulations of materials for fuel cell at different length scales from Ångstrom to meters. Theoretical studies of molecular mechanisms of proton transport (PT) in PEMs require quantum mechanical techniques, that is *ab initio* calculations based on density functional theory. These mechanisms determine the conductance of aqueous nanosized pathways in PEMs. At the same resolution, elementary steps of adsorption, surface diffusion, charge transfer, recombination, desorption proceed on the surfaces of nanoscale catalyst particles. Density functional theory calculations and kinetic modeling of reactivities based on Monte Carlo simulations or mean field approximations are applied to study stable conformations of supported nanoparticles as well as of the elementary processes on their surface.

These fundamental processes control the electrocatalytic activity of the catalyst surface. At the mesosopic scale, interactions between molecular components control the processes of structural formation that lead to random phase-segregated morphologies in membranes and CLs. Such complex processes can be studied by coarse-grained molecular dynamic (CGMD) simulations, and multiscale modeling

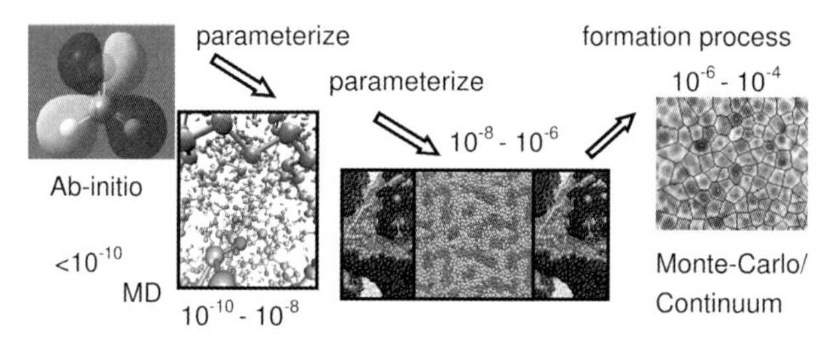


**Figure 15.2** Schematic multiscale strategy for simulation of materials for catalyst layers and polymer electrolyte membrane in polymer electrolyte fuel cells.

approaches. Complex morphologies of the emerging media can be related to relevant effective properties that characterize transport and reaction, using concepts from the theory of random heterogeneous media. Finally, conditions for stationary operation at the macroscopic device level can be defined and balance equations for involved species, that is electrons, protons, reactant gases, and water, can be established on the basis of fundamental conservation laws. Thereby, full relations between structure, properties, and performance could be established, which in turn would allow to predict architectures of materials and operating conditions that optimize fuel cell operation.

## 15.3 Proton Transport in PEMs

*Ab initio* MD calculations by Tuckerman *et al.* have unraveled molecular details of the structural diffusion of excess protons in bulk water [14–16]. In PEMs, conditions for genuine bulk-water-like PT are, however, hardly ever encountered [17,18]. Similar to PT in biophysical systems, rates of PT in PEMs are strongly affected by confinement of water in nanochannels, electrostatic effects at interfaces, and desolvation phenomena [19–25]. In PEMs, the interfaces between charged polymeric side chains and water account for differences in membrane morphology, stability, state of water, and proton-conductive abilities. It is still unclear whether the observed increase in the activation energy of PT from 0.12 eV at high levels of hydration to $>$0.35 eV at lowest water uptake of PEM [26] is due to a change in the molecular mechanism of proton mobility, a morphological transition or both [27,28].

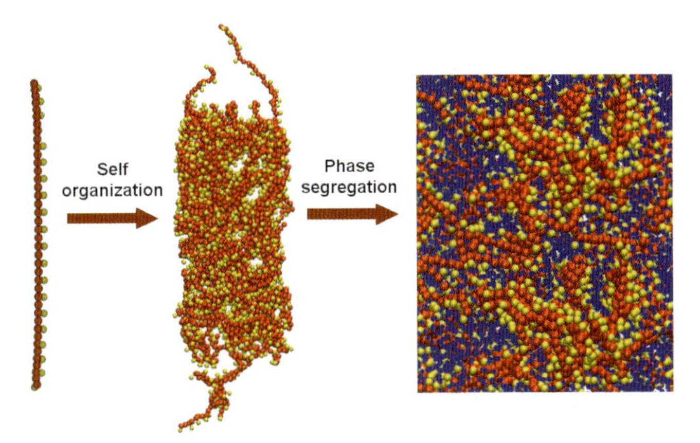


**Figure 15.3** Structural evolution of polymer electrolyte membranes. (Left) The primary coarse-grained structure of the Nafion-type ionomer with hydrophobic backbone (red) and hydrophilic side chains (yellow). (Middle) Polymeric aggregates with complex interfacial structure. (Right) The heterogeneous membrane morphology at the macroscopic scale consisting of randomly interconnected phases of the aggregates and water-filled viods.

Currently the most widely used and tested PEMs are perfluorinated sulfonated polymers. The base polymer, shown schematically in Figure 15.3, consists of Teflon-like backbones with randomly attached pendant side chains, terminated by charged hydrophilic segments (usually sulfonic acid groups, $-SO_3H$) [29–31].

Rationalizing the dependence of membrane conductivity on water content is a multiscale problem. Figure 15.3 depicts the three major levels in the structural evolution from primary chemical architecture to polymer aggregates at the nanaometer scale to random heterogeneous morphology at the macroscale.

When exposed to water, self-organization of polymer backbones leads to the formation of a hydrophobic skeleton that consists of interconnected elongated fibrillar or lamellar aggregates [32–34]. The $-SO_3H$ groups dissociate and release protons as charge carriers into the aqueous subphase that fills the void spaces between polymer aggregates. Polymer side chains remain fixed at the surfaces of those aggregates where they form a charged, flexible interfacial layer. The structure of this interface determines the stability of PEMs, the state of water, the strength of interactions in the polymer–water–ion system, vibration modes of side chains, and mobilities of water molecules and protons. The role of acid-functionalized surface groups on prolon conduction mechanisms in water filled nanopores of PEMs was explained by molecular modeling [22, 23].

Proton conductivity depends on the distribution and structure of water and dynamics of protons and water molecules at multiple scales. In order to describe the conductivity of membrane, one needs to take into account explicit polymer–water interactions at molecular level, interfacial phenomena at polymer–water interfaces and in pores at mesoscopic scale, and the statistical geometry and percolation properties of aqueous domains at macroscopic scale.

The complications for the theoretical description of PT in the interfacial region stem from fluctuations of the side chains, their random distributions at polymeric aggregates, and their partial penetration into the bulk of water-filled pores. The importance of an appropriate flexibility of hydrated side chains has been explored recently in extensive molecular modeling studies.

Empirical valence bond (EVB) approaches, introduced by Warshel *et al.* [35–37], are an effective way for incorporating environmental effects on bond-breaking and bond-making in solution. They are based on parameterizations of empirical interactions between reactant states, product states, and where appropriate a number of intermediate states. The interaction parameters, corresponding to off-diagonal matrix elements of the classical Hamiltonian, are calibrated using *ab initio* potential energy surfaces in solution and relevant experimental data.

Voth *et al.* [38, 39] and Kornyshev *et al.* [40, 41] adopted EVB-based models to study the effect of confinement in nanometer-sized pores and the role of acid-functionalized polymer walls on solvation and transport of protons in PEMs. The calculations by the Voth *et al.* displayed a strong inhibiting effect of sulfonate ions on proton motions. The EVB model by Kornyshev *et al.* was specifically designed to study the effects of charge delocalization in $SO_3^-$ groups, side chain density, and fluctuations of side chains and head groups on proton mobility. It was found that proton mobility increases with increasing delocalization of the negative countercharge on $SO_3^-$. The motion of sulfonate groups increases the mobility of protons. Conformational motions of the side chains facilitate proton motion as well. EVB-based studies were able to explain qualitatively the increase in proton conductivity with increasing water content in PEMs.

EVB-based MD simulations as well as continuum dielectric approaches involve empirical correlations between the structure of acid-functionalized interfaces in PEMs and proton distributions and mobilities in aqueous domains. These approaches fail in describing proton conduction in PEMs under conditions of low hydration with less than two water molecules added per sulfonate group, where interfacial effects prevail.

*Ab initio* molecular level simulations in PEM modeling based on density functional theory were employed by Paddison and Elliott [42] in order to study side chain correlations and examine direct proton exchange between water of hydration and surface groups. Detailed *ab initio* calculations of hydrated polymeric fragments, including several side chains, by Paddison and Elliott are insightful in view of fundamental polymer–water interactions [42].

The regular structure of the crystal [43] provides a proper basis for controlled *ab initio* MD. The Vienna *Ab initio* Simulation Package (best known as VASP) based on density functional theory was used to study the dynamics in the system [44–47]. Overall, an MD trajectory of more than 200 ps was simulated. This time frame is still by far too short for direct observation of proton transfer events, which occur on timescales of more than 1 ns. Intermittent introduction of a proton-hole defect triggered the transition from the native crystal structure, with localized positions of protonic charges, to an activated state with two delocalized protons. One of these protons resides within a Zundel ion, $H_5O_2^+$, whereas the other one is accommodated between two $SO_3^-$ groups, which approach each other at hydrogen-bond distance. The formation of the latter sulfonate O···H···O complex requires a considerable rearrangement of the crystal structure. The two almost simultaneously formed proton complexes stabilize the intermediate state. The energy of formation for the defect state is $\sim$0.3 eV. These calculations suggest that an appropriate flexibility of anionic side chains could be vital for high proton mobility in PEMs under conditions of minimal hydration and high anion density. Furthermore, a drift of the Zundel-ion was observed that indicates its possible role as a relay group for proton shuttling between hydronium ions or sulfonate anions.

Comparison of such calculations with the experimental data (e.g., NMR, IR, pulse experiments, and scanning electrochemical microscopy) on molecular mobilities in PEMs and on conductivities (e.g., Arrhenius plots) could help to rationalize the effects of interfacial structure and segmental motions of side chains on PT.

## 15.4
## Water Transport in Hydrated Nafion Membrane

Transport properties of hydrated Nafion membranes were also studied by means of both experiments and full-atomistic MD simulations [15–25, 48]. Zawodzinski *et al.* [49] determined proton diffusion coefficients at various levels of hydration from conductivity measurements and water diffusion coefficients from NMR measurements. The authors concluded that structural diffusion becomes more significant at higher hydration level due to more bulk-like behavior of water. The diffusion coefficients of water molecules obtained by Pivovar [50] and using quasielastic neutron scattering (QENS) are considerably greater than those reported by Gebel and Diat [33] using NMR technique. A recent QENS study [51] of water diffusion in hydrated Nafion-112 based on Gaussian statistical analysis of QENS data has concluded that there are different populations of diffusing proton and water with different characteristic timescales differ by a factor of 50 at all hydration levels. Most of recent experimental and computational studies of water and PT in model hydrated Nafion confirm that the exact mechanism of water transport is still unknown, mostly because of a lack of morphology and dynamic picture of the hydrated Nafion at long length- and timescales.

Mesoscale models in a CG level are able to capture the morphology at the longer time- and length scales. CG models based on dynamic self-consistent mean field (SCMF) theory have recently been developed to emerge the hydrated structure of Nafion at several water contents [13, 14]. In the SCMF calculations, Nafion is represented by a set of thread-like side chains. Each side chain and backbone is constructed of a number of CG segments (beads), representing several atomic groups [13]. The interaction parameters between CG beads are either generated using classical MD or determined by Flory–Huggins parameters [13, 14, 52, 53]. In general, variation of the hydration level at a fixed temperature leads to structural reorganization of the phase-separated morphology [13, 14]. It was shown that at low water content ($\lambda < 6$), the isolated hydrophilic domains are spherical, whereas at higher water content ($\lambda > 8$), the domains deform into elliptical shape [13].

Dissipative particle dynamics (DPD) was also applied to predict mesoscopic structure of hydrated Nafion membrane, using a CG model for Nafion [54]. In DPD simulations, the time evolution for a set of interacting particles is governed by Newton's equations [36, 37]. The force acting on a particle is given by a conservative force, dissipative force, pair-wise random force, and a binding spring force. Recent DPD simulations provide microsegregated structure of hydrated Nafion at various water contents [55]. By increasing water content, the morphologies of the membrane show a percolation-type transition from isolated hydrophilic clusters to the 3D network of irregular channels, forming a hydrophilic subdomain. In spite of many advantages, DPD and SCMF are not able to accurately predict physical properties that rely upon time correlation functions (e.g., diffusion), making them less applicable to extract structural-related transport properties of the phase-segregated membrane.

## 15.5
## Atomistic MD Simulations of CL

In a standard fabrication process, Pt particles are deposited on carbon surfaces by reducing $H_2PtCl_6$ agents on the carbon surface [56, 57]. Carbon particles (Vulcan or Ketjen by Tanaka) are dispersed in a tetrahydrofuran solution of $H_2PtCl_6$ by an ultrasonic treatment, whereas formic acid is added as the reducing agent. Filtering the solution collects the dispersed Pt/carbon particles.

In MD simulations, the molecular adsorption concept is used to interpret the Pt–C interactions during the fabrication processes. The Pt complexes are mostly attached to the hydrophilic sites on carbon particles, viz carbonyl or hydroxyl groups [58]. The adsorption is based on both physical and chemical adsorption. Carbon particle preparations, impregnation, and reduction are three main steps of the catalyst preparation. The point of zero charge determines the pH range at which the impregnation step should be carried out. This point is an important parameter in catalyst preparation.

Extensive simulation studies based on MD calculations have focused on the Pt nanoparticles adsorbed on carbon in the presence or absence of ionomers [59–63]. Lamas and Balbuena performed classical MD simulations on a simple model for the interface between graphite-supported Pt nanoparticles and hydrated Nafion. In MD studies of CLs, the equilibrium shape and structure of Pt clusters are usually simulated using the embedded atom method. Semiempirical potentials such as the many-body Sutton–Chen (SC) potential [64] are popular choices for the close-packed metal clusters. Such potential models include the effect of the local electron density to account for many-body terms. The SC potential for Pt–Pt and Pt–C interactions provides a reasonable description of bulk and small cluster properties of Pt clusters. The potential energy in the SC potential is expressed by

$$U = \varepsilon_{pp} \sum_{1}^{N} \left[ \frac{1}{2} \sum_{j \neq i}^{N} \left( \frac{\sigma_{pp}}{r_{ij}} \right)^{n} - c\sqrt{\rho_i} \right], \quad \rho_i = \sum_{j \neq i}^{N} \left( \frac{\sigma_{pp}}{r_{ij}} \right)^{m}, \tag{15.1}$$

where for Pt–Pt interaction, $\varepsilon_{pp} = 0.019833\,\text{eV}$, $\sigma_{pp} = 3.92\,\text{Å}$, $n = 10$, $m = 8$, and $c = 34.408$. The first term is a pair-wise repulsive potential and the second term represents the metallic bonding energy associated with the local electron density. The weak interaction between a Pt cluster and a graphite substrate is defined by 6–12 Lennard–Jones (LJ) interactions with $\varepsilon_{Pt-C} = 0.022\,\text{eV}$ and $\sigma_{Pt-C} = 2.905\,\text{Å}$ in Eq. (15.1). In order to obtain realistic results, the size of Pt cluster system should be sufficiently large. The LJ parameters for Pt–Pt interactions are obtained by fitting the LJ equation to SC potential with the values of $\varepsilon_{Pt-Pt} = 2336\,K$ and $\sigma_{Pt-Pt} = 2.41\,\text{Å}$. These simulations based on adapted LJ interactions from SC potential allow one to account for Pt cluster–cluster interactions as well. Nafion is represented by oligomer models interacting through Dreiding force field [65, 66], which is composed by a series of bond angle and dihedrals and nonbonded (LJ and electrostatic) interactions. A two-site model and single-point charge (SPC) models are used for oxygen and water, respectively [67]. The simulation setup used in Ref. [61] contained three $Pt_{256}$ clusters (each of ~1.6 nm in diameter) deposited over a position-restrained graphite substrate, modeled by an AB arrangement of two graphite sheets. MD simulations have been used to examine the properties of Pt nanoparticles supported by graphite [59, 68]. Diffusion of Pt nanoparticles on graphite has been investigated, and high diffusion coefficients, on the order of

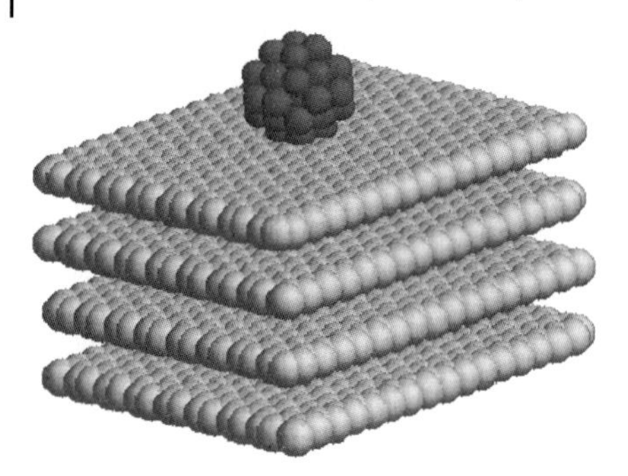

**Figure 15.4** Initial configuration of a small platinum nanocluster deposited on a static graphite substrate.

$10^{-5}$ cm$^2$ s$^{-1}$, have been predicted [58]. Some studies focused on the morphology of supported nanoclusters. It has been predicted that supported and unsupported metal nanoparticles of similar sizes have different dynamical behavior at different temperatures.

The morphology and mobility of metal nanoparticles on their support can also be altered by MD simulations. When deposited on carbon nanotubes, Pt particles diffuse at a slower rate than on graphite, which decreases sintering and should lead to increased long-term performance in catalytic applications [68]. Figure 15.4 shows a simulation setup in which four graphite sheets are parallel to each other and are perpendicular to the $z$-axis. Normal periodic boundary conditions are maintained in the $x$ and $y$ directions. This simulation system consists of multiple Pt clusters on the model graphite surface. Several simulations were carried out to test the relative stability of various cluster symmetries for different cluster structure and sizes.

The MD simulations in Ref. [60] showed how dynamics properties of water in the vicinity of the catalyst surface change as a function of Nafion contents. Observed water structures ranged from the ones similar to those in the bulk membrane region to somewhat disconnected set of clusters. Water dynamics varies according to the contribution from three different mechanisms as a function of the water content (or Nafion content). The first mechanism emerges at low water contents and is due to strong interactions between the water and side chain groups. The second mechanism presents at intermediate water contents where a high ratio of water molecules is located in the water clusters, and in bridges where water molecules have high mobility. The third mechanism relates to the behavior of bulk water and is observed when the MEA is flooded.

## 15.6
## Self-Organization in PEMs and CLs at the Mesocopic Scale

Mesoscale simulations can describe the morphology of heterogeneous materials and rationalize their effective properties beyond length- and timescale limitations of atomistic simulations. In this section, we describe how mesoscale computational tools can be used to investigate the self-organization phenomena during fabrication of CLs in PEFCs [69]. Such studies will help us to understand structure-related properties of CLs and membrane for further developing and designing novel materials for PEFC. Improving the fabrication of CL provides opportunities to improve the Pt catalytic utilization by increasing the interfacial area between Pt particles and ionomer [70–72]. This is achieved by mixing ionomer with dispersed Pt/C catalysts in the ink suspension prior to deposition to form a CL. The solubility of the ionomer depends upon the choice of a dispersion medium. This influences the microstructure and pore-size distribution of the CL [72]. The random morphology, in turn, has a marked impact on transport and electrochemical properties of the emerging composite medium. In principle, only catalyst particles at Pt–water interfaces are electrochemically active. Electrochemical reactions, thus, occur on the walls of wetted pores inside and between agglomerates as well as at interfaces between Pt and wetted ionomer. Relative contributions of these distinct types of interfaces will depend on the corresponding interfacial areas. Evidently, high catalytic yields hinge on compositions and porous structures with well-attuned pore-size distributions and wetting properties of the pore network. Agglomerates are usually referred to as the building blocks of CL at the mesoscopic scale. They consist of carbon/Pt particles and are vital for the catalyst utilization and the overall voltage efficiency of the layer [69–71, 73, 74]. This section presents a recently introduced mesoscale computational method to evaluate key factors during the fabrication of CL. These simulations rationalize structural factors such as pore sizes, internal porosity, and wetting properties of internal/external surfaces of such agglomerates. They help elucidating whether ionomer is able to penetrate into primary pores inside agglomerates [75–77]. Moreover, dispersion media with distinct dielectric properties can be evaluated in view of capabilities for controlling sizes of carbon/Pt agglomerates, ionomer domains, and the resulting pore network topology. These insights are highly valuable for the structural design of CLs with optimized performance and stability.

The performance of CL strongly depends on several fabrication parameters such as applicable solvent, particle sizes of primary carbon powders, wetting properties of carbon materials, and composition of the CL ink [69–71]. These factors determine the complex interactions between Pt/carbon particles, ionomer molecules, and solvent molecules and, therefore, control the CL formation process.

Advanced experimental and theoretical tools can address the correlation of transport-reaction processes with structural details at meso-to-micro- and down to atomistic scale. CGMD techniques can describe the system at the micro-to-meso level, while still being able to capture the morphology at long time- and length scales [78–80]. The application of CG models, however, requires special care. Due to the reduced number of degrees of freedom, CG simulations may not be able to accurately predict physical properties that directly rely upon time correlation functions (e.g., diffusion) [79].

A significant number of mesoscale computational approaches have been employed to understand the phase-segregated morphology and transport properties of water-swollen Nafion membranes [27, 28, 40, 78, 79, 81–83]. Because of computational limitations, full atomistic models are not able to probe the random morphology of these systems. Structural complexity is even more pronounced in CLs since they consist of a mixture of Nafion ionomer, Pt clusters supported on carbon particles, solvent, and water. Independent computational strategies are generally needed in order to simulate the microstructure formation in CLs. For instance, the structure of the ionomer phase in CLs cannot be trivially inferred from that in membrane simulations, as there are distinct correlations between Nafion, water, and carbon particles in CLs. The computational approach based on CGMD simulations is developed in two major steps. In the first step, Nafion chains, water and hydronium molecules, other solvent molecules, and carbon/Pt particles are replaced by the corresponding spherical beads with predefined subnanoscopic length scale. In the second step, parameters of renormalized interaction energies between the distinct beads are specified.

We consider four main types of spherical beads: polar, nonpolar, apolar, and charged beads [84]. Clusters including a total of four water molecules or three water molecules plus a hydronium ion are represented by polar beads of radius 0.43 nm. The configuration of a Nafion and carbon particles in simulation box is shown in Figure 15.5. A side chain unit in Nafion ionomer has a molecular volume of 0.306 $nm^3$, which is comparable to the molecular volume of a four-monomeric unit of polytetrafluoroethylene (0.325 $nm^3$) [78]. Therefore, each four-monomeric unit ($-[-CF_2-CF_2-CF_2-CF_2-CF_2-CF_2-CF_2-CF_2-]-$) and each side chain (represented by a charged bead) are CG as spherical beads of volume 0.315 $nm^3$ ($r = 0.43$ nm). The hydrophobic Nafion backbone is replaced by a CG chain of 20 apolar beads, as illustrated in Figure 15.5.

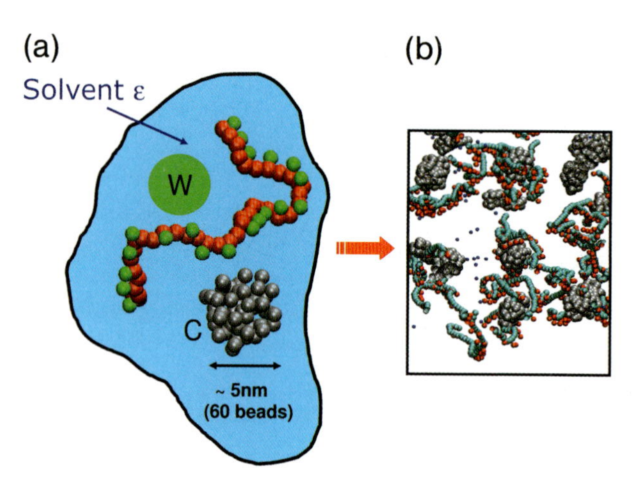


**Figure 15.5** (a) Coarse-grained model of Nafion with 20-unit oligomer, length ~30 nm; carbon particles, diameter ~5 nm; and water. Initial simulation box had a size of 50 × 50 × 50 $nm^3$ and includes 52 carbon particles, 72 Nafion oligomers (38% wt), 1440 CG hydronium ions, and 13 $H_2O$ per side chain (λ). Panel (b) shows a snapshot of the simulation box after the equilibration.

Carbonaceous particles can be CG in various ways, building upon a new technique called multiscale coarse graining [85–87]. In this method, the CG potential parameters are systematically obtained from atomistic-level interactions [88]. Using this technique, a model was built for semispherical carbon particles based on CG sites in the C60 system.

The interactions between nonbonded beads are modeled by the LJ potential. In this potential, the effective bead diameter ($r_{ij}$) is 0.43 nm for side chain, backbone, and water beads. The strength of interaction is limited to five possible values ranging from weak (1.8 kJ $mol^{-1}$) to strong (5 kJ $mol^{-1}$) beads [84]. The electrostatic interactions between charged beads are described by the Columbic interaction with relative dielectric constant $\varepsilon_r = 20$ in order to include screening. The effect of solvent is incorporated by changing $\varepsilon_r$ as well as by varying the degree of dissociation of Nafion side chains. Interactions between chemically bonded beads (e.g., in Nafion chains) are modeled by harmonic potentials for the bond length and bond angle, where the force constants are $K_{bond} = 1250$ kJ $mol^{-1}$ $nm^{-2}$ and $K_{angle} = 25$ kJ $mol^{-1}$ angle, respectively. The size of the simulation box can vary from $50 \times 50 \times 50$ $nm^3$ to $500 \times 500 \times 500$ $nm^3$, depending on system and composition.

The snapshot of the final microstructure was analyzed in terms of density map profiles, RDFs, pore-size distributions, and pore shapes. Figure 15.6 visualized a snapshot of the carbon–Nafion–water–solvent (CNWS) composite. Carbon particles are assumed to be hydrophobic. The interaction parameters of the carbon particles are selected to mimic the properties of VULCAN-type C/Pt particles. They

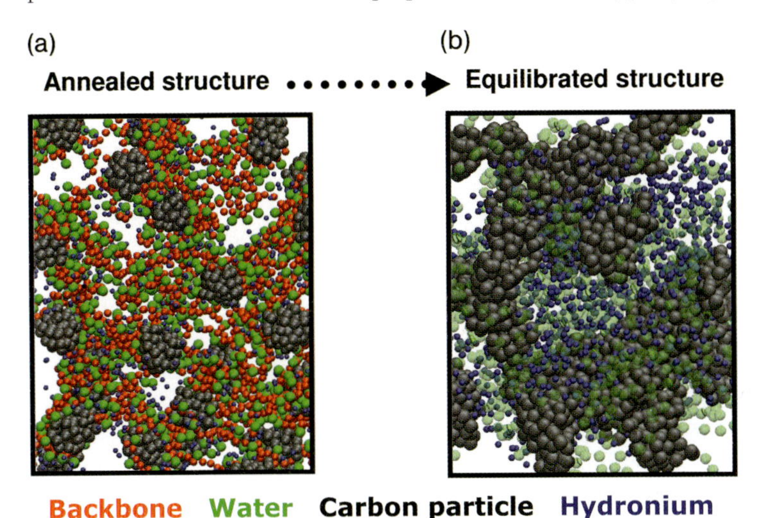


**Figure 15.6** (a) Equilibrium annealed and equilibrated structures of a catalyst blend composed of carbon (black), Nafion (red), water (green), and implicit solvent. Hydrophilic domain is removed in (b) for better visualization. Reproduced from Malek *et al.* [11], Copyright © 2007 with permission from ACS. (See Color Insert.)

are hydrophobic, with a repulsive interaction with water and Nafion side chains and semiattractive interactions with other carbon particles as well as with Nafion backbones.

Structural analysis based on site–site RDFs is fully discussed in Ref. [11]. The RDFs reveal general features of phase segregation. There is a strong correlation between carbon particles. Ionomer forms a structure that is most likely random. As expected, hydrated proton (H) and water (W) behave similarly. The correlation between hydrophilic species (H and W) and ionomer (N) is significantly stronger than that between those species and carbon (C). The autocorrelation functions $g_{SS}$ and $g_{HH}$ exhibit a similar structure as $g_{WW}$. This indicates a strong clustering of side chains and hydronium ions due to the aggregation and folding of polymer backbones.

This CGMD helped consolidating the main features of microstructure formation in CLs of PEFCs. CGMD simulations showed that the final microstructure depends on carbon particle choices and ionomer–carbon interactions. While ionomer side chains are buried inside hydrophilic domains with a weak contact to carbon domains, the ionomer backbones are attached to the surface of carbon agglomerates. They form a randomly interconnected proton-conducting network. The correlation between hydrophilic species and ionomer is significantly stronger than that between those species and carbon particles. A layer of ionomer of ~10 nm thickness with well-packed morphology around Pt/C clusters was evident in the presence of polar solvents. Note that the compactness of carbon agglomerates is affected by the polarity of solvent.

The evolving structural characteristics of the CL are particularly important for further analysis of transport of protons, electrons, reactant gas ($O_2$), and water as well as the distribution of electrocatalytic activity at Pt–water interfaces. In principle, such mesoscale simulation studies allow relating these properties to the selection of solvent, carbon particles (sizes and wettability), catalyst loading, and level of membrane hydration in the CL. There is still a lack of explicit experimental data, with which we can compare our results directly. Versatile experimental techniques are still needed to study structural characterization of phases and interfaces, carbon–ionomer phase correlation, and particle–particle interactions. Qualitative and quantitative agreements with experimental data will corroborate predictive capabilities of these calculations.

CGMD simulations can also predict the mesoscopic structure of the hydrated Nafion membrane with different water contents [92]. Membrane simulations were performed with different water contents of 3, 6, 12, and 19 wt% (unphysical). The mesoscopic structure of the hydrated membrane is visualized in Figure 15.7, revealing a sponge-like structure. Water beads together with hydrophilic beads of Nafion side chain form aggregated clusters, which are embedded in the hydrophobic phase of the Nafion backbones. Our analysis shows that hydrophilic subphase is composed of a 3D network of irregular channels. The channel size increases with water content from 1 through 2 to 4 nm at 3, 6, and 19 wt.% water, respectively. The site–site RDF obtained from CGMD simulations match perfectly to those from the atomistic MD simulations. The RDF between the side chain beads and the

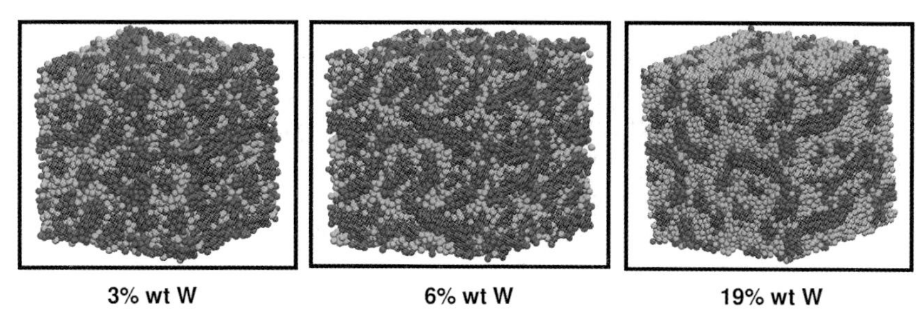


**Figure 15.7** Snapshots of the final microstructure in hydrated Nafion membrane at different water contents. Hydrophilic domain (water, hydronium ions, and side chains) is shown in green, whereas hydrophobic domain is in red. (See Color Insert.)

components of the mixture show that side chains are surrounded with water and hydrated protons. Water and hydrated protons show similar interactions with side chain and ionomer backbone.

## 15.7 Concluding Remarks

Design and development of advanced materials plays a key role in vigorous efforts to enhance performance and cost-effectiveness of PEFCs [89]. The practice over the last two decades has shown that the dramatic improvements in power density, durability, and cost would be impossible without a breakthrough in the concept of proton-conducting membranes and CLs. Moreover, all components of operational PEFCs have to cooperate well in order to optimize the interplay of transport and reaction. Current membranes and CLs are fabricated as random heterogeneous media. Their properties evolve on a hierarchy of scales, from Ångstrom to meters, exhibiting complex morphologies and an intricate interplay of fundamental processes. Evidently, the rational design of functionally optimized materials demands strategies that are based on systematic understanding provided by physical theory, molecular modeling, and physical modeling of performance at the macroscopic device level.

Here we focused on the main challenges in molecular modeling of transport processes in polymeric membranes and catalyst layers of polymer Electrolyte Fuel Cells. Recent efforts in theory, modeling, and diagnostics have explicitly identified constraints and reserves for improvements of the current design of CLs and membranes in PEFCs.

Current calculations exploiting DFT and CPMD methods are on the way to establishing fundamental steps that could be of similar significance as the transformation between $H_9O_4^+$ and $H_5O_2^+$ for PT in bulk water. The packing density of protogenic surface groups transpires as the critical parameter for achieving high mobility of protons at lowly hydrated interfaces. If the packing density is too low, surface groups are only weakly correlated, prohibiting direct proton transfer between them. If the density is too high, strong hydrogen bonds will suppress the dynamics of protons and water at the interface. A critical separation that could represent a good compromise of long-range correlations and flexibility is ~7 Å. Further important criteria, which depend on the chemical architecture of the polymer (backbones, side chains, and acid head groups), include the ease of formation and breaking of H-bonds and the conformational flexibility of side chains and acid groups.

Systematic studies in theory and molecular modeling on proton mobilities at 2D hydrated structures with controlled molecular architectures could identify collective coordinates (or reaction coordinates) and transition pathways at the interface. Differences in activation energies and rates of proton transfer, calculated with appropriate sampling techniques (e.g., transition path sampling), could be evaluated in relation to distinct polymer constituents, acid head groups, lengths and densities of side chains, and levels of hydration. New mechanisms could be exploited together with opportunities to control membrane structure via block-copolymer self-assembly [90,91]. Moreover, theory and modeling efforts should also address PT in anhydrous systems, which offer promising alternatives for membrane materials.

At the nanopartile scale, combined computational and theoretical methods (DFT, OF-DFT, kinetic Monte Carlo simulations, MD simulations) are needed to overcome the size and timescale limitations of conventional DFT techniques. Simulations of physical properties of realistic Pt–C nanoparticle systems could provide interaction parameters needed in models of self-organization phenomena (agglomeration and phase segregation). Theoretical models of electrokinetic mechanisms, on the other hand, are needed to establish trends between catalyst structure and oxygen reduction kinetics.

We have discussed statistical computational and continuum approaches to modeling of microstructure formation and self-organization phenomena in porous CLs and membrane of PEFC. Dynamic simulations of structural changes during formation processes lend themselves equally well to the study of mechanisms underlying structural degradation. CGMD simulations of self-organization in CLs suggest that the resulting structures are inherently unstable. Applicable solvents with different dielectric constants correspond to different stable conformations in terms of agglomerate sizes, sizes of ionomer domains, and pore space morphology. The replacement of low-dielectric solvents used during fabrication of CLs by water as the working liquid in the operating fuel cell will thus destabilize the initially formed structures, causing a drift toward a new stable conformation.

## References

1 K. Malek, A.P.J. Jansen, C. Li, R.A. van Santen, **2007**. Enantioselectivity of Immobilized Mn–salen complexes: A Computational Study, *J. Catal.* 246, 127–135.

2 W. Vielstich, A. Lamm, H.A. Gasteiger (Eds.), **2003**. *Handbook of Fuel Cells: Fundamentals, Technology and Applications*, Wiley-VCH, Weinheim.

3 A.Z. Weber, J. Newman, **2004**. *Chem. Rev.* 104, 4679.
4 M. Eikerling, **2006**. *J. Electrochem. Soc.* 153, E58.
5 F.J. Keil, R. Krishna and M.-O. Coppens, *Rev. Chem. Engng*, **2000**, 16, 71.
6 W. Vielstich, A. Lamm, H. Gasteiger (Eds.), **2003**. *Handbook of Fuel Cells: Fundamentals, Technology, Applications*, Wiley-VCH, Weinheim.
7 M. Eikerling, A.A. Kornyshev, A.R Kucernak, **2006**. *Phys. Today* 59, 38.
8 M. Eikerling, A.A. Kornyshev, A.A. Kulikovsky, **2007**. Encyclopedia of electrochemistry, In: A.J. Bard and M. Stratmann (Eds.), *Electrochemical Engineering*, Vol. 5, volume edited by D.D. Macdonald and P. Schmuki, chapter 8.2, pp. 447–543, Wiley-VCH, Weinheim, **2007**.
9 S. Gottesfeld, T.A. Zawodzinski, **1997**. In: R.C. Alkire, H. Gerischer, D.M. Kolb, C.W. Tobias (Eds.), *Advances in Electrochemical Science and Engineering*, Vol. 5, pp. 195–301, Wiley-VCH, Weinheim.
10 E.J. Carlson, **2006**. *DOE Hydrogen Program Annual Merit Review*.
11 M. Eikerling, K. Malek, Q. Wang, **2008**. Catalyst Layer Modelling: Structure, Properties, and Performance, in *PEM Fuel Cells Catalysts and Catalyst Layers–Fundamentals and Applications*, J.J, Zhang, (Ed.), pp. 381–446.
12 M. Eikerling, A.A. Kornyshev, A.A. Kulikovsky, **2005**. *Fuel Cell Review* 1, 15.
13 D. Wang, J.S. Wainright, U. Landau, R.F. Savinell, **1996** *J. Electrochem. Soc.* 143, 1260.
14 D. Marx, M.E. Tuckerman, J. Hutter, M. Parinello, **1999**. *Nature* 397, 601.
15 M.E. Tuckerman, K. Laasonen, M. Sprik, M. Parrinello, **1995**. *J. Phys. Chem.* 99, 5749.
16 M.E. Tuckerman, K. Laasonen, M. Sprik, M. Parrinello, **1995**. *J. Chem. Phys.* 103, 150.
17 M.K. Petersen, F. Wang, N.P. Blake, H. Metiu, G.A. Voth, **2005**. *J. Phys. Chem. B* 109, 3727.
18 A. Roudgar, S.P. Narasimachary, M. Eikerling, **2006**. *J. Phys. Chem. B* 110, 20469.
19 G. Hummer, C. Dellago, **2006**. *Phys. Rev. Lett.* 97, 245901.
20 S. Braun-Sand, M. Strajbl, A. Warshel, **2004**. *Biophys. J.* 87, 2221.
21 A.Y. Mulkidjanian, J. Heberle, D.A. Cherepanov, **2006**. *Biochim. Biophys. Acta* 1757, 913.
22 M. Eikerling, A.A. Kornyshev, **2001**. *J. Electroanal. Chem.* 502, 1.
23 P. Commer, A.G. Cherstvy, E. Spohr, A.A. Kornyshev, **2002**. *Fuel Cells* 2, 127.
24 M. Eikerling, A.A. Kornyshev, A.M. Kuznetsov, J. Ulstrup, S. Walbran, **2001**. *J. Phys. Chem. B* 105, 3646.
25 M. Kato, A.V. Pisliakov, A. Warshel, **2006**. *Proteins Struct., Funct., Genet.* 64, 829.
26 M. Cappadonia, J.W. Erning, S.M.S. Niaki, U. Stimming, **1995**. *Solid State Ionics* 77, 65.
27 A. Vishnyakov, A.V. Neimark, **2001**. *J. Phys. Chem. B* 105, 9586.
28 A.S. Ioselevich, A.A. Kornyshev, J.H.G. Steinke, **2003**. *J. Phys. Chem. B* 108, 11953.
29 M. Eikerling, A.A. Kornyshev, E. Spohr, **2008**. In: Proton-Conducting Polymer Electrolyte Membranes: Water and Structure in Charge, *Advances in Polymer Science, Polymers in Fuel Cells 215*, edited by G.G. Scherer, Springer, Berlin/Heidelberg, 15.
30 D. Seeliger, C. Hartnig, E. Spohr, **2005**. *Electrochim. Acta* 50, 4234.
31 S. Tanimura, T. Matsuoka, **2005**. *J. Polym. Sci. B* 42, 1905.
32 G. Gebel, **2000**. *Polymer* 41, 5829; L. Rubatat, G. Gebel, O. Diat, **2004**. *Macromolecules* 37, 7772.
33 G. Gebel, O. Diat, **2005**. *Fuel Cells* 5, 261.
34 R.R. Netz, D. Andelmann, **2003**. *Phys. Rep.* 380, 1.
35 A. Warshel, R.M. Weiss, **1980**. *J. Am. Chem. Soc.* 102, 6218.
36 A. Warshel, **1991**. *Computer Modeling of Chemical Reactions in Enzymes and in Solutions*, Wiley, New York.

37 J. Åqvist, A. Warshel, **1993**. *Chem. Rev.* 93, 2523.
38 M.K. Petersen, G.A. Voth, **2006**. *J. Phys. Chem. B* 110, 18594.
39 M.K. Petersen, F. Wang, N.P. Blake, H. Metiu, G.A. Voth, **2005**. *J. Phys. Chem. B* 109, 3727.
40 S. Walbran, A.A. Kornyshev, **2001**. *J. Chem. Phys.* 114, 10039.
41 E. Spohr, P. Commer, A.A. Kornyshev, **2002**. *J. Phys. Chem. B* 106, 10560.
42 S.J. Paddison, J.A. Elliott, **2006**. *Phys. Chem. Chem. Phys.* 8, 2193.
43 J.B. Spencer, J.O. Lundgren, **1973**. *Acta Cryst. B.* 29, 1923.
44 G. Kresse, J. Hafner, **1993**. *Phys. Rev. B* 47, 558.
45 G. Kresse, J. Hafner, **1994**. *Phys. Rev. B* 49, 14251.
46 G. Kresse, J. Furthmüller, **1996**. *Phys. Rev. B* 54, 11169.
47 G. Kresse, J. Hafner, **1994**. *J. Phys.: Condens. Matter.* 6, 8245.
48 N.P. Blake, **2007**. *J. Phys. Chem. B* 111, 2490.
49 T.A. Zawodzinski, M. Neeman, L.O. Sillerud, S. Gottesfeld, **1991**. *J. Phys. Chem.* 95, 6040.
50 A.A. Pivovar, **2005**. *J. Phys. Chem. B* 109, 785.
51 J.-C. Perrin, S. Lyonnard, F. Volino, **2007**. *J. Phys. Chem. C* 111, 3393.
52 R.D. Groot, P.B. Warren, **1997**. *J. Chem. Phys.* 107, 4423.
53 R.D. Groot, **2003**. *J. Chem. Phys.* 118, 11265–11277.
54 S. Yamamoto, S.A. Hyodo, **2003**. *Polym. J. (Tokyo, Jpn.)* 35, 519.
55 A. Vishnyakov, A.N. Neimark, **2005**. Final Report for US Army Research Office, DAAD190110545.
56 I. Moriguchi, F. Nakahara, H. Furukawa, H. Yamada T. Kudo, **2004**. *Electrochem. Solid State Lett.* 7, A221.
57 H. Yamada, H. Nakamura, F. Nakahara, J. Moriguchi, T. Kudo, **2007**. *J. Phys. Chem.* 111, 227.
58 X. Hao, *et al* **2003**. *J. Colloid Interfaces Sci.* 267, 259.
59 E.J. Lamas, P.B. Balbuena, **2003**. *J. Phys. Chem. B* 107, 11682.
60 S.P. Huang, P.B. Balbuena, **2002**. *Mol. Phys.* 100, 2165.
61 J. Chen, K.-Y. Chan, **2005**. *Mol. Simul.*, 31, 527.
62 E.J. Lamas, P.B. Balbuena, **2006**. *Electrochem. Acta.* 51, 5904.
63 P.B. Balbuena, E.J. Lamas, Y. Wang, **2005**. *Electrochem. Acta* 50, 3788.
64 A.P. Sutton, J. Chen, **1990**. *Philos. Mag. Lett.* 61, 139.
65 S.S. Jang, V. Molinero, T. Çadin, W.A. Goddard III, **2004**. *J. Phys. Chem. B* 108, 3149.
66 W. Goddard, B. Merinov, A. van Duin, T. Jacob, M. Blanco, V. Molinero, S.S. Jang, Y.H. Jang, **2006**. *Mol. Sim.* 32, 251.
67 R.C. Reid, J.M. Prausnitz, B.E. Poling, **1987**. *The Properties of Gases and Liquids*, McGraw-Hill, Boston.
68 B.H. Morrow, A. Striolo, **2007**. *J. Phys. Chem. C* 111, 17905.
69 K. Malek, M. Eikerling, Q. Wang, T. Navessin, Z. Liu, **2007**. *J. Phys. Chem. C* 111, 13627.
70 M. Uchida, Y. Aoyama, E. Eda, A. Ohta, **1995**. *J. Electrochem. Soc.* 142, 463.
71 M. Uchida, Y. Aoyama, E. Eda, A. Ohta, **1995**. *J. Electrochem. Soc.* 142, 4143.
72 M. Uchida, Y. Fuuoka, Y. Sugawara, N. Eda, Ohta, A. **1996**. *J. Electrochem. Soc.* 143, 2245.
73 M. Eikerling, A.A. Kornyshev, **1998**. *J. Electroanal. Chem* 453, 89.
74 M. Eikerling, **2006**. *J. Electrochem. Soc.* 153, E58.
75 R. Fernandez, P. Ferriera-Aparicio, L. Daza, **2005**. *J. Power. Sources* 151, 18.
76 P. Gode, F. Jaouen, G. Lindbergh, A. Lundblad, G. Sundholm, **2003**. *Electrochem. Acta* 48, 4175.
77 Q. Wang, M. Eikerling, D. Song, Z. Liu, **2004**. *J. Electroanal. Chem.* 573, 61.
78 J.T. Wescott, Y. Qi, L. Subramanian, T.W. Capehart, **2006**. *J. Chem. Phys.* 124, 134702.
79 D.Y. Galperin, A.R. Khokhlov, **2006**. *Macromol. Theory Simul.* 15, 137.
80 R.D. Groot, P.B. Warren, **1997**. *J. Chem. Phys.* 107, 4423.
81 A. Vishnyakov, A.V. Neimark, **2000**. *J. Phys. Chem.* 104, 4471.

82 D.A. Mologin, P.G. Khalatur, A.R. Kholhlov, **2002**. *Macromol. Theory Simul.* 11, 587.

83 P.G. Khalatur, S.K. Talitskikh, A.R. Khokhlov, **2002**. *Macromol. Theory. Simul.* 11, 566.

84 S.J. Marrink, A.H. de Vries, A.E. Mark, **2007**. *J. Phys. Chem. B* 111, 7812.

85 S. Izvekov, A. Violi, **2006**. *J. Chem. Theory Comput.* 2, 504.

86 S. Izvekov, A. Violi, **2005**. *J. Phys. Chem. B* 109, 2469.

87 S. Izvekov, A. Violi, G.A. Voth, **2005**. *J. Phys. Chem. B* 109, 17019.

88 G.A. Voth, **2005**. *J. Phys. Chem. B* 109, 17019.

89 A.S. Arico, P. Bruce, B. Scrosatti, J.-M. Tarascon, W. van Schalwijk, **2005**. *Nature Mater.* 4, 366.

90 S. Förster, M. Anotnietti, **1998**. *Adv Mater* 10, 195.

91 O. Ikkala, G. ten Brinke, **2002**. *Science* 295, 2407.

92 K. Malek, M. Eikerling, Q. Wang, Z. Liu, S. Otsuka, K. Akizuki, M. Abe, **2008**. *J. Chem. Phys.* 129, 20472.

# Part IV
# Catalytic Applications

# 16
# Application of the DFT Method to the Study of Intramolecular Palladium Shifts in Aryl and Polyaryl Complexes

*Alain Dedieu, and Antonio J. Mota*

## 16.1
## Introduction

Reactions that involve the catalytic transformation of the rather inert C–H bond are currently the subject of intense research. In many instances, palladium complexes have turned out to be quite efficient to reach this goal [1, 2]. Activation and functionalization of C–H bond can be achieved through several ways, going from the mere C–H bond oxidative addition to multistep processes. Among the various reactions, which have been recently investigated, the through-space shift of palladium between two carbon atoms, associated to a simultaneous C–H bond activation [3–19], is quite intriguing from a mechanistic point of view. As shown in Scheme 16.1, two different *intramolecular* pathways may be envisioned *a priori*: (1) an oxidative addition/reductive elimination two-step process in which a hydridopalladium(IV) intermediate is involved or (2) a one-step process in which the oxidation state +2 of the palladium atom is retained. In this case, it could be considered either as a metathesis reaction of the C–H $\sigma$ bond with the Pd–C bond or as a concomitant 1,$n$ shift of both the hydrogen and palladium atoms. This process, which is important from a synthetic point of view, because it allows the attachment of groups in positions that are otherwise difficult to substitute by usual chemical reactions [19], has been reported for 1,4 migrations either between two $sp^2$ [4–9, 12, 15, 16, 18] or between an $sp^3$ and an $sp^2$ carbon atoms [3, 9, 11, 13]. There has been also a report of a 1,5 vinyl-to-aryl (i.e., $sp^2$-to-$sp^2$) shift [14]. Yet the experimental studies could not assess unambiguously which of the two pathways sketched in Scheme 16.1 is operative [16, 18, 19]. We have recently addressed this mechanistic issue via DFT-B3LYP calculations, and we will review here the corresponding results for 1,$n$ aryl-to-aryl ($n$ = 3–6) [20, 21] and 1,$n$ aryl-to-alkyl ($n$ = 3 and 5) [22] Pd shifts.

*Computational Methods in Catalysis and Materials Science.* Edited by Rutger A. van Santen and Philippe Sautet

ISBN: 978-3-527-32032-5

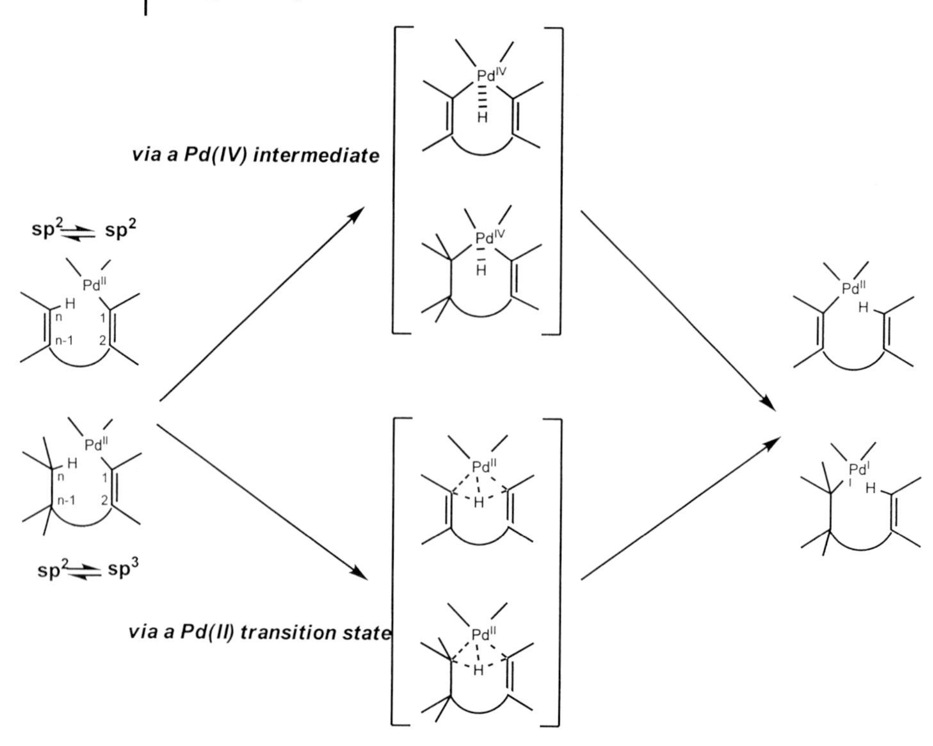


**Scheme 16.1** General pathways for the $sp^2$–to-$sp^2$ and $sp^2$–to-$sp^3$ Pd shift.

## 16.2 Computational Details

The calculations were carried out at the DFT-B3LYP level [23–25] with the Gaussian 03 program [26]. The systems under investigation were monopalladated bromo derivatives of the type shown in the retrosynthetic pathways of Scheme 16.2. They have been found or should result from an initial oxidative addition to palladium of the C–Br bond of the corresponding monobromated derivatives. For the product of this oxidative addition, various structures can be anticipated depending on the coordination sphere around the palladium atom, and in particular on the number of phosphine ligands. We investigated this point rather thoroughly in our initial study that pertained to the 1,5 vinyl-to-aryl shift [20] and found that the monophosphine species were energetically favorable. In particular, we showed that the transition states calculated for the vinyl-to-aryl Pd shift are higher in energy in the biphosphine complexes compared with the monophosphine complexes [20]. These monophosphine complexes, although being formally three-coordinated, will in most cases fill the fourth valence of palladium through an agostic-type interaction, either with the C–H or the C=C bonds of one of the pendant aromatic rings or with a C–H bond of the neighboring alkyl chain (except in the methyl case, *vide infra*). The involvement of three-coordinated monophosphine complexes as intermediates is also well documented experimentally [11, 27–36] and theoretically [29, 37–40].

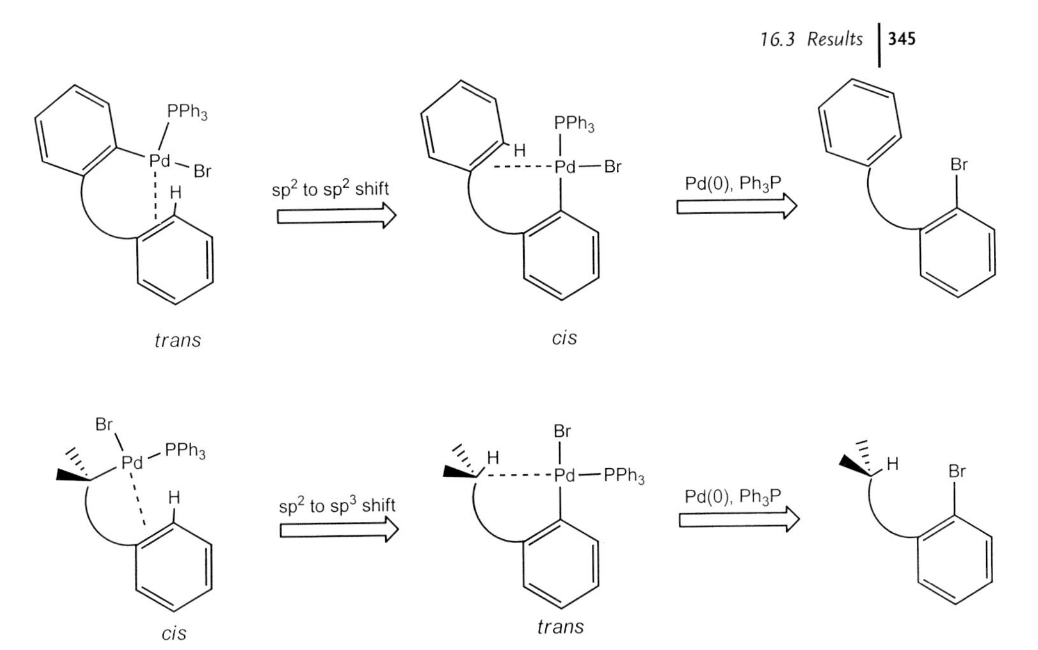


**Scheme 16.2** Retrosynthetic pathways from the bromo derivatives.

The *cis/trans* label that will be used throughout this chapter refers to the relative position of the bromine atom and the carbon atom covalently bound to palladium. Note that the transfer will connect a *cis* structure to a *trans* structure and vice versa. Since both types of transfer led to similar conclusions, we will focus here on the transfer that starts from the most stable isomer which is the *trans* isomer in the aryl-to-aryl case and the *cis* isomer in the aryl-to-alkyl case. The triphenylphosphine ligand $PPh_3$ was modeled by $PH_3$. The geometries were fully optimized by the gradient technique. Details about the basis sets used and the accuracy of the calculations (geometries and thermodynamic values) may be found in the original publications [20–22]. Here we may simply stress that the computed geometrical parameters are in agreement with related experimental structures of bromo palladium complexes [27–29].

## 16.3
## Results

### 16.3.1
### The 1,3 Pd Migration: Naphthalene (1) and Toluene (2) Systems

The two migrations – 1,3 Pd migration in naphthalene (1) and toluene (2) – are sketched in Scheme 16.3, and the key structures of the reaction paths of lowest energy (reactant, transition state, and product) are shown in Figure 16.1.

They are characterized by one single transition state of high energy, as confirmed by an IRC calculation. Quite interestingly this transition state is best viewed as

PPh$_3$ | Pd—Br | H — **1**

Br | Pd—PPh$_3$ | H — **2**

$\Delta$H (TS$^{IV}$) = +46.6 $\quad$ $\Delta$H (TS$^{IV}$) = +42.0

Scheme 16.3.

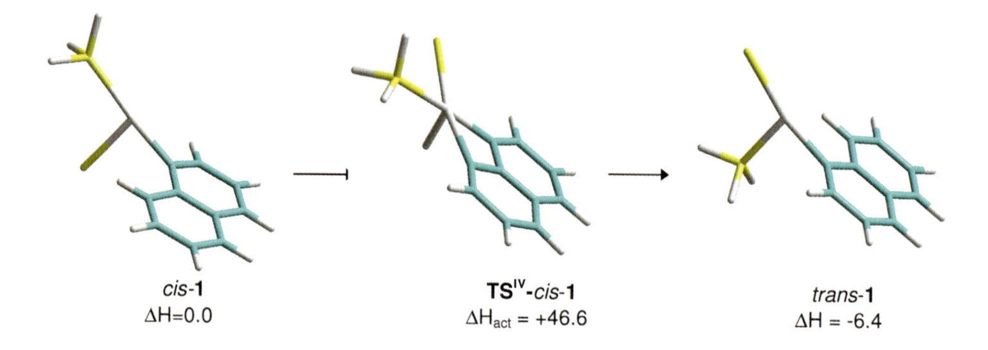


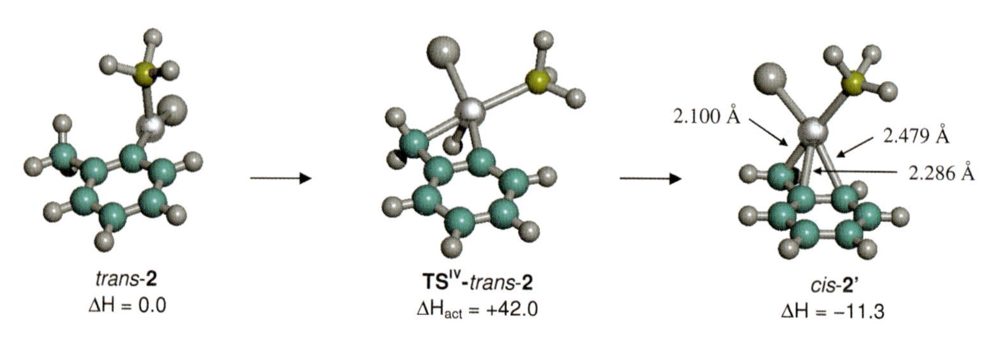


**Figure 16.1** Optimized structures for the *cis*-**1** → *trans*-**1** and *trans*-**2** → *cis*-**2′** 1,3 migration process. The enthalpy values are in kcal mol$^{-1}$.

being formally a d$^6$ Pd(IV) complex. This somewhat unexpected conclusion was drawn from the consideration of the geometrical parameters and from the value computed for the imaginary frequency: The Pd–H$_{transferred}$ distance is quite short, 1.54 Å in both cases. The distances between the transferred hydrogen atom and the two carbon atoms involved in the transfer are rather long, ranging between 2.0 and 2.5 Å. The Pd atom lies roughly in the plane formed by the organic moiety and the Br and PH$_3$ ligands, with the hydrogen atom at the apex of this plane. The system therefore adopts a geometry that is close to that of a square pyramid. Finally, the imaginary frequency is quite low, $335i\ cm^{-1}$ for **TS$^{IV}$**-*cis*-**1** and $557i\ cm^{-1}$ for **TS$^{IV}$**-*cis*-**2**. Thus, the transfer is best referred to as an oxidative hydrogen migration (OHM), a denomination which has been coined by Oxgaard et al. [41]. They encountered this type of transfer in their study of the mechanism of homogeneous Ir(III)-catalyzed arylation of olefins, where a $d^4$ Ir(V) transition state involving five covalent bonds was found [41a]. OHM transition states implying other transition metals have also been recognized recently [42]. Yet, that an OHM mechanism, i.e., a one-step Pd(IV) mechanism, can be at work rather than the two-step Pd(IV) mechanism or the one-step Pd(II) mechanism of Scheme 16.1, was not expected at the onset of our studies.

An additional and interesting feature that is displayed in the case of the 1,3 Pd migration in the palladium tolyl complex **2** is that the product is an $\eta^3$-benzyl complex: the three Pd–C(benzyl) bond distances (see Figure 16.1) are quite comparable to known $\eta^3$-benzyl Pd bond lengths [43–45]. The expected $\eta^1$ structure was found to be a transition state for the torsion about the $C_{aromatic}$–$C_{methyl}$ bond, well above the $\eta^3$ structure by 18.9 kcal mol$^{-1}$. As a result of the stabilization brought by the trihapto coordination mode in the product, the reaction is relatively exothermic, −11.3 kcal mol$^{-1}$ instead of −6.4 kcal mol$^{-1}$ in the case of the $sp^2$-to-$sp^2$ shift.

Finally, it is worth mentioning that the barrier is quite high in both reactions, see Scheme 16.3. This is because of the ring tension that the cyclometalated species experiences in two transition states, the C–Pd–C angle and the C $\cdots$ C distance for the carbon atoms involved in the 1,3 shift amounting to 67° and 2.30 Å, respectively, in both cases.

### 16.3.2
### The 1,4 Pd Migration: Phenanthrene (3), Biphenyl (4), Methylnaphthalene (5), and Ethylbenzene (6) Systems

A conclusion that may be drawn from Section 16.3.1 is that *three* rather than two intramolecular mechanisms can be possibly at work in the 1,*n* Pd migrations. The three possibilities have indeed been found for the 1,4 migrations, depending on the nature of the organic moiety and, more specifically, on the rigidity of the $C_4$ backbone involved in the transfer. In systems **3** and **5**, this backbone will remain roughly planar, owing to the presence of the fused aromatic rings. Hence, one finds a single Pd(IV) transition state characteristic of the OHM mechanism, as in systems **1** and **2**. The barriers are much lower, however, 24.4 kcal mol$^{-1}$ for the $sp^2$-to-$sp^2$ transfer (system **3**) and 27.7 kcal mol$^{-1}$ for the $sp^2$-to-$sp^3$ transfer (system **5**). We traced, in part, this relatively low barrier to the relief of the ring tension in the organic moiety, which is now a five-member metalla ring (compare, for instance, the values of the C–Pd–C angle and the C $\cdots$ C distance in **TS$^{IV}$**-*cis*-3 which amount 83° and 2.73 Å, respectively, to the values reported above for **TS$^{IV}$**-*cis*-**1**) [21]. In contrast to the systems **3** and **5**, the systems **4** and **6**, which do not have the constraint of the rigid $C_4$ backbone, can

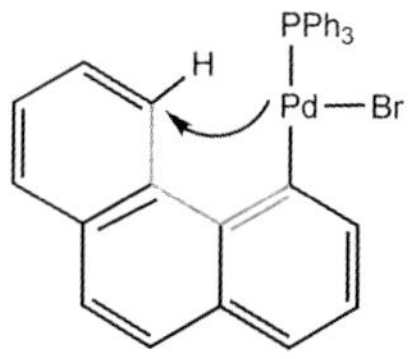

3

ΔH ($TS^{IV}$) = +24.4

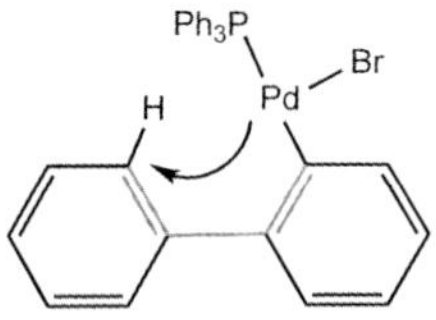

4

ΔH ($TS^{IV}$) = +27.5

ΔH ($TS^{II}$) = +29.0

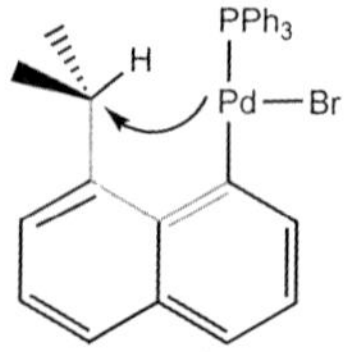

5

ΔH ($TS^{IV}$) = +27.7

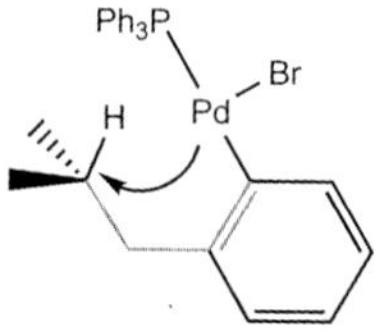

6

ΔH ($TS^{IV}$) = +29.2

ΔH ($TS^{II}$) = +30.5

$Inter^{IV}$, ΔH($TS_{max}$) = +31.9

**Scheme 16.4.**

undergo 1,4 shift not only via one-step Pd(IV) pathway (OHM mechanism) but also via one-step Pd(II) mechanism and – for **6** – via multistep pathway involving Pd(IV) *intermediates* (see Figure 16.2). The three pathways are competitive, with comparable energy barrier. (See Scheme 16.4), the one-step Pd(IV) mechanism being slightly favored.

At this stage, one should stress that, for the $sp^2$-to-$sp^3$ 1,4 Pd shift, the primary product of the reaction *cis*-**6′**, in which the palladium atom is coordinated to a terminal carbon atom of the alkyl chain, is not the most stable structure on the whole potential energy surface: it can rearrange easily to a quite stable $\eta^3$-benzyl structure, *cis*-**6′b**, see Figure 16.2. Such a rearrangement has been previously encountered in metal alkyl complexes and referred to as a "chain running" mechanism [46–49]. The overall reaction leading from the reactant *trans*-**6′** to the $\eta^3$-benzyl product *cis*-**6′ b** is therefore a [1, 4 + 1, 2] tandem Pd migration. Although being conceptually very simple, the 1,2 Pd/H interchange is mechanistically rather complex, involving several rotations, $\beta$-elimination, and insertion steps, all characterized by low-enthalpy barriers. The reader can find a detailed description on this in our original publication [22, 50].

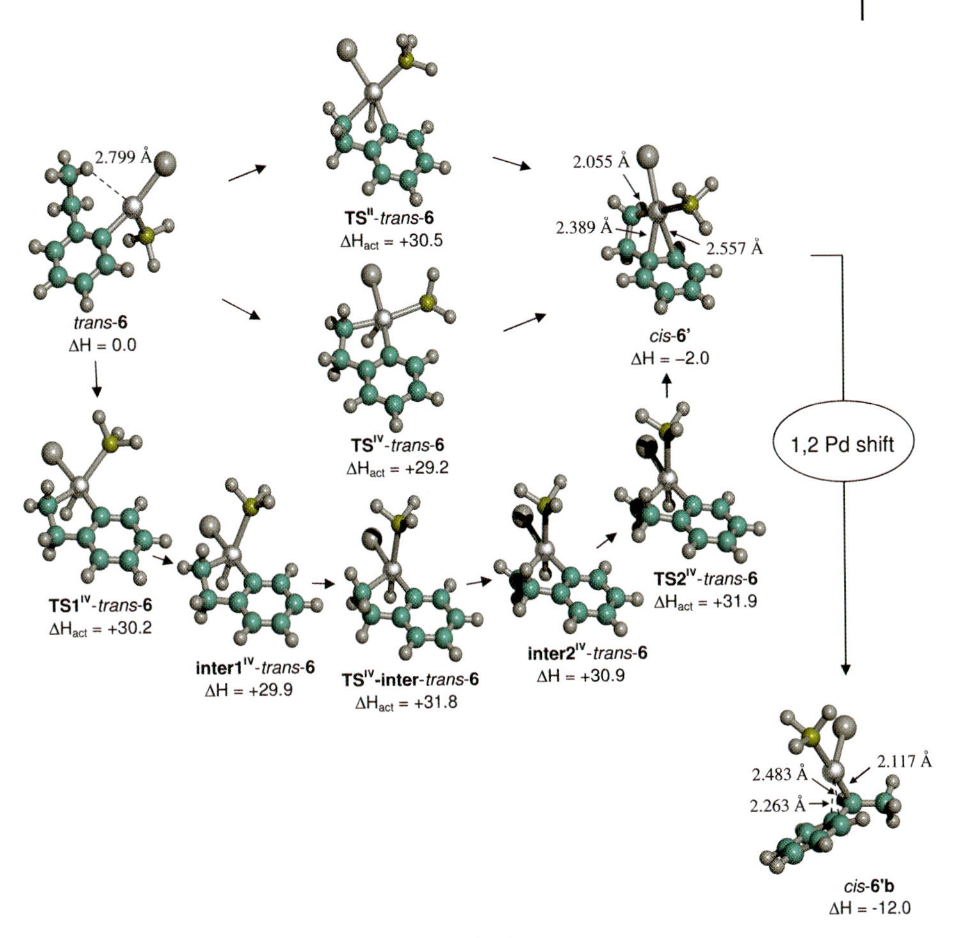


**Figure 16.2** Optimized structures of the Pd(II) and Pd(IV) pathways for the *trans*-**6** → *cis*-**6′** → *cis*-**6′ b** [1, 4 + 1, 2] migration process. The enthalpy values are in kcal $mol^{-1}$.

## 16.3.3
## The 1,5 Pd Migration: Benzo[c]phenanthrene (7), Benzylbenzene (8), and Propylbenzene (9) Systems

In contrast to the previous cases, the most stable transition state for the 1,5 migration, either from $sp^2$ to $sp^2$ or from $sp^2$ to $sp^3$ carbon atoms, turned out to be the Pd(II) transition state (see Scheme 16.5). In the case of benzo[c]phenanthrene system 7, no other transition state could be found. This is due to the fact that the square pyramidal conformation, which is characteristic of a Pd(IV) transition state, cannot be achieved owing to the geometrical requirements of the benzo[c]phenanthrene moiety: having two external phenyl rings coplanar with the palladium atom would lead to a Pd–C bond distance of the order of 1.7–1.8 Å, which is clearly too short.

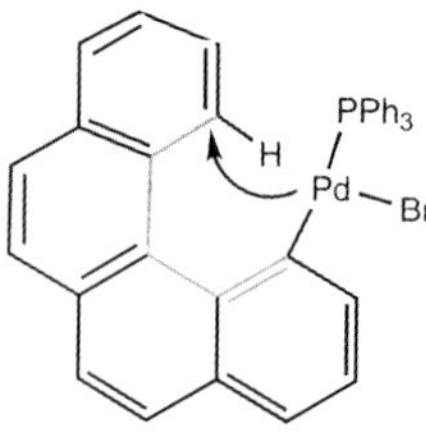

**7**

$\Delta H\ (TS^{II}) = +27.7$

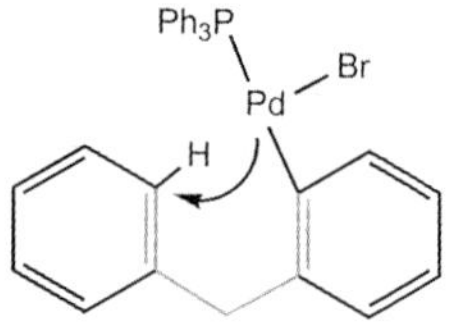

**8**

$\Delta H\ (TS^{II}) = +23.3$

$Inter^{IV}$, $\Delta H(TS_{max}) = +36.2$

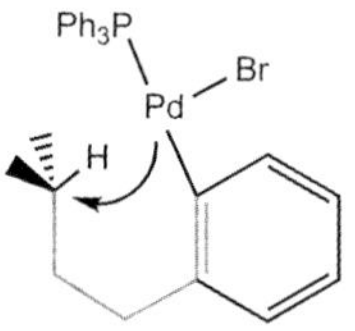

**9**

$\Delta H\ (TS^{II}) = +25.4$

$\Delta H\ (TS^{IV}) = +30.9$

$Inter^{IV}$, $\Delta H(TS_{max}) = +33.0$

Scheme 16.5.

On the other hand, for the benzyl- and propylbenzyl systems **8** and **9**, Pd(IV) routes were also obtained, of the oxidative addition/reductive elimination type in both cases and of the OHM type in system **9** only. For these two systems, a square pyramidal conformation is possible. At this stage, it is also interesting to compare the geometries of **TS**$^{II}$-*cis*-**7**, **TS**$^{II}$-*cis*-**8**, and **TS**$^{II}$-*cis*-**9** (see Figure 16.3). One salient feature of these transition states is the conformation of the organic moieties directly involved in the transfer, which are no longer coplanar with the $PdBr(PH_3)$ unit but adopt somehow a clamping arrangement around palladium.

As far as the energetics are concerned, these 1,5 migration processes turn out to be rather facile, the lowest enthalpy barrier for each system (which corresponds to the Pd(II) route) ranging between 17.8 and 25.4 kcal mol$^{-1}$.

### 16.3.4
### The 1,6 Pd Migration: Phenethylbenzene (10) and Styryl Benzene (11) Systems

In phenethylbenzene (**10**) and styryl benzene (**11**) systems that correspond to either having an additional methylene group in the alkyl chain binding the two

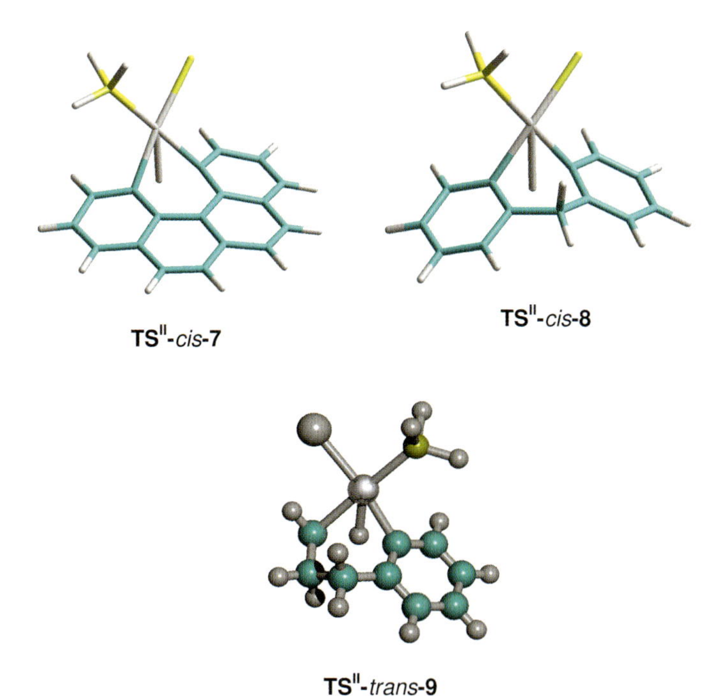


**Figure 16.3** Optimized structures of **TS$^{II}$**-*cis*-**7**, **TS$^{II}$**-*cis*-**8**, and **TS$^{II}$**-*trans*-**9**.

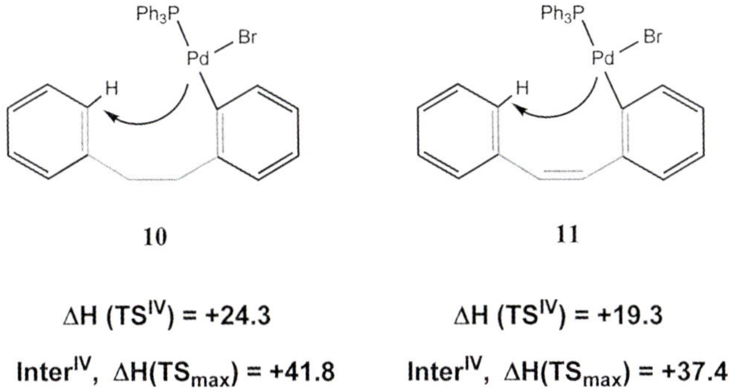

$Ph_3P$ Pd Br H

**10**

$\Delta H$ ($TS^{IV}$) = +24.3

Inter$^{IV}$, $\Delta H(TS_{max})$ = +41.8

$Ph_3P$ Pd Br H

**11**

$\Delta H$ ($TS^{IV}$) = +19.3

Inter$^{IV}$, $\Delta H(TS_{max})$ = +37.4

**Scheme 16.6.**

phenyl groups (the phenylbenzene system, see **10** in Scheme 16.6) or replacing the ethyl bridge of **10** by an ethenic bridge (see **11**), the favored pathway is the Pd(II) single-step process: the corresponding barriers are low and range from 19 to 24 kcal mol$^{-1}$ (enthalpy value). We refer the reader to the original publication for a more detailed analysis of this process and of the Pd(IV) pathways that are of the oxidative addition/reductive elimination type only and much more energy demanding (with barriers of the order of $\geqslant$ 35 kcal mol$^{-1}$) [21].

## 16.4
## Discussion

### 16.4.1
### How to Rationalize These Results?

The importance of geometric factors has been already described. But one would like to delineate more precisely the factors that govern the preference for one pathway over the others. To do so, we chose to analyze a 1,$n$ migration between two *disconnected* organic molecules. This allowed looking at the Pd/H interchange in an unconstrained system, both sides being able to adopt the more adequate position with respect to the Pd atom. The system under study was therefore the $PdBr(PH_3)(C_2H_3)(C_2H_4)$ complex *cis*-**12**, in which an hydrogen atom is transferred to the vinyl group and exchanges with palladium (see Figure 16.4). The calculations led to two single-step transition states, formally of Pd(II) and Pd(IV) character. The corresponding structures and energies are displayed in Figure 16.4. A first feature of this figure is that in the Pd(II) transition state $\mathbf{TS^{II}}$-*cis*-**12**, the transferred hydrogen is located precisely in the C–Pd–C plane. Moreover, the two vinyl groups are not coplanar with the $PdBr(PH_3)$ unit (the Br–Pd–C–H and P–Pd–C–H dihedral angles amounting to 66° and 78°, respectively) but adopt a structure that is reminiscent of their structure in *cis*-**12** and *trans*-**12**, so as to clamp the transferred hydrogen. We have already noticed this feature in $\mathbf{TS^{II}}$-*cis*-**7**, $\mathbf{TS^{II}}$-*cis*-**8**, and $\mathbf{TS^{II}}$-*cis*-**9**. The Pd(IV) transition state $\mathbf{TS^{IV}}$-*cis*-**12**, on the other hand, has the transferred hydrogen atom placed almost apically with respect to the plane defined by the ligands of the metal. Also note that the two $C_2H_3$ groups are now more coplanar with the $PdBr(PH_3)$ group (Figure 16.4). Finally, the single imaginary frequency is much lower in$\mathbf{TS^{IV}}$-*cis*-**12** compared with $\mathbf{TS^{II}}$-*cis*-**12**, 240$i$ $cm^{-1}$ instead of 1232$i$ $cm^{-1}$. All these features are therefore characteristic of a transition state for an oxidative hydrogen addition mechanism.

The orbital analysis is consistent with this picture. In $\mathbf{TS^{II}}$-*cis*-**12** transition state, the three occupied orbitals that have a significant weight on the transferred hydrogen all correspond to a bonding combination between the hydrogen s orbital and the occupied orbitals belonging to the $\sigma$ framework of the two $C_2H_3$ fragments (see Figure 16.5). One should note that these bonding interactions are helped by the clamping arrangement of the two vinyls. There is also some contribution from a nonbonding (with respect to the spectator ligands) d orbital on palladium. Thus, the transfer does correspond to a through-space proton transfer between the two carbon atoms, somewhat assisted by the metal. As mentioned earlier, this proton transfer takes place in the $PdC_2$ plane. We have also provided a rationalization of this feature on the basis of the isolobal analogy of such transition states with $CH_5^+$ in the $C_s$ geometry [20]. We also found that the formal oxidation state of +2 for $\mathbf{TS^{II}}$-*cis*-**12** is well reflected in the nature of its lowest unoccupied molecular orbital (LUMO) which is – as expected in a square planar Pd(II) complex – the $d_{x^2-y^2}$ orbital (somewhat antibonding with the ligand $\sigma$ orbitals), the $d_{z^2}$ orbital being one of the four doubly occupied metal d orbitals. In the $\mathbf{TS^{IV}}$-*cis*-**12** transition

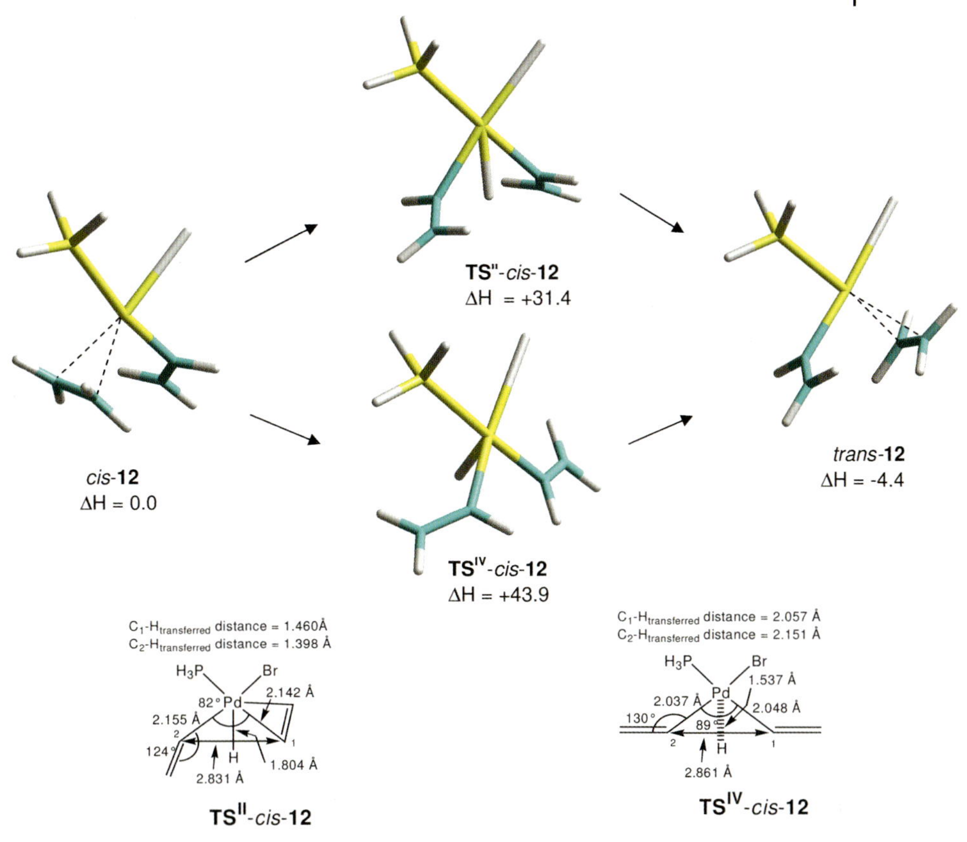


**Figure 16.4** Optimized structures for the *cis*-**12** → *trans*-**12** 1,∞ migration process. The enthalpy values are in kcal mol$^{-1}$.

state, there are two occupied orbitals with a significant weight on the transferred hydrogen (Figure 16.6). They both involve a bonding combination of the s orbital of the transferred hydrogen with the palladium $d_{z^2}$-type orbital. The LUMO of the system is made of the corresponding antibonding combination between $d_{z^2}$ and $s_H$, and the next LUMO is the antibonding combination of $d_{x^2-y^2}$ with the ligand $\sigma$ orbitals. Thus, the system has an electronic structure that indeed corresponds to that of a square pyramid $d^6$ complex (i.e., a Pd(IV) complex), the transferred hydrogen being formally a hydride.

As far as the activation energies are concerned, because **12** is devoid of any geometrical or steric constraints, the Pd(II) pathway is favored over the Pd(IV) pathway by about 12 kcal mol$^{-1}$ ($\Delta H$ values). This is consistent with the known difficulty of palladium complexes to achieve a +4 oxidation state. Yet the energy demand is quite high in both pathways. We trace this feature to a relatively strong stabilization of the *cis*-**12** species, for which there is no spatial restrictions to accommodate simultaneously the ethylene and the vinyl ligands.

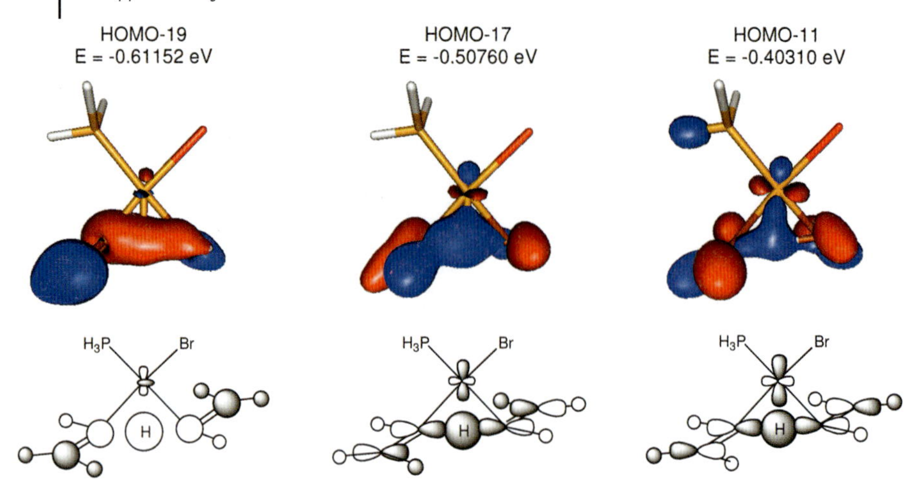


**Figure 16.5** Orbitals contributing to the transferred hydrogen in **TS**$^{II}$-*cis*-**12**.

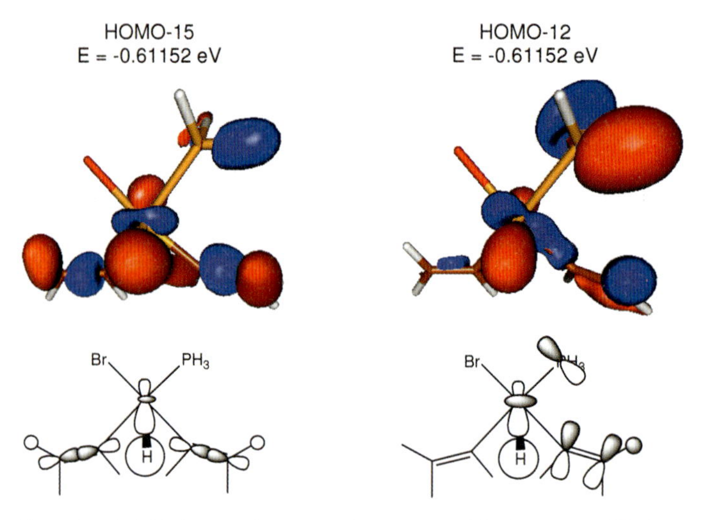


**Figure 16.6** Orbitals contributing to the transferred hydrogen in **TS**$^{IV}$-*cis*-**12**.

Thus, the above analysis shows clearly the interplay between electronic and geometric factors [51]. We have mentioned that a Pd(II) transition state is helped by a clamping arrangement of the two organic moieties, in contrast to the Pd(IV) transition state for which a planar arrangement is more favorable. Thus, **1**, **2**, **3**, and **5**, which cannot achieve this clamping arrangement, are characterized by the existence of a Pd(IV) transition state only, whereas for **4**, **6**, **8**, **9**, **10**, and **11**, both Pd(II) and Pd(IV) transition states can be located. The ease of the Pd(II) pathway in these systems will result from the balance between the propensity of the two organic moieties to achieve this clamping arrangement and the ability of the metallic fragment to fill the fourth coordination intramolecularly either via a weak agostic interaction with a C–H bond (systems **4**, **6**, **7**, and **9**) or via a stronger

$\pi$ interaction with a C–C bond (systems **8**, **10**, and **11**). The Pd(IV) transition states are generally much higher in energy, as expected for +4 oxidation state, the only exceptions being **3**, **4**, **6**, and **9**. This is due to the fact that in these systems, the coplanar arrangement of the two organic moieties in the transition state is easily achieved, the ring tension in the metallacycle is not too large (contrary to **1**, **2**, **10**, and **11**), and the reactant is not stabilized too much because it experiences only a C–H agostic interaction, which is relatively weak.

### 16.4.2
### Are the $sp^2$-to-$sp^3$ Shifts Different from the $sp^2$-to-$sp^2$ Shifts? Concluding Remarks

We may finally emphasize that the conclusions obtained for the shift between two $sp^2$ carbon atoms generally hold for the shift between an $sp^2$ and an $sp^3$ carbon atom, in particular with regard to the preference the Pd(II) route over the routes that involve either a Pd(IV) transition state (the OHM route) or a Pd(IV) intermediate (the oxidative addition/reductive elimination route). This is best seen in Schemes 16.3–16.5, where the $sp^2$-to-$sp^2$ and $sp^2$-to-$sp^3$ cases have been deliberately drawn next to each other. Within the Pd(IV) route, the preference for the one-step pathway over the two-step pathway has its origin in structural requirements and is therefore more system dependent. A single-step Pd(IV) transfer will take place when the transferred hydrogen has the possibility to interact simultaneously with both organic sides. Conversely, if the system cannot set up such a structure due to geometric requirements (as in systems **6**, **8**, **9**, **10**, and **11**), one finds an intermediate connected by two transition states in which the hydrogen atom interacts more with the carbon atom to which it will be transferred than with the other.

Finally, we hope that this chapter illustrates clearly the usefulness of standard DFT calculations to unravel rather intricate mechanisms in the field of organometallic reactivity and homogeneous catalysis, how they can provide a quite detailed knowledge on the geometric and electronic structures of intermediates and transition states. The calculations have been used not only to rationalize experimental observations but also to make proposals such as the possible occurrence of 1,6 Pd shifts, for which low barriers have been computed.

## References


1 For some recent references, see: (a) Tsuji, J. *Palladium Reagents and Catalysis: New Perspectives for the 21st Century*; John Wiley & Sons: New York, **2004**; (b) *Handbook of Organopalladium Chemistry for Organic Synthesis*; Negishi, E., Ed; John Wiley & Sons: New York, **2002**; (c) see also chapters 5, 6, 8–10, 13, and 15 in *Metal-Catalyzed Cross-Coupling Reactions*, 2nd ed; de Meijere A.; Diederich, F., Eds; Wiley-VCH: Weinheim, Germany, **2004**; Chapter 5, pp. 217–317.

2 See also: (a) Negishi, E., Copéret, C., Ma, S., Liou, S.-Y., Liu, F. *Chem. Rev.* **1996**, 365, (b) Herrmann, W.A., Öfele, K., Preysing, D.v., Schneider, S.K. *J. Organomet.* **2003**, *687*, 229;

(c) Palladium Chemistry in 2003: Recent Developments, special issue of *J. Organomet. Chem.*, G. Bertrand Ed.; (d) Zeni, G., Larock, R.C. *Chem. Rev.* **2004**, *104*, 2285; (e) Tietze, L.F., Ila, H., Bell, H.P. *Chem. Rev.* **2004**, *104*, 3453; Dupont, J., Consort, C.S., Spencer, J. *Chem. Rev.* **2005**, *105*, 2557; (f) Zeni, G., Larock, R.C. *Chem. Rev.* **2006**, *106*, 4644; (g) Yin, L., Liebscher, J. *Chem. Rev.* **2007**, 133; (h) Alberico, D., Scott, M.E., Lautens, M. *Chem. Rev.* **2007**, 174.

3 Wang, L., Pan, Y., Jiang, X., Hu, H. *Tetrahedron Lett.* **2000**, *41*, 725.

4 Tian, Q., Larock, R.C. *Org. Lett.* **2000**, *2*, 3329.

5 Larock, R.C., Tian, Q. *J. Org. Chem.* **2001**, *66*, 7372.

6 Karig, G., Moon, M.-T., Thasana, N., Gallagher, T. *Org. Lett.* **2002**, *4*, 3115.

7 Campo, M.A., Larock, R.C. *J. Am. Chem. Soc.* **2002**, *124*, 14326.

8 Campo, M.A., Huang, Q., Yao, T., Tian, Q., Larock, R.C. *J. Am. Chem. Soc.* **2003**, *125*, 11506.

9 Huang, Q., Fazio, A., Dai, G., Campo, M.A., Larock, R.C. *J. Am. Chem. Soc.* **2004**, *126*, 7460.

10 Huang, Q., Campo, M.A., Yao, T., Tian, Q., Larock, R.C. *J. Org. Chem.* **2004**, *69*, 8251.

11 Barder, T.E., Walker, S.D., Martinelli, J.R., Buchwald, S.L. *J. Am. Chem. Soc.* **2005**, *127*, 4685.

12 Zhao, J., Larock, R.C. *Org. Lett.* **2005**, *7*, 701.

13 Zhao, J., Campo, M., Larock, R.C. *Ang. Chem. Int. Ed.* **2005**, *44*, 1873.

14 Bour, C., Suffert, J. *Org. Lett.* **2005**, *7*, 653.

15 Masselot, D., Charmant, J.P.H., Gallagher, T. *J. Am. Chem. Soc.* **2006**, *128*, 694.

16 Zhao, J., Larock, R.C. *J. Org. Chem.* **2006**, *71*, 5340.

17 Singh, A., Sharp, P.R. *J. Am. Chem. Soc.* **2006**, *128*, 5998.

18 Campo, M., Zhang, H., Yao, T., Ibdah, A., McCulla, R.D., Huang, Q., Zhao, J., Jenks, W.S., Larock, R.C. *J. Am. Chem. Soc.* **2007**, *129*, 6298.

19 Ma, S., Gu, Z. *Ang. Chem. Int. Ed.* **2005**, *44*, 7512.

20 Mota, A.J., Dedieu, A., Bour, C., Suffert, J. *J. Am. Chem. Soc.* **2005**, *127*, 7171.

21 Mota, A.J., Dedieu, A. *Organometallics* **2006**, *25*, 3130.

22 Mota, A.J., Dedieu, A. *J. Org. Chem.* **2007**, *72*, 9669.

23 Becke, A.D. *Phys. Rev. A* **1988**, *38*, 3098.

24 Lee, C., Yang, W., Parr, R.G. *Phys. Rev. B* **1988**, *37*, 785.

25 Becke, A.D. *J. Chem. Phys.* **1993**, *98*, 5648.

26 Gaussian 03, Revision B.04. Frisch, M.J., Trucks, G.W., Schlegel, H.B., Scuseria, G.E., Robb, M.A., Cheeseman, J.R., Montgomery, Jr., J.A., Vreven, T., Kudin, K.N., Burant, J.C., Millam, J.M., Iyengar, S.S., Tomasi, J., Barone, V., Mennucci, B., Cossi, M., Scalmani, G., Rega, N., Petersson, G.A., Nakatsuji, H., Hada, M., Ehara, M., Toyota, K., Fukuda, R., Hasegawa, J., Ishida, M., Nakajima, T., Honda, Y., Kitao, O., Nakai, H., Klene, M., Li, X., Knox, J.E., Hratchian, H.P., Cross, J.B., Adamo, C., Jaramillo, J., Gomperts, R., Stratmann, R.E., Yazyev, O., Austin, A.J., Cammi, R., Pomelli, C., Ochterski, J.W., Ayala, P.Y., Morokuma, K., Voth, G.A., Salvador, P., Dannenberg, J.J., Zakrzewski, V.G., Dapprich, S., Daniels, A.D., Strain, M.C., Farkas, O., Malick, D.K., Rabuck, A.D., Raghavachari, K., Foresman, J.B., Ortiz, J.V., Cui, Q., Baboul, A.G., Clifford, S., Cioslowski, J., Stefanov, B., Liu, B.G., Liashenko, A., Piskorz, P., Komaromi, I., Martin, R.L., Fox, D.J., Keith, T., Al-Laham, M.A., Peng, C.Y., Nanayakkara, A., Challacombe, M., Gill, P.M.W., Johnson, B., Chen, W., Wong, M.W., Gonzalez, C., Pople, J.A., Gaussian, Inc., Pittsburgh PA, **2003**.

27 Alcazar-Roman, L.M., Hartwig, J.F., Rheingold, A.L., Liable-Sands, L.M., Guzei, I.A. *J. Am. Chem. Soc.* **2000**, *122*, 4618.

28 Yin, J., Buchwald, S.L. *J. Am. Chem. Soc.* **2002**, *124*, 6043.
29 Stambuli, J.P., Incarvito, C.D., Bühl, M., Hartwig, J.F. *J. Am. Chem. Soc.* **2004**, *126*, 1184.
30 Paul, F., Patt, J., Hartwig, J.F. *J. Am. Chem. Soc.* **1994**, *116*, 5969.
31 Hartwig, J.F., Paul, F. *J. Am. Chem. Soc.* **1995**, *117*, 5373.
32 Louie, J., Hartwig, J.F. *J. Am. Chem. Soc.* **1995**, *117*, 11598.
33 Galardon, E., Ramdeehul, S., Brown, J.M., Cowley, A., Hii, K.K., Jutand, A. *Ang. Chem. Int. Ed.* **2002**, *41*, 1760.
34 Stambuli, J.P., Bühl, M., Hartwig, J.F. *J. Am. Chem. Soc.* **2002**, *124*, 9346.
35 Strieter, E.R., Blackmond, D.G., Buchwald, S.L. *J. Am. Chem. Soc.* **2003**, *125*, 13978.
36 Hills, I.D., Fu, G.C. *J. Am. Chem. Soc.* **2004**, *126*, 13178.
37 Cundari, T.R., Deng, J. *J. Phys. Org. Chem.* **2005**, *18*, 417.
38 Ahlquist, M., Fristrup, P., Tanner, D., Norrby, P.-O. *Organometallics* **2006**, *25*, 2066.
39 Lam, K.C., Marder, T.B., Lin, Z. *Organometallics* **2007**, *26*, 758.
40 Braga, A.A.C., Ujaque, G., Maseras, F. *Organometallics* **2006**, *25*, 3647. See the discussion herein about the difficulty to evaluate the free energy in solution of bimolecular reactions via standard DFT/PCM calculations.
41 (a) Oxgaard, J., Muller, R.P., Goddard III, W.A., Periana, R.A. *J. Am. Chem. Soc.* **2004**, *126*, 352, (b) Oxgaard, J., Periana, R.A., Goddard III, W.A. *J. Am. Chem. Soc.* **2004**, *126*, 11658.
42 For a review of OHM transition states, see also Lin, Z. *Coord. Chem. Rev.* **2007**, *251*, 2280.
43 Ariafard, A., Lin, Z. *J. Am. Chem. Soc.* **2006**, *128*, 13010.
44 Johns, A.M., Utsunomiya, M., Incarvito, C.D., Hartwig, J.F. *J. Am. Chem. Soc.* **2006**, *128*, 1828.
45 Gatti, G., Lopez, J.A., Mealli, C., Musco, A. *J. Organomet. Chem* **1994**, *483*, 77.
46 Rix, F.C., Brookhart, M., White, P.S. *J. Am. Chem. Soc.* **1996**, *118*, 2436.
47 Schmidt, G.F., Brookhart, M. *J. Am. Chem. Soc.* **1985**, *107*, 1443.
48 Brookhart, M., Volpe, A.F., Lincoln, D.M., Hovarth, I.T., Millar, J.M. *J. Am. Chem. Soc.* **1990**, *112*, 5634.
49 See also (a) Larock, R.C., Lu, Y., Bain, A.C., Russell, C.E. *J. Org. Chem.* **1991**, *56*, 4589, (b) Albeniz, A.C., Espinet, P., Lin, Y.-S. *Organometallics*, **1997**, *16*, 4138.
50 **6′b** can also be obtained from **6** via a direct 1,3 Pd migration. This process is, however, characterized, like the other 1,3 migrations, by a high barrier. Thus, any apparent 1,3 Pd shift in ethylbenzene is most likely the result of a [1,4 + 1,2] tandem reaction. More generally, one may also anticipate that a 1,*n* Pd migration in alkylbenzene derivatives could be followed by a series of 1,2 shifts, provided that such shifts would be allowed by either the geometrical constraints of the system or the nature of the alkyl moiety. See ref [22] for a discussion of this issue in the light of the experimental results.
51 In a recent paper, see reference [18], Larock et al. have also stressed the importance of steric hindrance.

# 17
# Combining Electronic Structure Calculations and Spectroscopy to Unravel the Structure of Grafted Organometallic Complexes

*Raphael Wischert, Christophe Copéret, Françoise Delbecq, and Philippe Sautet*

## 17.1
## Introduction

Chemical processes in industry heavily rely on heterogeneous catalysts because of their several advantages (more efficient process: separation of products, recycling, and regeneration). While it is still necessary to improve these systems (in terms of energy efficiency and selectivity), especially in the context of sustainable development, the ill-defined nature of their active sites impedes a rational approach to the development of better processes through structure–reactivity relationships. With this aim, surface organometallic chemistry (SOMC) has emerged in the last decades, and it is directed toward the generation of heterogeneous catalysts via a molecular approach in order to combine the advantages of homogeneous and heterogeneous catalysis [1]. Therefore, one approach involves obtaining well-defined surface complexes by chemically grafting organometallic complexes on supports, typically oxides such as silica and alumina. One of the key aspects of this approach is to characterize these systems at a molecular level; therefore, it involves the use of spectroscopic methods, either well known from homogeneous catalysis or specific to the solid state, such as IR, solid-state NMR, and extended X-ray absorption fine structure spectroscopy (EXAFS). This approach has already proven to be successful in many cases, such as in the area of olefin metathesis [2], hydrogenation [3, 4], and polymerization. It has even permitted the discovery of novel reactions such as depolymerization [5–7] and the metathesis of alkanes [8–10]. More recently, this approach has also used computational chemistry as a tool to help in the characterization and understanding of the reactivity of these systems. This allows the exploitation of a strong synergy and complementarities between spectroscopy (such as nuclear magnetic resonance or infrared vibrational characterization), molecular simulations, and catalytic reactivity studies.

Over the years, these techniques have been able to show that the surface of a support acts as a mono- or polydentate ligand for the metal center, depending on its chemical reactivity toward the complex. The support should therefore not be regarded as inert but as playing an active chemical and structural role. This becomes particularly obvious when identical complexes grafted on different supports display

*Computational Methods in Catalysis and Materials Science.* Edited by Rutger A. van Santen and Philippe Sautet

ISBN: 978-3-527-32032-5

different catalytic activities and selectivities. For instance, in several cases, the choice of an alumina support can even become essential: (i) well-defined tungsten carbyne complexes are active alkane metathesis catalyst precursors when supported on alumina, but inactive when supported on silica [11] and (ii) $Zr(CH_2{}^tBu)_4$ (where $^tBu$ stands for $[-C(CH_3)_3]$) provides an efficient polymerization catalyst when grafted on alumina, while it displays no activity when grafted on silica [12–14]. In this chapter, we will present the role that a combined theoretical and experimental approach has played in order to obtain a detailed molecular understanding of alumina-supported systems, using the two above-mentioned examples: $Zr(CH_2{}^tBu)_4$ and $W(\equiv C^tBu)(CH_2{}^tBu)_3$ grafted on partially dehydroxylated $\gamma$-alumina [15].

First, a brief discussion of the theoretical methods will be presented (Section 17.2), followed by a description of the alumina model (Section 17.3), and the determination of alumina-supported chemical structures resulting from the grafting of organometallic molecular complexes on alumina (Section 17.4).

## 17.2 Methods

Density functional theory (DFT, cf. Chapter 2) calculations on a $\gamma$-$Al_2O_3$ model were performed using the generalized gradient approximation functional PW91 of Perdew and Wang [16] for exchange and correlation, and a periodic description (cf. Chapter 4), as implemented in the VASP code [17, 18]. Atoms were described by an all-electron, frozen core approach using the projector augmented wave (PAW) technique [19, 20]. With these potentials, a cut-off energy of 275 eV was found to be sufficient for a converged total energy; a mean variation of 1 meV/atom is observed when the cut-off energy is increased up to 400 eV. The Brillouin zone integration is converged with a $3 \times 3 \times 1$ k-point grid generated by the Monkhorst–Pack algorithm [21]. Vibrational frequencies were calculated in the harmonic approximation by numerical evaluation of the Hessian matrix. An anharmonicity term of 80 $cm^{-1}$, previously calculated for hydroxyl groups on boehmite, was added to all values [22]. Because the calculation of NMR parameters is not implemented in VASP, model clusters were extracted from the periodic model and chemical shifts were computed using the GIAO method implemented in the Gaussian03 package [23] at the DFT/B3LYP level (cf. Chapter 4). For hydrogen and carbon atoms, the IGLO-II basis set [24] was chosen along with the LANL2DZ basis set [25–27] for all other atoms, with *d* polarization functions added on Al and O atoms. Chemical shifts are reported relative to the shielding of tetramethylsilane that is calculated at the same level.

## 17.3 Modeling $\gamma$-Alumina

Of the known transition aluminas, $\gamma$-alumina is the industrially most important ones with key applications in refining and petrochemistry [28, 29]. It is used as a catalyst itself, for example in the Claus process, or more frequently, as a catalyst support. In the latter case, it often plays more than the role of a support, and in some cases, it can even be considered as a co-catalyst (*vide infra*). Its use as a support is due to several important properties: high surface area (100–200 $m^2\ g^{-1}$), high thermal stability, and good mechanical properties [30]. Moreover, in contrast to supports such as silica, its surface chemistry is more complex. This is due to the presence of Al atoms with different coordination numbers, which leads to various types of Brønsted and Lewis acid sites (*vide infra*).

### 17.3.1
### Modeling the Structure of γ-Alumina Bulk Material

The precise crystallographic nature of $\gamma$-$Al_2O_3$ is still under intense discussion as its structural investigation is hindered by its poor crystallinity. $\gamma$-$Al_2O_3$ has been typically described as a defective spinel-like structure (model of Knözinger and Ratnasamy [31]). While this model is still widely accepted by the experimental and theoretical community, it imposes arbitrarily constraints on the type and number of tetrahedral interstices present in the bulk material.

More recently, a new model has been proposed on the basis that $\gamma$-$Al_2O_3$ is a metastable transition alumina usually obtained by dehydration of boehmite ($\gamma$-AlOOH) at 700°C or by flame pyrolysis. The model is the result of a detailed theoretical study of this dehydration process [32, 33]. The topotactic transformation of boehmite into $\gamma$-$Al_2O_3$ (i.e., the crystallographic parameters are inherited from the starting material) involves (i) hydrogen transfer; (ii) collapse of the $\gamma$-AlOOH sheets and structural shearing; and (iii) Al migration from octahedral to tetrahedral interstices. Out of a wide range of Al atom distributions, the most stable structure contains 25% of tetrahedral Al, in agreement with $^{27}$Al NMR data [34] (see Figure 17.1). This model has several other advantages. It features a slightly distorted cubic face-centered sublattice (*fcc*) of oxygen atoms, which is consistent with XRD data [35]. Furthermore, the calculated cell volume is accurate within the limits of the GGA–DFT approach; the distortion of the cell in the *c* direction is well-reproduced and the bulk modulus is in considerably better agreement with the experimental value than that obtained for a spinel-based model [36]. For instance, in order to match the experimental powder X-ray and neutron diffractions patterns [35], supplementary spinel sites must be assumed in the spinel-based models in order to achieve a Rietveld refinement [32, 37]. Finally, the stability per $Al_2O_3$ unit is higher in the non-spinel model.

For these reasons, we chose the model system of Krokidis and Digne for our study. It is described by a unit cell containing 8 $Al_2O_3$ units (Figure 17.1), which is a good compromise between structural reliability and system sizes adapted to *ab initio* calculations. However, such a periodic model does not account for the poor crystallinity and size of the alumina (nano)particles.

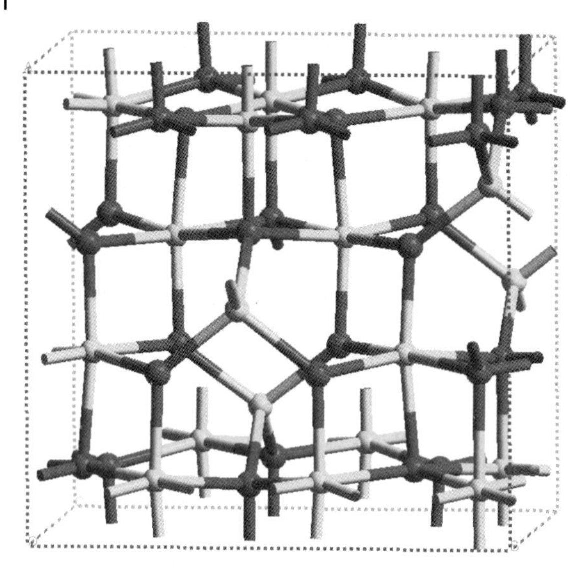

**Figure 17.1** Unit cell of $\gamma$-$Al_2O_3$ according to Krokidis et al. [32] Oxygen atoms are indicated by black and aluminum atoms by grey balls. The unit cell contains 25% of tetrahedral Al and 75% of octahedral Al.

### 17.3.2
### Modeling the Surface of γ-Alumina Including Hydration Behavior

From this model, it is possible to investigate the surface composition and properties of hydrated alumina [33]. The (110) facet is predominant on $\gamma$-$Al_2O_3$ crystallites according to neutron diffraction (83%) [38] and electron microscopy data (70%) [39]. As mentioned above, the transformation of boehmite to $\gamma$-$Al_2O_3$ is topotactic, which implies that the particle morphology of alumina is directly inherited from its precursor [29, 39]. On the basis of this finding, the model predicts 74% of (110) surface, along with 16% (100) and 10% (111) surface respectively, which is in excellent agreement with the experimental values. Therefore, the (110) surface was chosen for the theoretical investigation of our $\gamma$-$Al_2O_3$-supported systems. The bare (110) surface (noted as *s*0) exposes 1 $Al_{III}$ site originating from bulk tetrahedral Al, and 3 $Al_{IV}$ sites originating from bulk octahedral Al per unit cell (see Figure 17.2(a)). In the calculations, the (110) surface was represented by a four-layer periodic slab of unit formula $Al_{16}O_{24}$ with cell vectors of $8.1 \times 8.4 \times 23$ Å, a vacuum slab preventing artificial interactions between unit cells in the periodic description of the system. In order to reproduce the bulk material properties, the bottom two layers were kept fixed during the calculations, while the top layers were allowed to relax.

For a realistic description of the surface, it was of utmost importance to include a description of the surface hydration under working conditions. Therefore Digne

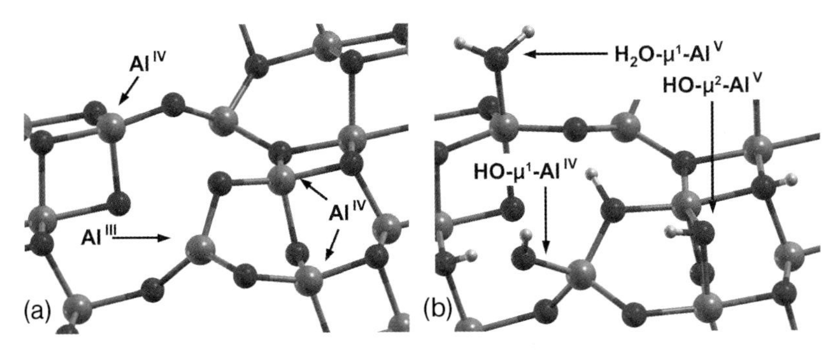


**Figure 17.2** Relaxed configurations of $\gamma$-$Al_2O_3$ (110) surface. (a) fully dehydroxylated s0 surface. (b) trihydrated s3 surface. Black balls: oxygen atoms; gray balls: aluminum atoms; and white balls: hydrogen atoms. The bottom two layers of the four-layer periodical slab have been omitted for clarity.

et al. studied the temperature-dependent surface hydroxyl coverage of the (110) $\gamma$-$Al_2O_3$ surface [33]. This work was further refined to the experimental conditions of our grafting experiments by Joubert et al., who simulated temperature-programmed desorption of water by first principles calculations and found that it was essentially a nonactivated process, thereby validating the thermodynamic (temperature–surface hydroxylation) relationship postulated by Digne. At a simulated dehydroxylation temperature of 770 K and a pressure of $10^{-5}$ Torr (as in the experiment), a surface covered by one water molecule in each surface unit cell ($s1$, 3 OH $nm^{-2}$), dissociated on an $Al_{III}$ site, is found to predominate, along with a small proportion of fully dehydroxylated ($s0$) surface [40]. However, it should be taken into account that different water pressures exist in solids due to the local curvature of the pores [41]. Because of the high porosity of the material, $s0$- and $s1$-type surfaces will likely be favored on exposed outer areas, whereas $s2$- and $s3$-type surfaces (see Figure 17.2(b)), covered by 2 or 3 molecules of water, respectively, will more likely occur in the less accessible pores, which, however, represent the major part of the surface area.

As a matter of fact, the experimental IR spectrum is best reproduced by the s3 surface. As depicted, five types of OH-groups are present; however, we focus on the following three types relevant for grafting molecules, as explained below: (i) HO-$\mu^1$-$Al_{IV}$, which is also present on the s1 surface, is formed by dissociation of one water molecule on an $Al_{III}$ defect site; (ii) $H_2O$-$\mu^1$-$Al_V$, a water molecule coordinated to an $Al_{IV}$ site; and (iii) HO-$\mu^2$-($Al_V$,$Al_V$), a bridging OH-group formed by dissociation of a third water molecule on two vicinal $Al_{IV}$. The OH coverage on s3 is 8.8 OH/$nm^{-2}$. Taking into account the experimentally determined percentage of the (110) facet (*vide supra*), this amounts to 6.2–7.3 OH $nm^{-2}$, which is to be compared with the experimental values of 2.5–5 OH $nm^{-2}$, as determined by H/D exchange [42], gravimetric studies [43] and pulse flow methods [44, 45] for $\gamma$-$Al_2O_3$ dehydrated at 500°C.

In the region above 3500 $cm^{-1}$, HO-$\mu^1$-$Al_{IV}$ exhibits the highest frequency (3842 $cm^{-1}$), followed by the bridging HO-$\mu^2$-$Al_V$,$Al_V$ group (3707 $cm^{-1}$), and the coordinated water molecule $H_2O$-$\mu^1$-$Al_V$ (3616 $cm^{-1}$). However, it should be noted that for a complete description of the five IR bands widely reported in the literature for partially dehydroxylated $\gamma$-$Al_2O_3$ (see Morterra [46] for a review), contributions from all surface orientations have to be taken into account (see Ref. [33] for further details).

In order to evaluate the reactivity of the OH-groups, the adsorption energy of pyridine on the different hydroxyl sites was calculated; the higher the adsorption energy, the higher the Lewis acidity toward pyridine, i.e., the capacity of the proton to interact with this Lewis base without breaking the O–H bond: HO-$\mu^1$-$Al_{IV}$, having the lowest Al coordination, was found to be the most reactive toward calculated pyridine adsorption, followed by HO-$\mu^2$-($Al_V$,$Al_V$) and $HO_2$-$\mu^1$- $Al_V$. The other OH-types, namely, –OH,–groups on Al sites with higher coordination or those involved in hydrogen bonds, did not show any reactivity toward pyridine and therefore were not considered as grafting sites [15].

We may conclude this part with some final remarks on the choice of our model surfaces. As mentioned before, the complete description of the IR spectrum of an alumina sample is not possible with a single surface model. The *s*3 surface was, therefore, chosen as the best average model for all hydroxyl types present on the surface; all OH-groups of *s*1 and *s*2 are present and the IR spectrum is well reproduced. The study of the $Zr(CH_2{}^tBu)_4$ and the $W(\equiv C^tBu)$ $(CH_2{}^tBu)_3$ systems focuses on the *s*3 and the fully dehydroxylated *s*0 surfaces as the latter features low coordination Al sites which were experimentally shown to be present on $\gamma$-$Al_2O_{3\text{-}(500)}$. Indeed, hydrogen is split at room temperature on low coordination $Al_{III}$ and $Al_{IV}$ sites (cumulated amount of 0.069 sites $nm^{-2}$), methane at higher temperatures (100–150°C) selectively on the more reactive $Al_{III}$ Lewis acidic sites, which is fully consistent with calculated reaction and activation energies [40].

## 17.4
## Understanding the Structure of Surface Species Resulting from Grafting of Molecular Organometallic Complexes on γ-Alumina

### 17.4.1
### Grafting of $Zr(CH_2{}^tBu)_4$ on γ-Alumina

The reaction of $Zr(CH_2{}^tBu)_4$ (**1**) with $\gamma$-$Al_2O_3$ partially dehydroxylated at 500°C, $\gamma$-$Al_2O_{3\text{-}(500)}$, yields a solid, which contains 2.8%wt Zr and $11 \pm 2$ C per grafted Zr, along with 2 equivalents of 2,2-dimethylpropane per grafted Zr. Monitoring the same reaction by IR spectroscopy reveals that the OH-bands at 3795, 3776, and 3730 $cm^{-1}$ have mainly disappeared, whereas the band at 3695 $cm^{-1}$ is only partially consumed, which is consistent with the presence of less accessible OH-groups. While these findings are consistent with the formation of a bisaluminoxy surface species (**2**, see Scheme 17.1), $^{13}C$ CP-MAS solid-state NMR spectroscopy of the corresponding compound selectively $^{13}C$ labeled on the carbon attached to the

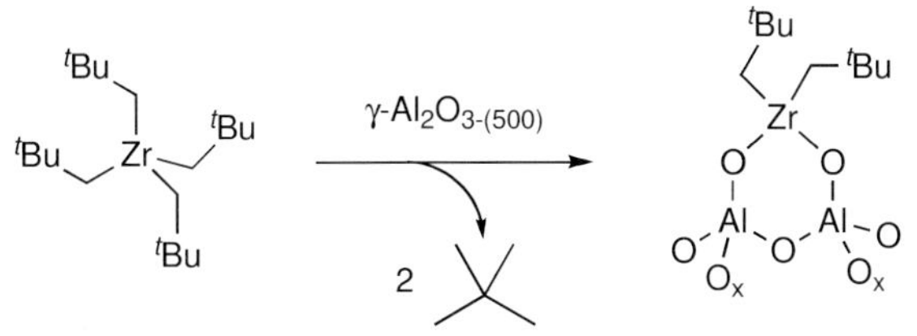

**Scheme 17.1** Proposed surface species for the reaction of $Zr(CH_2{}^tBu)_4$ (**1**) with $\gamma$-$Al_2O_{3\text{-}(500)}$.

metal, clearly shows that several species are formed, as evidenced by three major signals at 26, 84, and 99 ppm (not observed on the unlabeled compound).

Further investigation of the thus-formed surface species was undertaken by studying the chemical grafting, the structure of surface species, and their corresponding spectroscopic signatures by comparing experimental and calculated data.

First, the chemical grafting of $Zr(CH_3)_4$, (**1m**) taken as a model compound (as an approximation to $Zr(CH_2{}^tBu)_4$, which is computationally demanding for a full study because quadruple unit cells are required) is calculated to be highly exothermic on all OH sites (see Table 17.1), with the highest value associated with the grafting on the most acidic site, HO-$\mu^1$-$Al_{IV}$. For instance, grafting on the bridging HO-$\mu^2$- ($Al_V$,$Al_V$) or on the coordinated water molecule $H_2O$-$\mu^1$-$Al_V$ is around 15 kJ $mol^{-1}$ less favorable, as predicted by the weaker Lewis acidic character of the protons in this case.

In term of grafting mechanism, the reaction of $Zr(CH_3)_4$ on a HO-$\mu^1$-$Al_{IV}$ site yields a monoaluminoxy species (**3m**) and one molecule of $CH_4$, which is associated with a highly exothermic reaction ($\Delta_r E = -201$ kJ $mol^{-1}$) and a relatively low activation energy (37 kJ $mol^{-1}$). The grafting mechanism corresponds to the reaction of a Zr–C bond and an O–H bond via $\sigma$-bond metathesis, as evidenced

**Table 17.1** Adsorption energies ($E_{ads}$) of pyridine and reaction energies $\Delta_r E$ of the model compounds $Zr(CH_3)_4$ (**1m**) and $W(\equiv CCH_3)(CH_3)_3$ (**7m**) on –OH–groups of the s3 surface. $\Delta_r E = E_{tot} - (E_{s3} + E_{molecule})$.

| | | $\Delta_r E$ (kJ $mol^{-1}$)[b] | |
|---|---|---|---|
| **OH-group** | $E_{ads}$ (kJ $mol^{-1}$)[a] | $Zr(CH_3)_4$ | $W(\equiv CCH_3)(CH_3)_3$ |
| HO-$\mu^1$-$Al_{IV}$ | −40 | −201 | −150 |
| HO-$\mu^2$-($Al_V$,$Al_V$) | −31 | −186 | −139 |
| $H_2O$-$\mu^1$-$Al_V$ | −26 | −187 | −130 |
| HO-$\mu^3$-($Al_V$,$Al_{VI}$,$Al_{VI}$) | No adsorption | | |
| HO-$\mu^2$-($Al_{IV}$,$Al_V$) | No adsorption | | |

[a] Adsorption energy for pyridine.
[b] Reaction energies for $Zr(CH_3)_4$ (**1m**) and $W(\equiv CCH_3)(CH_3)_3$ (**7m**).

by the quasilinear transition state (C–H–O). The monaluminoxy species **3m** is surrounded by Lewis acidic Al sites and hydroxyl groups and therefore can further react via (i) transfer of a methyl group to an adjacent Al Lewis acid site, an exothermic process (98 kJ mol$^{-1}$) associated with a very low barrier (15 kJ mol$^{-1}$), yielding a partially cationic species $Al_sOZr(CH_3)_2(\mu\text{-}CH_3)Al_s$ (**4m**, see Scheme 17.2) with a $CH_3$ ligand bridging between the Zr centre and an $Al_V$ site and further stabilized by an oxygen of an aluminoxane bridge, and alternatively, (ii) the reaction of the monoaluminoxy surface complex **3m** with a second HO-$\mu^1$-$Al_{IV}$ site yielding a second equivalent of methane and a neutral bisaluminoxy species $(Al_sO)_2Zr(CH_3)_2$ (**2m**). Similar to the first grafting step, the barrier for this second $\sigma$-bond metathesis is low (20 kJ mol$^{-1}$); however, the reaction is less favored ($\Delta_r E = -154$ kJ mol$^{-1}$), probably due to geometric constraints because no perfect tetrahedral arrangement of the ligands is possible. Note that this bipodal species **2m** can rearrange via an exothermic (80 kJ mol$^{-1}$) and nearly barrierless process ($\Delta E^{\ddagger} = 5$ kJ mol$^{-1}$) into $(Al_sO)_2Zr(CH_3)(\mu\text{-}CH_3)(Al_s)$ (**5m**), featuring a bridging $CH_3$ ligand. This species can also be formed through the reaction of the partially cationic complex **4m** with an adjacent HO-$\mu^1$-$Al_{IV}$ site, releasing a second equivalent of methane. This step is again associated with an exothermic process ($\Delta_r E = -136$ kJ mol$^{-1}$) and a small activation barrier (14 kJ mol$^{-1}$). In contrast, the formation of the trisaluminoxy species, $(Al_sO)_3Zr(CH_3)$ (**6m**), along with a third equivalent of methane is not favorable: (i) a third $\sigma$-bond metathesis from **5m**, involving the bridging $CH_3$ ligand and a bridging HO-$\mu^2$-($Al_V$,$Al_V$) is associated with a high barrier of 150 kJ mol$^{-1}$, despite the high exothermicity of the process (−90 kJ mol$^{-1}$ from **5m**), and (ii) a two-step reaction that involves the reformation of **2m** also requires a high activation energy (90 kJ mol$^{-1}$). Changing the methyl ligands of the model system to the neopentyl ligands (used experimentally) has no significant effect on the calculated activation energies of the elimination processes. However, the formation of a bridged neopentyl complex is endoenergetic on going from **3** to **4** ($\Delta_r E = +35$ kJ mol$^{-1}$), probably because of the bulkiness of the neopentyl ligand. However, it remains exoenergetic on going from the bisaluminoxy species $(Al_sO)_2Zr(CH_2{}^tBu)_2$ (**2**) to ({$[(Al_sO)_2Zr(CH_2{}^tBu)]^+$ $[(^tBuCH_2)(Al_s)]^-$} (**5**) ($\Delta_r E = -23$ kJ mol$^{-1}$), which is, in this case, a true cationic complex, where the neopentyl group is fully shifted to an adjacent Lewis acidic site (see Figure 17.3). Overall, these computational studies suggest that grafting takes place via two successive $\sigma$-bond metathesis reactions on adjacent OH sites to form two alkane molecules and a neutral bisaluminoxy species $(Al_sO)_2Zr(CH_2{}^tBu)_2$ (**2**), which ultimately evolves to the true cationic surface complex {$[(Al_sO)_2Zr(CH_2{}^tBu)]^+$ $[(^tBuCH_2)(Al_s)]^-$} (**5**), a neopentyl ligand having shifted to an adjacent Lewis acidic site.

The validation of the proposed surface species {$[(Al_sO)_2Zr(CH_2{}^tBu)]^+$ $[(^tBuCH_2)(Al_s)]^-$} (**5**) was further pursued by comparing the experimental and calculated spectroscopic signatures. The calculated IR frequencies of remaining OH found in **5** show that they are not perturbed by the grafted species, which is consistent with the experimental observations. Chemical shifts were calculated on a model cluster extracted from the optimized geometry of **5**. In the proposed major surface complex **5**, the calculated chemical shifts of the methylene carbons bonded to Zr

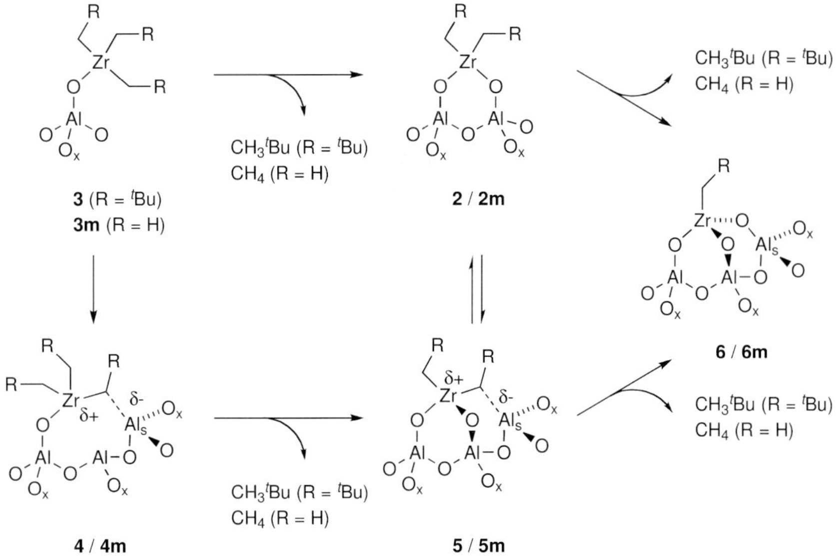

3 (R = $^tBu$)
3m (R = H)

$CH_3{}^tBu$ (R = $^tBu$)
$CH_4$ (R = H)

2 / 2m

$CH_3{}^tBu$ (R = $^tBu$)
$CH_4$ (R = H)

6 / 6m

4 / 4m

$CH_3{}^tBu$ (R = $^tBu$)
$CH_4$ (R = H)

5 / 5m

$CH_3{}^tBu$ (R = $^tBu$)
$CH_4$ (R = H)

**Scheme 17.2** Pathways evolving from the monopodal species **3**/**3m**. Stabilizing interactions from surface oxygen atoms are not shown here.

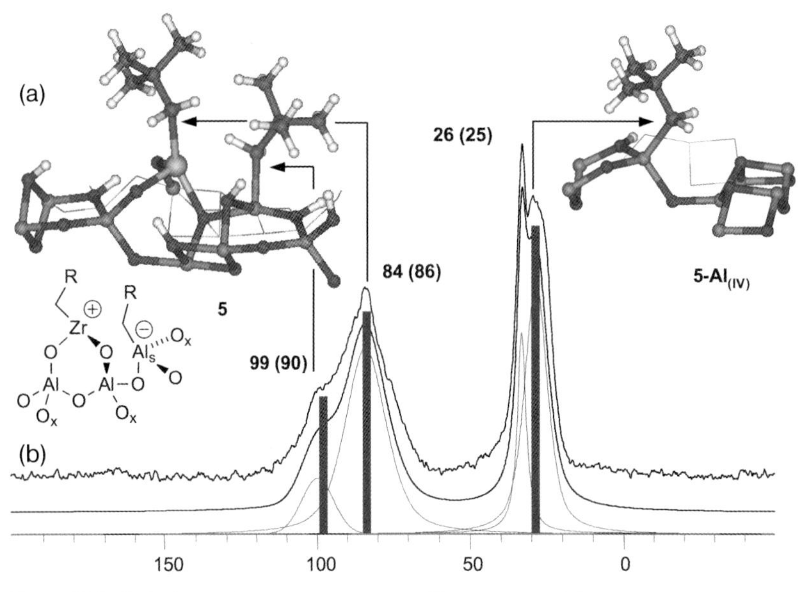


**Figure 17.3** Cationic species $\{[(Al_sO)_2Zr(CH_2{}^tBu)]^+ [Al_s(CH_2{}^tBu)]^-\}$ (**5**) and alkylaluminate species $[\mathbf{Al_{IV}}(CH_2{}^tBu)]^-$ (**5**-$Al_{IV}$) resulting from the grafting of $Zr(CH_2{}^tBu)_4$ (**1**) on $\gamma$-$Al_2O_{3\text{-}(500)}$. Calculated $^{13}C$ chemical shifts are given in parentheses and compared to the $^{13}C$ CP-MAS (10 kHz) solid-state NMR spectrum (33% labeled $^{13}C$ on $CH_2$) of $Zr(CH_2{}^tBu)_4$ on $\gamma$-$Al_2O_{3\text{-}(500)}$. Adapted from Ref. [47].

```
     H
    /
   O         □
   |         |
---O-Al--O--Al-O---
    / \     / \
  Ox   O  Ox   O
```

**Scheme 17.3** $Al_2O_3$ surface acting as a bifunctional support.

and $Al_{VI}$ are 90 and 86 ppm, respectively, which can readily explain two out of three experimental NMR signals. Note, however, that it is not possible to explain the third signal at 26 ppm by using this model. Taking into account the presence of Al Lewis acidic defect sites (see Section 17.3) that are not present in the s3 model, it is also possible to transfer the neopentyl ligand from Zr onto an adjacent $Al_{III}$ defect site. In this case, the calculated NMR chemical shift of the tetrahedral alkylaluminate species $[Al_{IV}(CH_2{}^tBu)]^-$ (**5** – $\mathbf{Al_{IV}}$, see Figure 17.3) is 25 ppm, which is in excellent agreement with the third observed NMR signal.

Overall, from experimental and computational studies, the grafting of $Zr(CH_2{}^tBu)_4$ on $\gamma$-$Al_2O_3$ probably generates a cationic surface species like **5**. While this type of surface species is possible on partially dehydroxylated alumina (or on silica–alumina), it cannot be generated on pure silica because this support completely lacks Lewis acidic sites. The presence of a strongly electrophilic cationic Zr center, having a free coordination site in the case of $Zr(CH_2{}^tBu)_4$ supported on $\gamma$-alumina, is fully consistent with its high polymerization activity in comparison with the neutral species obtained on silica. This also shows that alumina should be considered as a bifunctional surface; –OH-groups playing the role of anchoring sites for organometallic complexes and the Lewis acidic sites acting as a co-catalyst by forming the active cationic species (Scheme 17.3) [47].

## 17.4.2
## Grafting of $W(\equiv C^tBu)(CH_2{}^tBu)_3$ on $\gamma$-Alumina

The reaction of $W(\equiv C^tBu)(CH_2{}^tBu)_3$ (**7**) with $\gamma$-$Al_2O_{3\text{-}(500)}$ generates 0.9 equivalent of 2,2-dimethylpropane per grafted W, and the resulting solid contains 3.8%wt W and $14.7 \pm 2$ C per grafted W, as determined by elemental analysis. These findings are consistent with the formation of a monoaluminoxy surface species $(Al_sO)W(\equiv C^tBu)(CH_2{}^tBu)_2$ (**8**, see Scheme 17.4). Moreover, monitoring this reaction by IR spectroscopy reveals that upon reaction of $W(\equiv C^tBu)(CH_2{}^tBu)_3$ on $\gamma$-$Al_2O_{3-(500)}$, the OH-bands at 3795, 3776, and 3730 $cm^{-1}$ have disappeared, while a large OH band appears at 3650 $cm^{-1}$. Additionally, the $^{13}C$ CP-MAS spectrum of the labeled species is noteworthy, displaying broad intense peaks at 318, 103, and 95 ppm, in contrast to sharp peaks at 318 and 95 ppm in the case of the silica-supported species [11, 48]. Note that these results are in contrast to what is observed for Zr: formation of a bisaluminoxy surface species (**5**), presence of an intense signal at 26 ppm associated with $[Al_{IV}(CH_2{}^tBu)]^-$(**5** – $\mathbf{Al_{IV}}$) and the disappearance of most OH-groups (*vide supra*). These findings indicate that the grafting of Zr and W molecular complexes on alumina is probably very different, and this has also been investigated in greater detail via calculations.

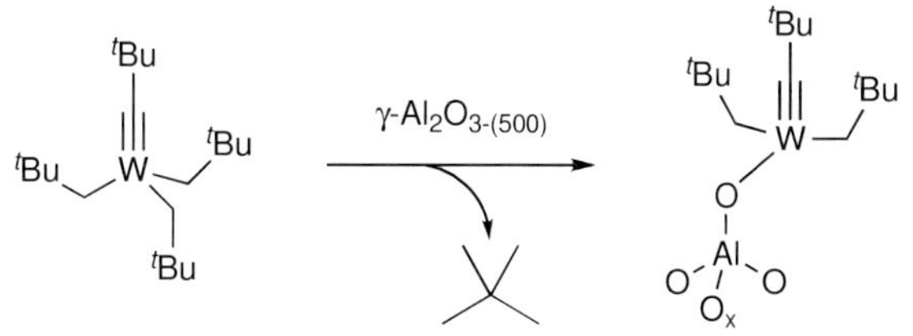

**Scheme 17.4** Proposed surface species for the reaction of $W(\equiv C^tBu)(CH_2{}^tBu)_3$ (**7**) with $\gamma$-$Al_2O_{3\text{-}(500)}$.

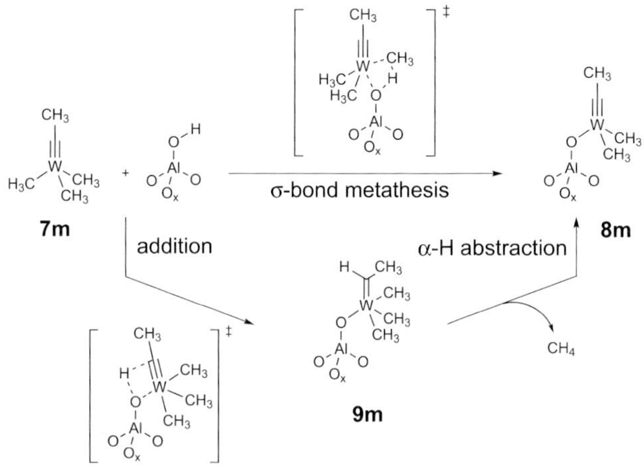

**Scheme 17.5** Two different pathways for the initial grafting step of $W(\equiv CCH_3)Me_3$ (**7m**) on $\gamma$-$Al_2O_{3\text{-}(500)}$.

First of all, as observed for Zr, the chemical grafting of the model complex $W(\equiv CCH_3)(CH_3)_3$ (**7m**, see Scheme 17.5) is calculated to be highly exothermic on all OH sites, but less than for Zr (see Table 17.1), the highest value being associated with the grafting on the most acidic site, HO-$\mu^1$-$Al_{IV}$. In this case, two grafting mechanisms have been considered: (i) $\sigma$-bond metathesis between $Al_sO{-}H$ and $W{-}CH_3$ of $W(\equiv CCH_3)(CH_3)_3$ (**7m**), yielding $(Al_sO)W(\equiv CCH_3)(CH_3)_2$ (**8m**), and (ii) addition of the hydroxyl group onto the carbyne forming an intermediate carbene (**9m**), which generates methane and the same species after $\alpha$-H abstraction.

The first pathway is significantly easier with an activation energy of 81 kJ mol$^{-1}$ compared to 126 kJ mol$^{-1}$ for the addition of $Al_sOH$ onto the carbyne. This is also consistent with what has been observed experimentally for other systems, i.e., alkylidyne ligands are not involved in the grafting process [49], in contrast to alkyl and alkylidene substituents [49–52]. Moreover, structures involving cationic forms were calculated to be disfavored (>49 kJ mol$^{-1}$), in contrast to Zr. Furthermore, the elimination of a second equivalent of methane by reaction of **8m** with an adjacent $Al_sOH$ group to form the bisaluminoxy species $(Al_sO)_2W(\equiv CCH_3)(CH_3)$ (**10m**, not depicted) is calculated to be thermodynamically favored by 90 kJ mol$^{-1}$. Note,

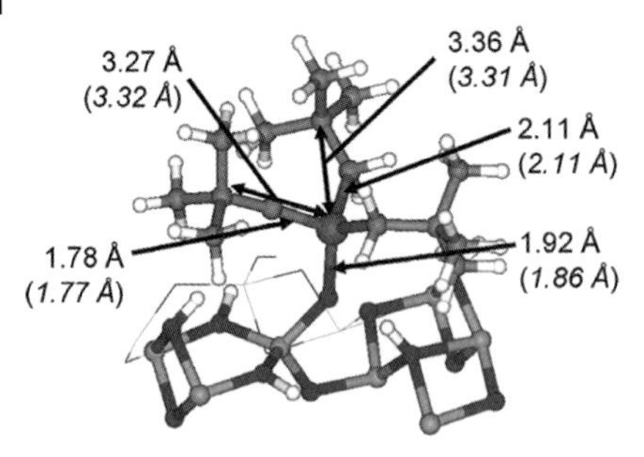


**Figure 17.4** Calculated bond lengths compared to EXAFS data (in italic letters) for the grafted tungsten complex (**8**). Reproduced from Ref. [15].

however, that this is considerably less than that required for the first elimination step. Additionally, the barrier for the formation of **10m** starting from **8m** is slightly lower (70 kJ $mol^{-1}$) than that for the initial grafting reaction (81 kJ $mol^{-1}$), but it becomes significantly higher (97 kJ $mol^{-1}$) when the Me groups of the model complex are replaced with the bulky neopentyl ligands of the real system. Compared to the Zr system, all these activation barriers are high, and this being the principal structural difference between both systems, the presence of the carbyne ligand is probably responsible for the observed differences in reactivity. In fact, the complex has to adopt a penta-coordinated structure in the transition state of the $\sigma$-bond metathesis, which is disfavored as deformation of the complex and the presence of a $\sigma$-donor (oxygen of OH-group) lead to the distortion of the $\pi$-bonding system of the carbyne ligand. So far, elemental analysis, thermodynamic, and kinetic considerations clearly point toward a single monoaluminoxy surface species **8**, which is further supported by an excellent agreement between the calculated bond distances of **8** and experimental EXAFS data (see Figure 17.4) [11]. However, how to reconcile these data with the complex NMR spectrum of **8**?

First, a hint for the interpretation of this apparent discrepancy was given by the IR spectrum of the grafted complex, which interestingly featured the appearance of a broad OH band at 3650 $cm^{-1}$; this data is in good agreement with the interaction of OH-groups with the surface species. Therefore, several isomeric species of **8**, featuring OH-groups in interaction were searched, and the corresponding OH stretching frequencies were calculated. For instance, when HO-$\mu^2$-($Al_V$,$Al_V$) interacts with the carbynic carbon of **8** ($d_{(H-C)} = 2.78$ Å), the frequency for the OH-group is shifted to a lower value by 70 $cm^{-1}$. A much smaller shift (12 $cm^{-1}$) is calculated for an isomer of **8**, in which the same proton interacts with the methylene carbon of the neopentyl group. Note that no shift was found for the bisaluminoxy species $(Al_sO)_2W(\equiv CCH_3)(CH_3)$ (**10m**) ($d_{(H-C)} = 3.40$ Å).

The calculated IR shift obtained for grafted complexes interacting with remaining OH-groups is fully consistent with the observed IR data, and this indicates that such interactions probably occur for these supported species. With these interactions in mind, $^{13}C$ NMR shifts, including the real ligands, were calculated for the two

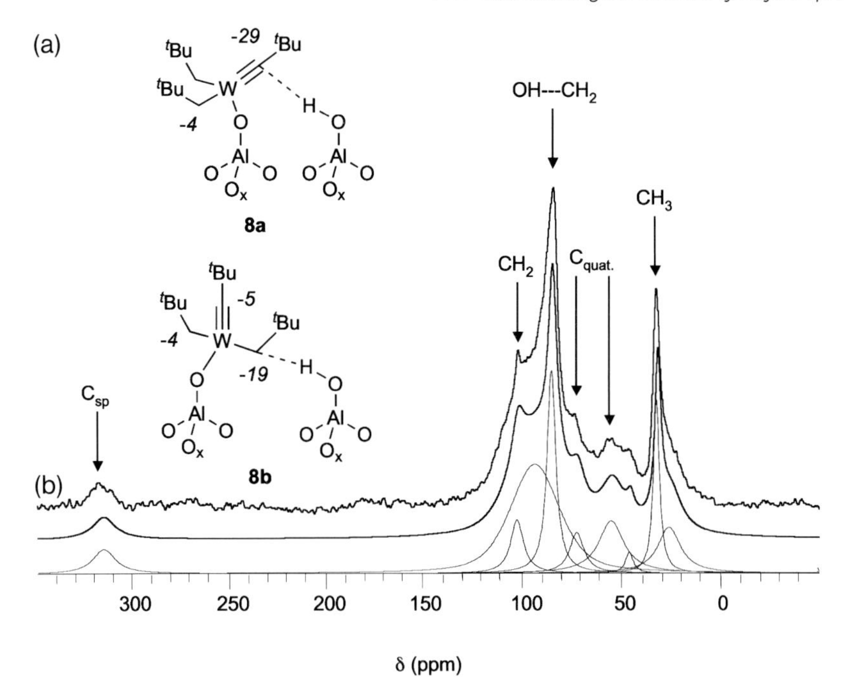


**Figure 17.5** (a) Possible interactions between hydroxyl groups of the surface and the ligands of $(Al_VO)W(\equiv C^tBu)(CH_2{}^tBu)_2$. (**8**). Italic numbers correspond to $\Delta\delta$ between the calculated $^{13}C$ chemical shift values for **8a**/**8b** and those for the isolated molecular complex (**7**). (b) $^{13}C$ CP-MAS (10 kHz) solid-state NMR spectrum (30% labeled $^{13}C$ on $CH_2$) of $W(\equiv C^tBu)(CH_2{}^tBu)_3$ (**7**) on $\gamma$-$Al_2O_{3-(500)}$.

isomers of **8** mentioned previously: one showing an interaction of a vicinal OH-group with the carbyne ligand (**8a**, see Figure 17.5) as suggested by the experimental and calculated IR data, and the second featuring an interaction between a vicinal OH-group and one of the neopentyl ligands (**8b**). The calculated shifts for the methyl groups are very accurate and do not differ significantly from the calculated values for the molecular complex, as observed in the experiment. In the case of **8a**, a significant shift upfield (−29 ppm) is found for the carbynic carbon interacting with the OH-group, while the methylene carbon is barely (−4 ppm) affected. Likewise, a remarkable shift (−19 ppm) is calculated for the interacting methylene carbon in **8b**, while it is only −4 ppm for the methylenic carbons of the "free" neopentyl ligands and −5 ppm for the noninteracting carbynic carbon. In conclusion, these values demonstrate that interactions with surrounding hydroxyl groups lead to a considerable modification of the chemical shifts of the affected atoms. Overall, this data (mass balance analysis, EXAFS, calculations) shows that it is very likely that there is the formation of a single species by reaction of **7** with $\gamma$-$Al_2O_3$. However, the remaining OH-groups surrounding this system perturb it significantly, which gives rise to complex IR and NMR spectra.

## 17.5 Conclusion

Surface organometallic chemistry heavily relies on mass balance analysis and a combination of advanced spectroscopic tools, and this has allowed a molecular understanding of the structure of surface species for silica-supported systems. However, in this chapter, we have shown that even with all these techniques, obtaining a molecular understanding is somewhat more difficult in the case of alumina-supported systems. This is due to alumina, a very complex support in term of surface sites: various types of OH–groups and Lewis acidic sites are present simultaneously. Here, we have shown through two case studies that the tight combination of experimental and theoretical investigations, especially by using thermodynamic, kinetic, and spectroscopic considerations, allows for such a molecular understanding.

Clearly, computational studies belong to the toolbox of surface organometallic chemistry to help the characterization of surface species. Even though not discussed here, computational studies will be of great value in obtaining structure–reactivity relationships in heterogeneous catalysis and in developing more predictive approaches to catalyst design.

## References

1 C. Copéret, M. Chabanas, R. Petroff Saint-Arroman, J.-M. Basset, *Angew. Chem.* **2003**, *115*, 164–191.

2 C. Copéret, *Dalton Trans.* **2007**, 5498–5504.

3 T.J. Marks, *Acc. Chem. Res.* **1992**, *25*, 57–65.

4 C. Copéret, *Handbook of Homogeneous Hydrogenation*, Vol. 1, Wiley-VCH, Weinheim, **2007**, pp. 111–151.

5 V.R. Dufaud, J.M. Basset, *Angew. Chem., Int. Ed.* **1998**, *37*, 806–810.

6 C. Lecuyer, F. Quignard, A. Choplin, D. Olivier, J.-M. Basset, *Angew. Chem., Int. Ed. Engl.* **1991**, *30*, 1660–1661.

7 M. Chabanas, V. Vidal, C. Copéret, J. Thivolle-Cazat, J.M. Basset, *Angew. Chem., Int. Ed.* **2000**, *39*, 1962–1965.

8 V. Vidal, A. Theolier, J. Thivolle-Cazat, J.-M. Basset, *Science* **1997**, *276*, 99–102.

9 J.-M. Basset, C. Copéret, D. Soulivong, M. Taoufik, J. Thivolle-Cazat, *Angew. Chem., Int. Ed.* **2006**, *45*, 6082–6085.

10 C. Thieuleux, A. Maraval, L. Veyre, C. Copéret, D. Soulivong, J.-M. Basset, G.J. Sunley, *Angew. Chem., Int. Ed.* **2007**, *46*, 2288–2290.

11 E. Le Roux, M. Taoufik, C. Copéret, A.d. Mallmann, J. Thivolle-Cazat, J.-M. Basset, B.M. Maunders, G.J. Sunley, *Angew. Chem., Int. Ed.* **2005**, *44*, 6755–6758.

12 J.W. Collette, C.W. Tullock, R.N. MacDonald, W.H. Buck, A.C.L. Su, J.R. Harrell, R. Mulhaupt, B.C. Anderson, *Macromolecules* **1989**, *22*, 3851–3858.

13 C.W. Tullock, R. Mulhaupt, S.D. Ittel, *Makromol. Chem. Rapid Commun.* **1989**, *10*, 19.

14 C.W. Tullock, F.N. Tebbe, R. Mülhaupt, D.W. Ovenall, R.A. Setterquist, S.D. Ittel, *J. Polym. Sci., Part A: Polym. Chem.* **1989**, *27*, 3063–3081.

15 J. Joubert, F. Delbecq, P. Sautet, E.L. Roux, M. Taoufik, C. Thieuleux, F. Blanc, C. Copéret, J. Thivolle-Cazat, J.-M. Basset, *J. Am. Chem. Soc.* **2006**, *128*, 9157–9169.

16 J.P. Perdew, J.A. Chevary, S.H. Vosko, K.A. Jackson, M.R. Pederson, D.J. Singh, C. Fiolhais, *Phys. Rev. B: Condens. Matter* **1992**, *46*, 6671–6687.

17 G. Kresse, J. Furthmüller, *Phys. Rev. B: Condens. Matter* **1996**, *54*, 11169–11186.

18 G. Kresse, J. Furthmüller, *Comput. Mater. Sci.* **1996**, *6*, 15–50.

19 P. Blöchl, C. Först, J. Schimpl, *Bull. Mater. Sci.* **2003**, *26*, 33–41.

20 P.E. Blöchl, *Phys. Rev. B: Condens. Matter* **1994**, *50*, 17953–17979.

21 H.J. Monkhorst, J.D. Pack, *Phys. Rev. B: Condens. Matter* **1976**, *13*, 5188–5192.

22 P. Raybaud, M. Digne, R. Iftimie, W. Wellens, P. Euzen, H. Toulhoat, *J. Catal.* **2001**, *201*, 236–246.

23 Gaussian 03, Revision C.02, M.J. Frisch, G.W. Trucks, H.B. Schlegel, G.E. Scuseria, M.A. Robb, J.R. Cheeseman, J.A. Montgomery, Jr., T. Vreven, K.N. Kudin, J.C. Burant, J.M. Millam, S.S. Iyengar, J. Tomasi, V. Barone, B. Mennucci, M. Cossi, G. Scalmani, N. Rega, G.A. Petersson, H. Nakatsuji, M. Hada, M. Ehara, K. Toyota, R. Fukuda, J. Hasegawa, M. Ishida, T. Nakajima, Y. Honda, O. Kitao, H. Nakai, M. Klene, X. Li, J.E. Knox, H.P. Hratchian, J.B. Cross, V. Bakken, C. Adamo, J. Jaramillo, R. Gomperts, R.E. Stratmann, O. Yazyev, A.J. Austin, R. Cammi, C. Pomelli, J.W. Ochterski, P.Y. Ayala, K. Morokuma, G.A. Voth, P. Salvador, J.J. Dannenberg, V.G. Zakrzewski, S. Dapprich, A.D. Daniels, M.C. Strain, O. Farkas, D.K. Malick, A.D. Rabuck, K. Raghavachari, J.B. Foresman, J.V. Ortiz, Q. Cui, A.G. Baboul, S. Clifford, J. Cioslowski, B.B. Stefanov, G. Liu, A. Liashenko, P. Piskorz, I. Komaromi, R.L. Martin, D.J. Fox, T. Keith, M.A. Al-Laham, C.Y. Peng, A. Nanayakkara, M. Challacombe, P.M.W. Gill, B. Johnson, W. Chen, M.W. Wong, C. Gonzalez, J.A. Pople, Gaussian, Inc., Wallingford CT, **2004**.

24 W. Kutzelnigg, U. Fleischer, M. Schindler, *NMR Basic Principles and Progress* (EDS.: P. Diehl, E. Fluck, H. Günther, R. Kosfeld, J. Seelig), Springer, Berlin, Heidelberg, **1990**, p. 165.

25 P.J. Hay, R.W. Willard, *J. Chem. Phys.* **1985**, *82*, 270–283.

26 W.R. Wadt, P.J. Hay, *J. Chem. Phys.* **1985**, *82*, 284–298.

27 P.J. Hay, R.W. Willard, *J. Chem. Phys.* **1985**, *82*, 299–310.

28 F. Schüth, K. Unger, *The Handbook of Heterogeneous Catalysis*, Vol. 1 (G. Ertl, H. Knözinger, J. Weitkamp), Wiley-VCH, Weinheim, **1997**, pp. 79–80.

29 P. Euzen, P. Raybaud, X. Krokidis, H. Toulhoat, J.-L. Le Loarer, J.-P. Jolivet, C. Froidefond, *Handbook of Porous Materials* (F. Schüth, K. Sing, J. Weitkamp), Wiley-VCH, Weinheim, **2002**, p. 1591.

30 A. Zecchina, D. Scarano, S. Bordiga, G. Spoto, C. Lamberti, *Adv. Catal.*, Vol. 46 (H. Knözinger, B.C. Gates), Academic Press, San Diego, **2001**, pp. 265–397.

31 H. Knözinger, P. Ratnasamy, *Catal. Rev. Sci. Eng.* **1978**, *17*, 31–70.

32 X. Krokidis, P. Raybaud, A.E. Gobichon, B. Rebours, P. Euzen, H. Toulhoat, *J. Phys. Chem. B* **2001**, *105*, 5121–5130.

33 M. Digne, P. Sautet, P. Raybaud, P. Euzen, H. Toulhoat, *J. Catal.* **2004**, *226*, 54–68.

34 M.H. Lee, C.-F. Cheng, V. Heine, J. Klinowski, *Chem. Phys. Lett.* **1997**, *265*, 673–676.

35 R.S. Zhou, R.L. Snyder, *Acta Crystallogr. Sect. B: Struct. Sci.* **1991**, *47*, 617–630.

36 G. Gutiérrez, B. Johansson, *Phys. Rev. B: Condens. Matter* **2002**, *65*, 104202–104209.

37 C. Wolverton, K.C. Hass, *Phys. Rev. B: Condens. Matter* **2000**, *63*, 024102–024116.

38 J.P. Beaufils, Y. Barbaux, *J. Chim. Phys. Phys.-Chim. Biol.* **1981**, *78*, 347–352.

39 P. Nortier, P. Fourre, A.B.M. Saad, O. Saur, J.C. Lavalley, *Appl. Catal.* **1990**, *61*, 141–160.

40 J. Joubert, A. Salameh, V. Krakoviack, F. Delbecq, P. Sautet, C. Copéret, J.-M. Basset, *J. Phys. Chem. B* **2006**, *110*, 23944–23950.
41 L.D. Gelb, K.E. Gubbins, R. Radhakrishnan, M. Sliwinska-Bartkowiak, *Rep. Progr. Phys.* **1999**, 1573.
42 B.A. Hendriksen, D.R. Pearce, R. Rudham, *J. Catal.* **1972**, *24*, 82–87.
43 J.B. Peri, *J. Phys. Chem.* **1965**, *69*, 211–219.
44 L. Nondek, *React. Kinet. Catal. Lett.* **1975**, *2*, 283–289.
45 Z. Vít, J. Vala, J. Málek, *Appl. Catal.* **1983**, *7*, 159–168.
46 C. Morterra, G. Magnacca, *Catal. Today* **1996**, *27*, 497–532.
47 J. Joubert, F. Delbecq, C. Copéret, J.-M. Basset, P. Sautet, *Top. Catal.* **2008**, *48*, 114–119.
48 E. Le Roux, M. Taoufik, M. Chabanas, D. Alcor, A. Baudouin, C. Copéret, J. Thivolle-Cazat, J.-M. Basset, A. Lesage, S. Hediger, L. Emsley, *Organometallics* **2005**, *24*, 4274–4279.
49 M. Chabanas, C. Copéret, J.-M. Basset, *Chem. Eur. J.* **2003**, *9*, 971–975.
50 V. Dufaud, G.P. Niccolai, J. Thivolle-Cazat, J.-M. Basset, *J. Am. Chem. Soc.* **1995**, *117*, 4288–4294.
51 M. Chabanas, E.A. Quadrelli, B. Fenet, C. Copéret, J. Thivolle-Cazat, J.-M. Basset, A. Lesage, L. Emsley, *Angew. Chem., Int. Ed.* **2001**, *40*, 4493–4496.
52 E. Le Roux, M. Chabanas, A. Baudouin, A. de Mallmann, C. Copéret, E.A. Quadrelli, J. Thivolle-Cazat, J.-M. Basset, W. Lukens, A. Lesage, L. Emsley, G.J. Sunley, *J. Am. Chem. Soc.* **2004**, *126*, 13391–13399.

# 18
# Physical and Chemical Properties of Oxygen at Vanadium and Molybdenum Oxide Surfaces: Theoretical Case Studies

*Klaus Hermann*

## 18.1
## Introduction

Transition metal oxides are extremely interesting and important due to their physical and chemical properties [1–3]. Physicists find these materials attractive as they show phase transitions with exciting structural, electronic, and magnetic behavior [1]. Some materials are also known in connection with high temperature superconductivity or for exciting optical properties [1, 3]. Amongst these, vanadium and molybdenum oxides form an important group of materials due to their large variety in crystal structures and physical/chemical properties. They cover a wide range of electronic properties from metals to semiconductors and insulators and are, therefore, studied and used in many fields of technological applications. Examples are vanadium-oxide-based electrical and optical switching devices, write-erase media, light detectors, temperature sensors, infrared spatial light modulators [4–9], and even vanadia constituents of surfaces of medical Ti–Al–V implants [10]. On the other hand, molybdenum trioxide forms n- and p-type semiconductor phases [11], which are of technological interest.

Chemists find transition metal oxides attractive since many of them exhibit high catalytic activity [2]. In fact, most of the present day heterogeneous catalysts of industrial relevance contain reactive sites that include transition metal oxides. Here vanadium and molybdenum oxides are of particular interest and widely exploited. Their catalytic behavior is determined to a large extent by the mobility of surface/lattice oxygen as well as by different reactivity depending on the orientation of the crystal faces [2, 12–15]. The specific properties of both oxides in combination with other elements such as bismuth, cobalt, aluminum, or alkali metals, lead to their use as active and selective catalysts in many reactions belonging to redox processes including those where oxygen is involved as well as those where only hydrogen participates. Examples are oxidation, ammoxidation, and dehydrogenation of hydrocarbons, oxidation of $SO_2$ to $SO_3$, naphthalene or oxylene to phthalic anhydride and more recently *n*-butane to maleic anhydride [16]. Vanadia-based catalysts seem also promising for the oxidation of toluene to

*Computational Methods in Catalysis and Materials Science.* Edited by Rutger A. van Santen and Philippe Sautet

ISBN: 978-3-527-32032-5

benzaldehyde, methanol to formaldehyde and to methyl formate, as well as for the removal of $NO_x$ by selective reduction with $NH_3$ [4, 17]. Molybdenum oxides are used commercially, either in pure form or together with other elements, as catalysts for isomerization or polymerization processes, for the production of formaldehyde and acrylonitrile [2, 12, 13], as well as for the partial oxidation of hydrocarbons and alcohols [18–26]. Despite the importance of vanadium and molybdenum oxides as catalysts, many details of their catalytic behavior at an atomic level are still far from being understood [4, 12, 16, 27]. This is mainly due to the large structural complexity of these materials as well as to their diverse electronic behavior which makes quantitative studies, both experimental and theoretical [28], rather challenging.

In this chapter, we will discuss theoretical concepts and results concerning vanadium and molybdenum oxide surfaces which can be used to interpret and understand catalytic behavior of these systems at an atomic scale. While the two elements form different classes of oxides, their surfaces exhibit structural and electronic similarities connected with microscopic surface binding which suggest a common theoretical treatment. In general, theoretical methods that have been used to study oxide surfaces can be classified according to the approximations made in the system geometry where local cluster and repeated slab models are applied at present. Local cluster models assume that the physical/chemical behavior at selected surface sites can be described approximately by finite sections cut out from the region near the oxide surface. These sections (so-called surface clusters) are treated as fictitious molecules with or without additional boundary conditions to take the effect of electronic coupling with the environment (so-called electronic embedding) into account. Their electronic structure can be calculated by modern quantum chemical methods. In particular, *ab initio* density functional theory (DFT) techniques [29, 30] have proven to be quite successful in recent work. On the other hand, repeated slab models [31] are based on the assumption of an exact two-dimensionally periodic arrangement of all atoms and molecules at the oxide surface. Thus, a slab of full translational periodicity and finite thickness is used to describe the surface system approximately. For computational convenience, surface slabs are repeated perpendicular to their surfaces with vacuum separating adjacent slabs (so-called repeated slab geometry). This yields an altogether three-dimensionally periodic system with a large unit cell (supercell) which can be studied by modern bulk methods of solid state theory. Here applications have been restricted in almost all cases to *ab initio* DFT methods [32–35].

As a result of space limitations, the following discussion cannot cover all aspects of vanadium and molybdenum oxide surfaces connected with catalytic behavior or other properties. Therefore, we will restrict ourselves to the geometric and electronic behavior of bulk and surface oxygen which is of catalytic relevance. Here different theoretical concepts will be discussed and illustrated by example results from recent work where we point out both differences and similarities between the oxides of the two chemical elements.

## 18.2 Vanadium Oxide

### 18.2.1 Vanadium Oxide Bulk Structure

Bulk vanadium oxides as single crystals can be characterized in their geometric structure, apart from their quantitative lattice definition by lattice and lattice basis vectors, by the occurrence of specific elementary $VO_x$ building units. Here the four single valence types of vanadium oxide, VO, $V_2O_3$, $VO_2$, $V_2O_5$, shown in Figure 18.1, contain a common octahedral $VO_6$ unit. These units differ by their distortion (given by neighboring V–O distances and O–V–O angles) as well as by their relative arrangement in the crystal (i.e., links between $VO_6$ corners, edges, or faces) as visualized by the balls-and-sticks models in Figure 18.1.

The **monoxide, VO**, forms a cubic rocksalt lattice [36] with two atoms (1 V + 1 O) in the unit cell containing regular $VO_6$ octahedra of distances $d_{V\text{–}O} = 2.06$ Å and angles $\angle$(O–V–O) = 90°. Here adjacent $VO_6$ units are connected by their edges where oxygen centers share six octahedra each and are 6-fold coordinated with respect to their vanadium neighbors (see Figure 18.1(a)). In this oxide, vanadium assumes its lowest formal oxidation state +2 yielding a Mott–Hubbard metal with the highest V 3d occupation in the valence band region [37]. The **sesquioxide,**

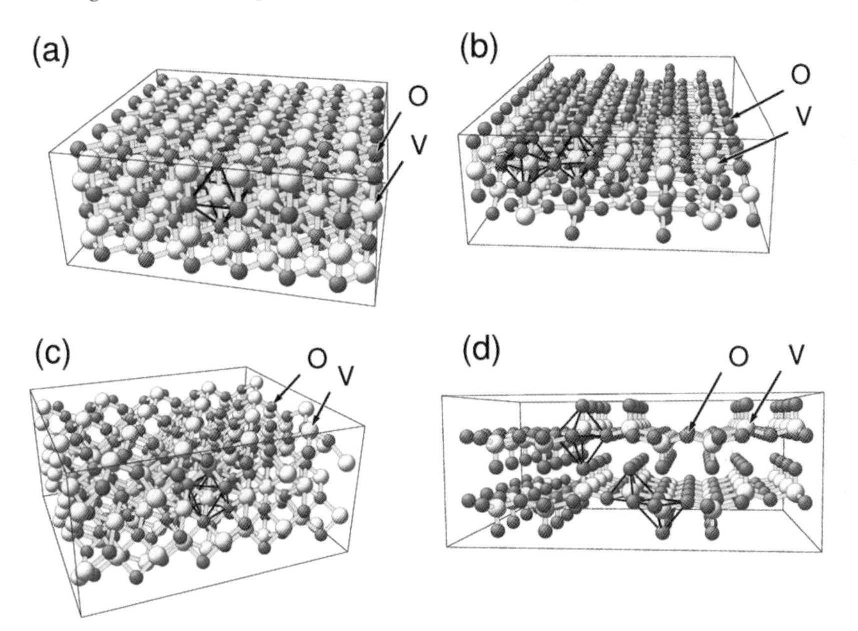


**Figure 18.1** Geometric structure of (a) cubic bulk VO ((100) netplane stacking), (b) monoclinic bulk $VO_2$ ((011) stacking), (c) trigonal bulk $V_2O_3$ ((0001) stacking), and (d) orthorhombic bulk $V_2O_5$ ((010) stacking). Vanadium and oxygen atoms are shown as shaded large and small balls, respectively. Octahedral $VO_6$ and bipyramidal $V_2O_8$ units are indicated by black lines.

$V_2O_3$, forms a monoclinic phase [38] below 150–170 K and a trigonal corundum phase [39] above this temperature with ten atoms (4 V + 6 O) in the unit cell. In the trigonal lattice the $VO_6$ octahedra are distorted with three distances $d_{V-O}$ each at 2.02 Å and 2.13 Å, respectively, and angles ∠(OVO) varying between 82° and 97°. Adjacent $VO_6$ units are connected by their edges or corners (see Figure 18.1(c)). The oxygen centers share four octahedra each and are 4-fold coordinated with respect to their vanadium neighbors with distances varying by 5% in the trigonal lattice. Here vanadium assumes a formal oxidation state +3, where the monoclinic low-temperature phase is an antiferromagnetic insulator while the trigonal phase at higher temperatures is metallic [40] with a Mott–Hubbard transition separating the phases. The **dioxide, $VO_2$**, exists as a tetragonal rutile crystal above 337 K [41,42] with six atoms (2 V + 4 O) in the unit cell. This structure is distorted slightly by shear and doubles its unit cell (4V + 8O) becoming monoclinic below 337 K [43]. In the tetragonal lattice the regular $VO_6$ octahedra are elongated slightly along their 4-fold axes with four distances $d_{V-O}$ of 1.90 Å and two of 1.95 Å and all angles ∠(OVO) = 90°. Adjacent $VO_6$ units are connected by their edges or corners (see Figure 18.1(b)). The oxygen centers share three octahedra each and are 3-fold coordinated with respect to their vanadium neighbors with distances varying by 3%. In both structures, vanadium assumes a formal oxidation state +4 resulting in a semiconductor (monoclinic crystal) or a metal (tetragonal crystal) [44,45] with reduced V 3d occupation compared with $V_2O_3$. The **pentoxide, $V_2O_5$**, is described by an orthorhombic layer-type lattice [46] with 14 atoms (4 V + 10 O) in the unit cell. Here the $VO_6$ octahedra are distorted substantially with one small distance of $d_{V-O}$ = 1.59 Å and one very large, $d_{V-O}$ = 2.79 Å, while the other V–O distances range between 1.78 Å and 2.02 Å. Angles ∠(OVO) involving the vanadyl oxygen at short distance amount to 105° while those including the farthest oxygen range between 73° and 77°. Adjacent $VO_6$ units are connected by edges or corners (see Figure 18.1(d)), where three oxygen centers share three and two octahedra, respectively. As a result, there is singly, 2-, and 3-fold coordinated oxygen in the crystal. The large distortion of the octahedra suggests alternatively the use of $V_2O_8$ bipyramids as building units to describe the lattice structure which reflects the layer-type lattice more appropriately (see Figure 18.1(d)). Here the oxide layers lie parallel to the (010) netplanes of the lattice following the nomenclature used, e.g., in Ref. [47]. Note that, depending on the choice of the orthorhombic crystal axes, the layer netplane orientation may also be denoted by (001). The latter corresponds to an interchange of the orthorhombic lattice vectors $\underline{b}$ and $\underline{c}$ as proposed in Ref. [48]. In $V_2O_5$, vanadium assumes its highest formal oxidation state, +5, yielding in a small-gap semiconductor [49] with O 2sp-type valence and V 3d-type conduction bands [50, 51].

Apart from the single valence vanadium oxides discussed above there is a multitude of mixed valence oxides of vanadium, which can result from different growth conditions or may be due to structural changes during catalytic reactions. When oxide surfaces are exposed to catalytic reduction processes then oxygen vacancies can be formed [1–3, 12, 14, 52]. If the vacancy concentration at the crystal surface goes beyond a critical value vacancies may propagate into the substrate which will give rise to crystallographic shear planes [1, 14] resulting in vacancy annihilation. As a consequence, the crystal can be considered as being composed of slabs of mostly undistorted parent structure separated by shear planes. If the spacing between these planes is regular a mixed valence oxide of crystalline order is created. Depending on the actual spacing different intermediate oxides may be formed which are usually described as homologous series. Examples are Magneli phases, which form series $V_nO_{2n-1}$ or $V_nO_{2n+1}$ [53]. In this chapter, we will restrict our discussion of theoretical concepts to pentoxide, $V_2O_5$, and the sesquioxide, $V_2O_3$, for which extended studies have been performed [28].

## 18.2.2
## Characterizing Oxygen at the $V_2O_5(010)$ Surface

As discussed earlier, vanadium oxide, when combined with other elements such as bismuth, cobalt, aluminum, or alkali metals, forms an essential component in selective catalysts for many reactions where examples are the reduction of $NO_x$ with $NH_3$ [54, 55] or the selective oxidation of hydrocarbons [56]. These reactions involve surface oxygen of the substrate which is incorporated into the reacting species or acts as an intermediate reactive site. Since vanadium oxide surfaces contain oxygen sites of different geometry and coordination, suggesting different chemical behavior [57, 58], it is very important to know which of these sites are involved in a particular reaction step. An answer of this question will help to elucidate microscopic reaction details [59].

So far, theoretical studies on vanadium oxide surfaces have focused exclusively on single crystal surfaces which are described by low Miller indices and are believed to be energetically favorable. These surfaces are most easily accessible by theory due to their relatively simple geometry although their relevance as to catalytic activity has been doubted. In this section we focus on the (010) oriented surface of vanadium pentoxide, $V_2O_5$, which has been examined extensively by theory and experiment [27]. The orthorhombic lattice of bulk $V_2O_5$ is layer type with weakly coupling physical layers stacked along the (010) direction. (Please note that, as mentioned earlier, we use in this section crystal axes and Miller indices as introduced in Ref. [49].) Thus, the corresponding (010) surface is determined in its structure by local V–O binding rather than by stacking of closely packed crystal planes, as found for compact oxides. This results in an altogether rough surface shape. The physical layers of $V_2O_5$ consist of six (010) atom layers ($4 \times$ O, $2 \times$ V) offering, altogether, three structurally different oxygen atoms, terminal (vanadyl) oxygen, O(1), coordinated to one vanadium atom by a short bond and bridging oxygen, O(2)/O(3), coordinated to two or three vanadium atoms. As a result, there are five different oxygen atoms at the ideal bulk terminated $V_2O_5(010)$ surface as indicated in Figure 18.2. Terminal vanadyl oxygen atoms O(1) are located above vanadium atoms. Oxygen atoms O(2), O(2′) bridge two vanadyl groups pointing into the bulk and sticking out of the surface while oxygen atoms O(3), O(3′) are connected to three vanadyl groups. In addition, the surface exposes bare vanadium

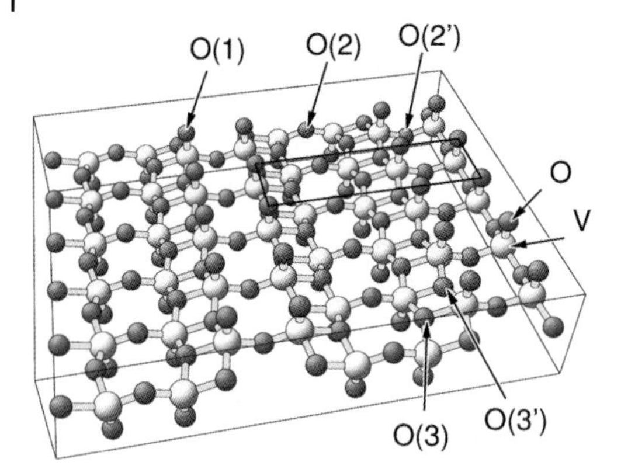


**Figure 18.2** Geometric structure of the (010) oriented single crystal surface of orthorhombic $V_2O_5$. Vanadium and oxygen atoms are shown as shaded large and small balls, respectively, with connecting sticks indicating nearest-neighbor relations. Differently coordinated surface oxygen atoms, O(1–3, 2′, 3′), are labeled and the rectangular cell of the vanadium/oxygen netplanes is sketched accordingly.

atoms (vanadyl VO groups with oxygen pointing into the substrate) connected with O(2) and O(3, 3′) atoms.

There are several physical parameters which may be used to discriminate between differently coordinated oxygen in vanadium oxides and at their surfaces where one is based on the notion that oxygen in different atomic/ionic neighborhoods assumes a different atom charge. Here it must be emphasized that the assignment of atomic charges to atoms inside a compound, molecular or solid state type, can never be taken quantitatively. Due to the quantum mechanical nature of these systems a physically strict separation of an electron charge into atomic contributions is impossible by definition. Atom charges are usually connected qualitatively with experimental evidence, such as spectroscopic peak shifts in core electron photoemission (ESCA, XPS), or they are derived in theoretical work from empirical or mathematical recipes rather than from physical axioms.

The oldest conceptual definition, the **formal valence charge** of an atom, is based on the electron shell occupation of the corresponding free atom together with relative electron affinities and ionization potentials of different atoms in the compound. As an example, bulk $V_2O_5$ contains four vanadium and ten oxygen atoms in its elementary (neutral) building block. The neutral V and O atoms are described by electron configurations $1s^2\ 2s^2\ 2p^6\ 3s^2\ 3p^6\ 3d^3\ 4s^2$ and $1s^2\ 2s^2\ 2p^4$, respectively. Due to its large relative electronegativity, oxygen is believed to accept two electrons to form a rare gas (neon) closed shell configuration $1s^2\ 2s^2\ 2p^6$ while vanadium donates five electrons to assume a rare gas (argon) closed shell configuration $1s^2\ 2s^2\ 2p^6\ 3s^2\ 3p^6$. This yields formal valence charges $Q^{val}$ of +5 for vanadium and −2 for oxygen. Clearly, the definition of formal valence charges is rather approximate and cannot, for example, account for fine details such as distinguishing between differently coordinated atom species in a compound.

Within the cluster framework the electronic structure and derived properties of the oxide systems can be evaluated by standard quantum chemical methods where *ab initio* DFT methods [30] have been quite successful also for large size clusters. These quantum chemical methods use almost exclusively the linear-combination-of-atomic-orbitals (LCAO) approximation to represent one-electron functions $\varphi_i(\underline{r})$, Kohn–Sham orbitals within the DFT concept, by an ansatz

$$\varphi_i(\underline{r}) = \sum_A \sum_\kappa c_{iA\kappa} \cdot \gamma_{A\kappa}(\underline{r}), \tag{18.1}$$

where the summation runs over all atoms $A$ and atom-centered basis functions $\gamma_{A\kappa}(\underline{r})$. The latter can be numeric or analytic functions, for example Slater exponential [60] or (contracted) Gaussian functions [61]. Further, the $c_{iA\kappa}$ in (18.1) are mixing coefficients determined by solutions of the N-electron equations. The atom-based representations (18.1) can also be used to define atom charges. The normalization of all one-electron functions $\varphi_i(\underline{r})$ requires that

$$1 = \langle \varphi_i | \varphi_i \rangle = \int \varphi_i^*(\underline{r}) \varphi_i(\underline{r}) d^3 r = \sum_A \sum_\kappa \sum_{A'} \sum_{\kappa'} c_{iA\kappa}^* c_{iA'\kappa'} \cdot S_{A\kappa A'\kappa'} \tag{18.2}$$

with

$$S_{A\kappa A'\kappa'} = \int \gamma_{A\kappa}^*(\underline{r}) \gamma_{A'\kappa'}(\underline{r}) d^3 r \tag{18.3}$$

denoting the so-called overlap integrals which arise due to the nonorthogonality of basis functions from different atom centers. Then the total number of electrons in a cluster, $N_{el}$, can be written as a sum over all occupied orbitals $\varphi_i(\underline{r})$ yielding

$$N_{el} = \sum_i^{occ} \langle \varphi_i | \varphi_i \rangle = \sum_A \left\{ \sum_i^{occ} \sum_\kappa \sum_{A'} \sum_{\kappa'} c_{iA\kappa}^* c_{iA'\kappa'} \cdot S_{A\kappa A'\kappa'} \right\} \tag{18.4}$$

or

$$N_{el} = \sum_A q_A^{Mull} \quad \text{with} \quad q_A^{Mull} = \sum_i^{occ} \sum_\kappa \sum_{A'} \sum_{\kappa'} c_{iA\kappa}^* c_{iA'\kappa'} \cdot S_{A\kappa A'\kappa'}. \tag{18.5}$$

Thus $N_{el}$ can be decomposed mathematically into atom-centered contributions $q_A^{Mull}$, which, when combined with the nuclear charges $Z_A$ of the atom centers, yield atom charges according to

$$Q_A^{Mull} = Z_A - q_A^{Mull}. \tag{18.6}$$

This definition of atom charges, first proposed by Mulliken [62] assumes that charge contributions resulting from two-center terms ($A \neq A'$) in (18.4) are distributed evenly between the corresponding atoms which is somewhat arbitrary. Alternative definitions, also starting from Eq. (18.4), have been given by **Löwdin** [63] and are subject to similar mathematical arbitrariness as **Mulliken charges**.

As an illustration of different charge definitions we consider a $V_{10}O_{31}H_{12}$ cluster (see Figure 18.3), which represents local sections of a $V_2O_5$ slab with (010)

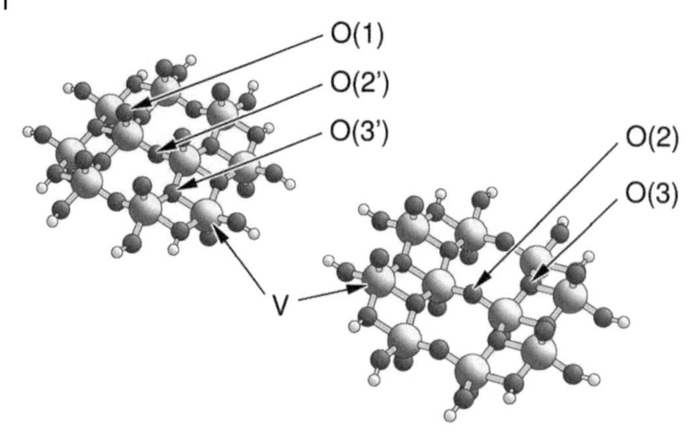


**Figure 18.3** Geometric structure of the $V_{10}O_{31}H_{12}$ cluster representing local sections of the $V_2O_5$ (010) substrate (one crystal layer). Vanadium and oxygen atoms are shown as shaded large and small balls, respectively, while very small balls refer to hydrogen atoms (saturators) at the cluster periphery. The cluster is shown from both sides exposing different oxygen sites labeled accordingly.

**Table 18.1** Atom charges of the $V_{10}O_{31}H_{12}$ cluster modeling local sections of $V_2O_5$ (010) slabs.[a]

| Site | $Q^{val}$ | $Q^{Mull}$ | $Q^{Löw}$ | $Q^{Badr}$ |
|---|---|---|---|---|
| V | +5 | +1.43 | +0.87 | +2.08 |
| O(1) | −2 | −0.28 | −0.20 | −0.58 |
| O(2, 2′) | −2 | −0.62 | −0.41 | −0.91 |
| O(3, 3′) | −2 | −0.79 | −0.52 | −1.07 |

[a] For a definition of the formal valence, Mulliken, Löwdin, and Bader charges, $Q^{val}$, $Q^{Mull}$, $Q^{Löw}$, $Q^{Badr}$, see text. All data refer to V, O surface sites near the cluster center. The differently coordinated oxygen species O(1), O(2, 2′), O(3, 3′) are indicated in Figure 18.3. All values are given in atomic units.

orientation where hydrogen is used to saturate dangling V–O bonds at the cluster periphery. This cluster has, amongst others, been successfully applied in many cluster studies on chemical and physical properties of $V_2O_5$ substrate [28]. Table 18.1 lists atom charges for differently coordinated oxygen in the $V_{10}O_{31}H_{12}$ cluster (see Figure 18.3) obtained by *ab initio* DFT calculations [64] using the gradient corrected RPBE (revised Perdew–Burke–Ernzerhof [65]) functional. First, the differences between the computed Mulliken charges, $Q^{Mull}$, and the Löwdin charges, $Q^{Löw}$, of the same atom are quite sizeable which makes their quantitative meaning somewhat doubtful. Further, both Mulliken and Löwdin charges for the same atom are dramatically smaller in absolute value compared with the corresponding formal valence charge. This reflects the fact that the definition of formal valence charges assumes an ionic compound ignoring covalent binding which reduces charging in the substrate. It also illustrates the quantitative uncertainty that is attached to atom charges. However, the qualitative picture given in Table 18.1 does not depend on the actual charge definition. In all cases, vanadium becomes positively and oxygen negatively charged in $V_2O_5$ substrate which is intuitively obvious. In addition, the oxygen charges scale with the coordination where singly coordinated vanadyl oxygen O(1) carries the smallest negative charge while 3-fold bridging oxygen O(3), O(3′) is the most strongly negative.

Table 18.1 contains also atom charges $Q^{\text{Badr}}$ from a **Bader charge** analysis [66] for the $V_{10}O_{31}H_{12}$ cluster, where charges are evaluated by spatial integration of charge densities. This requires a partitioning of space into atom cells $C_A$ assigned to one atom each such that each point $\underline{r}$ in space belongs to exactly one atom cell $C_A$. Then the total number of electrons, obtained by integration of the charge density $\rho(\underline{r})$ over the whole space can be written as

$$N_{\text{el}} = \int \rho(\underline{r}) d^3r = \sum_A \int_{C_A} \rho(\underline{r}) d^3r = \sum_A q(C_A), \tag{18.7}$$

where the quantities $q(C_A)$ can be interpreted as atom specific electron charges and

$$Q_A^{\text{Vol}} = Z_A - q(C_A) \tag{18.8}$$

as total atom charges. Obviously, the values of $Q_A^{\text{Vol}}$ depend on the shape of the cells $C_A$ where different choices have been suggested. These range from purely geometric partitioning by space filling polyhedral cells (Voronoi or Wigner-Seitz cells) about atom centers at $\underline{R}_A$ defined by

$$\underline{r} \in C_A \quad \text{if} \quad |\underline{r} - \underline{R}_A| \leqslant |\underline{r} - \underline{R}_B| \quad \text{for all} \quad B \neq A \tag{18.9}$$

to more sophisticated choices involving the spatial variation of the electron density $\rho(\underline{r})$ as, for example, in Bader's topological analysis [66] where boundaries of adjacent cells $C_A$, $C_B$ are defined by the constraint

$$\nabla \rho(\underline{r}) \cdot \underline{n} = 0, \tag{18.10}$$

with $\underline{n}$ denoting the normal vector of the boundary at $\underline{r}$. This yields the atom charges $Q^{\text{Badr}}$ given in Table 18.1. Obviously, the definitions of atom charges by spatial integration are conceptionally different from Mulliken or Löwdin charge analyses and, therefore, give different numerical results, as shown in Table 18.1, while the qualitative results are identical. The definitions of space-integrated charges rely only on the electron charge density and the cell shapes but not on wavefunction details, such as basis function representations. Therefore, atom charges by spatial integration can be obtained for finite clusters and extended solids on the same footing.

While atom charges can be used to qualitatively discriminate between differently coordinated oxygen species in the $V_2O_5$ substrate in theory an immediate experimental discrimination based on atom charges is not available since it is impossible to directly measure atom charges. Therefore, indirect experimental evidence has to be applied when differently coordinated oxygen needs to be identified. One possible method makes use of the fact that core electron ionization of atoms depends also on their charge state. More generally, atoms in different electronic or geometric environments inside a compound are expected to yield different core ionization potentials which can be measured using X-ray photoemission spectroscopy (XPS). This is, in fact, one of the factors determining chemical peak shifts considered in electron spectroscopy for chemical analysis (ESCA) [67]. In the present context, it may be used to discriminate differently coordinated oxygen in the $V_2O_5$ substrate. However, the assignment of ionization peaks to specific oxygen species requires substantial theoretical support.

The basic process of core electron photoemission of an atom is described by a highly energetic X-ray photon impinging on the atom and exciting one of its highly localized core electrons from a bound into an unbound state, i.e., removing the electron and leaving the ionized atom behind. Thus, a theoretical treatment must consider the atom in its electronic ground state, determined by an N-electron ground state wavefunction $\Phi_{el}^{N}(\underline{r}_1, \ldots, \underline{r}_{i-1}, \underline{r}_i, \underline{r}_{i+1}, \ldots, \underline{r}_N)$ with a total energy $E_{tot}^{N}$, and in its ionized state determined by an $(N-1)$-electron wavefunction $\Phi_{el}^{N-1,i}(\underline{r}_1, \ldots, \underline{r}_{i-1}, \underline{r}_{i+1}, \ldots, \underline{r}_N)$ with a total energy $E_{tot}^{N-1,i}$, where $i$ labels the photoelectron which is removed. Then the ionization potential $IP(i)$, i.e., the smallest energy required to remove the photoelectron from the atom, is given by

$$IP(i) = E_{tot}^{N-1,i} - E_{tot}^{N} \, . \tag{18.11}$$

Depending on the approximations made to evaluate the many-electron wavefunctions and corresponding total energies one can distinguish different scenarios. First, one may assume that removing the photoelectron will not affect the other electrons which remain in the atom. In a theory where many-electron wavefunctions are represented by one-electron functions (orbitals) $\varphi_j(\underline{r}), j = 1, \ldots, N$, with $\varphi_i(\underline{r})$ denoting the photoelectron orbital, this scenario can be accounted for by evaluating the ionized state wavefunction $\Phi_{el}^{N-1,i}(\underline{r}_1, \ldots, \underline{r}_{i-1}, \underline{r}_{i+1}, \ldots, \underline{r}_N)$ and its total energy $E_{tot}^{N-1,i}$, using the orbitals $\varphi_j(\underline{r})$ obtained for the ground state $\Phi_{el}^{N}(\underline{r}_1, \ldots, \underline{r}_{i-1}, \underline{r}_i, \underline{r}_{i+1}, \ldots, \underline{r}_N)$. This so-called frozen orbital approximation, yielding ionization potentials $IP^{frozen}(i)$, ignores electronic relaxation of the atom after the photoelectron has been removed. For wavefunctions determined within the Hartree–Fock approach [68] the frozen orbital approximation is also known as Koopmans approximation.

In an improved description of electronic relaxation can be included in different ways. The most accurate method is to calculate the $(N-1)$-electron wavefunction $\Phi_{el}^{N-1,i}(\underline{r}_1, \ldots, \underline{r}_{i-1}, \underline{r}_{i+1}, \ldots, \underline{r}_N)$ and its total energy $E_{tot}^{N-1,i}$ by solving the appropriate $N$-electron equations of the ionized state self-consistently. This yields ionization potentials $IP^{relaxed}(i)$ which include the full electronic relaxation of the atom in response to its ionization and which are smaller than the corresponding frozen orbital values $IP^{frozen}(i)$. An alternative, more approximate method to include electronic relaxation in the evaluation of ionization potentials is available within the DFT scheme, where the corresponding many-electron states are represented by Kohn–Sham orbitals $\varphi_j^{KS}(\underline{r}), j = 1, \ldots, N$. According to Slater's transition state theory [69, 70] it is assumed that total energies of an electron system can be written as expansions involving occupation numbers $n_j$ of the Kohn–Sham orbitals

$$E_{\text{tot}} = E_{\text{tot}}(n_1, \ldots, n_N) = E_{\text{tot}}^0 + \sum_{j=1}^{N} \alpha_j n_j + \sum_{j=1}^{N} \sum_{k=1}^{N} \beta_{jk} n_j n_k + \cdots . \tag{18.12}$$

Then the total energy of the ground state is given by

$$E_{\text{tot}}^N = E_{\text{tot}}(n_1 = 1, \ldots, n_N = 1), \tag{18.13a}$$

while the energy of an ionized final state with an electron in orbital $\varphi_i^{\text{KS}}(\underline{r})$ removed corresponds to

$$E_{\text{tot}}^{N-1,i} = E_{\text{tot}}(n_1 = 1, \ldots, n_{i-1} = 1, n_i = 0, n_{i+1} = 1, \ldots, n_N = 1), \tag{18.13b}$$

which according to (18.11), (18.12) yields an ionization potential

$$IP(i) = E_{\text{tot}}^{N-1,i} - E_{\text{tot}}^N = -\alpha_i - \beta_{ii} - \sum_{j=1, j \neq i}^{N} (\beta_{ij} + \beta_{ji}) - \cdots . \tag{18.14}$$

This can be connected with Kohn–Sham eigenvalues $\varepsilon_i$ which appear, together with the orbitals $\varphi_i^{\text{KS}}(\underline{r})$, as solutions of the Kohn–Sham equations. According to Janak [71] the eigenvalues $\varepsilon_i$ can, considering (18.10), be written as

$$\varepsilon_i = \frac{\partial E_{\text{tot}}(n_1, \ldots, n_N)}{\partial n_i} = \alpha_i + 2\beta_{ii} n_i + \sum_{j=1, j \neq i}^{N} (\beta_{ij} + \beta_{ji}) n_j + \cdots , \tag{18.15}$$

depending on the occupation number $n_i$ of orbital $\varphi_i^{\text{KS}}(\underline{r})$. Thus, if in expansion (18.12) energy terms higher than quadratic in the occupation numbers are neglected, a comparison of (18.14) and (18.15) yields the approximate ionization potential within Slater's transition state theory

$$IP^{\text{TS}}(i) = -\varepsilon_i \quad \text{for} \quad n_i = \frac{1}{2}, n_j = 1, j \neq i . \tag{18.16}$$

Therefore, the solution of the Kohn–Sham equations with a half-filled core orbital, $n_i = 1/2$, can be used to determine an approximate ionization potential for that core electron based on the corresponding Kohn–Sham eigenvalue $\varepsilon_i$.

Clearly, the above discussion of atom ionization can be generalized by replacing atomic wavefunctions and orbitals with the corresponding wavefunctions of extended electronic systems like clusters, molecules, or solids. Then differently coordinated atom species inside a compound may be identified experimentally by their ionization properties. This brings us back to the initial problem of discriminating different oxygen species in $V_2O_5$ substrate. Table 18.2 lists computed ionization potentials for different oxygen sites in the $V_{10}O_{31}H_{12}$ cluster, see Figure 18.3, representing a local section of $V_2O_5$ substrate with (010) orientation. The values are obtained from *ab initio* DFT calculations [72]. Obviously, the absolute values of the computed *IP* values vary considerably between different approximations, which

**Table 18.2** Computed ionization potentials for different oxygen sites of the $V_{10}O_{31}H_{12}$ cluster [72].[a]

| Site | $IP^{frozen}$ | $IP^{TS}$ | $IP^{relaxed}$ |
|---|---|---|---|
| O(1) | 566.95 | 537.48 | 535.63 |
| O(2, 2′) | 566.59 | 537.20 | 535.43 |
| O(3, 3′) | 566.47 | 537.30 | 535.52 |

[a] For a definition of the frozen orbital, transition state, and fully relaxed ionization potentials, $IP^{frozen}$, $IP^{TS}$, $IP^{relaxed}$, see text. The differently coordinated oxygen species O(1), O(2, 2′), O(3, 3′) are indicated in Figure 18.3. All values are given in eV.

are explained by the different amount of electronic relaxation accounted for by the approximations. However, for a given approximation the *IP* values for the three types of oxygen species differ by less than 0.5 eV. This has to be compared with the experimental results from photoemission spectroscopy (XPS) on a clean $V_2O_5(010)$ surface [73], where only one 1.5 eV wide ionization peak at $536.4 \pm 0.8$ eV (including a workfunction of 6.4 eV for the $V_2O_5(010)$ surface [59]) with no further structure is observed. While the experimental and the computed fully relaxed IPs values are quite similar, a discrimination of differently coordinated oxygen on the basis of the XPS data does not seem to be possible.

Alternative experimental methods that may allow identifying and discriminating different atom species in solids and at surfaces make use of the valence electron structure of these systems. Here ultraviolet photoemission spectroscopy (UPS) is a promising candidate. As mentioned earlier, vanadium pentoxide is a small gap semiconductor. The 5.5 eV wide valence band region is found to be dominated by oxygen 2sp-type electron states determining V–O valence bonds. Since the binding situation of each oxygen depends on its local coordination, the corresponding O 2sp-derived valence orbitals are expected to differ in their energetic positions inside the valence band region. This can be determined theoretically by atom-resolved valence densities-of-states (PDOS) where extensive cluster studies [59] have shown that vanadyl oxygen O(1), see Figure 18.3, derived orbitals are focused energetically in the center of the valence band region while contributions from bridging oxygen O(2), O(3) are spread over the whole band region. As a result, the total valence density-of-states (DOS) is peaked at the center of the valence band region and is consistent with angle-resolved ultraviolet photoemission (UPS) experiments probing the valence DOS of a freshly cleaved $V_2O_5$ (010) surface [59] shown in Figure 18.4. Thus, experimental UPS data together with theoretical DOS analyses may be used to identify specific oxygen species at the $V_2O_5$ (010) surface but the analysis is not fully conclusive.

Spectroscopies that combine information about the electronic core with that of valence and conduction band regions turn out to be the most promising in

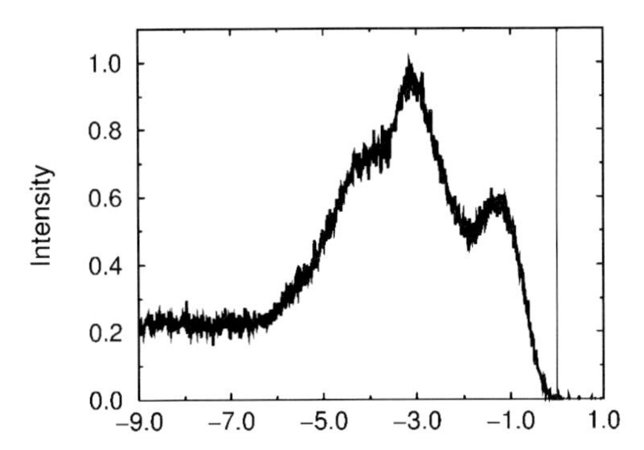


**Figure 18.4** Angle-resolved He-II ultraviolet photoemission spectrum of a $V_2O_5(010)$ surface sample taken at normal incidence [59].

analyzing differently coordinated atom species inside the oxide substrate. Here near-edge X-ray absorption fine structure (NEXAFS) spectroscopy, discussed in detail also in Chapter 12, is a prominent example. The basic X-ray absorption process is described by an X-ray photon impinging at an atom and exciting a core electron from its ground state to unoccupied final states where the final states may be bound valence or Rydberg-type states or may be unbound resonances. The excitations are determined by dipole transitions with the corresponding transition probabilities given, in a simple one-electron theory, by dipole matrix elements between initial and final state orbitals, $\varphi_{\text{init}}(\underline{r})$ and $\varphi_{\text{fin}}^{(k)}(\underline{r})$, respectively. This yields absorption intensities $I(E_k, \underline{e}_{\text{pol}})$ for each transition depending on the transition energy $E_k$ and on the polarization vector of the incoming photon, $\underline{e}_{\text{pol}}$, where

$$I(E_k, \underline{e}_{\text{pol}}) = \alpha\, E_k \cdot (\underline{e}_{\text{pol}}\underline{m}^{(k)})^2 \tag{18.17}$$

with the dipole transition moment vector defined by

$$\underline{m}^{(k)} = \langle \varphi_{\text{init}} | q\underline{r} | \varphi_{\text{fin}}^{(k)} \rangle = \int \varphi_{\text{init}}^{*}(\underline{r})\; q\underline{r}\; \varphi_{\text{fin}}^{(k)}(\underline{r})\; d^3 r \tag{18.18}$$

and $\alpha$ denoting a global scaling constant. The complete excitation spectrum is then given by the sum over all possible excitation processes starting from an initial orbital $\varphi_{\text{init}}(\underline{r})$, i.e., by

$$I(E, \underline{e}_{\text{pol}}) = \sum_k I(E_k, \underline{e}_{\text{pol}})\; \delta(E - E_k) = \alpha \sum_k E_k \cdot (\underline{e}_{\text{pol}}\underline{m}^{(k)})^2 \cdot \delta(E - E_k), \tag{18.19}$$

where $\delta(E - E_k)$ is the Dirac delta function. In general, the excitation spectrum (18.19) will depend on the relative orientation of the polarization vector $\underline{e}_{\text{pol}}$ with respect to the different dipole transition moment vectors $\underline{m}^{(k)}$ yielding polarization angle dependent spectra for systems which are fixed in space, such as solids or adsorbate molecules in unique positions at the substrate surface. However, for randomly oriented molecules in gas phase or at surfaces the angle dependence has to be summed over all orientations which results in angle-integrated excitation spectra given by

$$I_{\text{avg}}(E) = \int I(E, \underline{e}_{\text{pol}}) d\Omega = \frac{2\pi}{3} \alpha \sum_k E_k \cdot (\underline{m}^{(k)})^2 \cdot \delta(E - E_k). \tag{18.19a}$$

In an improved treatment the dipole transitions have to be described as excitations of an $N$-electron system where the rearrangement of all electrons accompanying the transition (electronic relaxation) has to be accounted for. This results in dipole transition moment vectors evaluated with the corresponding $N$-electron wavefunctions, $\Phi_{\text{init}}(\underline{r}_1, \ldots, \underline{r}_N)$ and $\Phi_{\text{fin}}^{(k)}(\underline{r}_1, \ldots, \underline{r}_N)$, i.e.,

$$\underline{m}^{(k)} = \left\langle \Phi_{\text{init}} | \sum_{i=1}^{N} q\underline{r}_i | \Phi_{\text{fin}}^{(k)} \right\rangle = \int \Phi_{\text{init}}^* \cdot \sum_{i=1}^{N} q\underline{r}_i \cdot \Phi_{\text{fin}}^{(k)} d^3 r_1 \cdots d^3 r_N \tag{18.20}$$

to be used for the spectrum function (18.19). Thus, the theoretical determination of X-ray absorption spectra (XAS) requires an accurate $N$-electron theory which is provided by quantum chemical methods for clusters and by solid state methods for extended systems. In particular, DFT methods have been applied extensively to evaluate theoretical spectra for molecular and surface systems [74] in comparison with the experimental NEXAFS data. For a detailed discussion and references see Chapter 12. Within the DFT scheme Slater's transition potential approximation [69, 70] has proven to be extremely successful in obtaining theoretical spectra in close agreement with experiment. Here the $N$-electron wavefunctions of the initial and final excited states are described by Kohn–Sham orbitals $\varphi_j^{\text{KS}}(\underline{r}), j = 1, \ldots, N$, which are solutions of the Kohn–Sham equations where the corresponding core orbital, $\varphi_{\text{init}}^{\text{KS}}(\underline{r})$, characterizing the core electron to be excited is half filled, setting $n_{\text{init}} = 1/2$, yielding the transition state (and the approximate ionization potential $-\varepsilon_{\text{init}}$ corresponding to $\varphi_{\text{init}}^{\text{KS}}(\underline{r})$). Then possible final state orbitals $\varphi_{\text{fin}}^{\text{KS},(k)}(\underline{r})$ to be used for the calculation of dipole transition moment vectors (18.18) are taken from the set of unoccupied orbitals of the transition state calculation and, in a first approximation, corresponding excitation energies are obtained from differences of Kohn–Sham eigenvalues, $E_k = \varepsilon_k - \varepsilon_{\text{init}}$. Thus, Slater's transition potential approach allows an evaluation of the complete excitation spectrum for a given core orbital based on final state orbitals of only one self-consistent calculation. However, the treatment does not fully account for electronic relaxation in the core hole excited state yielding excitation and ionization energies which are too large. This can be corrected by shifting all excitation energies $E_k$ by the difference of the ionization potential evaluated with the transition potential method, i.e., equal to $-\varepsilon_{\text{init}}$, and the corresponding value from $\Delta$Kohn–Sham ($\Delta$SCF) calculations where total energies of the initial ground state and the core hole state, with $n_{\text{init}} = 0$, are compared. The global shifting assumes that the relaxation corrections vary only slightly in the energy region near the ionization threshold which seems reasonable. This procedure is approximate but avoids having to calculate fully relaxed final states for each excitation separately which may become very tedious for large systems. For a discussion of more sophisticated strategies to correct for electron relaxation see Chapter 12.

The theoretical excitation spectrum (18.19) can be extended further by including unbound final state resonances which lie energetically above the ionization threshold. This is feasible by applying a so-called double basis set technique [75] where very diffuse basis functions are used to describe the spatial variation of the resonances inside the core electron region in an approximate way. (Note that according to (18.18) and (18.19) the excitation spectrum requires a reliable representation of the resonance orbitals only within the core electron region, which is feasible with diffuse localized basis functions while the resonances themselves are not localized.) The procedure is known to lead to reliable core excitation spectra up to 10 eV above ionization threshold.

The use of X-ray absorption spectroscopy (XAS) for analyzing differently coordinated atom species in oxides will be illustrated in the following by considering oxygen at the $V_2O_5$ (010) surface [72, 76]. Figure 18.5 compares experimental polarization-resolved NEXAFS spectra of the $V_2O_5$(010) surface [77] with the corresponding theoretical total O 1s core level excitation spectra computed [72] for the

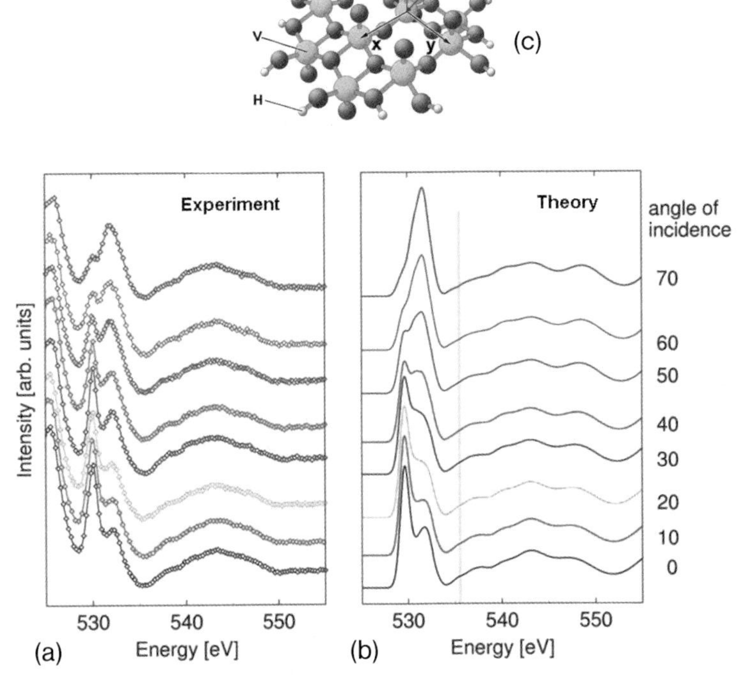


**Figure 18.5** Comparison of experimental polarization-resolved NEXAFS spectra of the $V_2O_5$(010) surface (a) [77] with corresponding theoretical total O 1s core excitation spectra (b) calculated for the $V_{10}O_{31}H_{12}$ cluster shown at the top [72]. The spectra refer to different angles of incidence $\Theta$ (c) of the photon beam with respect to the surface normal as indicated in the cluster sketch. The vertical gray line in the right figure, near 535 eV, indicates the ionization threshold.

$V_{10}O_{31}H_{12}$ cluster shown at the top of the figure. The spectra refer to different polarization directions $\underline{e}_{pol}$ of the incoming photon beam and have been converted to polar angles of incidence $\Theta$ between $0^\circ$ (normal photon incidence) and $70^\circ$ (almost grazing incidence) taken with respect to the surface normal ($z$ axis), where an azimuthal angle $\varphi$ corresponding to a beam variation (and a polarization direction) along the $xz$ plane was chosen. (The corresponding Cartesian coordinate system is sketched in the cluster picture at the top of Figure 18.5.) The comparison of the spectra requires an additional step in the theoretical treatment. According to (18.19) the theoretical spectra are based on computed discrete excitations and need to be convoluted by broadening functions, replacing the Dirac delta functions $\delta(E - E_k)$ by localized continuous functions which account for lifetime broadening and instrumental resolution effects in the experiment. This is usually performed in an empirical way by applying Gaussian broadening with energy dependent widths [78] in the theoretical spectra following common wisdom amongst the synchrotron radiation community.

A comparison between the experimental and theoretical spectra of Figure 18.5 shows an overall very good agreement for all photon angles $\Theta$ where the agreement is excellent for normal photon incidence, $\Theta = 0^\circ$, corresponding to $x$ polarization. A detailed analysis of the final state orbitals involved in the core excitations shows, first, that the broad peak in the continuum region near 545 eV is assigned to transitions from O 1s core to O 3p orbital resonances. (The ionization threshold lies at 535 eV, see vertical gray line in Figure 18.5, right graphics.) Second, the two-peak structure found in the energy range between 529 and 536 eV is characterized by excitations from O 1s core to final state orbitals, which are antibonding mixtures of O 2p and V 3d contributions. Both the energetic positions and the intensities of these two experimental peaks are reproduced quite nicely by the calculated spectrum. The experimental spectra include excitation peaks at energies lower than 529 eV, which do not appear in the theoretical spectra. These peaks are caused by excitations of V 2p core electrons to V 3d final state orbitals and require a theoretical treatment of vanadium core electrons including relativistic spin–orbit coupling which is not yet available in the present theory.

The theoretical spectra can be decomposed into contributions from 1s core excitations at differently coordinated oxygen in the oxide which is of particular interest in the energy region between 529 and 536 eV and allows a discrimination of the oxygen species in $V_2O_5$ based on angle-dependent peak intensities. This is shown in Figure 18.6, which compares experimental and theoretical spectra for two photon incidence angles, $\Theta = 0^\circ, 70^\circ$ where the theoretical spectra are decomposed into contributions from the three oxygen species O(1), O(2), and O(3). Obviously, the partial spectra between 529 and 536 eV, all exhibiting a two-peak structure, differ dramatically. This can be understood by the symmetries of the final state orbitals which depend on the oxygen center where the core excitation happens. Vanadyl oxygen O(1) binds with one vanadium along the $z$ direction (see Figure 18.5(c)). As a consequence, final state orbitals, described as antibonding mixtures of O 2p and V3d contributions, can be classified, according to symmetry, by either O $2p_x$ ($2p_y$) mixing with V $3d_{xz}$ ($3d_{yz}$) or by O $2p_z$ mixing with V $3d_{z^2}$ where the latter

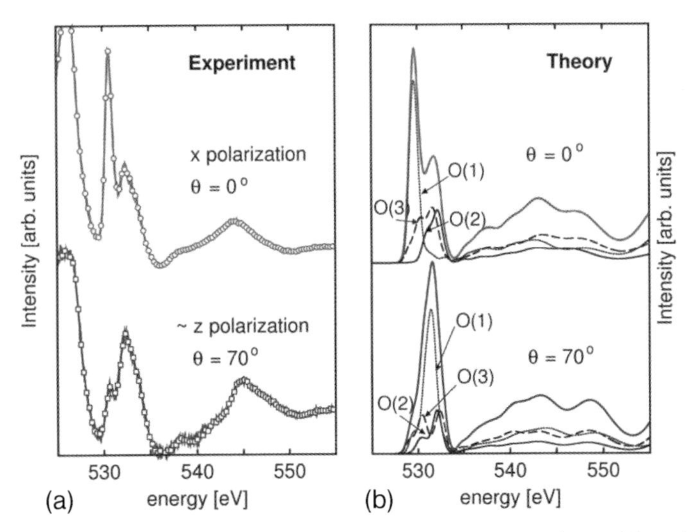


**Figure 18.6** Comparison of experimental (a) and theoretical (b) polarization-resolved O 1s NEXAFS spectra of the $V_2O_5(010)$ surface [72, 77]. The spectra refer to two characteristic photon polarization directions (converted to angles of incidence $\Theta$), $\Theta = 0^\circ$ (*x* polarization), $= 70^\circ$ (nearly *z* polarization), see Figure 18.5. The theoretical excitation spectra are obtained by calculations using the $V_{10}O_{31}H_{12}$ cluster of Figure 18.5(c), where the total spectra are decomposed into contributions due to the differently coordinated oxygen, O(1), O(2), and O(3), shown by dotted, dashed, and dash-dotted lines, respectively.

is more strongly antibonding and, thus, higher in energy. This characterizes the two peaks at 529.5 and 531.4 eV of the O(1)-derived partial excitation spectrum. The low-energy peak at 529.5 eV is analyzed as being due to O $2p_x$ – V $3d_{xz}$ and O $2p_y$ – V $3d_{yz}$ type final state orbitals where an example of the latter is shown as a shaded contour plot along the *yz* plane, see orbital (1) in Figure 18.7. Thus, the corresponding dipole transition moment vectors $\underline{m}^{(k)}$ defined by (18.19) lie in the *xy* plane of the cluster, i.e., parallel to the $V_2O_5$ (010) surface, which, according to (18.17), yields maximum transition probability and hence largest peak intensity for photon polarization parallel to the surface, termed "*x* polarization" ($\Theta = 0^\circ$) in Figure 18.6. By the same reasoning the peak intensity is expected to be smallest for polarization normal to the surface, approximately described by "$\sim z$ polarization" ($\Theta = 70^\circ$) in Figure 18.6. On the other hand, the high-energy peak at 531.4 eV is found to be due to O $2p_z$ – V $3d_{z^2}$ type final state orbitals where an example is shown as a shaded contour plot along the *yz* plane, see orbital (2) in Figure 18.7. Therefore, the moment vectors $\underline{m}^{(k)}$ point along the z direction of the cluster, i.e., normal to the $V_2O_5$ (010) surface, and, according to (18.17), maximum transition probability (i.e., largest peak intensity) is obtained for photon polarization normal to the surface, $\Theta = 90^\circ$ (approximated by $\Theta = 70^\circ$ in Figure 18.6). In contrast, polarization parallel to the surface, corresponding to $\Theta = 0^\circ$, must result in very small (zero) intensity as confirmed in Figure 18.6.

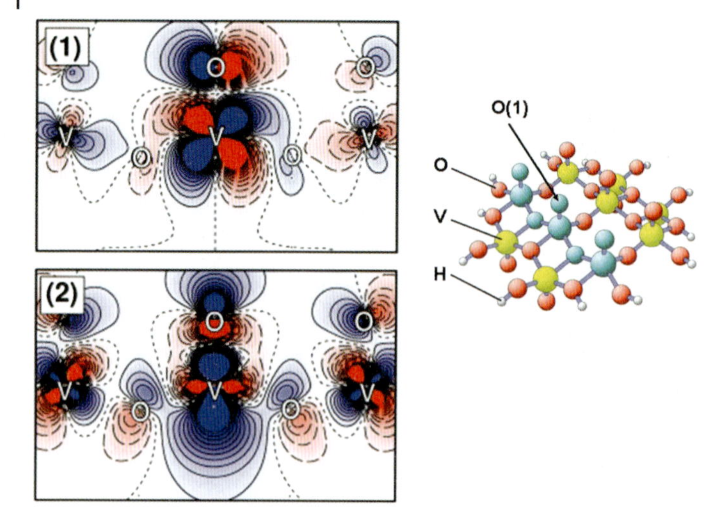


**Figure 18.7** Shaded contour plots of selected final state orbitals, (1) and (2), referring to O 1s core level excitation at an O(1) atom of the $V_{10}O_{31}H_{12}$ surface cluster sketched to the right, see text. Full (long dashed) contour lines denote positive (negative) orbital values while short dashed lines belong to zero values. The amount of shading illustrates the absolute contour values. The plotting plane extends along *xz* through O(1) and its V neighbor indicated by green painted atoms in the cluster sketch.

The O(2)-derived partial excitation spectrum in the energy range between 529 and 536 eV is noticeably different from that for O(1). The partial spectrum given in Figure 18.6 shows also a two-peak structure where, however, the structure is less pronounced with the high-energy peak dominating and being less dependent on photon incidence angles. This can be rationalized, as for O(1) core excitations, by the symmetry of the final state orbitals. Bridging oxygen O(2) binds with two vanadium atoms along the $x$ direction with some tilt toward the $z$ axis, see Figure 18.5. Therefore, final state orbitals are characterized as O $2p_y$ – V $3d_{xy}$ and O $2p_x$ – V $3d_{x^2-y^2}$ mixtures but also O $2p_z$ – V $3d_{z^2}$ mixtures appear as a result of the V–O(2) bond geometry. Examples of characteristic orbitals are shown as shaded contour plots along the $xy$ plane in Figure 18.8. The high-energy peak in the O(2)-derived partial excitation spectrum is determined by core excitations involving final state orbitals of both O $2p_x$ – V $3d_{x^2-y^2}$, see orbital (2) in Figure 18.8, and O $2p_z$ – V $3d_{z^2}$ shape resulting in dipole transition moment vectors $\underline{m}^{(k)}$ which point along the $x$ and the $z$ axis, respectively. This leads to transition probabilities and hence to peak intensities where the two orbital contributions complement each other and the combined peak intensity becomes almost independent of the photon incidence angle Θ which is obvious in Figure 18.6.

Similar arguments hold for the low-energy peak in the O(2)-derived partial excitation spectrum, which as a consequence of the interplay between excitations involving final state orbitals of O $2p_y$–V $3d_{xy}$, see orbital (1) in Figure 18.8, and O $2p_z$ – V $3d_{z^2}$ character, decreases weakly with increasing Θ. Finally, Figure 18.6 shows that the O(3)-derived partial excitation spectrum in the energy range between

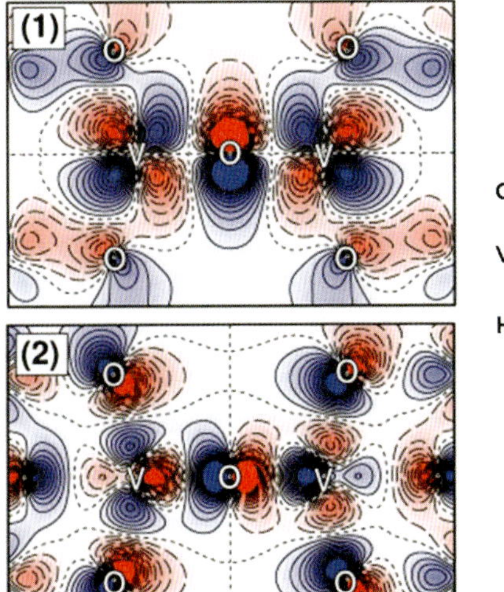


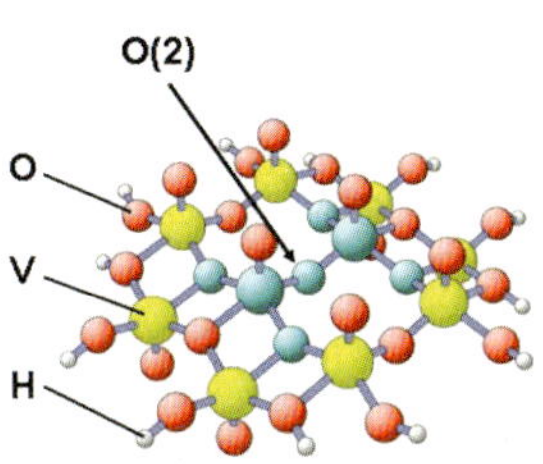


**Figure 18.8** Shaded contour plots of selected final state orbitals, (1) and (2), referring to O 1s core level excitation at an O(2) atom of the $V_{10}O_{31}H_{12}$ surface cluster sketched to the right, see text. For a definition of the contours and shading, see Figure 18.7. The plotting plane extends along *xy* through O(2) and close to its two V neighbors indicated by green painted atoms in the cluster sketch.

529 and 536 eV differs clearly from those for O(1) and O(2). This can also be explained by the symmetry of the corresponding final state orbitals of bridging oxygen O(3) binding with three vanadium atoms as discussed elsewhere [72]. Altogether, the low-energy peak in the experimental NEXAFS spectrum for $x$ polarization in Figure 18.6 can be clearly assigned to singly coordinated vanadyl oxygen O(1), while the high-energy peak originates from core excitations at the bridging oxygen centers O(2), O(3). The corresponding analysis for $y$ polarization [72] shows that the high-energy peak of the two-peak structure contains mainly contributions referring to the 3-fold coordinated center O(3). This illustrates that O 1s NEXAFS spectroscopy can be used to discriminate between differently coordinated oxygen in $V_2O_5$.

### 18.2.3
### Oxygen Vacancies at the $V_2O_5$(010) Surface

Oxygen vacancies at oxide surfaces and in their bulk are of major importance for the physical and chemical behavior of these systems. Examples are point defects in silicon dioxide or oxygen vacancies in MgO, which can accommodate localized electron states observed as color centers [79]. In many catalytic reactions such as hydrocarbon oxidation the efficiency of vanadium-oxide-based catalysts depends strongly on their ability to provide surface oxygen as a reactant. Therefore, theoretical studies of physical and chemical (catalytic) properties of the differently coordinated oxygen at vanadium oxide surfaces can help to understand details of

**Table 18.3** Oxygen vacancy energies, $E^r_{vac}$, $E^f_{vac}$, for the O(1–3, 2′, 3′) sites at the $V_2O_5(010)$ surface with and without vacancy-induced relaxation [80].[a]

| Vacancy site | $E^r_{vac}$ | $E^f_{vac}$ | $\Delta Q^r(V)$ | $\Delta Q^f(V)$ |
|---|---|---|---|---|
| O(1) | 5.48 | 7.20 | −0.18 | −0.26 |
| O(2, 2′) | 7.23, 6.17 | 8.00, 7.81 | −0.16, −0.06 | −0.30, −0.35 |
| O(3, 3′) | 6.18, 6.00 | 6.92, 7.01 | −0.19, −0.32 | −0.33, −0.33 |

[a]The energies are obtained for model clusters $V_{20}O_{61}H_{24}/V_{20}O_{62}H_{24}$ (two $V_2O_5$ layers, see text). The table includes vacancy-induced atom charge differences of the vanadium center(s) nearest to the oxygen vacancy, $\Delta Q^r(V)$ (relaxed geometry), $\Delta Q^f(V)$ (frozen geometry), from Mulliken analyses. All energies are given in eV, charges in atomic units per atom.

corresponding reaction steps. Here we focus on the $V_2O_5(010)$ surface which has been studied extensively [28, 80–83] as a model system.

As mentioned earlier, bulk vanadium pentoxide contains three structurally different oxygen species which yield five different oxygen sites, O(1–3, 2′, 3′), at the $V_2O_5(010)$ surface (see Figure 18.3). Thus, the surface offers five different types of surface oxygen vacancies which can be studied theoretically by removing corresponding oxygen species from the surface clusters and re-evaluating the electronic structure. Frozen vacancy states are obtained from calculations on the vacancy clusters where all atom positions are kept frozen at the geometry of the initial surface cluster. In addition, relaxed vacancy states are determined by geometry optimizations of the corresponding vacancy cluster allowing all cluster atoms near the vacancy to rearrange until the lowest total energy is obtained. As an illustration, Table 18.3 lists results from theoretical cluster studies [80, 82] on different oxygen vacancies. The calculations use vacancy clusters $V_{20}O_{61}H_{24}$ derived from the substrate cluster $V_{20}O_{62}H_{24}$ representing two $V_2O_5$ layers, where the two layer sections themselves are identical with $V_{10}O_{31}H_{12}$ shown in Figure 18.3, and placed at a perpendicular distance corresponding to the interlayer separation in bulk $V_2O_5$. Oxygen vacancy energies $E^r_{vac}$ (relaxed geometry) and $E^f_{vac}$ (frozen geometry) are determined by appropriate total energy differences for the clusters with and without vacancy, i.e., by

$$E^x_{vac} = E^x_{tot}(V_{20}O_{61}H_{24}) + E_{tot}(O) - E_{tot}(V_{20}O_{62}H_{24}), \quad x = r, f \qquad (18.21)$$

using the ground state of atomic oxygen as a reference. (Taking 1/2 $O_2$ as a reference would reduce the energies in Table 18.3 by 2.57 eV.) All vacancy energies (from DFT calculations applying the RPBE functional [82]) are found to be rather large of which the value for O(1) is the smallest. Vacancy-induced geometric relaxation decreases the frozen geometry energies $E^f_{vac}$, where the decrease is the largest for O(1). This is explained by the specific relaxation at the O(1) site which involves oxygen of the second $V_2O_5$ layer as will be discussed below. Obviously, oxygen is bound very strongly to its substrate environment and may not be removed in one single step during an oxidation reaction. However, several theoretical studies

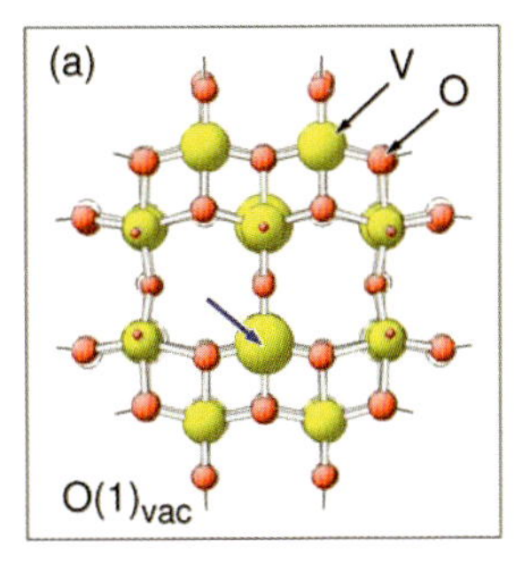


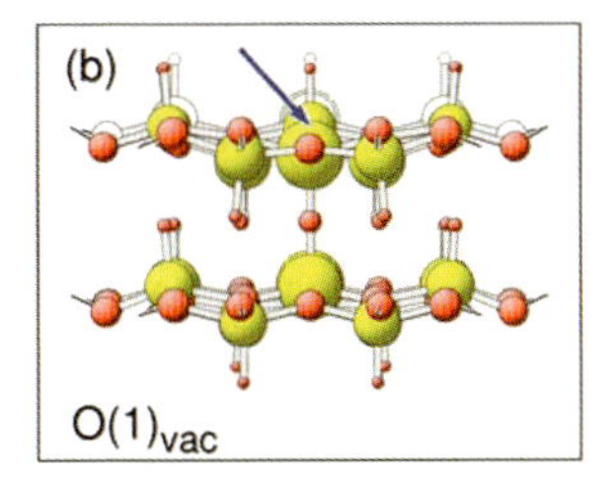


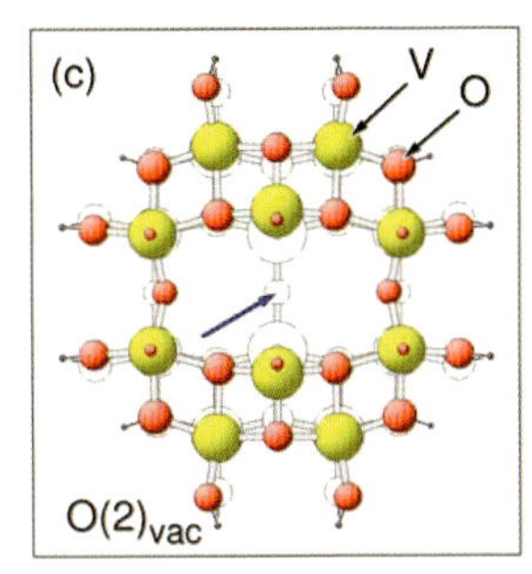


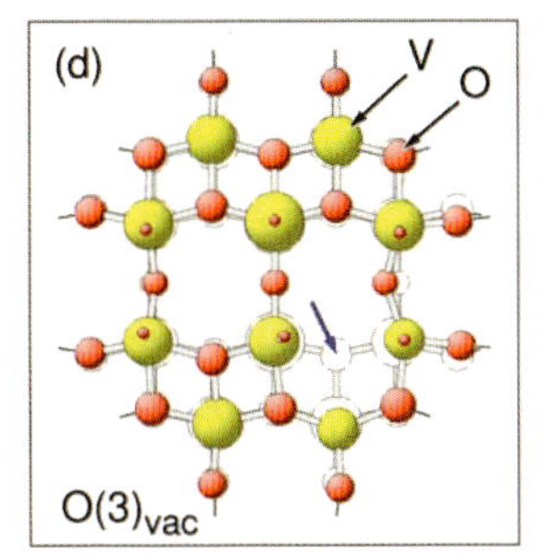


**Figure 18.9** Relaxed geometry of the differently coordinated oxygen vacancies for (a, b) vanadyl O(1), (c) bridging O(2), and (d) bridging O(3) modeled by $V_{20}O_{61}H_{24}$ clusters (two $V_2O_5(010)$ layers, see text) [80, 82]. The views are perpendicular (plots a, c, d) and parallel (plot b) to the $V_2O_5(010)$ surface, respectively. Cluster atoms are shown by shaded balls where ball radii represent atom charges, see text. Red (yellow) shading refers to negative (positive) charge while the radii give the amount of charge. The white balls behind each relaxed cluster describe the geometry without vacancy. The vacancy locations are indicated by blue arrows.

indicate strongly that preadsorbed atoms or molecules, such as hydrogen [82,83], oxygen [28,80], CO [83], or $NH_x$ [84], can weaken surface binding of the oxygen and make it available for oxidation processes.

Figures 18.9(a)–(d) visualizes geometric consequences of substrate relaxation as well as charge transfer due to vacancy formation at different oxygen sites of the $V_2O_5(010)$ surface where vacancy clusters $V_{20}O_{61}H_{24}$ (representing two $V_2O_5$ layers, see above) are used as models [80]. All cluster atoms are displayed as shaded balls where ball radii represent atom charges as obtained by Mulliken populations. As a main result, geometric relaxation is always found to be locally confined. When oxygen is removed from a (singly coordinated) vanadyl O(1) site, see Figures 18.9(a) and (b), relaxation causes the vanadium atom below the vacancy site (indicated by a blue arrow) to move perpendicular to the surface toward the substrate by about 0.5 Å. As a result, the vanadium penetrates further into the substrate and can bind with the nearest vanadyl oxygen O(1) of the second substrate layer to form a V–O–V bond bridge between the first and second layer. This interlayer V–O–V bond bridge is found to be very similar in its electronic structure with the V–O(2)–V bridge of the first layer as confirmed by bond order and charge analyses [80]. Thus, vacancy formation at the first layer of the $V_2O_5(010)$ surface can increase the electronic coupling with the second layer. This may be the starting point of major surface relaxation and may result eventually in a reconstruction and changed stoichiometry of the whole surface region. If oxygen is removed from a (doubly coordinated) bridging O(2) site, see Figure 18.9(c), the strongest relaxation shifts occur at the two vanadium atoms adjacent to the vacancy which move laterally by 0.6 Å such that the vacancy opening is enlarged. However, the lattice topology of the $V_2O_5(010)$ surface is conserved suggesting that a single O(2) vacancy will not introduce major surface restructuring. Finally, geometric relaxation near the O(3) vacancy, shown in Figure 18.9(d), is also characterized by mainly lateral shifts of the atoms adjacent to the vacancy increasing the vacancy opening.

A closer inspection of the ball radii in Figures 18.9(a)–(d) reveals that the positive charges of the vanadium atoms closest to each oxygen vacancy are smaller than the corresponding values of the cluster without the vacancy. This is also obvious from the atom charge differences $\Delta Q(V)$ listed in Table 18.3 and can be explained by electronic consequences of the vacancy formation. When oxygen, residing as a negatively charged species at the $V_2O_5$ surface, is removed from the surface as a neutral atom the remaining negative charge is redistributed in the vicinity of the vacancy and affects the local electronic structure. An analysis of the electronic structure of bulk $V_2O_5$ [50] shows that the 5.5 eV wide valence band region of this small-gap semiconductor, see Figure 18.10, is characterized mainly by oxygen 2p-type electron states while the energy range above the 2 eV wide gap is described by (unoccupied) vanadium 3d-type electron states. As a consequence, electron transfer due to oxygen vacancy formation can be thought of conceptionally as

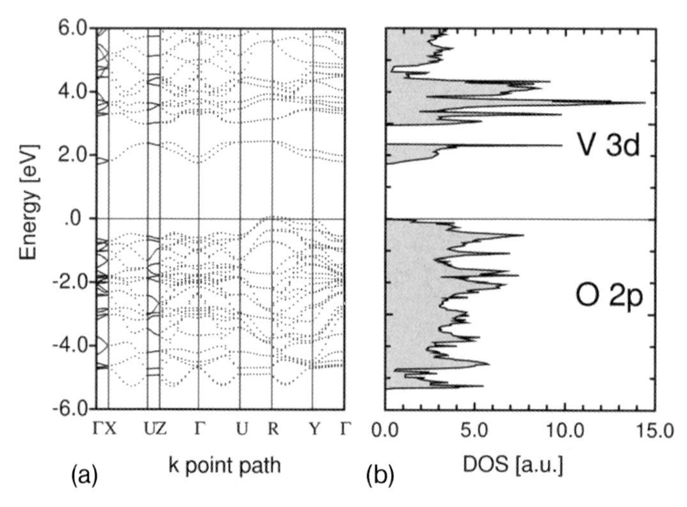


**Figure 18.10** Band structure (a) and total density-of-states (DOS) (b) of bulk $V_2O_5$ [50]. The energy bands are shown for characteristic linear paths connecting high symmetry points of the irreducible part of the orthorhombic Brillouin zone. All energies $\varepsilon(k)$ are taken with respect to that of the highest occupied level. The DOS is given in states per unit volume and per eV.

occupying V 3d-type electron states increasing the electron count at the vanadium centers adjacent to the oxygen vacancy. This reduces the positive charges of the corresponding vanadium atoms and explains the decreased ball radii due to vacancy formation in Figures 18.9(a)–(d). The electronic process of increasing local vanadium 3d occupation is actually observed in ultraviolet photoemission (UPS) experiments [73] and gives a clear microscopic picture of vacancy-induced chemical reduction of the metal sites.

### 18.2.4
### Termination of the $V_2O_3(0001)$ Surface

Vanadium sesquioxide, $V_2O_3$, is characterized in its bulk structure by a rather densely packed lattice, monoclinic [38] or trigonal [39], in contrast to the open layer structure found for the previously discussed $V_2O_5$. This suggests different possible geometric models for the surface termination of $V_2O_3$ substrate which can influence the catalytic behavior of this material in different ways. In the simplest case of $V_2O_3$ single crystals, densely packed low-Miller-index netplanes may be assumed to terminate corresponding surfaces where, however, changed local binding can influence the surface geometry. So far, experimental studies have been reported only for surfaces of $(10\bar{1}2)$ orientation (corresponding to the cleavage plane of bulk $V_2O_3$) and of (0001) orientation (corresponding to the basal plane) [3, 27, 85]. In this section we will focus on (0001) oriented surfaces which have been grown as thin films on alumina [86] and on $Cu_3Au(111)$ substrate [87] to examine catalytic activity.

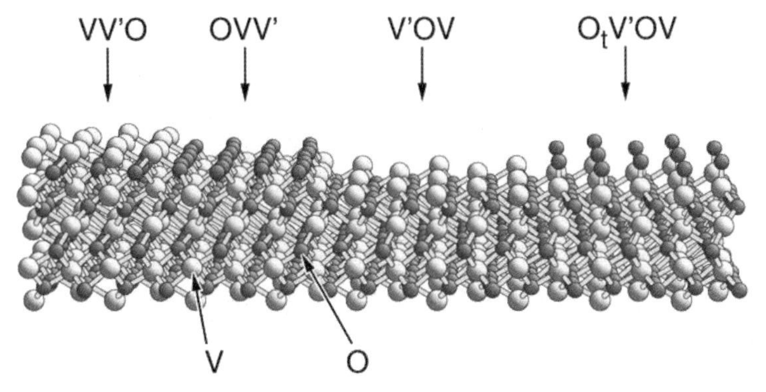


**Figure 18.11** Perspective view of the rhombohedral $V_2O_3(0001)$ surface for different terminations (from left to right): full metal termination VV′O, oxygen termination OVV′, half metal layer termination V′OV, and vanadyl termination $O_tV'OV$, see text. Vanadium and oxygen atoms are shown as shaded large and small balls, respectively, and labeled accordingly.

The trigonal corundum lattice of bulk $V_2O_3$ is described in its ideal structure along the (0001) direction by three different types of hexagonal netplanes, two vanadium planes (V, V′ of equal atom density) very close together and one oxygen plane (O,

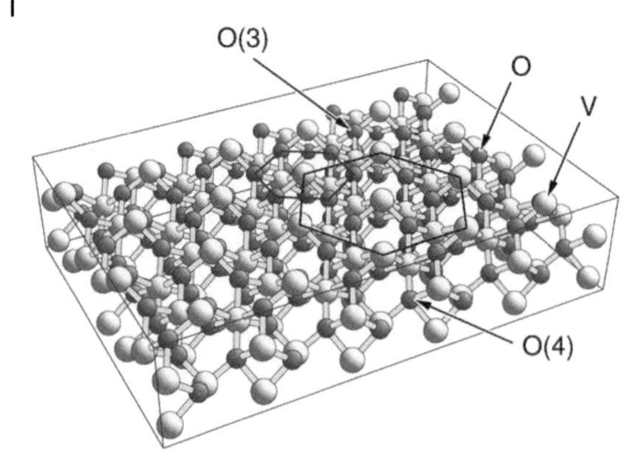


**Figure 18.12** Geometric structure of the (0001) oriented single crystal surface of trigonal $V_2O_3$. for a surface termination V′OV. Vanadium and oxygen atoms are shown as shaded large and small balls, respectively, with connecting sticks indicating nearest-neighbor relations. Differently coordinated surface oxygen atoms, O(3, 4), are labeled and the hexagonal cells of the vanadium and oxygen netplanes are sketched accordingly.

with three times the atom density of each vanadium plane) in between, which yields a stacking sequence ... OVV′OVV′O.... This offers three different surface terminations of the ideal bulk, VV′O with the two metal layers at the top, OVV′ with the oxygen netplane as the topmost layer, and the half metal termination V′OV with only one of the two metal layers at the top. (The nomenclature of these intrinsic bulk terminations, sketched in Figure 18.11, refers to the element composition of the topmost netplanes.) The latter, V′OV, reflects the ion arrangement with the smallest average Madelung potential at the surface corresponding to a surface arrangement of one hexagonal V plane on top of a hexagonal O plane (containing 3-fold coordinated oxygen O(3) in contrast to 4-fold coordinated O(4) in the bulk) as shown in Figure 18.12. This surface geometry has been confirmed in quantitative LEED experiments [88] under ultrahigh vacuum conditions, where major relaxation of the top layers along the surface normal has been observed and described by theory [89–91]. The geometric arrangement is very similar to the surface structure found experimentally for the $Cr_2O_3(0001)$ surface [92, 93].

Theoretical studies on the geometry of differently terminated $V_2O_3(0001)$ surfaces have been performed using DFT methods applying both periodic [91] and cluster models [89, 90], where good agreement was found between the different approaches. As an example, Table 18.4 lists interlayer distances $z_{ij}$ for the topmost five layers of the V′OV terminated surface obtained from cluster model calculations [89, 90] including geometry optimizations to represent layer relaxation without reconstruction. A comparison of the results of Table 18.4 and their graphic representation in Figure 18.13 evidences that the dominant relaxation effect occurring at the clean V′OV terminated $V_2O_3(0001)$ surface is described by the topmost vanadium layer V′(3) (with 3-fold coordinated vanadium) moving inward by 0.30 Å with respect to the underlying oxygen layer O(3). In addition, the two subsurface

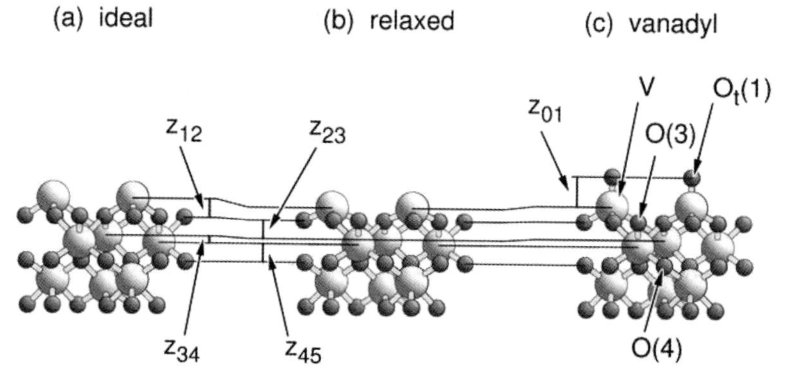


**Figure 18.13** Geometric details of the $V_2O_3(0001)$ surface for the (a) ideal V′OV bulk termination, (b) V′OV termination with the topmost four layers relaxed, (c) vanadyl terminated (relaxed) $O_tV'OV$ termination. The different sketches for a view parallel to the surface refer to calculations [89, 90] on embedded surface clusters $V_{11}O_{33}H_{33}$ (a, b) and $V_{11}O_{36}H_{33}$ (c), respectively, where saturator hydrogen atoms are hidden. The different vanadium and oxygen atoms as well as interlayer distances $z_{ij}$ are labeled accordingly. Corresponding layers are connected by horizontal lines to illustrate the relaxation effect, see text.

**Table 18.4** Interlayer distances $z_{ij}$ of the the V′OV terminated $V_2O_3$ (0001) surface (a) for ideal bulk termination, (b) with the topmost four layers relaxed, (c) with additional oxygen on top (vanadyl $O_tV'OV$ termination).[a]

| | (a) V′OV ideal | (b) V′OV relaxed | (c) $O_tV'OV$ relaxed |
|---|---|---|---|
| $z_{01}$ {$O_t(1) - V'(3)$} | – | – | 1.586 |
| $z_{12}$ {$V'(3)–O(3)$} | 0.984 | 0.684 | 0.815 |
| $z_{23}$ {$O(3)–V(6)$} | 0.984 | 0.983 | 0.932 |
| $z_{34}$ {$V(6)–V'(6)$} | 0.363 | 0.270 | 0.341 |
| $z_{45}$ {$V'(6)–O(4)$} | 0.984 | 0.955 | 0.955 |

[a]The distances are sketched in Figure. 18.13. The results are taken from calculations [89, 90] on embedded surface clusters $V_{11}O_{33}H_{33}$ (a, b) and $V_{11}O_{36}H_{33}$ (c), see text. All distances are given in Å.

vanadium layers V(6), V′(6) (with 6-fold coordinated vanadium) move closer together by 0.09 Å. This relaxation scheme influences the charge distribution in the surface layers. Mulliken charge analyses show [89] that the atoms of the topmost vanadium layer V′(3) increase their positive charges due to relaxation resulting in an electron charge flow into the bulk. This charge flow, enhancing the surface charging slightly, is distributed over several sublayers, affecting each layer only little.

Depending on preparation and growth conditions, thin $V_2O_3$ films can form rather complex surface structures which have been studied extensively by surface tunneling microscopy (STM) [85, 91]. Here we mention only the existence of vanadyl groups at the V′OV terminated surface after oxygen exposure which have been considered to be of catalytic importance [94] and have been identified by experiment [86, 95]. These VO groups exist at the intrinsic $V_2O_5(010)$ surface as discussed before but not at other ideal bulk terminated $V_xO_y$ surfaces. Therefore, in the early literature on properties of vanadia catalysts samples of unknown V/O stoichiometry where often incorrectly claimed to be "$V_2O_5$" because vanadyl groups were observed by infrared absorption.

Vanadyl groups at the V′OV terminated $V_2O_3(0001)$ surface have been simulated by stabilizing oxygen on top of vanadium atoms V′(3) of the topmost layer yielding a $O_t$V′OV termination as shown in Figure 18.11. Obviously, the added singly coordinated vanadyl oxygen $O_t(1)$ must influence the intrinsic geometric and electronic structure of the V′OV terminated surface. As an illustration, Table 18.4 includes interlayer distances $z_{ij}$ for the topmost five layers and the vanadyl oxygen layer of the $O_t$V′OV termination. As before, the values have been evaluated in cluster model studies [90] including layer relaxation. The calculations yield quite strongly bound vanadyl oxygen $O_t(1)$ above the V′(3) centers at a distance of 1.59 Å, which reflects a typical V–O distance of vanadyl found at the $V_2O_5(010)$ surface [81]. As a result, the vanadium atoms change their coordination from 3 to 4, yielding V′(4) which affects their geometric properties. A comparison of corresponding interlayer distances with and without the vanadyl oxygen, given in Table 18.4 and sketched in Figure 18.13, shows clearly that the presence of the oxygen removes a major part of the initial layer relaxation at the V′OV terminated $V_2O_3(0001)$ surface. In particular, the inward relaxation of the topmost V′ layer is compensated by 45% and the deeper layers assume also positions closer to the unrelaxed bulk geometry. This can be explained by the fact that the additional vanadyl bonds at the V′ atoms increase the total number of V–O bonds per V′ atom, which weakens the electronic coupling of V′ with the substrate. As a consequence, corresponding bond lengths and the interlayer distance $z_{12}$ are increased. Mulliken atom charge analyses show [90] that the added vanadyl oxygen $O_t$ assumes a negative charge while the vanadium V′(4) directly underneath looses charge due to the electron transfer. As a result, vanadyl formation leads to an additional surface dipole which may be observed in workfunction experiments.

As discussed above, the intrinsic V′OV termination of the $V_2O_3(0001)$ surface offers two different types of oxygen, 3-fold coordinated O(3) in the second layer near the surface and 4-fold coordinated O(4) in the 5th layer which may be considered bulk type, see Figure 18.13. Since O(3) type oxygen is assumed to be involved in catalytic reactions happening at the surface it is important to be able to identify this species in the experiment. As discussed in great detail in Section 18.2.2 a very promising method for this purpose is based on O 1s core excitation spectroscopy available with NEXAFS measurements. Therefore, O 1s core excitation spectra have been calculated for differently coordinated oxygen near the $V_2O_3(0001)$ surface [96, 97] following exactly the same approach discussed in Section 18.2.2.

Figure 18.14 compares experimental polarization-resolved NEXAFS spectra of the $V_2O_3(0001)$ surface [98] with the corresponding theoretical total O 1s core level

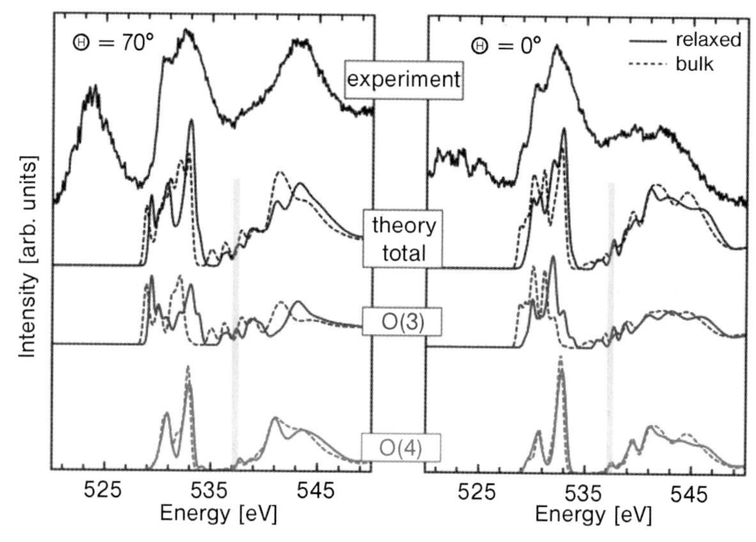


**Figure 18.14** Comparison of experimental polarization-resolved O K-edge NEXAFS spectra [98] with theoretical O 1s core excitation spectra calculated for a $V_{11}O_{33}$ cluster [96] shown in Figure 18.13. The spectra refer to two different angles of photon beam incidence, $\Theta = 70^\circ$, $0^\circ$, at the V′OV bulk terminated $V_2O_3(0001)$ surface. The total theoretical spectra are decomposed into contributions from differently coordinated oxygen, O(3) and O(4). Full lines denote spectra for the fully relaxed surface while dashed lines show results for ideal bulk termination. The vertical gray lines indicate the oxygen 1s ionization threshold.

excitation spectra computed for the $V_{11}O_{33}$ cluster shown in Figure 18.13 [96,97]. The spectra refer to two different polarization directions $\underline{e}_{pol}$ of the incoming photon beam where the $\underline{e}_{pol}$ directions have been converted to polar angles of incidence $\Theta = 70^\circ$ (near grazing photon incidence) and $0^\circ$ (normal incidence) taken with respect to the surface normal. (It can be shown that, as a result of the surface symmetry and domain formation, the spectra do not depend on the azimuthal angle $\varphi$ [97].) The theoretical spectra, below the experimental spectra, are given for the V′OV terminated surface both including surface relaxation, full lines, and for the ideal bulk termination, dashed lines. The comparison between the experimental and theoretical total spectra of Figure 18.14 shows an overall good agreement for the two photon angles Θ. The analysis of the final state orbitals involved in the core excitations yields qualitatively the same results as found for the $V_2O_5$ surface discussed in Sec. 18.2.2. First, the broad peak in the continuum region near 543 eV is due to transitions from O 1s core to O 3p orbital resonances. (The ionization energy amounts to 537 eV as indicated by vertical gray lines in Figure 18.14.) Second, the multipeak structure in the energy range between 529 and 537 eV is assigned to excitations from O 1s core to final state orbitals described as antibonding mixtures of O 2p and V 3d contributions. Here the energetic positions and the intensities agree rather nicely between theory and experiment. However, the multipeak structure in the theoretical spectrum is not well resolved in the experiment. The experimental spectra include excitation peaks at energies lower than 529 eV, which are not included in the theoretical spectra. These peaks have been also observed in the case of $V_2O_5$, see Figure 18.5, and are of the same origin. They are assigned to excitations of V 2p core electrons to V 3d final state orbitals and require a theoretical treatment of vanadium core electrons including relativistic spin–orbit coupling which is not yet available in the present theory.

The theoretical total spectra in Figure 18.14 are decomposed into contributions from 1s core excitations at differently coordinated oxygen, surface-type O(3) and bulk-type O(4), which is interesting in the energy range between 529 and 537 eV and may allow a discrimination of the two oxygen species. In this range the peak structure referring to O(3)-derived excitations is wider than that for O(4). In addition, the O(3)-derived peaks start at a lower energy than those for O(4) and the peak distributions as well as their angle dependence differ substantially. Clearly, these differences originate from details of the local binding environment of the two types of oxygen which manifest themselves also in the excited final state orbitals determining the excitation spectrum by their excitation energies $E_k$ and by corresponding dipole transition matrix elements $\underline{m}^{(k)}$. However, the local binding of O(3) with its three vanadium neighbors in a quasi-planar geometry and of O(4) with its four vanadium neighbors in a distorted tetrahedral arrangement is quite complex. Therefore, simple geometric arguments concerning the shape of the final state orbitals, which have been very successful for $V_2O_5$ discussed above, are not adequate to characterize the differences between the two atom-derived spectra in greater detail. An inspection of the theoretical core excitation spectra with and without surface relaxation in Figure 18.14 evidences quite interesting differences. First, the partial spectra for O 1s excitation at the surface-type O(3) species shift to higher energies by 0.6 to 1.1 eV, which translates to the same shift in the total spectra yielding better agreement between theory and experiment. This can be explained by the increased charging of the $V_2O_3(0001)$ surface discussed above, which accumulates small amounts of negative charge at the O(3) atoms. The relaxation-induced spectral shift does not occur for the bulk-type O(4) species, see Figure 18.14, which is expected since the binding environment of these atoms is not affected by relaxation.

Experimental O K-edge NEXAFS spectra for the $V_2O_3(0001)$ surface after exposure to oxygen are assumed to be influenced by the presence of vanadyl groups discussed above. Figure 18.15 compares experimental polarization-resolved NEXAFS spectra of the $V_2O_3(0001)$ surface after oxygen exposure with the corresponding theoretical total O 1s core level excitation spectra computed [97] for a $V_{11}O_{36}$ cluster modeling a section at the $O_tV'OV$ terminated surface (see Figures 18.11 and 18.13). The spectra refer to two different polar angles of incidence $\Theta = 70^\circ, 0^\circ$ of the incoming photon beam analogous to the data for the V′OV termination shown in Figure 18.14. Further, the theoretical spectra, below the experimental spectra, are given for the relaxed surface geometry taken from the calculations. The comparison between the experimental and theoretical total spectra of Figure 18.15 shows overall

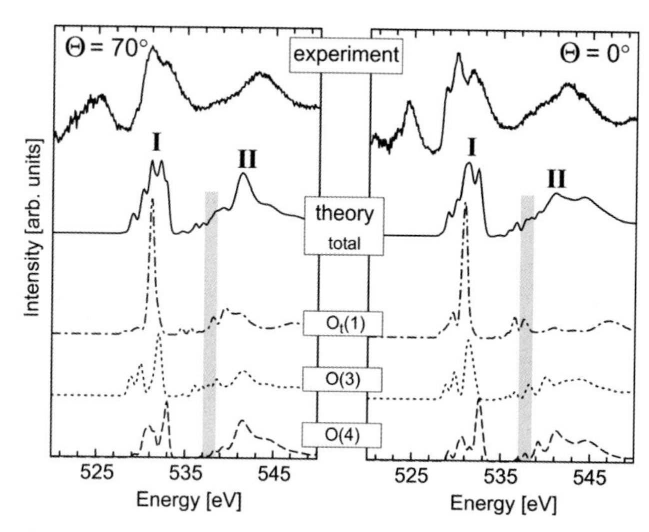


**Figure 18.15** Comparison of experimental polarization-resolved O K-edge NEXAFS spectra [97] with theoretical O 1s core excitation spectra for two different angles of photon beam incidence, $\Theta = 70^\circ, 0^\circ$, at the $O_tV'OV$ terminated $V_2O_3(0001)$ surface. The total theoretical spectra are decomposed into contributions from differently coordinated oxygen, vanadyl $O_t(1)$ (dash-dotted), O(3) (dotted), and O(4) (dashed). The vertical gray lines indicate the corresponding oxygen 1s ionization threshold.

reasonable agreement for the two photon angles Θ. As discussed before, the experimental spectra can be divided roughly into three energy regions and interpreted using final state orbital analyses completely analogous to the discussion for the V′OV termination.

Figure 18.15 includes partial spectra calculated for 1s core excitations at differently coordinated oxygen, vanadyl $O_t(1)$, surface-type O(3), and bulk-type O(4), which yield major intensity in the energy range between 527.5 and 535 eV. In this range the peak structures referring to O(3) and O(4)-derived excitations are similar to those of the corresponding partial spectra for the V′OV terminated $V_2O_3(0001)$ surface without vanadyl groups. This suggests that the presence of the vanadyl groups at the $O_tV'OV$ terminated surface influences the local electronic structure near the surface-type O(3) and the bulk-type O(4) only little. The partial spectra due to core excitations at the singly coordinated vanadyl oxygen $O_t(1)$ exhibit one major peak with a small low-energy satellite in the energy range between 529 and 533 eV, where the intensity of the main peak depends very little on the photon incidence angle Θ while the small satellite increases in intensity with decreasing Θ. The latter can be explained in analogy to the O(1)-derived low-energy peak observed for the $V_2O_5(010)$ surface, see Sec. 18.2.2, by the symmetry of the involved final state orbitals characterized as antibonding $O\,2p_{x,y}$–$V\,3d_{xz,yz}$ mixtures. In contrast, the dominant peak at 531 eV is found to originate mainly from two O 2p–V 3d final state orbitals very close in energy [97], where the corresponding dipole transition matrix elements $\underline{m}^{(k)}$, determining the peak intensity, are almost perpendicular to each other. As a consequence, the corresponding intensity contributions complement each other resulting in only a weak dependence on the photon incidence angle $\Theta$. This is consistent with detailed analyses of the experimental polarization-resolved NEXAFS spectra [97], which also a show a dominant peak near 531 eV depending only weakly on $\Theta$. Thus, it can be used to identify singly coordinated vanadyl groups at the $V_2O_3(0001)$ surface.

## 18.3 Molybdenum Oxide

### 18.3.1 Molybdenum Oxide Bulk Structure

Bulk molybdenum oxides as single crystals exhibit in their basic structure interesting similarities compared with vanadium oxides. There are, however, only two different single valence type molybdenum oxides, $MoO_2$, and $MoO_3$, shown in Figure 18.16. These crystals are characterized in their geometric structure quantitatively by lattice and lattice basis vectors, but also qualitatively by the occurrence of specific elementary $MoO_x$ building units. They may be described by octahedral $MoO_6$ and tetrahedral $MoO_4$ units which differ by their distortion (determined by neighboring Mo–O distances and O–Mo–O angles) as well as by their relative arrangement (i.e., links between corners, edges, or faces) in the crystal.

The **dioxide, $MoO_2$**, crystallizes in a monoclinic lattice that deviates only slightly from the tetragonal rutile structure and is characteristic for several early transition metal dioxides like $VO_2$. Its elementary cell contains four chemical $MoO_2$ units [99] and almost equal Mo–O bond lengths describe the octahedral Mo environments. The $MoO_6$ units form rows in which they are connected by edges and the octahedra of adjacent rows are linked by corners. Further, every second row is rotated by $90^\circ$ about the (011) direction, completely analogous to what has been discussed for $VO_2$ (see Figure 18.1(b)). Bulk $MoO_2$ contains only one type of oxygen species, O(3) linking three metal centers. However, at the (011) oriented surface there is both 2- and 3-fold coordinated oxygen, O(2), O(3) where O(2) bridges asymmetrically between two Mo centers at distances $d_{Mo\text{–}O} = 1.74, 2.25$ Å while O(3) is connected with three Mo centers by two equal ($d_{Mo\text{–}O} = 1.95$ Å) and one longer distance ($d_{Mo\text{–}O} = 2.33$ Å). In $MoO_2$, molybdenum appears in a formal oxidation state +4 where the valence band structure is determined by O 2sp and Mo 4d-type electron states exhibiting a highly interesting metal-insulator transition [100]. The **trioxide, $MoO_3$**, appears in an orthorhombic $\alpha$-phase [101] at low-pressure and a monoclinic $\beta$-phase at high pressure [102]. The orthorhombic lattice contains four elemental $MoO_3$ units (4 Mo + 12 O) in the unit cell. It has a layered structure with weakly coupling bilayers parallel to the (010) crystal plane. Each bilayer consists of two interleaved planes of corner-linked distorted $MoO_6$ octahedra (see Figure 18.16, one bilayer is marked by shaded atom balls) where octahedra from adjacent planes share edges. The octahedra are described by Mo–O distances $d_{Mo\text{–}O} = 1.67, 1.73, 2 \times 1.95, 2.25$, and 2.33 Å where the largest distance refers to

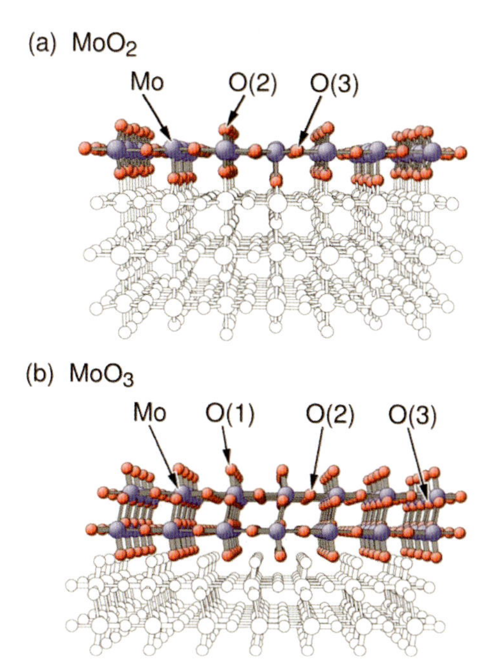


**Figure 18.16** Geometric structure of (a) monoclinic $MoO_2$ with netplane stacking along the (011) direction, (b) orthorhombic $MoO_3$ with netplane stacking along the (010) direction. Molybdenum and oxygen atoms are shown as large and small balls, respectively, where those of the topmost layers are emphasized by shading. Molybdenum and differently coordinated oxygen atoms, O(1), O(2), O(3), are labeled accordingly.

oxygen and molybdenum in different planes. In $MoO_3$, molybdenum assumes its highest formal oxidation state, +6, yielding a semiconductor with dominantly O 2sp-type valence and Mo 4d-type conduction bands [103].

Apart from the two single valence molybdenum oxides, there are many mixed valence oxides of molybdenum resulting from different growth conditions or due to stoichiometric changes during catalytic reactions connected with oxygen vacany formation, which affects also structural properties. As examples we mention only the **homologous series** $Mo_nO_{3n-1}$ [104] and $Mo_nO_{3n-2}$ [105], where the ratio $Mo^{6+}/Mo^{4+}$ equals $(n-1)/1$ and $(n-2)/2$, respectively. In the following, we will restrict our discussion of theoretical concepts to molybdenum trioxide, $MoO_3$, which has been studied extensively due to its importance for catalytic applications [12].

### 18.3.2
### Characterizing Oxygen at the $MoO_3$ (010) Surface

In the introduction of this chapter it has been mentioned that molybdenum oxides are used commercially as catalysts for the partial oxidation of hydrocarbons and alcohols. In these reactions surface oxygen of the substrate is incorporated into the

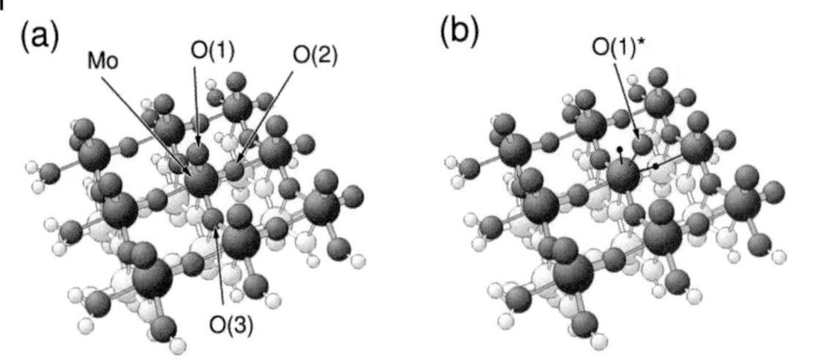


**Figure 18.17** Geometric structure of (a) the $Mo_{15}O_{56}H_{22}$ cluster representing a bi-layer section at the $MoO_3(010)$ surface, (b) Geometric structure of the $Mo_{15}O_{55}H_{22}$ cluster modeling (relaxed) oxygen vacancies O(1) and O(2). Molybdenum and oxygen atoms are shown as large and small balls, respectively, while saturator hydrogen at the cluster periphery is represented by very small balls. In addition, all atoms of the topmost layer are emphasized by shading. Molybdenum and differently coordinated oxygen atoms, O(1), O(2), O(3), O(1)* are labeled accordingly, see text.

reactant resulting in an oxidized species which desorbs from the surface leaving an oxygen vacancy behind. Since molybdenum oxide surfaces contain oxygen sites of different geometry and coordination, suggesting different chemical behavior [28], it is very important to identify oxygen sites that participate in a particular reaction step. This can help to understand details of catalytic reactions on a microscopic scale. The problem of identifying differently coordinated oxygen in molybdenum oxide is, from a methodological point of view, completely analogous to that for vanadium oxide discussed in Section 18.2.2 and theoretical/experimental methods discussed there can be applied equally well to molybdenum oxide.

So far, theoretical studies on molybdenum oxide surfaces have considered single crystal surfaces of (010) and (100) orientation only [28] where, in this section, we will focus on (010) surfaces. The orthorhombic lattice of bulk $MoO_3$ is layer type with weakly coupling physical bilayers stacked along the (010) direction, which is the preferred cleavage plane direction of the crystal. Molybdenum atoms as such are not exposed at the surface while there are three types of surface oxygen differing in their coordination as indicated in Figures 18.16(b) and 17(a). Oxygen appears as terminal (molybdenyl) oxygen, O(1), coordinated to one molybdenum center at a distance $d_{Mo\text{–}O} = 1.67$ Å, as bridging oxygen, O(2), coordinated asymmetrically to two Mo centers at distances $d_{Mo\text{–}O} = 1.74$, 2.25 Å, and as bridging oxygen, O(3), coordinated to three Mo centers with two equal ($d_{Mo\text{–}O} = 1.95$ Å) and one longer distance ($d_{Mo\text{–}O} = 2.33$ Å), see also Figure 18.17(a).

Theoretical cluster studies on the electronic structure of the $MoO_3(010)$ surface have been performed for different size clusters by *ab initio* DFT calculations [28, 106, 107]. As an illustration, Table 18.5 lists atom charges for the $Mo_{15}O_{56}H_{22}$ cluster shown in Figure 18.17(a) representing a local section of a $MoO_3(010)$ bilayer, where hydrogen is used to saturate dangling Mo–O bonds at the cluster periphery. The results are obtained from Mulliken analyses in DFT calculations [108] using the gradient corrected RPBE [65] functional. As in the case of vanadium oxide, the

**Table 18.5** Formal valence charges $Q^{val}$ and atom charges $Q^{Mull}$ from Mulliken populations for the $Mo_{15}O_{56}H_{22}$ cluster modeling local sections at the $MoO_3(010)$ surface [108].[a]

| Site | $Q^{val}$ | $Q^{Mull}$ |
|---|---|---|
| Mo | +6 | 2.20 |
| O(1) | −2 | −0.48 |
| O(2) | −2 | −0.73 |
| O(3) | −2 | −1.00 |

[a]The Mulliken charges refer to Mo, O sites near the cluster center. The differently coordinated oxygen species O(1), O(2), O(3) are indicated in Figure. 18.17(a). All values are given in atomic units.

metal species is positively and oxygen negatively charged in $MoO_3$. Further, the oxygen charges scale with the coordination where singly coordinated molybdenyl oxygen O(1) carries the smallest negative charge while 3-fold bridging oxygen O(3) is the most strongly negative. However, all calculated Mulliken charges differ substantially from corresponding formal valence charges also given in Table 18.5 which substantiates the various caveats attached to different atom charge definitions discussed in Section 18.2.2.

As mentioned earlier, atom charges can provide a theoretical method to qualitatively discriminate between the differently coordinated oxygen in the $MoO_3$ substrate. However, this discrimination cannot be verified by experiment since it is impossible to directly measure atom charges. Therefore, indirect experimental evidence has to be used when differently coordinated oxygen needs to be identified. NEXAFS spectroscopy discussed in Section 18.2.2 has proven to be quite successful for vanadium oxide. Therefore, it will be also applied to $MoO_3$ substrate in the following.

Figure 18.18(a) compares the experimental polarization-resolved NEXAFS spectra of the $MoO_3(010)$ surface with the corresponding theoretical total O 1s core level excitation spectra computed for the $Mo_{15}O_{56}H_{22}$ cluster shown in Figure 18.17(a) [106]. The experimental spectra refer to two different photon polarization directions $\underline{e}_{pol}$ corresponding to polar angles of incidence $\Theta = 0°$ (normal photon incidence) and 70° (near-grazing incidence) taken with respect to the surface normal. For normal photon incidence the agreement between theory and experiment is very good in the energy region up to ionization threshold at 536 eV, which exhibits the most prominent spectral features. Above the ionization threshold, theoretical and experimental data seem to refer to different broadening but the spectra remain in agreement. For near-grazing incidence, $\Theta = 70°$, the agreement between theoretical and experimental peak intensities is less satisfactory which may be explained by corrections required in the determination of the experimental polar angle $\Theta$. Fresnel diffraction effects at the surface affect the

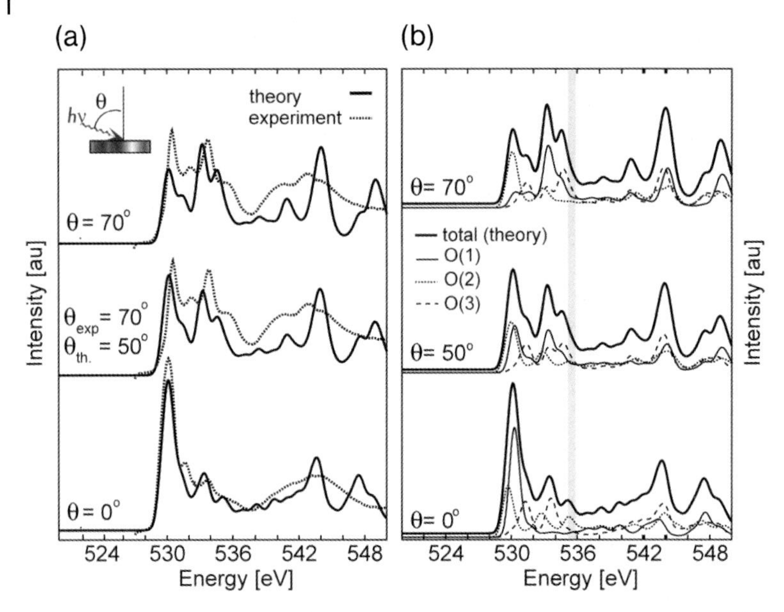


**Figure 18.18** (a) Comparison of experimental polarization-resolved NEXAFS spectra of the $MoO_3(010)$ surface (dotted lines) [106] with the corresponding theoretical total O 1s core excitation spectra (solid lines). The spectra refer to different angles of incidence $\Theta$ of the photon beam with respect to the surface normal as indicated in the inset. (b) Decomposition of the theoretical NEXAFS spectra into contributions from the three differently coordinated oxygen sites, O(1–3) for photon incidence angles $\Theta = 0^\circ$, $50^\circ$, and $70^\circ$. The vertical gray line indicates the position of the computed ionization threshold.

incoming photon beam such that the effective beam angle $\Theta$ deviates from that determined by macroscopic beam adjustment and given in Figure 18.18(a). This is quite suggestive from a comparison of the experimental spectrum for $\Theta = 70^\circ$ with the theoretical result for $\Theta = 50^\circ$ in Figure 18.18(a), which yields rather good agreement. A theoretical analysis of the final state orbitals participating in the core excitations shows that the broad peak in the continuum region near 544 eV is due to transitions from O 1s core to O 3p orbital resonances. In addition, the multipeak structure found in the energy range between 528 and 536 eV is characterized by excitations from O 1s core to final state orbitals, which are antibonding mixtures of O 2p and Mo 4d contributions.

Figure 18.18(b) shows decompositions of the theoretical polarization-resolved NEXAFS spectra into individual contributions from the three non-equivalent oxygen sites, O(1), O(2), O(3), for photon incidence angles $\Theta = 0^\circ, 50^\circ, 70^\circ$ [106]. The corresponding partial spectra depend on $\Theta$ in a very similar way as has been found for the $V_2O_5(010)$ surface discussed in Section 18.2.2, where the underlying mechanisms are identical. The theoretical spectra for core excitation at O(1) and O(2) are, in the energy region between 530 and 533 eV, dominated by two peaks each originating from O 1s core excitations to antibonding O 2p–Mo 4d orbitals. In both cases the peak at lower (higher) energy refers to final state orbitals where the O 2p admixture points perpendicular (parallel) to the corresponding O–Mo bond. Obviously, the different O 2p orientation affecting the antibonding character of these final state orbitals explains the energetic peak sequence. In addition, the symmetry of the O 2p admixture can describe the angle dependence of the corresponding peaks completely analogous to what has been discussed in detail for the oxygen-derived core excitation spectra of the $V_2O_5(010)$ surface, see Section 18.2.2. For core excitations at molybdenyl oxygen O(1), see Figure 18.17(a), the low-energy peak refers to final state orbitals where the corresponding dipole matrix elements point roughly parallel to the surface. Thus, the NEXAFS peak is the largest at normal incidence of the photon beam and becomes the smallest at grazing incidence. In contrast, the final state orbitals causing the high-energy peak result in dipole matrix elements pointing nearly perpendicular to the surface such that this NEXAFS peak is the smallest at normal incidence but the largest at grazing incidence. Core excitations at the asymmetric bridging site O(2), see Figure 18.17(a), follow the same scheme as described for O(1) except that the O–Mo bond at the excitation site points approximately parallel to the surface. As a result, dipole matrix elements pointing both parallel and perpendicular to the surface determine the intensities of the O(2)-derived NEXAFS peaks, which can explain the weaker angle dependence of these peaks [106]. Finally, core excitations at the symmetric bridging site O(3), see Figure 18.17(a), are characterized completely analogous to those at O(1) and O(2) resulting, however, in a somewhat more complex angle dependence of the corresponding NEXAFS peaks due to the more complicated coordinative binding at the O(3) site. Altogether, the theoretical spectra for core excitation at the three differently coordinated oxygen sites show pronounced differences in their dependence on photon incidence angle in the energy region between 530 and 533 eV. This can be made use of in the analysis of corresponding experimental polarization-resolved NEXAFS spectra allowing a discrimination of different oxygen species by experiments analogous to the findings for the $V_2O_5(010)$ surface.

### 18.3.3 Oxygen Vacancies at the $MoO_3(010)$ Surface

As mentioned earlier, the efficiency of oxide surfaces for catalytic reactions depends strongly on their ability to provide surface oxygen as a reactant. Therefore, studies of physical and chemical parameters of surface oxygen including oxygen vacancy formation are essential for a detailed understanding of corresponding reaction steps. In this section we will discuss only oxygen vacancies at the $MoO_3(010)$ surface, which have been examined in different theoretical studies [28, 106–108].

The three structurally different oxygen species in bulk $MoO_3$ yield three different oxygen sites, O(1–3), at the $MoO_3(010)$ surface as shown in Figure 18.17(a). As a consequence, the surface allows for three types of surface oxygen vacancies. These can be studied theoretically by evaluating the electronic structure of appropriate vacancy clusters where vacancy states of both frozen and relaxed geometry can be considered, cp. Section 18.2.3. Corresponding vacancy energies have been

**Table 18.6** Oxygen vacancy energies, $E^r_{vac}$, $E^f_{vac}$, for the O(1–3) sites at the $MoO_3$ (010) surface with and without vacancy-induced relaxation [108].[a]

| Vacancy site | $E^r_{vac}$ | $E^f_{vac}$ | $\Delta Q^r$(Mo) | $\Delta Q^f$(Mo) |
|---|---|---|---|---|
| O(1) | 5.36 | 7.33 | −0.32 | −0.26 |
| O(2) | 5.36 | 6.74 | −0.32 | −0.78 |
| O(3) | 6.45 | 6.50 | −0.40 | −0.48 |

[a]The energies are obtained for model clusters $Mo_{15}O_{55}H_{22}/Mo_{15}O_{56}H_{22}$. The table includes vacancy-induced atom charge differences of the molybdenum center(s) nearest to the oxygen vacancy, $\Delta Q^r$(Mo) (relaxed geometry), $\Delta Q^f$(Mo) (frozen geometry), from Mulliken analyses, see text. All energies are given in eV and charges in atomic units per atom.

computed for the three different oxygen sites in the cluster $Mo_{15}O_{56}H_{22}$ shown in Figure 18.17(a). The results, referring to DFT calculations [82, 108] using the ground state of atomic oxygen as a reference, are listed in Table 18.6 and yield very similar results compared with the data obtained for $V_2O_5$ shown in Table 18.3. All oxygen vacancy energies for $MoO_3$(010) are found to be quite large, 5.4–6.5 eV ($E^r_{vac}$ values, relaxed geometry) and 6.5–7.3 eV ($E^f_{vac}$ values, frozen geometry), where amongst relaxed vacancy energies the value for O(1) (and O(2), see below) is the smallest. Thus, oxygen is bound very strongly to its $MoO_3$ substrate environment and may not be removed in one single step during an oxidation reaction.

There is one interesting difference between molybdenum and vanadium oxide concerning the geometric relaxation as a result of oxygen vacancy formation. In $V_2O_5$ the 2-fold bridging oxygen O(2) is located at the lateral midpoint between its two vanadium neighbors with equal V–O distances of $d_{V\text{–}O} = 1.78$ Å. As a result, geometric relaxation connected with O(2) vacancy formation is distributed evenly between the two vanadium environments as discussed in Section 18.2.3 (see Figure 18.9). In contrast, $MoO_3$ contains 2-fold bridging oxygen O(2) which is placed asymmetrically between its two molybdenum neighbors with quite different Mo–O distances, $d_{Mo\text{–}O} = 1.73$ Å and $= 2.25$ Å. In fact, the O(2) species at the $MoO_3$(010) surface may be viewed almost as singly coordinated oxygen where, however, the corresponding Mo–O bond points parallel to the surface rather than normal as for molybdenyl O(1). As a result, local geometric relaxation for O(1) and O(2) vacancy formation yields the same final vacancy geometry shown in Figure 18.17(b). In this geometry a singly coordinated oxygen, O(1)*, originating from the site complementary to where the oxygen was removed, O(2) or O(1), stabilizes in the intermediate region between the two sites forming a bond which is comparable in strength to the initial O(1)–Mo bond (with the same bond length, $d_{Mo\text{–}O} = 1.67$ Å) at an angle of about 45° with respect to the (010) surface (see Figure 18.17(b)). (For a discussion of the existence of a strongly tilted O(1)* species see also [109].) This explains the identical data for relaxed oxygen vacancy energies $E^r_{vac}$ and charge differences $\Delta Q^r$(Mo) given in Table 18.6 for the energetically most favorable O(1) and O(2) vacancy formation.

Singly coordinated oxygen, O(1)* near the O(1) and O(2) vacancies may be identified in polarization-resolved NEXAFS spectra of the oxygen deficient $MoO_3$(010) surface. As a result of the similarity in the local binding environment, core excitations originating at the O(1)* atom of the oxygen deficient surface are found to involve very similar final state orbitals compared with excitations at O(1) atoms, namely antibonding O 2p–Mo 4d mixtures of different symmetry. These excitations yield a two-peak structure in the energy region between 529 and 534 eV of the theoretical core excitation spectrum in close analogy with the O(1)-derived spectra of Figure 18.18(b) discussed in Section 18.3.2. However, the tilted orientation of the O(1)*–Mo bond compared with O(1)–Mo and O(2)–Mo leads to differently oriented dipole transition matrix element vectors $\underline{m}^{(k)}$ and, hence, according to (18.19) to a different dependence of the excitation spectrum on the photon polarization direction. This is illustrated in Figure 18.19 comparing theoretical polarization-resolved spectra for O(1) and O(2)-derived core excitations at the clean $MoO_3$(010) surface, evaluated for a $Mo_{15}O_{56}H_{22}$ cluster, with spectra for O(1)*-derived core excitations

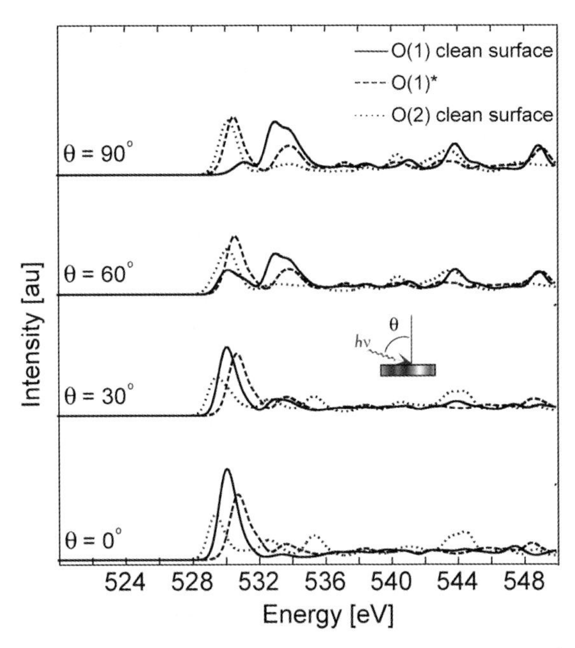


**Figure 18.19** Comparison of theoretical polarization-resolved O 1s NEXAFS spectra at O(1) and O(2) sites of the clean $MoO_3$(010) surface with those of the O(1)* site near O(1) and O(2) vacancies at the defective surface, see Figure 18.17(b) and text. The results are obtained from calculations using embedded clusters $Mo_{15}O_{56}H_{22}$ and $Mo_{15}O_{55}H_{22}$, respectively [106].

near a relaxed O(1)/O(2) vacancy at the surface, modeled by a $Mo_{15}O_{55}H_{22}$ cluster [106]. Obviously, the dependence of the O(1)*-derived spectra on the photon incidence angle Θ, shown in Figure 18.19 for Θ between 0° (normal photon incidence) and 90° (grazing incidence), is rather different and less pronounced compared with those for O(1), O(2). Therefore, O(1)* contributions should be observable in polarization-resolved NEXAFS spectra and could provide a spectroscopic fingerprint for O(1) and O(2) vacancies at the oxygen deficient $MoO_3$(010) surface. So far, NEXAFS experiments along these lines, which could also test the existence of a strongly tilted O(1)* species [109] have not been performed.

## References


1 C.N.R. Rao and B. Raven, *Transition Metal Oxides*, Wiley-VCH, New York, Weinheim, Cambridge, **1995**.

2 H.K. Kung, in *"Transition Metal Oxides: Surface Chemistry and Catalysis"*, B. Delmon, J.T. Yates (Eds.), Studies in Surface Science and Catalysis, Vol. 45, Elsevier, Amsterdam, **1989**.

3 V.E. Henrich and P.A. Cox, *The Surface Science of Metal Oxides*, University Press, Cambridge, **1994**.

4 B. Grzybowska-Swierkosz, F. Trifiro, J.C. Vedrine (Eds.), *J. Appl. Catal.* **157** (**1997**) 1; and references therein.

5 E.E. Chain, *Appl. Opt.* **30** (**1991**) 2782.

6 L.A. Gea, L.A. Boatner, J.D. Budai, and R.A. Zuhr, in *"Ion–Solid Interactions for Materials Modification and Processing Materials"*, D.B. de Poker, D. Ila, Y.T. Cheng, L.R. Harriott, T.W. Sigmon (Eds.), Research Society, Boston, **1995**, p. 215.

7 D. Yin, N. Xu, J. Zhang, and X. Zheng, *J. Phys. D* **29** (**1996**) 1051.

8 V.L. Galperin, I.A. Khakhaev, F.A. Chudnovskii, and E.B. Shadrin, *Sov. Tech. Phys. Lett.* **18** (**1992**) 329.

9 C.E. Lee, R.A. Atkins, W.N. Gibler, and H.F. Taylor, *Appl. Opt.* **28** (**1991**) 4511.

10 G.C. Granqvist, *Solid State Ionics* **70** (**1994**) 678.

11 A.K. Pandit, M. Prasad, T.H. Ansari, R.A. Singh, and B.M. Wanklyn, *Solid State Comm.* **80** (**1991**) 125.

12 E.R. Braithwaite, J. Haber (Eds.), *"Molybdenum: An Outline of its Chemistry and Uses"*, Studies in Inorganic Chemistry, Vol. 19, Elsevier, Amsterdam, **1994**.

13 R. Pearce, W.R. Patterson (Eds.), *Catalysis and Chemical Processes*, Wiley-Halsted, New York, **1981**.

14 A. Bielanski and J. Haber, *Oxygen in Catalysis*, Marcel-Dekker, New York, **1991**.

15 J. Haber in *"New Developments in Selective Oxidation by Heterogeneous Catalysis"*, P. Ruiz, B. Delmon (Eds.), Studies in Surface Science and Catalysis, Vol. 72, Elsevier, Amsterdam, **1992**.

16 see e. g. G. Ertl, H. Knözinger, and J. Weitkamp, *Handbook of Heterogeneous Catalysis*, VCH/Wiley Publishing, New York, **1997**.

17 B. Grzybowska-Swierkosz, J. Haber (Eds.), *Vanadia Catalysts for Processes of Oxidation of Aromatic Hydrocarbons*, PWN Publishers, Warsaw, **1984**.

18 J. Haber and E. Lalik, *Catal. Today* **33** (**1997**) 119.

19 K. Bruckman, R. Grabowski, J. Haber, A. Mazurkiewicz, J. Sloczynski, and T. Wiltowski, *J. Catal.* **104** (**1987**) 71.

20 R.K. Grasseli and J.D. Burrington, *Adv. Catal.* **30** (**1981**) 133.

21 J.C. Volta and J.L. Portefaix, *Appl. Catal.* **18** (**1985**) 1.

22 M. Abon, J. Massardier, B. Mingot, J.C. Volta, N. Floquet, and O. Bertrand, *J. Catal.* **134** (**1992**) 542.

23 M.A. Banares, J.L.G. Fierro, and J.B. Moffat, *J. Catal.* **142** (**1993**) 406.

24 A. Parmaliana and F. Arena, *J. Catal.* **167** (**1997**) 75.

25 A. Bielanski and M. Najbar, *Appl. Catal.* **157** (**1997**) 223.

26 A. Baiker and D. Gasser, *Z. Phys. Chem.* **149** (**1986**) 119.

27 see e. g. *"The Chemical Physics of Solid Surfaces"*, Vol. 9, Oxide Surfaces, D.P. Woodruff (Ed.), Elsevier, Amsterdam, **2001**.

28 K. Hermann and M. Witko in *"The Chemical Physics of Solid Surfaces"*, Vol. 9, Oxide Surfaces, D.P. Woodruff (Ed.), Elsevier, Amsterdam, **2001**, p. 136.

29 J.K. Labanowski, J.W. Andzelm (Eds.), *Density Functional Methods in Chemistry*, Springer, New York, **1991**.

30 see e. g. W. Koch and M.C. Holthausen, *A Chemist's Guide to Density Functional Theory*, Wiley-VCH, Weinheim, **2000**.

31 J.R. Chelikowski. M. Schlüter, S.G. Louie, and M.L. Cohen, *Solid State Comm.* **17** (**1975**) 1103.

32 P. Blaha, K.H. Schwarz, P. Sorantin, and S.B. Trickey, *Comp. Phys. Comm.* **59** (**1990**) 399.

33 M. Petersen, F. Wagner, L. Hufnagel, M. Scheffler, P. Blaha, and K.H. Schwarz, *Comp. Phys. Comm.* **126** (**2000**) 294.

34 G. Kresse and J. Furthmüller, *Phys. Rev. B* **54** (**1996**) 11169.

35 G. Kresse and D. Joubert, *Phys. Rev. B* **59** (**1999**) 1758.

36 R.E. Lohman, C.N.R. Rao, and J.M. Honig, *J. Phys. Chem.* **73** (**1969**) 1781.

37 A. Neckel, P. Rastl, R. Eibler, P. Weinberger, and K.H. Schwarz, *J. Phys.* **9** (**1976**) 579.

38 P.D. Dernier and M. Marezio, *Phys. Rev. B* **2** (**1970**) 3771.

39 M.G. Vincent and K. Yvon, *Acta Crystal. A* **36** (**1980**) 808.

40 M. Catty and G. Sandrone, *Faraday Discuss.* **106** (**1997**) 189.

41 Y. Oka, T. Yao, and N. Yamamoto, *J. Solid State Chem.* **86** (**1990**) 116.

42 T. Yao, Y. Oka, and N. Yamamoto, *J. Solid State Chem.* **112** (**1994**) 196.

43 Y. Oka, T. Yao, N. Yamamoto, Y. Neday, and A. Hayashi, *J. Solid State Chem.* **105** (**1993**) 271.

44 R.M. Wentzcovitch, W.W. Schulz, and P.B. Allen, *Phys. Rev. Lett.* **72** (**1994**) 3389.

45 K. Hermann, A. Chakrabarti, A. Haras, M. Witko, and B. Tepper, *phys. stat. solidi (a)* **187** (**2001**) 137.

46 R. Enjalbert and J. Galy, *Acta Cryst. C* **42** (**1986**) 1467.

47 R. Ramirez, B. Casal, L. Utrera, and E. Ruiz-Hitzky, *J. Phys. Chem.* **94** (**1990**) 8960.

48 R.W.G. Wyckoff, *Crystal Structures*, Wiley, New York, **1965**.

49 N. Van Hieu and D. Lichtman, *J. Vac. Sci. Technol.* **18** (**1981**) 49.

50 A. Chakrabarti, K. Hermann, R. Druzinic, M. Witko, F. Wagner, and M. Petersen, *Phys. Rev. B* **59** (**1999**) 10583.

51 V. Eyert and K.H. Höck, *Phys. Rev. B* **57** (**1998**) 12727.

52 A.M. Chippindale and A.K. Cheetham in Ref. [12].

53 S. Anderson, B. Collen, U. Kuylenstierna, and A. Magneli, *Acta Chem. Scand.* **11** (**1957**) 1641.

54 J.H. Choi, S.K. Kim, and Y.C. Bak, *Korean J. Chem. Eng.* **18** (**2001**) 719.

55 P. Forzatti, *Appl. Catal. A: General* **222** (**2001**) 221.

56 E.A. Mamedov and V.C. Corberan, *Appl. Catal. A: General* **127** (**1995**) 1.

57 B. Tepper, B. Richter, A.C. Dupuis, H. Kuhlenbeck, C. Hucho, P. Schilbe, M.A. bin Yarmo, and H.-J. Freund, *Surf. Sci.* **496** (**2002**) 64.

58 V.A. Ranea, J.L. Vicente, E.E. Mola, P. Arnal, H. Thomas, and L. Gambaro, *Surf. Sci.* **463** (**2000**) 115.

59 K. Hermann, M. Witko, R. Druzinic, A. Chakrabarti, B. Tepper, M. Elsner, A. Gorschlüter, H. Kuhlenbeck, and H.-J. Freund, *J. Electron Spectrosc. Relat. Phenom.* **98/99** (**1999**) 245.

60 R. McWeeny, *Acta Cryst.* **6** (**1953**) 631.

61 S.F. Boys, *Proc. Roy. Soc. A* **200** (**1950**) 542.

62 R.S. Mulliken, *J. Chem. Phys.* **23** (**1955**) 1833, 1841, 2388, 2343.

63 P.O. Löwdin, *Adv. Phys.* **5** (**1956**) 111.

64 K. Hermann, unpublished.

65 B. Hammer, L.B. Hansen, and J.K. Nørskov, *Phys. Rev. B* **59** (**1999**) 7413.

**66** R. Bader, *Atoms in Molecules: A Quantum Theory*, Oxford University Press, London, **1994**.

**67** K. Siegbahn, *Rev. Mod. Phys.* **54** (**1982**) 709; and references therein.

**68** A. Szabo and N.S. Ostlund, *Modern Quantum Chemistry: Introduction to Advanced Electronic Structure Theory*, Macmillan, New York, **1982**.

**69** J.C. Slater, *Adv. Quant. Chem.* **6** (**1972**) 1.

**70** J.C. Slater and K.H. Johnson, *Phys. Rev. B* **5** (**1972**) 844.

**71** J.F. Janak, *Phys. Rev. B* **18** (**1978**) 7165.

**72** C. Kolczewski and K. Hermann, *Surf. Sci.* **552** (**2004**) 98.

**73** Z.M. Zhang and V.E. Henrich, *Surf. Sci.* **321** (**1994**) 133.

**74** L. Triguero, L.G.M. Pettersson, and H. Ågren, *Phys. Rev. B* **58** (**1998**) 8097.

**75** H. Ågren, V. Carravetta, O. Vahtras, and L.G.M. Pettersson, *Theor. Chem. Acc.* **97** (**1997**) 14.

**76** C. Kolczewski and K. Hermann, *J. Chem. Phys.* **118** (**2003**) 7599.

**77** B. Richter and H. Kuhlenbeck, private communication.

**78** C. Kolczewski, R. Püttner, O. Plashkevych, H. Ågren, V. Staemmler, M. Martins, G. Snell, A.S. Schlachter, M. Sant'Anna, G. Kaindl, and L.G.M. Pettersson, *J. Chem. Phys.* **115** (**2001**) 6426.

**79** G. Paccioni in *"The Chemical Physics of Solid Surfaces"*, Vol. 9, Oxide Surfaces, D.P. Woodruff (Ed.), Elsevier, Amsterdam, **2001**, p. 94

**80** R. Druzinic, Ph.D. Thesis, Free University, Berlin (**2000**).

**81** K. Hermann, M. Witko, and R. Druzinic, *Faraday Discuss.* **114** (**1999**) 53.

**82** K. Hermann, M. Witko, R. Druzinic, and R. Tokarz, *Appl. Phys. A* **72** (**2001**) 429.

**83** C. Friedrich, Ph.D. Thesis, Free University, Berlin (**2004**).

**84** M. Gruber, to be published.

**85** S. Surnev, M.G. Ramsey, and F.P. Netzer, *Prog. Surf. Sci.* **73** (**2003**) 117.

**86** N. Magg, J.B. Giorgi, Th. Schroeder, M. Bäumer, and H.-J. Freund, *J. Phys.Chem. B* **106** (**2002**) 8756.

**87** H. Niehus, R.P. Blum, and D. Ahlbehrendt, *Surf. Rev. Lett.* **10** (**2003**) 353.

**88** Y. Romanyshyn, M. Naschitzki, H. Kuhlenbeck, and H.-J. Freund, in preparation.

**89** I. Czekaj, K. Hermann, and M. Witko, *Surf. Sci.* **525** (**2003**) 33.

**90** I. Czekaj, K. Hermann, and M. Witko, *Surf. Sci.* **545** (**2003**) 85.

**91** G. Kresse, S. Surnev, J. Schoiswohl, and F.P. Netzer, *Surf. Sci.* **555** (**2004**) 118.

**92** F. Rohr, M. Bäumer, H.-J. Freund, J.A. Mejias, V. Staemmler, S. Müller, L. Hammer, and K. Heinz, *Surf. Sci.* **372** (**1997**) L291; *Surf. Sci.* **389** (**1997**) 391.

**93** C. Rebhein, N.M. Harrison, and A. Wander, *Phys. Rev. B* **54** (**1996**) 14066.

**94** M. Inomata, A. Miyamoto, and Y. Murakami, *J. Catal.* **62** (**1980**) 140.

**95** S. Surnev, G. Kresse, M. Sock, M.G. Ramsey, and F.P. Netzer, *Surf. Sci.* **495** (**2001**) 91.

**96** C. Kolczewski and K. Hermann, *Theor. Chem. Acc. (Theor. Chim. Acta)* **114** (**2005**) 60.

**97** C. Kolczewski, K. Hermann, S. Guimond, H. Kuhlenbeck, and H.-J. Freund, *Surf. Sci.* **601** (**2007**) 5394.

**98** A.C. Dupuis, M. Abu Haija, B. Richter, H. Kuhlenbeck, and H.-J. Freund, *Surf. Sci.* **539** (**2003**) 99.

**99** B.G. Brant and A.C. Skapski, *Acta Chem. Scand.* **21** (**1967**) 661.

**100** V. Eyert, R. Horny, K.H. Hock, and S. Horn, *J. Phys., Condens. Matter* **12** (**2000**) 4923.

**101** L. Kihlborg, *Ark. Kemi* **21** (**1963**) 357, 443, 461; *Acta Chem. Scand.* **17** (**1963**) 1485.

**102** J.B. Parrise, E.M. McCarron III, R. Von Dreele, and J.A. Goldstone, *J. Solid State Chem.* **93** (**1991**) 193.

**103** A. Papakondylis and P. Sautet, *J. Phys. Chem.* **100** (**1996**) 10681.

**104** M. Sato, M. Onoda, and Y. Matsuda, *J. Phys. C* **20** (**1987**) 4763.

**105** Z. Zhu, S. Chowdhary, V.C. Long, J.L. Musfeldt, H.J. Koo, M.H. Whangbo, X. Wei, H. Negishi, M. Inoue, J. Sarro, and Z. Fisk, *Phys. Rev. B* **61** (**2000**) 10057.

**106** M. Cavalleri, K. Hermann, S. Guimond, Y. Romanyshyn, H. Kuhlenbeck, and H.-J. Freund, *Catal. Today* **124** (**2007**) 21.

**107** R. Tokarz-Sobieraj, K. Hermann, M. Witko, A. Blume, G. Mestl, and R. Schlögl, *Surf. Sci.* **489** (**2001**) 107.

**108** X. Shi and K. Hermann, to be published.

**109** R. Coquet and D.J. Willock, *Phys. Chem. Chem. Phys.* **7** (**2005**) 3819.

# 19
# Modeling Catalytic Reactivity in Heterogeneous Catalysis

*Rutger A. van Santen*

## 19.1
## General Concepts

A catalytic reaction is composed of several reaction steps. Molecules have to adsorb the catalyst, become activated, and product molecules have to desorb. The catalytic reaction is a reaction cycle of elementary reaction steps. The catalytic center is regenerated after reaction. This is the basis to the key molecular principle of catalysis: the Sabatier principle.

According to this principle, the rate of a catalytic reaction has a maximum when the rate of activation and rate of product desorption balance.

The time constant of a heterogeneous catalytic reaction is typically a second. This implies that the catalytic event is much slower than diffusion ($10^{-6}$ s) or elementary reaction steps ($10^{-4}$–$10^{-2}$ s). Activation energies of elementary reaction steps are typically in the order of 100 kJ mol$^{-1}$. The overall catalytic reaction cycle is slower than elementary reaction steps because usually several reaction steps compete and surfaces tend to be covered with an overlayer of reaction intermediates.

Clearly catalytic rate constants are much slower than vibrational and rotational processes that take care of energy transfer between the reacting molecules ($10^{-12}$ s). For this reason, transition reaction rate expressions can be used to compute the reaction rate constants of the elementary reaction steps.

Eyring's transition-state reaction rate expression is

$$k_{\mathrm{TST}} = \Gamma \frac{Q^{\#}}{Q_0} e^{-\frac{E_b - E_0}{kT}}, \tag{19.1a}$$

$$Q = \prod_i \frac{e^{-\frac{1}{2}\frac{h\nu_i}{kT}}}{1 - e^{-\frac{h\nu_i}{kT}}}, \tag{19.1b}$$

where $Q^{\#}$ is the partition function of transition state and $Q_0$ is the partition function of ground state, $k$ is Boltzmann's constant, and $h$ Planck's constant.

*Computational Methods in Catalysis and Materials Science.* Edited by Rutger A. van Santen and Philippe Sautet

ISBN: 978-3-527-32032-5

From Eq. (19.1), the following expressions can be deduced for the activation energy and preexponent of the Arrhenius form of the reaction rate in Eq. (19.2a)

$$r = \nu_{\text{eff}} \cdot e^{-\frac{E_{\text{act}}}{kT}} \tag{19.2a}$$

with

$$\nu_{\text{eff}} = \Gamma \frac{ekT}{h} e^{\frac{\Delta S^{\#}}{kT}} \tag{19.2b}$$

$$\Delta S^{\#} = k \ln \frac{Q^{\#}}{Q_0} \tag{19.2c}$$

$$E_{\text{act}} = (E_b - E_0) + \frac{1}{2} h \left( \sum_i \nu_i^{\#} - \sum_t \nu_{i1}^{0} \right) + kT. \tag{19.2d}$$

The transition-state energy is defined as the saddle point of the energy of the system when plotted as a function of the reaction coordinates as illustrated in Figure 19.1.

$\Gamma$ is the probability that reaction coordinate passes the transition-state barrier when the system is in the activated state. It is the product of a dynamical correction and the tunneling probability. While statistical mechanics can be used to evaluate the preexponent and activation energy, $\Gamma$ has to be evaluated by molecular dynamics

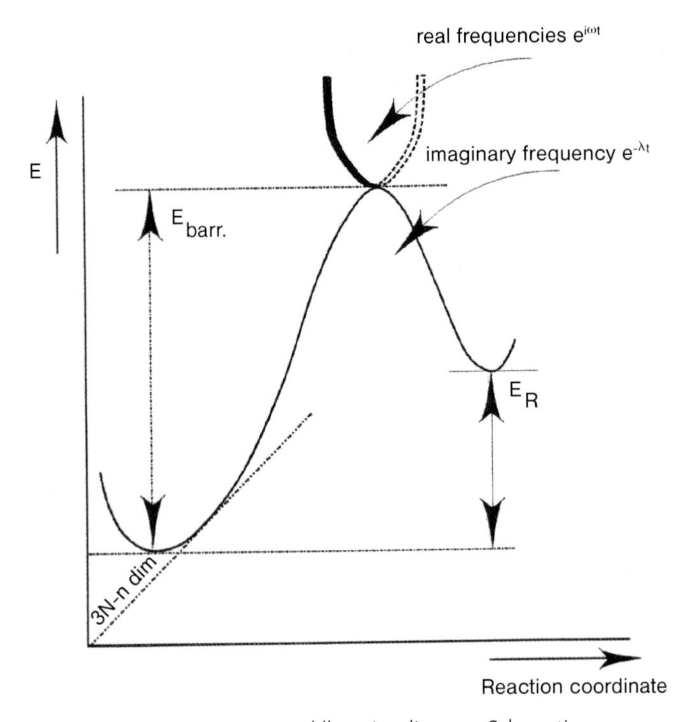


**Figure 19.1** Transition state saddle point diagram. Schematic representation of potential energy as a function of reaction coordinate.

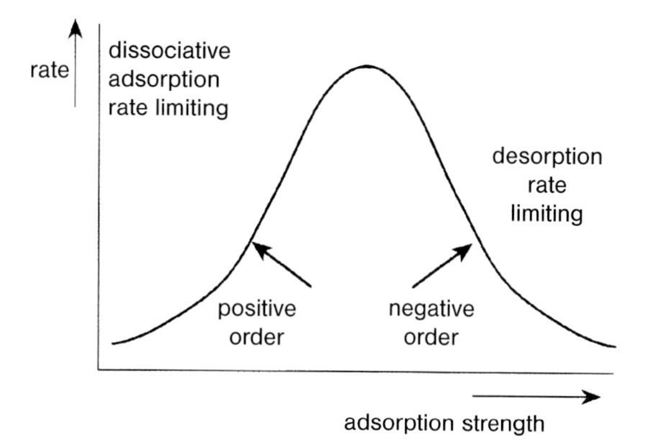


**Figure 19.2** Volcano plot illustrating Sabatier principle. Catalytic rate is maximum at optimum adsorption strength. At the left of the Sabatier maximum, rate has a positive order in reactant concentration, at the right of Sabatier maximum the rate has a negative order.

techniques because of the very short time scale of the system is in the activated state. For surface reactions not involving hydrogen $\Gamma$ is usually close to 1.

Most of the currently used computational chemistry programs provide energies and vibrational frequencies for ground as well as transition states.

A very useful analysis of catalytic reactions is provided for by the construction of so-called volcano plots (Figure 19.2). In a volcano plot, the catalytic rate of a reaction normalized per unit reactive surface area is plotted as a function of the adsorption energy of reactant, product molecule or reaction intermediates.

A volcano plot correlates a kinetic parameter as the activation energy with a thermodynamic parameter as the adsorption energy.

The maximum in the volcano plot corresponds to the Sabatier principle maximum where rate of activation of reactant molecules and desorption of product molecules balance.

## 19.2 Linear Activation Energy–Reaction Energy Relationships

The Sabatier principle deals with the relation of catalytic reaction rate and adsorption energies of surface reaction intermediates. A very useful relation is often also exists between the activation energy of elementary surface reaction steps, as adsorbate bond dissociation or adsorbed fragment recombination and corresponding reaction energies.

These give the Brønsted–Evans–Polanyi (BEP) relations.

For the forward dissociation reaction, the BEP relation is

$$\delta E^{\#}_{\text{diss}} = \alpha \delta E_{\text{react}}. \tag{19.3a}$$

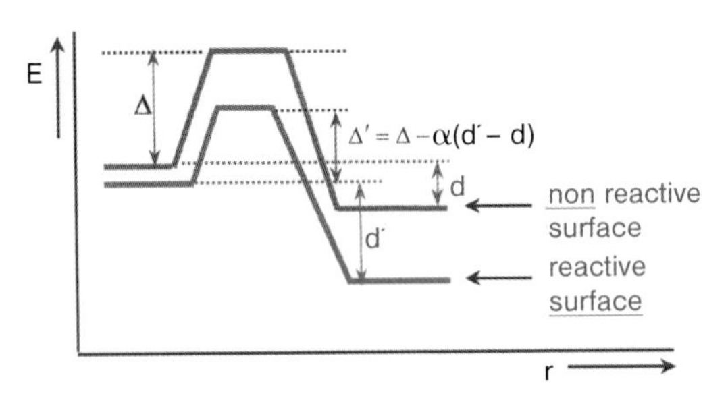


**Figure 19.3** Illustration of the BEP relation $\Delta' = \Delta - \alpha(d' - d)$.

For the backward recombination reaction, then Eq. (19.3b) has to hold:

$$\delta E^{\#}_{\text{rec}} = -(1 - \alpha)\delta E_{\text{react}}. \tag{19.3b}$$

Because of microscopic reversibility, the proportionality constants of the forward and backward reaction are related.

These relations are illustrated in Figure 19.3.

The original ideas of Evans and Polany [1] to explain such a linear relation between activation energy and reaction energy can be illustrated from a two-dimensional analysis of two crossing potential energy curves.

The two curves in Figure 19.4 represent the energy of a chemical bond that is activated before and after reaction. The difference between the location of the potential energy minima is the reaction coordinate $x_0$.

If one assumes the potential energy curves to have a similar parabolic dependence on the displacement of the atoms, a simple relation can be deduced between activation energy, the crossing point energy of the two curves, and the reaction energy.

$$E_{\text{act}} = E^{\circ}_{\text{act}}\left(\frac{\Delta E_r}{4E^{\circ}_{act}} - 1\right)^2 \tag{19.4a}$$

One then finds for $\alpha$

$$\alpha = \frac{\delta E_{\text{act}}}{\delta E_r} = \frac{1}{2}\left(1 + \frac{\Delta E_r}{4E^{\circ}_{\text{act}}}\right) \tag{19.4b}$$

One notes that the proportionality constant $\alpha$ depends on the reaction energy, $\Delta E_r$. Therefore, Eq. (19.4) is not strictly a linear relation between activation energy change and reaction energy. In the extreme limit of high exothermicity of the reaction energy $\alpha = 0$, and the crossing point of the two curves is at the minimum

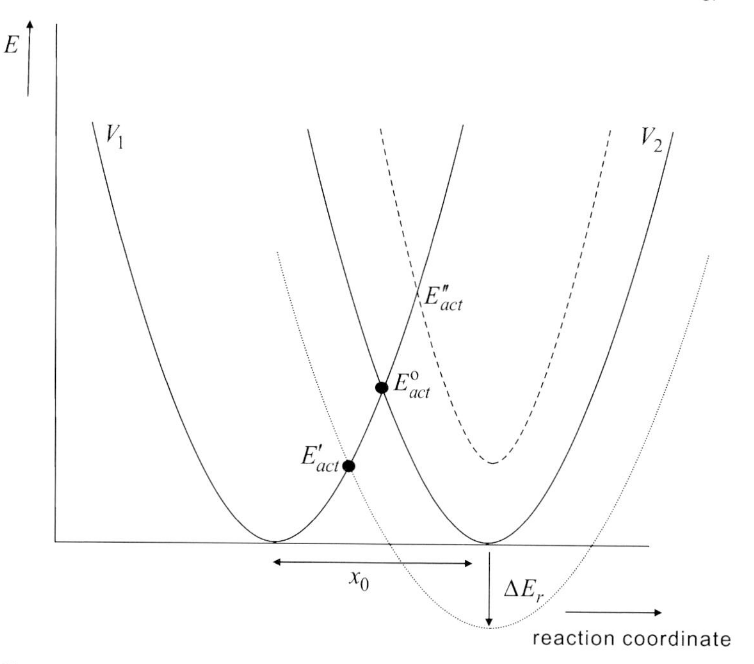


**Figure 19.4** Curve-crossing potential energy diagram of reacting system constant $k'$.

of curve $V_1$. In this case, the transition state is called early. Its structure is close to that of the reactant state.

In the limit of high endothermicity, $\alpha = 1$, and now the crossing point is close at the minimum of curve $V_2$. The transition-state structure is now close to that of the final state. The transition state can now be considered to be late.

This analyses is important, since it illustrates why $\alpha$ varies between 0 and 1. Often $\alpha$ is simply assumed to be equal 0.5.

The BEP relation is only expected to hold as long as one compares systems in which the reaction path of the reacting molecules is similar.

An illustration is provided by Figure 19.5 [2].

In this figure, the activation energies of $N_2$ dissociation are compared for different reaction centers: the (111) surface structure of a fcc crystal and a stepped surface.

Activation energies with respect to the energy of the gas-phase molecule are related to the adsorption energies of the $N$ atoms. As often found for bond-activating surface reactions, a value of $\alpha$ close to 1 is found. It implies that the electronic interactions between surface and reactant in the transition state and product state are similar. The bond strength of the chemical bond that is activated is substantially weakened due to the strong interaction with the metal surface. The structure of the transition state is close to that of the product state.

This is illustrated in Figure 19.6 for dissociation of CO [3].

As a consequence of the high value of $\alpha$, the proportionality constant of recombination is usually approximately 0.2, reflecting a weakening of the adatom surface bonds in transition state by this small amount. It implies that typically one

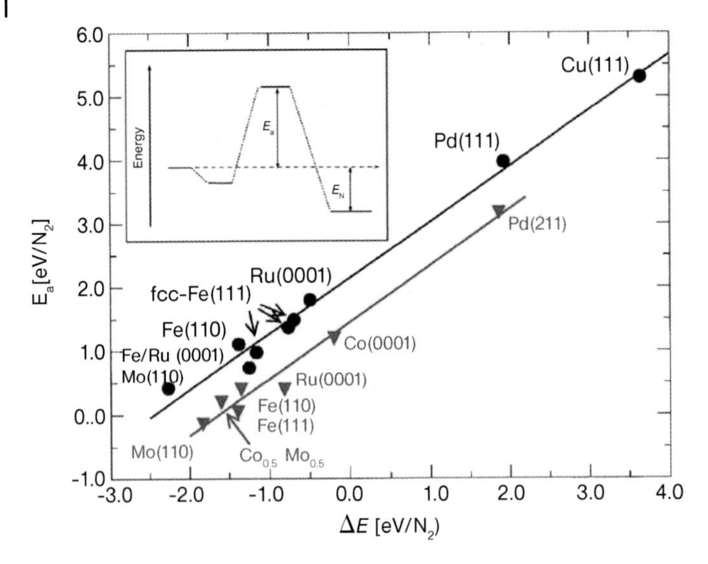


**Figure 19.5** Plot of computed reaction barriers for dissociation on $E_a$ for $N_2$ dissociation as a function of nitrogen atom adsorption energy on surface terrace and stepped surface [2]. The upper curve is for surface terrace of (111) type of fcc crystals, the lower curve presents data on the stepped surfaces.

on 6 surface bonds is broken in the transition state compared to the adsorption state of the two atoms before recombination.

## 19.3
## Micro-kinetic Expressions; Derivation of Volcano Curve

In microkinetics, overall rate expressions are deduced from the rates of elementary rate constants within a molecular mechanistic scheme of the reaction. We will use the methanation reaction as an example to illustrate the physical chemical basis to the Sabatier volcano curve. The corresponding elementary reactions are

$$\mathrm{CO_g} \underset{k_{\mathrm{diss}}}{\overset{k_{\mathrm{ads}}}{\rightleftarrows}} \mathrm{CO_{ads}} \tag{19.5a}$$

$$\mathrm{CO_{ads}} \xrightarrow{k_{\mathrm{diss}}} \mathrm{C_{ads}} + \mathrm{O_{ads}} \tag{19.5b}$$

$$\mathrm{C_{ads}} + 2\mathrm{H_2} \xrightarrow[r_H]{} \mathrm{CH_4} \nearrow . \tag{19.5c}$$

Surface carbon hydrogenation occurs through a sequence of hydrogenation steps in which $CH_{x,ads}$ species are formed with increasing hydrogenation. $r_H$, The rate of $C_{ad}$ hydrogenation depends implicitly on hydrogen pressure.

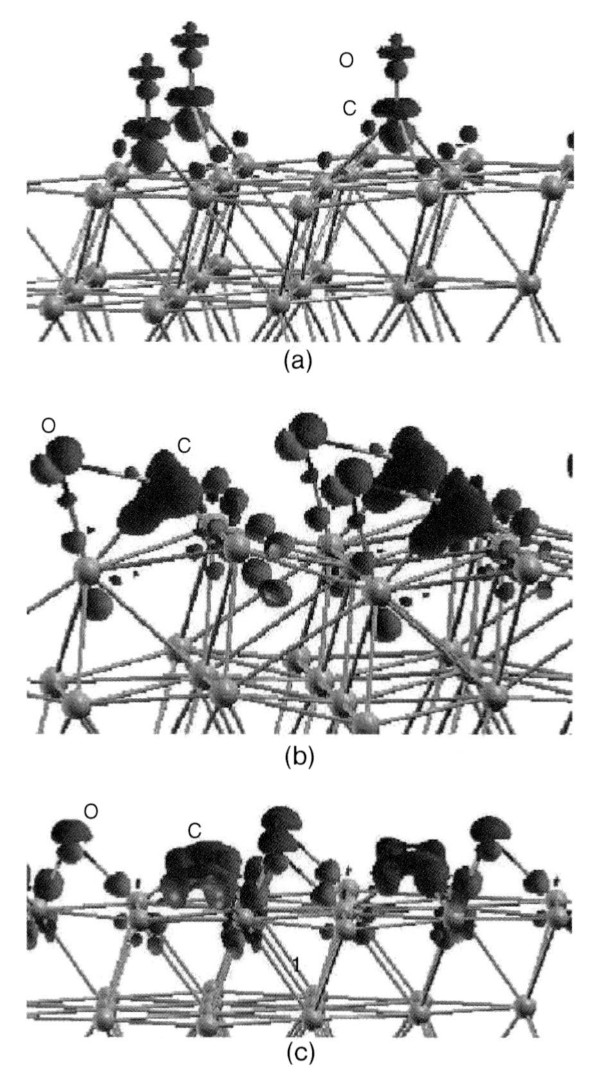


**Figure 19.6** Structures and electron density changes of dissociating CO on Ru(0001) surface: (a) adsorbed CO; (b) transition state for dissociation; and (c) dissociated state.

For Ru(0001), the corresponding reaction energy scheme is shown in Figure 19.7 [4].

The relative energies of the different reaction intermediates $C_{ads}$ or $CH_{ads}$ may strongly depend on the type of surface and metal. When for different surfaces or metals the relative interaction with $H_{ads}$ increases, $C_{ads}$ may for instance become more stable than CH. This is found for more coordinative unsaturated surfaces or more reactive metals.

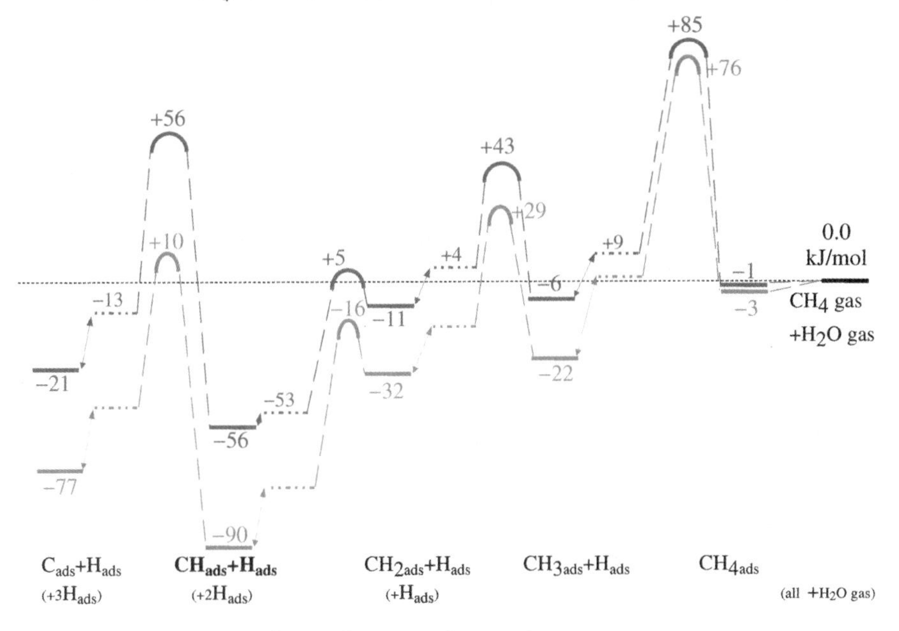


**Figure 19.7** Reaction scheme of reaction of $C_{ads}$ with $H_{ads}$ to produce $CH_4$ on Ru(0001) surface.

We will present expressions for reaction rates and steady-state concentrations using the simplifying assumption that $C_{ads}$ hydrogenation to $CH_4$ occurs in one reaction step.

We will also assume that $O_{ads}$ removal is fast and hydrogen adsorption is not influenced by the other adsorbates.

Then the activation energy for methane production from $C_{ads}$ is the overall activation energy for hydrogenation of $C_{ads}$ to $CH_4$, and Eq. (19.6) gives the rate of methane production:

$$R_{CH_4} = r_H \cdot \theta_c \tag{19.6}$$

with $\theta_C$. A closed expression for $\theta_C$ can be deduced:

$$\theta_c = 1 + \tfrac{1}{2}\lambda - \tfrac{1}{2}\sqrt{\lambda^2 + 4\lambda} \tag{19.7a}$$

$$\approx \frac{1}{1+\lambda} \tag{19.7b}$$

with

$$\lambda = \frac{r_H}{k_{diss}} \frac{\left(K_{ads}^{CO} \cdot [CO] + 1\right)^2}{\left(K_{ads}^{CO}[CO]\right)} \tag{19.8a}$$

$$= A\frac{r_H}{k_{diss}}. \tag{19.8b}$$

One notes that the coverage of $C_{ads}$ depends on two important parameters. The ratio $\rho$ of the rate of hydrogenation of $C_{ads}$ to give methane and the rate constant of CO dissociation:

$$\rho = \frac{r_H}{k_{diss}}, \tag{19.9}$$

and the equilibrium constant of CO adsorption, $K_{eq}^{CO}$. The coverage with $C_{ads}$ increases with decreasing value of $\rho$. This implies a high rate of $k_{diss}$ and slow rate of $C_{ads}$ hydrogenation. The strong pressure dependence of CO relates to the need of neighboring vacant sites for CO dissociation.

Beyond a particular value of $K_{eq}^{CO}$, the surface coverage with $C_{ads}$ will decrease because CO dissociation becomes inhibited.

In order to proceed, one needs to know the relation between the rate constants and reaction energies. This determines the functional behavior of $\rho$.

We will use the linear activation energy–reaction energy relationships as deduced from the BEP relation and write for expressions $k_{diss}$, $r_H$, and $\lambda$:

$$k_{diss} = \nu_0 e^{-\frac{E^0_{diss}}{kT}} \cdot e^{-\alpha \frac{E'_{ads}}{kT}} \tag{19.10a}$$

$$= \nu'_0 e^{-\alpha' \frac{E_{ads}}{kT}} \tag{19.10b}$$

$$r_H = r'_H e^{x \frac{E_{ads}}{kT}} \tag{19.10c}$$

$$\lambda = A \frac{r'_H}{\nu'_0} e^{(x+\alpha') \frac{E_{ads}}{kT}}. \tag{19.11}$$

The dissociation rate of $CO_{ads}$ will increase with increasing exothermicity of the reaction energy. As a measure, one can use the adsorption energy of the carbon atom.

From chemisorption theory, we know that adatom adsorption energies will decrease in a row of the group VIII metals when the position of the element moves to the right. The rate of hydrogenation of $C_{ads}$ will decrease with increasing adsorption energy of $C_{ads}$, and hence will decrease in the same order with element position in the periodic system.

We will now study the consequences of these BEP choices to the dependence of predicted rate of methane production on $E_{ads}$. Making the additional simplifying assumption that the adsorption energy parameters in Eqs. (19.10b) and (19.10c) are the same, one finds for the rate of methane production expression:

$$R_{CH_4} = C \frac{\lambda^{\frac{x}{x+\alpha'}}}{1+\lambda} \tag{19.12a}$$

with

$$C = r'_H \left( \frac{r'_H}{r'_0} A \right)^{-\frac{x}{x+\alpha'}}. \tag{19.12b}$$

In Eq. (19.12b), the constant $A$ depends on the equilibrium constant $K_{eq}^{CO}$. This will also vary with the adsorption energy of C or O, but will be much less

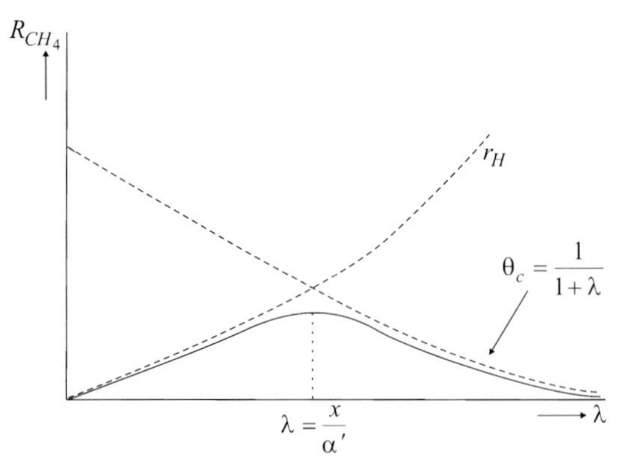


**Figure 19.8** Dependence of $R_{CH_4}$ on $\lambda$ (schematic).

sensitive to these variations than the activation energies of CO dissociation and hydrogenation [5].

The dependence of $R_{CH_4}$ on $\lambda$ is sketched in Figure 19.8.

Equation (19.12a) will only have a maximum as long as

$$\frac{x}{x+\alpha'} < 1. \tag{19.13}$$

Within our model, this condition is always satisfied. We find the interesting result that the Sabatier volcano maximum is found when $\lambda$ equals:

$$\lambda_{max} = \frac{x}{\alpha'}. \tag{19.14}$$

For $x = \alpha'$, the surface coverage $\theta_c$ equals $\frac{1}{2}$. At this coverage, the rates of dissociation and hydrogenation balance. Since the BEP parameter for hydrogenation is usually smaller than that for dissociation, the optimum surface coverage actually is expected to shift to one for the overall rate to be maximum.

The controlling parameters that determine the volcano curve are the BEP constants of $k_{diss}$ and $r_H$. It is exclusively determined by the value of $\rho$. It expresses the compromise of the opposing elementary rate events: dissociation versus product formation. The CO partial pressure determines the dependence of the rate on gas-phase pressure. It controls $\lambda$ through changes of parameter $A$. Note that the variation of $A$ would also give volcano-type behavior. This is obviously found when the rate is plotted as a function of gas-phase pressure. It relates to the blocking of surface sites by increasing adsorption of CO molecules. Volcano-type behavior is illustrated in Figure 19.9 for constant CO pressure.

Equation (19.8a) then allows deducing the optimum value of $E_{ads}$ of the Sabatier maximum rate. It can be deduced from Eq. (19.12a). The latter depends through $A$ on the CO pressure.

The volcano curve is bounded by the rate of dissociative adsorption of CO and hydrogenation of adsorbed carbon. This is illustrated in Figure 19.9.

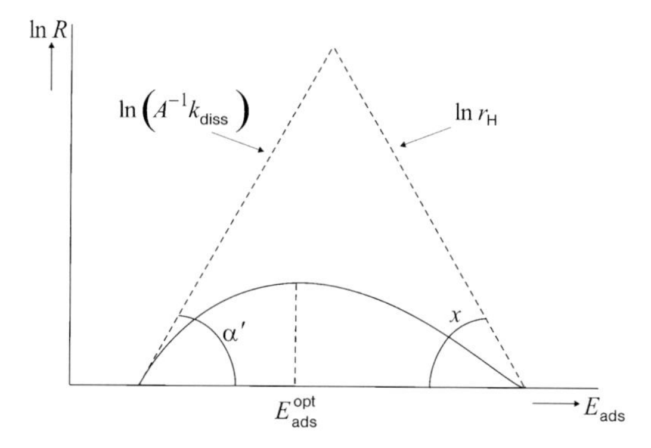


**Figure 19.9** Volcano curve dependence of rate of methanation R on $E_{ads}$ (schematic).

For relevant related treatments, see Ref. [6]. While the above discussion limits itself to the conversion of CO to a single product, the treatment can be easily extended to a selectivity problem.

Interestingly, one can easily deduce an expression of the relative rate of coke formation versus that of methanation. The rate of initial coke formation depends on the combination probability of carbon atoms and hence is given by

$$R_{C-C} = r_{CC} \cdot \theta_C^2. \tag{19.15}$$

The relative rate of coke versus rate of methane formation then follows from

$$\frac{R_{CC}}{R_{CH_4}} = \frac{r_{CC} \cdot \theta_C^2}{r_H \cdot \theta_C} \tag{19.16a}$$

$$= \frac{r_{CC}}{r_H} \cdot \theta_C. \tag{19.16b}$$

Interestingly as we have seen, this may have a maximum as a function of the metal–carbon bond energy.

The occurrence of a maximum depends on the BEP parameter $\alpha''$ of the C–C bond formation rate. Volcano-type behavior for the selectivity is found as long as Eq. (19.17a) is satisfied.

$$2x + \alpha > \alpha'' > x, \tag{19.17a}$$

then:

$$\lambda_{max}^{sel} = \frac{\alpha'' - x}{2x + \alpha - \alpha''}. \tag{19.17b}$$

$r_{CC}$ as well as $r_H$ decrease when the carbon adsorption energy increases. Volcano-type behavior of the selectivity to coke formation is found when the activation energy of C–C bond formation decreases faster with increasing metal–carbon bond energy than the rate of methane formation. Equation (19.16b) indicates that the rate of the nonselective C–C bond-forming reaction is low when $\theta_C$ is high and when the metal–carbon bond is so strong that methane formation wins from carbon–carbon bond formation. The other extreme is the case of very slow CO dissociation; then $\theta_C$ is so small that the rate of C–C bond formation is minimized.

This analysis indicates the importance of a proper understanding of BEP relations for surface reactions. It enables a prediction not only of conversion rates but also of selectivity trends.

## 19.4 Compensation Effect

For catalytic reactions and systems that are related through Sabatier-type relations based on kinetic relationships as expressed by Eqs. (19.6) and (19.7), one can also deduce that a so-called compensation effect exists. According to the compensation effect, there is a linear relation between the change in the apparent activation energy of a reaction and the logarithm of its corresponding preexponent in the Arrhenius reaction rate expression.

The occurrence of a compensation effect can be readily deduced from Eqs. (19.6) and (19.7). The physical basis to the compensation effect is similar to that of the Sabatier volcano curve. When reaction conditions or catalytic reactivity of a surface change, the surface coverage of the catalyst is modified. This change in surface coverage changes the rate through change in the reaction order of a reaction.

In Eq. (19.6), the surface coverage is given by $\theta_C$ and it is related to the parameter $\lambda$ of Eq. (19.8). Equation (19.6) can be rewritten to show explicitly its dependence on gas-phase concentration. Equation (19.18a) gives the result. This expression can be related to practical kinetic expressions by writing it as a power law as is done in Eq. (19.18b). Power law type rate expressions present the rate of a reaction as a function of the reaction order. In Eq. (19.18b), the reaction order is $m+n$ in $H_2$ and $-n$ in CO.

$$R_H = \frac{r_H k_{diss} K_{ads}^{CO}[CO]}{k_{diss} K_{ads}^{CO}[CO] + r_H \left(K_{ads}^{CO}[CO]+1\right)^2}, \tag{19.18a}$$

$$\approx k_H^{m+n} k_{diss}^{-n} (K_{ads}^{CO})^{-n} [H_2]^{m+n} [CO]^{-n} \quad -1 < n < 1. \tag{19.18b}$$

Power-law expressions are useful as long as the approximate orders on reactant concentration are constant over a particular concentration course. A change in the order of the reaction corresponds to a change in surface concentration of a particular reactant. A low reaction order implies usually a high surface concentration, and a low reaction order implies low surface reaction of the corresponding adsorbed intermediates. In order to deduce Eq. (19.18b) for the rate of surface carbon hydrogenation, the power law of Eq. (19.19) has been used.

$$r_H = k_H \cdot [H_2]^m. \tag{19.19}$$

From Eq. (19.18b), the apparent activation energy as well as preexponent can be readily deduced. They are given by Eq. (19.20).

$$E_{\text{app}} = mE_{\text{act}}^{\text{H}} + n\left\{E_{\text{act}}^{\text{H}} - E_{\text{act}}(\text{diss}) - E_{\text{ads}}(\text{CO})\right\},$$
$$= mE_{\text{act}}^{\text{H}} + n\Delta E_{\text{app}}, \quad (19.20)$$

$$\ln A_{\text{app}} = \ln \Gamma \cdot \frac{ekT}{h} + \frac{m}{k}\Delta S_{\text{act}}^{\text{H}} + \frac{n}{k}\Delta S_{\text{app}}, \quad (19.20b)$$

$$\Delta S_{\text{app}} = \Delta S_{\text{act}}^{\text{H}} - \Delta S_{\text{act}}(\text{diss}) - \Delta S_{\text{ads}}(\text{CO}). \quad (19.20c)$$

The orders of the reaction appear as coefficients of activation energies and adsorption energies and their corresponding entropies. For more detailed discussions, see Ref. [7].

A consequence of the compensation effect is the presence of an isokinetic temperature. For a particular reaction, the logarithm of the rate of a reaction measured at different conditions versus $\frac{1}{T}$ should cross at the same (isokinetic) temperature. This isokinetic temperature easily follows from Eq. (19.20) and is given by Eq. (19.21):

$$T_{\text{iso}} = \frac{\Delta E_{\text{app}}}{k\Delta S_{\text{app}}}. \quad (19.21)$$

It is important to realize that the compensation effect in catalysis refers to overall catalytic reactions.

Sometimes the activation energies of elementary reaction steps may also show a relationship between activation energy changes and activation entropies.

A reaction with a high activation energy tends to have a weaker interaction with the surface and hence will have enhanced mobility that is reflected in larger activation entropy. For this reason, the preexponents of surface desorption rate constants are $10^4$–$10^6$ larger than the preexponents of surface reaction rates.

In classical reaction rate theory expressions, this directly follows from the frequency–preexponent relationship:

$$k_{\text{class}} = \nu \frac{r_t^2}{r_i^2} e^{-\frac{E_{\text{act}}}{kT}}. \quad (19.22)$$

A high frequency of vibration between surface and reactant implies a strong bond, which will give a high activation energy. Increase of preexponent and corresponding activation energies hence counteract. Equation (19.22) is the rate expression for a weakly bonded complex. The bond frequency is $\nu$, and $r_{\text{i}}$ and $r_{\text{t}}$ are initial- and transition-state radii. Equation (19.22) is valid for a freely rotating diatomic complex.

Compensation-type behavior is quite general and has been extensively studied especially in transition metal catalysis [8a], sulfide catalysis [8b], and zeolite catalysis [7].

In the next section, we will present a short discussion of compensation-type behavior in zeolite catalysis.

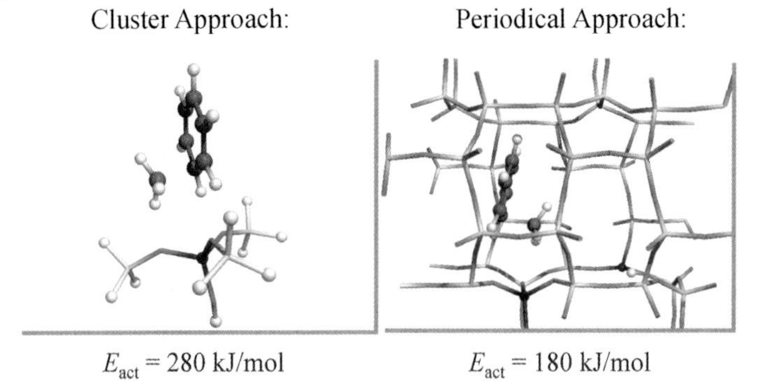


**Figure 19.10** Comparison of the activation energy of protonation of toluene by an acidic cluster versus activation of the same molecule by mordenite. DFT computed results [9a].

## 19.5
## Hydrocarbon Conversion Catalyzed by Zeolites

As further illustration of the compensation effect, we will use solid acid catalyzed hydrocarbon activation by microporous zeolites. A classical issue in zeolite catalysis is the relationship between overall rate of a catalytic reaction and the match of shape and size between adsorbate and zeolite micropore.

For a monomolecular reaction as the cracking of hydrocarbons by protonic zeolites, the rate expression is very similar to Eq. (19.6). The rate of reaction is now proportional to the concentration of molecules at the reaction center, the proton of the zeolite (Eq. (19.23a)).

$$r = k_{\mathrm{act}} \cdot \theta \tag{19.23a}$$

$$= k_{\mathrm{act}} \cdot \frac{K_{\mathrm{eq}}[C_{\mathrm{H}}]}{1 + K_{\mathrm{eq}}[C_{\mathrm{H}}]}. \tag{19.23b}$$

Assuming adsorption to behave according to the Langmuir adsorption isotherm gives Eq. (19.23b). Both the rate constant of proton activation $k_{\mathrm{act}}$ as well as the equilibrium constant of adsorption will depend on cavity details

Quantum-chemical studies have indeed shown that the presence of a surrounding cavity lowers barriers of charge separation that occur when a molecule is activated by zeolitic protons as shown in Figure 19.10 [9].

Presence of the zeolite cavity dramatically lowers the activation energy for protonation of toluene. It is mainly due to screening of charges in the transition state due to the polarizable lattice oxygen atoms. In the transition state, a positive charge develops on protonated toluene.

This reduction in activation energy will only happen when the structure of the transition-state complex fits well in the zeolite cavity. This is the case for the protonated toluene example in the zeolite Mordemite channel. The structure of the transition-state complex in the cluster simulation and zeolite can be observed to be very similar as in Figure 19.10.

The activation energy will be strongly increased when there is a mismatch between transition-state complex shape and cavity.

The rate constant then typically behaves as indicated in Eq. (19.24):

$$k_{act} = k_{act}^{n,st}\, e^{\frac{\Delta G_{st}}{kT}}, \tag{19.24}$$

where $\Delta G_{st}$ is the difference in free energy due to steric constants in reactant and transition state and $k^{n,st}$ is the rate constant of the nonsterically constrained reaction. The contribution of the steric component to the transition-state energy cannot be deduced accurately form DFT calculations because van der Waals energies are poorly computed. Force field methods have to be used to properly account for such interactions.

For adsorption in zeolites, the biased Monte Carlo method as developed by Smit, is an excellent method to determine the free energies of molecules adsorbed in zeolites [9b] (see also Chapter 7).

This method can be used to compute the concentration of molecules adsorbed in zeolites as we will discuss below.

We will use this method to deduce $\Delta G_{st}$ for hydrogen transfer reaction. The free energies of adsorption of reacting molecules as propylene and butane are compared with the free energies of reaction intermediate molecules that are analogous to the intermediates formed in the transition state. A C–C bond replaces the C–H–C bond. An example of such a transition state and analog intermediate is given in Figure 19.11.

In Table 19.1, a comparison is made of the differences in free energies for two different zeolites. Note the large repulsive energies computed for the intermediates and their sensitivity to zeolite structure.

An alternative view to interpret $\Delta G_{st}$ is the realization that reactants before they can be activated to a particular reaction have to be present in a conformation such that a particular reaction can occur. The actual activation of reacting molecule or molecules is not strongly affected by this state. Calculations on the activation of different isomers of xylene have indeed demonstrated that differences in the energies of the pretransition state configurations dominate the activation energy differences [10] and hence the Maxwell–Boltzmann term in Eq. (19.24) has to be interpreted as the relative probability that a particular intermediate pretransition state structure is realized in zeolite [11].

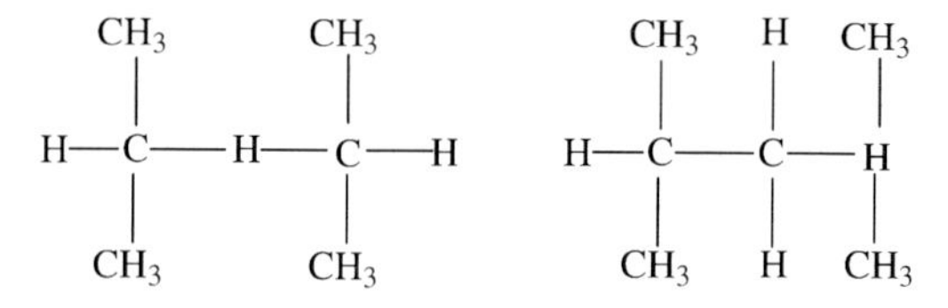


**Figure 19.11** Transition state for hydrogen transfer between propane and protonated propylene cation and its analog intermediate (schematic).

**Table 19.1** Configurationally biased Monte Carlo simulations of adsorption energies and reaction energies.

| | $\Delta H_{ads}$ (kJ mol$^{-1}$) | |
|---|---|---|
| | MFI | Chabasite |
| Propane | −41.0 | −34.6 |
| *n*-Butane | −44.1 | −47.0 |
| $C_6$ | −50.7 | −65.1 |
| $C_8$ | −49.9 | −43.1 |
| $\Delta E(C_6 - 2C_3)$ | +31.3 | + 4.1 |
| $\Delta E(C_8 - 2C_4)$ | +38.3 | +50.9 |

This is the reason that for complex cracking reactions in zeolites, the product pattern can be predicted from a simulation of the free energies of the corresponding intermediate molecules in the zeolite [11].

As long as there are no important steric contributions to the transition-state energies, the elementary rate constant of Eq. (19.23) does not sensitively depend on the detailed shape of the zeolite cavity. Then the dominant contribution is due to the coverage-dependent term $\theta$.

This has been demonstrated by a comparison of the cracking rates of small linear hydrocarbons in ZSM-5 [12] and also for reactions in different zeolites for hydroisomerization of hexane [13]. Differences in catalytic conversion appear to be mainly due to differences in $\theta$.

The apparent activation energies can be deduced from Eq. (19.23b). The corresponding expression is given by Eq. (19.25a).

$$E_{app} = E_{act} + E_{ads}(1-\theta) \tag{19.25a}$$

$$r \approx k_{act}\ K_{eq}^{(1-\theta)}\ [C_H]^{1-\theta} \qquad 0 < \theta < 1. \tag{19.25b}$$

In the absence of steric constraints in Eq. (19.25a), $E_{act}$ will not vary. $E_{ads}$ and $\theta$ are the parameters that significantly change with hydrocarbon chain length or zeolite.

Because the interaction of linear hydrocarbons is dominated by the van der Waals interaction with the zeolite, the apparent activation energies for cracking are decreasing linearly with chain length. In some cases, differences in the overall rate are not dominated by differences in the heat of adsorption but instead are dominated by differences in enthrones of adsorbed molecules.

One notes in Table 19.2 a uniform increase in the adsorption energies of the alkanes when the microspore size decreases (compare 12 ring channel zeolite MOR with 10 ring channel TON). However at the particular temperature of hydroisomerization, the equilibrium constant for adsorption is less in the narrow-pore zeolite than in the wide-pore system. This difference is due to the more limited mobility of the hydrocarbon in the narrow-pore material. This can be used to compute Eq. (19.23b) with the result that the overall hydroisomerization rate in

**Table 19.2** Calculated heats of adsorption and adsorption constants for various hydrocarbons in zeolites with different channel dimensions.

| | $\Delta H_{ads}$ kJ mol$^{-1}$ simulation | $K_{ads}$ ($T = 513$ K) mmol g Pa$^{-1}$ simulation |
|---|---|---|
| *n*-Pentane/TON | −63.6 | $4.8 \times 10^{-6}$ |
| *n*-Pentane/MOR | −61.5 | $4.8 \times 10^{-5}$ |
| *n*-Hexane/TON | −76.3 | $1.25 \times 10^{-5}$ |
| *n*-Hexane/MOR | −69.5 | $1.25 \times 10^{-4}$ |

the narrow-pore material is lower than that of the wide-pore material. This entropy difference dominated effect is reflected in a substantially decreased hydrocarbon concentration in the narrow-pore material.

## 19.6 Structure Sensitive and Non-sensitive Reactions

A classical issue in transition metal catalysis is the dependence of catalytic activity on changes in particle size of the metal clusters in the nanosize region.

As illustrated in Figure 19.12, three types of behavior can be distinguished.

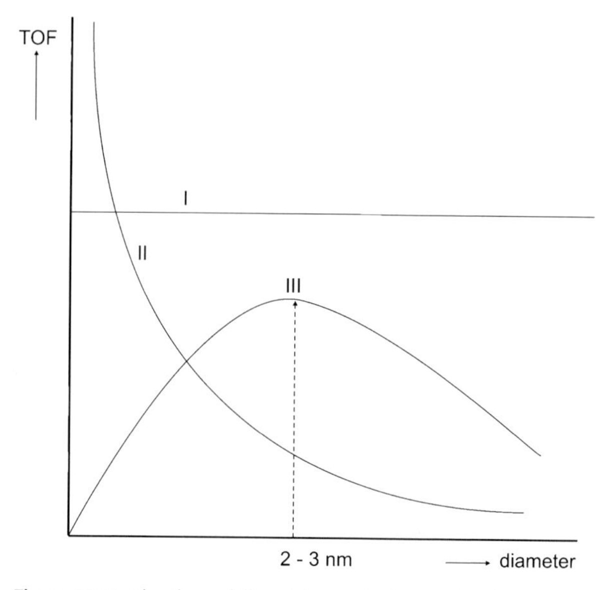


**Figure 19.12** The three different types of cluster size dependence of catalytic conversion. Rates are considered normalized per exposed surface atom.

The most significant surface feature that changes with metal particle size is the ratio of corner, edge, and terrace surface atoms.

The increase in the rate of Case II is related with an increase in the relative ratio of the edge and corner atoms over the decreasing number of terrace atoms. This increase in reactivity relates to the increased degree of coordinative unsaturation of the edge and corner atoms.

Important changes in the electronic structure occur. Electron delocalization decreases, which is reflected in a narrowing, especially, of the d-valence electron bands and a corresponding upward shift of the d-valence electron band center due to increased electron–electron interactions [5].

The decrease in bandwidth is proportional to

$$\sqrt{N_s}. \tag{19.26}$$

Within the tight-binding approximation, it implies a less electron localization energy

$$\Delta E_{\mathrm{loc}} \approx \left(\sqrt{N'_S} - \sqrt{N_S}\right)\beta \tag{19.27}$$

for the surface atom with the lowest number of nearest-neighbor surface atoms.

The dependence on electron localization energy can also be illustrated by using *bond order conservation* principle. This principle gives an approximate recipe to estimate changes in bond strength when coordination of a surface atom or adsorbate attachment changes [5, 16].

According to this principle, the valence of an atom is considered a constant. When more atoms coordinate to the same atom, the valence has to be distributed over more bonds and hence the strength per bond decreases. When the chemical bonds are equivalent, the bond strength of an individual bond $\varepsilon(n)$ depends on the following way on the corresponding bond strength of a complex with a single bond ($\varepsilon_0$):

$$\varepsilon(n) = \varepsilon_0\left(\frac{2n-1}{n^2}\right), \tag{19.28}$$

where $n$ is the total number of bonds with the metal atom. The surface atom metal coordination number is given by

$$N = n - 1. \tag{19.29}$$

As a consequence that activation energy for dissociation is lower when coordinative unsaturation of the surface atom increases.

Such dependence for the activation of a chemical bond as a function of surface metal atom coordinative unsaturation is typically found for chemical bonds of $\sigma$ character as the CH or C–C bond in an alkane. Activation of such bonds usually occurs atop of a metal atom. The transition-state configuration for dissociation of methane on a Ru surface illustrates this (Figure 19.13).

The data presented in Table 19.3 illustrate the dependence of the activation energy of methane on edge or corner (kink) atom position of some transition metal surfaces.

**Figure 19.13** Transition state configuration of methane activation on Ru(1120) surface.

**Table 19.3** Methane activation barrier as a function of edge and corner atoms.

| | |
|---|---|
| Ru(0001)[a] | 76 |
| Ru(1120)[b] | 56 |
| Rh(111)[c] | 67 |
| Rh step[c] | 32 |
| Rh kink[c] | 20 |
| Pd(111)[c] | 66 |
| Pd step[c] | 38 |
| Pd kink[c] | 41 |

[a] (0001) Ciobica et al. [4].
[b] (1120) Ciobica and van Santen [14].
[c] Liu, and Hu [15].

The BEP $\alpha$ value for methane activation is close to one. As a consequence, the BEP value for hydrogenation of adsorbed methyl, the reverse reaction, should be nearly zero.

The dependence on decreasing particle size that results for this recombination reaction is as Class I in Figure 19.11. The rate of this recombination reaction has become independent of surface atom coordination number. The differences between the activation energies for dissociation, hence, closely relate to the differences in energy of the adsorbed product fragments $methyl_{ads}$ and $H_{ads}$.

Class II and Class I behavior have been shown to be closely related and are complement to each other. In practical catalysis, Class I behavior is typically found for hydrogenation reactions.

Class III type behavior is representative of reactions in which $\pi$ bonds have to be broken. It is the typical behavior of reactions in which CO or $N_2$ bond activation is rate limiting.

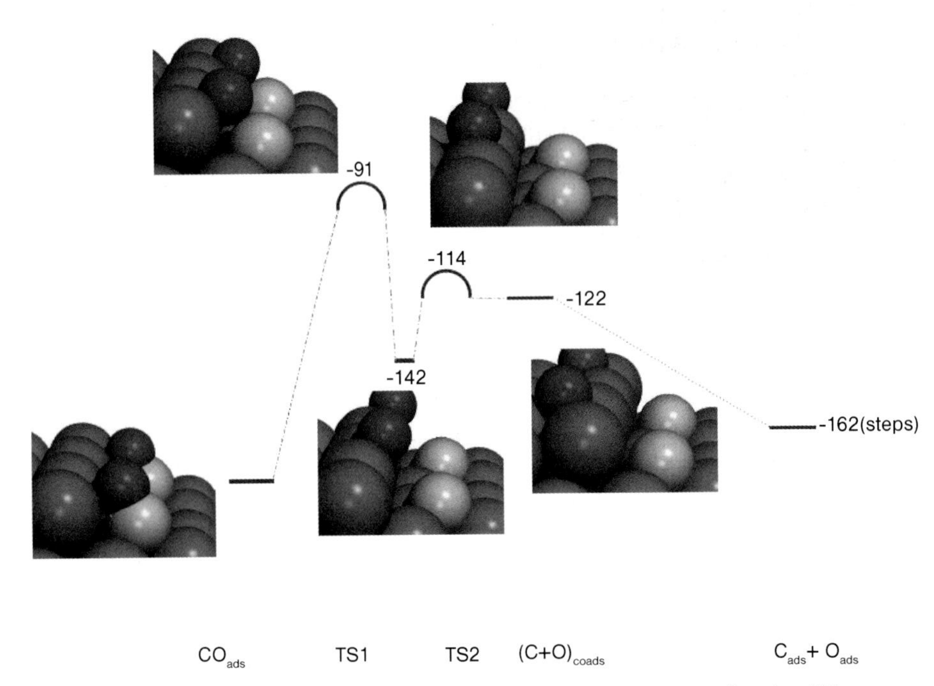


**Figure 19.14** Energetics and structure of CO dissociating on Ru step-edge site [17].

As we have already mentioned in the section on the BEP method, activation of such molecules depends strongly on the structure of the catalytically active center (see Figure 19.4). The structures of reactant, transition state, as well as product state at a step-edge site are shown for CO dissociation in Figure 19.14.

Surface step-edge sites have substantially lowered activation energies compared to the activation energies of the same dissociation reactions on surface terraces (compare 91 kJ mol$^{-1}$ in Figure 19.14 with 215 kJ mol$^{-1}$ on Ru(0001) surface). This lowered barrier is due to the several chemical bonding affecting factors. Due to multiple contact of the CO molecule at the step-edge site, there is substantially more back donation into the CO bond weakening $2\pi^*$ orbitals. As a consequence to stretch the CO bond to its transition state, only a small extension is required. Thirdly, in the transition state, the oxygen and carbon atoms do not share bonding with the same surface metal atom. Such sharing which is an important reason for enhanced barrier energies at terraces (due to the bond order conservation principle).

As can also be observed from Figure 19.5, the activation barriers at step sites are reduced compared to those at terraces whereas the adsorption energies of the fragment atoms do not change. Different as for activation of $\sigma$ type bonds activation of $\pi$ bonds proceeds through elementary reaction steps for which there

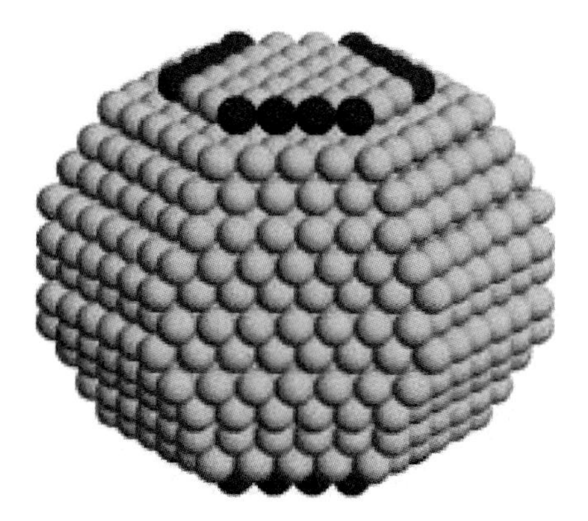

**Figure 19.15** Cubo octahedron with step-edge sites [18].

is no relation between reaction energy and activation barrier, when reaction site structure changes. The activation barrier for the forward dissociation barrier as well as the reverse recombination barrier is reduced for step-edge sites.

Interestingly, when the particle size of a metal nanoparticle becomes less than approximately 2 nm, terraces become so small that they cannot anymore support the presence of step-edge site metal atom configurations. This can be observed form Figure 19.15, which shows a cubo-octahedron just large enough to support a step-edge site.

Class III type behavior is a consequence of this impossibility to create step-edge type sites on smaller particles. Larger particles will also support the step-edge sites. Details may vary. Surface-step directions can have a different orientation and also the coordinative unsaturation of the atoms that participate in the ensemble of atoms that form the reactive center. This will enhance the activation barrier compared to that on the smaller clusters. Recombination as well dissociation reactions of $\pi$ molecular bonds will show Class III type behavior.

The different BEP behavior for the activation of $\sigma$ versus $\pi$ bonds, basic to the very different Class I and Class II particle size dependence compared to Class III particle size dependence, is summarized in Figure 19.16.

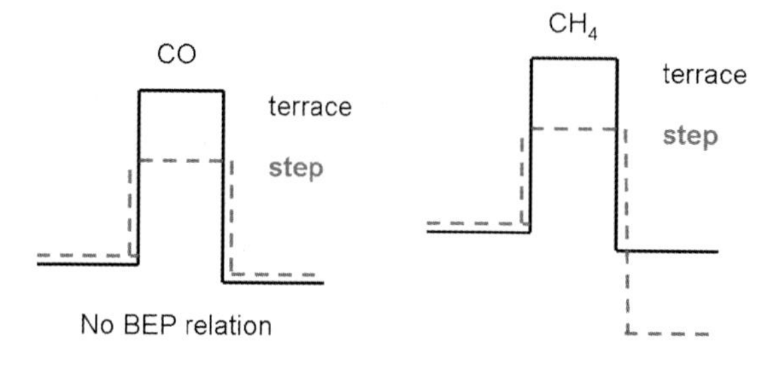


**Figure 19.16** Methane versus CO activation (schematic).

While Class I and Class II behavior are intrinsically related through microscopic reversibility, Class III type behavior implies that there is no BEP relation between the changes in activation energy with structure.

When the selectivity of a reaction is controlled by differences in the way molecules are activated on different sites, the probability of the presence of different sites becomes important. An example again can be taken from the activation of CO. For the methanation, activation of the CO bond is essential. This will proceed with low barriers at step-edge type sites. If one is interested in the production of methanol, catalytic surfaces are preferred that will not allow for easy CO dissociation. This will be typically the case for terrace sites. The selectivity of the reaction to produce methanol will then be given by an expression as Eq. (19.30a):

$$S = \frac{x_1 r_1}{x_1 r_1 + x_2 r_2} \tag{19.30a}$$

$$\frac{r_1}{r_2} = \frac{r_{CH_3OH} \cdot \theta_{CO}}{k_{diss}\theta_{CO}(1-\theta)} \approx \frac{r_{CH_3OH}}{k_{diss}}(1 + K_{ads}^{CO}[CO]). \tag{19.30b}$$

In this expression, $x_1$ and $x_2$ are the fractions of terrace versus step-edge sites, $r_1$ is the rate of conversion of adsorbed CO to methanol on a terrace site, and $r_2$ is the rate of CO dissociation at a step-edge type site. Increased CO pressure will also enhance the selectivity because it will block dissociation of CO.

## 19.7 Summary

Using microkinetic expressions, we have discussed the most important catalytic concepts that describe heterogeneous catalytic reactions. We have related these concepts with the energies, entropies, and transition-state features that are accessible by current state-of-the-art DFT techniques.

While it is very useful to relate reaction mechanistic proposals with catalytic kinetics, one has to be aware that DFT-predicted energies typically have an error of at least 10 kJ mol$^{-1}$.

Predictive kinetics requires accuracies that are an order of magnitude more precise. There are many examples that predict overall kinetics quite accurate. This is then due to a fortuitous cancellation of errors that needs to be understood.

We did not discuss extensively the consequences of lateral interactions of surface species adsorbed in adsorption overlayers. They lead to changes in the effective activation energies mainly because of consequences to the interaction energies in co-adsorbed pretransition states. At lower temperatures, it can also lead to surface overlayer pattern formation due to phase separation. Such effects cannot be captured by mean-field statistical methods as the microkinetics approaches but require treatment by dynamic Monte Carlo techniques as discussed in Chapter 10.

## References

1 E. Polanyi, *Trans. Far. Soc.* **34**, 11 (**1936**).

2 S. Dahl, A. Logadottir, C.J.H. Jacobsen, J.K. Nørskov, *Appl. Catal. A: General* **222**, 19 (**2001**).

3 S.G. Shetty, A.P.J. Jansen, R.A. van Santen, *J. Phys. Chem. C* **112**, 14027, (**2008**).

4 I.M. Ciobica, F. Frechard, R.A. van Santen, A.W. Kleijn, J. Hafner, *J. Phys. Chem. B.* **104**, 3364 (**2000**).

5 R.A. van Santen, M. Neurock, *Molecular Heterogeneous Catalysis*, Wiley-VCH, Wienhiem (**2006**).

6 (a) T. Bligaard, J.K. Nørskov, S. Dahl, J. Matthiesen, C.M. Christensen, J. Sehested, *J. Catal.* **222**, 206 (**2004**). (b) S. Kaszetelan, *Appl.Catal. A: General* **83**, L1–L6 (**1992**).

7 G.C. Bond, M.A. Keane, H. Kral, J.A. Lercher, *Catal. Rev.-Sci. Eng.* **42** (3), 323 (**2000**).

8 (a) T. Bligaard, K. Honkala, A. Logadottir, J.R. Nørskov, S. Dahl, C.J.H. Jacobson, *J. Phys. Chem. B* **107**, 9325 (**2007**).
(b) H. Toulhoat, P. Raybaud, *J. Catal.* **216**, 63 (**2003**).

9 (a) X. Rozanka, R.A. van Santen, in *"Computer Modelling of Microporous Materials"* (Eds. C.R.A. Catlow, R.A. van Santen, B. Smit) Elsevier, The Netherlands, Chapter 9 (**2004**). (b) B. Smit, in *"Computer Modelling of Microporous Materials* (Eds. C.R.A. Catlow, R.A. van Santen, B. Smit) Elsevier, The Netherlands, Chapter 2 (**2004**).

10 A.M. Vos, X. Rozanska, R.A. Schoonheydt, R.A. van Santen, F. Hutschka, J. Hafner, *J. Am. Chem. Soc.* **123**, 2799 (**2001**).

11 B. Smit, T.L.M. Maesen, *Nature* **451**, 671 (**2008**).

12 (a) W.O. Haag, in *"Zeolites and Related Microporous Materials: State of the Art 1994"* (Eds. J. Weitkamp, H.G. Karge, H. Pfeifer, W. Hölderick ), Elsevier, Amsterdam, p. 1375 (**1994**).(b) Th.F. Narbeshuber, H. Vinety, J. Lercher, *J. Catal.* **157**, 338 (**1995**).

13 F.J.J.M.M. de Gauw, J. van Grondelle, R.A. van Santen, *J. Catal.* **206**, 295 (**2002**).

14 I.M. Ciobica, R.A. van Santen, *J. Phys. Chem. B* **106**, 6200 (**2002**).

15 Z.P. Liu, P. Hu, *J. Am. Chem. Soc.* **125**, 1958 (**2003**).

16 E. Shustorovich, *Adv. Catal.* **37**, 101 (**1990**).

17 I.M. Ciobica, R.A. van Santen, *J. Phys. Chem. B* **107**, 3803 (**2003**).

18 K. Honkala, A. Hellman, I.N. Remediakis, A. Logadottir, A. Carlsson, S. Dahl, C.H. Christensen, J.K. Nørskov, *Science* **307**, 555 (**2005**).

# 20
# Conclusion: Challenges to Computational Catalysis

*Rutger A. van Santen and Philippe Sautet*

## 20.1
## Introduction

Computational chemistry has essentially two important contributions in the field of chemistry. First, it enables the calculation of spectroscopic properties of model systems proposed by the experimental scientists. In this role, it assists spectroscopic studies to analyze the spectra and assign chemical intermediates in the systems. Second, by calculating energies it allows to predict the relative stability of different arrangements and to simulate reaction pathways.

Methods that accurately reproduce spectroscopic properties from calculations on model systems that realistically represent the atomistic nature as well as the extent of the system are extremely useful. Representative for this approach are the techniques currently used routinely to analyze EXAFS spectra. In this book, the chapters on STM and XAS are representative of such applications.

Clearly new algorithms that improve the speed and accuracy of such calculations are very useful. Also methods that predict the relative stability of particular phases are needed. This relates to the topic of catalysis in relation to the structure of the catalytically reactive site.

For the study of catalytic reactivity features, we need to compute the relative stability of complexes or surfaces in liquid or reactive gas environment as well as the rates of the reactions that occur on these surfaces. Especially in heterogeneous catalysis, methods are needed that are able to predict the following:

1. Reactivity and kinetic parameters that relate to activity and selectivity patterns of systems with extended surfaces, as transition metals or oxides.
2. Surface-phase composition as a function of gas- or liquid composition in contact with catalyst substrate. Surface structures may change as a function of coverage. There is a need to predict such structures as well as rates of change between metastable phases.

In both themes, progress has been impressive. Many studies now demonstrate the feasibility of an *ab initio* approach to catalytic kinetics. Especially in desulfurization

*Computational Methods in Catalysis and Materials Science.* Edited by Rutger A. van Santen and Philippe Sautet

ISBN: 978-3-527-32032-5

catalysis and also in oxidation catalysis, there is a firm basis for techniques that predict surface structure and surface composition as a function of environmental conditions.

Less advanced is our ability to predict and simulate catalyst synthesis. It relates to studies in the liquid phase of catalyst preparation precursors and their reaction to form the catalytically relevant phase.

Of especial interest are techniques that enable the prediction of the size and morphology of dispersed particles on supports or the prediction of micropore dimensions in microporous systems.

Progress in this area is urgently needed because of our increasing capabilities to relate catalyst performance to catalyst structure and composition.

## 20.2
## The Simulation of Catalytic Reactivity

It is important to realize that the overall time scale of catalytic events are typically sec -1 and the corresponding length scales those of a micron. There is a time and length scale gap between simulations at a molecular level and macroscopic catalytic performance.

On the molecular level state-of-the-art DFT methods are available that predict energies and transition states at a level (10–30 kJ $mol^{-1}$) relevant to mechanistic reactivity studies. Kinetic data can be extracted using reaction rate transition state theory expressions with quantum-chemical results as input. This is an example of using data relevant at a picosecond time scale to predict events occurring at a longer time scale. Dynamic corrections may be necessary by doing molecular dynamics simulations on configurations near the transition-state saddlepoint.

Similarly, the energetic and kinetic data extracted from electronic structure calculations through the use of reaction-rate expressions can be used as input to kinetic Monte Carlo methods to predict the overall rate of a catalysis reaction. The catalytic cycle is composed of many elementary reaction steps that take place in parallel at different positions of the catalyst. Very often, such methods require the use of a grid to represent the catalyst surface. New approaches have to be developed to improve this and incorporate displacement of the surface atoms as well.

In cases where local equilibrium can be assumed, molecular dynamics and equilibrium Monte Carlo methods can be combined as for instance is done to estimate diffusion constants in zeolites using CB-MC. There is a need to develop such combined scaling methods to a greater extent.

This is even more important as long as the quantum method used, such as DFT-GGA, does not provide a reliable description of van der Waals interactions often relevant for conformational changes. In such a case, molecular mechanics and quantum chemical procedures have to be combined.

A major issue for the *ab initio* approach to catalytic kinetics deals with the accuracy of currently used DFT-based methods. For reliable kinetic simulations, energies of reaction intermediates and their corresponding activation energies have to be predicted with 1 kJ $mol^{-1}$ accuracy. This is outside the current possibilities for systems of realistic size. Practically empirical adjustments are often made. There is, therefore, a great need for methods that improve the accuracy of computational techniques.

A related issue is the accurate prediction of the nonbonding van der Waals interactions as already mentioned above. DFT methods so far are not able to simulate such terms properly. Obviously this is a great limitation for its application to many systems where van der Waals interactions control the stereochemistry of reactions.

A very important issue is the incorporation of environmental effects in the simulations. This holds for solvent effects or structural effects as those of the cavities of a zeolite.

In solution, one needs to incorporate the disorder in the system and especially important the reactivity of the solvent itself. This has been demonstrated to be critically important in electrocatalysis. Methods such as CP-MD, discussed also in this book are available, but one has to be aware of the limitations of these techniques as far as van der Waals interactions are concerned.

In zeolites, the shape and size of the micropores have a large influence on the free energies of adsorbed molecules. This will not only affect the relative stability of reaction intermediates, but also transition states of elementary reactions taking place in the micropores. The dominating forces that control the interaction with the zeolite channel are the dispersive van der Waals interactions. A great challenge is to compute transition states in the cavity of zeolite-type structures including the differences that result from the van der Waals interaction.

## 20.3 The Structure of the Catalytic Complex or Surface

This is a field molecular quantum chemical methods, solid-state force field based simulations and molecular dynamics methods needed to be combined.

To predict the overlayer structure of surfaces and their potential reconstruction, not only stationary thermodynamic information is important. The nucleation process itself that initiates the phase transformations also needs to be understood. This latter is a virtually unexplored area.

STM enables a direct imaging of surfaces in the presence of the reactants, and combined with image simulations (see chapter 11) allows a detailed study of surface structural changes. In addition, *ab initio* thermodynamic approaches have been used with much success to predict the structure and composition of surfaces under a temperature and pressure of gas similar to catalytic conditions.

Now computational DFT models are available on hydrated surfaces of oxides used as the support of heterogeneous catalysts. This creates the possibility to explore the generation and reactivity of small metal particles as well as reactive oxide clusters or sulfide particles on such supports. The classical topic of metal-support effects is becoming within reach. Detailed atomistic investigations are now becoming possible.

Comparison with experiment can be achieved through computation of spectroscopic properties. Computations on (partially) hydrated oxide surfaces also enable the prediction of acidity, basicity as well as surface reactivity related with the presence of hydroxyl groups.

Finally it is of utmost importance to realize that the reactivity of adsorbates or adatoms on surfaces may be a strong function of overlayer concentration, as is for instance the case for ethylene hydrogenation or ethylene epoxidation. Again this is a topic that can now be explored more widely with the current status of computational techniques.

## 20.4 Catalyst Synthesis

The reaction intermediates and transformations of catalyst precursor during catalyst synthesis is almost an explored area. These systems and their transformation have to be studied in the water-phase. An important issue is the interaction of organic structure directing agents (SDA) molecules with inorganic cluster precursors.

For the preparation of nanoporous materials with pores of larger dimensions self-assembly of SDA lipophylic molecules is relevant. This implies simulations of molecular complexes on larger length and time scales. Here coarse-graining methods as described in this book can be usefully applied. Again there is a need to combine different methods, operational on different length scales, so as to describe properly some of the molecular interactions on a shorter scale. The structure of the complexes may also significantly depend on nonbonding interactions from fragments far from reactive centers. Synthesis itself may proceed on different length or time scales. For a control of the synthesis pathway at the nanoscale an understanding of the relevant molecular chemistry is essential. At the intermediate size level aggregates may be formed undergoing pre-annealing processes before crystallization finally occurs. These are complex reaction sequences in which again dynamical events of different time and length scales play a role. Colloid chemical aspects also start to become important.

## 20.5 Grand Challenges and New Developments

In heterogeneous catalysis only for very few reactions the mechanism is known in conclusive detail. One of the best understood systems are the zeolite proton catalyzed reactions as isomerization or cracking. Even for these reactions, there is still a debate how to understand in detail the role of zeolite lattice or cavity. This is not only due to the earlier mentioned difficulty to describe the van der Waals interactions, but also to the difficulty to describe chemical reactivity of molecules with many conformational options for parts of the molecule not involved in the reactive center. This question is even less resolved when cations are used as promoters of reactions in zeolites.

A major contribution to this field is the use of CB-MC methods, described in this book, that predict the conformation and chemical potentials of adsorbed phases as a function of pore filling. There is a need to improve these simulations using a combination of coarse graining with atomistic modeling. Potentials for complex systems of use in molecular dynamic simulations are not yet widely available. The reactive field methods (chapter 7) are an important improvement.

In metal catalysis the understood reactions at a molecular level are the hydrogenation of ethylene and the ammonia synthesis reaction. However, again, the fine details of coverage dependence, surface reconstruction, and generally the effect of particle size are little understood. Molecular models of the action of promoters are even more limited.

It is now time to computationally explore in full detail the integrated complex features that control heterogeneous reactions taking full account of the features mentioned above. There are many catalytic systems of great technical importance for which a wide body of knowledge has been collected, but for which still a detailed mechanistic understanding is still lacking. Critically important is an understanding of the relation between catalytic reactivity, catalyst composition and catalyst structure. The impact of computational studies may be expected to be very large to understand how the catalyst structure controls the reactivity, but also in return how the nature of the gas phase reactant might control the structure of the catalytic active site.

Relevant catalytic reactions of current great interest are the Fischer–Tropsch reaction and related reactions of importance to the Fischer–Tropsch process. Hydrocarbon activation in many catalytic systems is now open for theoretical exploration. An important new topic is the conversion reactions of biomass. Of increasing interest is the development and hence the modeling of catalysts that mimic biocatalytic systems.

Apart from the detailed study of specific systems and the development of combined scaling methods to execute these programs computationally there is a need for the formulation of analytical theories of catalysis. The Sabatier principle provides such a framework that relates overall catalytic activity to the interaction energy of reactants with the catalyst surface. Activation energy–free energy relations as the Brønsted–Eyring–Polanyi relation between activation energy and reaction energy of surface reactions are another example.

Theories are emerging that relate trends in activation energies as a function of substrate bond type with the topological structure of the reactive centers. Methods are available that the relate electronic structure of the catalyst with trends in interaction energies as well as activation energies. Bond-order conservation rule can also be used to study lateral interactions between coadsorbed overlayer species.

As a result modeling approaches using parameters available from such theories to make predictions and to design complex catalysts are becoming possible. This is of great importance for attempts to use computational combinatorial approaches to discover new catalyst formulations.

Finally, it is critically important to predict the structure of the complex phases that are present on supported catalysts. Simulation of spectra as well as determination of relative stability of proposed surface or cluster complexes is critically important. During reaction these complexes may dynamically change. The in-situ experimental characterization of these dynamical behaviors is extremely demanding and hence this creates important opportunities for theory to have a major impact in the future of catalysis.

# Subject Index



*Computational Methods in Catalysis and Materials Science.* Edited by Rutger A. van Santen and Philippe Sautet

ISBN: 978-3-527-32032-5

***n***



***o***